Dear Jackie, Merry Christmas, Happy Hannukah!

Use this book well, and
have fun in the stream!

Love,

Ethan

Stream Hydrology

Stream Hydrology

AN INTRODUCTION FOR ECOLOGISTS

NANCY D. GORDON

THOMAS A. McMAHON

BRIAN L. FINLAYSON

Software author Rory J. Nathan

Centre for Environmental Applied Hydrology
University of Melbourne

JOHN WILEY & SONS
Chichester · New York · Brisbane · Toronto · Singapore

Reprinted April 1993, September 1993, March 1994, September 1994

Paperback edition September 1994

Other Wiley Editorial Offices

John Wiley & Sons, Inc., 605 Third Avenue,
New York, NY 10158-0012, USA

Jacaranda Wiley Ltd, G.P.O. Box 859, Brisbane,
Queensland 4001, Australia

John Wiley & Sons (Canada) Ltd, 22 Worcester Road,
Rexdale, Ontario M9W 1L1, Canada

John Wiley & Sons (SEA) Pte Ltd, 37 Jalan Pemimpin #05-04,
Block B, Union Industrial Building, Singapore 2057

Library of Congress Cataloging-in-Publication Data:

Gordon, Nancy D.
Stream hydrology : an introduction for ecologists / Nancy D. Gordon, Thomas A. McMahon, Brian L. Finlayson : software written by Rory J. Nathan.
p. cm.
Includes bibliographical references and index.
ISBN 0-471-93084-9 (ppc.)
1. Rivers. 2. Hydrology. I. McMahon, T. A. II. Finlayson, Brian L. III. Title.
GB1205.G65 1992
551.48′3—dc20 91–26728
CIP

British Library Cataloguing in Publication Data:

A catalogue record for this book
is available from the British Library.

ISBN 0-471-93084-9 (ppc)
ISBN 0-471-95505-1 (paper)

Typeset in 10/12pt Times by Photo·graphics, Honiton, Devon
Printed and bound in Great Britain by
Biddles Ltd, Guildford and King's Lynn

Contents

Preface

In interdisciplinary applications of stream hydrology, biologists and engineers interact in the solution of a number of problems such as the rehabilitation of streams, the design of operating procedures and fishways for dams, the classification of streams for environmental values and the simulation of field hydraulic characteristics in laboratory flumes to study flow patterns around obstacles and organisms. One of the main purposes in writing this book was to help improve communication between the two disciplines and foster a sense of co-operation in these interdisciplinary efforts.

On the surface, the definitions of ecology and hydrology sound very similar: *ecology* is the study of the interrelationships between organisms and their environment and with each other, and *hydrology* is the study of the interrelationships and interactions between water and its environment in the hydrological cycle. In general, ecology is a more descriptive and experimental science and hydrology is more predictive and analytical. This fundamental difference influences the way streams are studied and perceived in the two disciplines.

For example, a diagram of an ecologist's view of a stream might appear as follows:

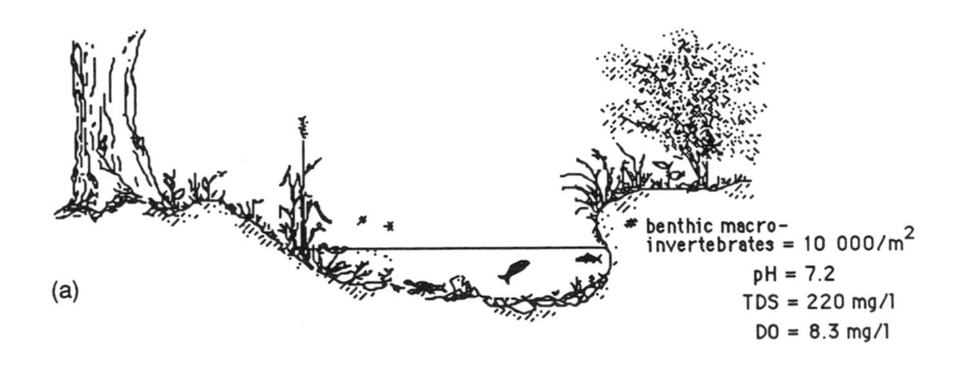

Here, the focus is on the aquatic biota, their interrelations, and the physical and chemical factors which affect them. An engineering hydrologist, on the other hand, might "view" the same stream much differently, perhaps more like this:

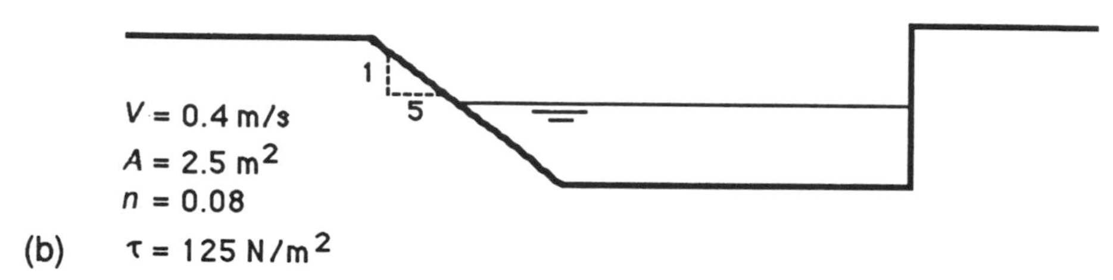

In this image, the physical dimensions of the stream have been simplified into a few numbers from which estimates can be made of how the stream will respond under different flow conditions.

Neither view is superior to the other; each represents only a fraction of "all there is to know" about the stream. Interdisciplinary interaction offers a way of merging the information contained in the different views into a more complete picture. It is often at the interface between disciplines, in fact, that new ideas are generated and progress made. Perhaps like a stereo pair, new "dimensions" will be revealed when the images are successfully superimposed.

The emphasis of this text is on the physical environment. Information has been drawn from the fields of geomorphology, hydrology and fluid mechanics, with examples given to highlight the information of biological relevance. Mathematical derivations have been omitted; instead, the intent was to provide an intuitive understanding of the principles, demonstrate their practical use and leave the mathematics to a computer. A software package, AQUAPAK, has been provided for this purpose (see Appendix 2). Omissions and simplifications were necessary in conveying the wide range of subject matter. We can only resort to the blanket statement that everything is more complicated than our description of it, and that ours is merely another "view" of streams.

A practical approach has been taken, with the chapter on field techniques forming a central part of the text. In other chapters, examples have been given so that the principles can be applied more readily. Field studies in the Acheron River Basin, located approximately 100 km to the east of Melbourne, provided information for examples throughout the book. We did this to maintain continuity, as well as to illustrate how we went about "getting to know" this river system.

The process of getting to know a stream is not unlike that of a doctor learning about a patient and his or her health. A conscientious doctor will look beyond the charts, images and the results of various tests to obtain a sense of what causes the patient's health to be what it is. In the same manner, hydrological data, aerial photographs, channel surveys and water quality analyses only measure "symptoms" of a stream's condition, and, as with human health, the underlying causes are complex and nebulous.

Just as patients are more than the sum of their connective tissues and blood vessels, streams, too, should be viewed "holistically" as a continuum from source to sea and as systems which interact with the surrounding environment. This book presents methods for "diagnosing" the physical condition of streams. Criteria for establishing what constitutes physical "health" are yet to be developed. As Leopold (1960) advocated over 30 years ago, benchmark stations free from grazing and other human influences are needed in order to evaluate the effects of humans on ecologic and geologic change. Interdisciplinary studies are essential for establishing these baseline conditions, for determining the sensitivity of a given stream to "stress" and for developing appropriate rehabilitation procedures to "cure" those streams which are found to be in poor condition.

ACKNOWLEDGEMENTS

The authors would like to express appreciation to a number of individuals for their contributions during the evolution of this text. Mr Andrew Douch deserves special credit for combining just the right amount of environmental science, artistic flair and skill with the Macintosh to create many of the drawings and diagrams in Chapters 4, 5 and 6. We are grateful to Dr Michael Keough of the Department of Zoology, University of Melbourne, for producing the realistic examples of Section 2.5. We would like to also thank the many individuals who reviewed sections of the text and provided thoughtful and valuable suggestions for improvement: our colleagues at the University of Melbourne, Mr George Benwell, Dr Juliet Bird, Dr Leon Bren, Dr Kien Gan, Dr Michael Keough, Mr Cliff Ogleby, Dr Ian O'Neill and Dr Bill Young; also Drs Leon Barmuta, Ian Campbell and P. Sam Lake of Monash University; Dr Richard Marchant of the Museum of Victoria, Dr Peter Jackson of the Queensland Department of Primary Industries, Drs Angela Arthington, Stuart Bunn and Brad Pusey of Griffith University, Professor Peter Brockwell of Colorado State University, Dr James Gore of Austin Peay State University, Dr Cliff Johnson of the US Agricultural Research Service, Dr Robert Milhous of the US Fish and Wildlife Service, Dr Ken Bencala and Dr Robert Jarrett of the US Geological Survey, and Dr Robert Newbury of Canada.

Comments and helpful information were also provided by Scott Seymour of the Dandenong Valley and Western Port Authority, Mr Steve Nolan of the Rural Water Commission and Jane Doolan and others of the Department of Conservation and Environment, Victoria. We are also grateful to the Rural Water Commission and the Bureau of Meteorology of Victoria for providing data for many of the examples and for the sample data files in AQUAPAK. Additionally, we are appreciative of the energy put into the project by the many students and other assistants who battled blackberries with us during stream surveys and helped with data analysis, literature review, computer programming and production of text and diagrams, in particular Lisa Langworthy, who helped with all of these tasks.

The senior author would like to give special recognition to several people who were constant sources of encouragement, inspiration, and advice. Special thanks go to Dr Robert Newbury for teaching us how to learn from streams and for help in focusing on what was important. She would like to thank Dr Sam Lake for his unending enthusiasm, for providing numerous reference materials, and for imparting ecological wisdom to the work. She is also most grateful to Dr Ian O'Neill for the many long hours he spent patiently explaining fluid mechanics to her, and for being pedantic enough to make her get it right. The sincerest of thanks go to her husband, Ralph, for his tolerance and support during the "panic" stages and for going to Australia with her; and to her parents, for allowing her to wade in puddles.

Finally, we are very appreciative of the professionalism and helpfulness of the people at John Wiley & Sons, especially Mrs Helen Bailey, Mrs Louise Portsmouth, Dr Lewis Derrick and Miss Claire Morrison, who greatly contributed to the quality of the final product.

This work was supported by funds provided by the Land and Water Resources Research and Development Corporation (formerly Australian Water Research Advisory Council).

COPYRIGHTED MATERIALS

We are grateful to the following organizations, individuals and publishers for permission to reproduce copyrighted material(s): Absoft Corporation (f77.rl file for Macintosh), American Association of Petroleum Geologists (Figure 7.18), American Statistical Association (Table 8.3), American Society of Civil Engineers (Figures 7.14 and 7.19), *Australian J. Mar. FW Res.* (Figure 8.13), Blackwell Scientific Publications and L. Barmuta (Figure 8.15), Butterworth-Heinemann (Figures 4.20, 9.11, 9.15 and 9.16), *Canadian J. Fish. Aq. Sci.* (Figure 9.2), *Canadian Water Resour. J.* (Figures 5.11 and 9.10), Chapman and Hall (Figure 5.19), Crown (State of Victoria) (Figures 3.4 and 3.5), Department of Water Resources Victoria (Figure 4.4), Diogenes Designs Ltd (Footrot Flats cartoon, Chapter 9), Director, Geological Survey of Victoria (Figure 3.6), D.R.F. Harleman (Figure 6.12), HarperCollins Publishers (Figure 5.36), Hemphill Publishing (Figures 5.29 and 5.36), Hodder and Stoughton Ltd (Figures 4.6 and 4.16), Houghton Mifflin Company (Table 6.3 and Figure 6.20), Institute of Hydrology (Figure 8.8), Institution of Engineers, Australia (Figure 3.7), John Wiley and Sons Ltd (Figures 4.2 and 9.6), John Wiley and Sons, New York (Figures 5.34, 7.5, and 9.3 and quote from Mark Twain, Chapter 8), *Journal of Geography in Higher Education* (Figures 5.24 and 5.28), Liverpool University Press (Table 1.4), Longman Group Ltd (Tables 1.2 and 5.5, and Figures 6.13, 6.27 and 7.20), Macmillan (Figure 4.3), McGraw-Hill (Tables 5.2, 6.1, 6.3, and 8.1, Figures 6.20, 6.23 and 6.36, and routine in program BACKWATR), Naval Ocean Systems Center, US Navy (Figure 6.21), *NZ J. Mar. FW Res.* (Figure 8.16), *Oecologia* and B. Statzner (Figure 6.4), Oxford University Press (Figure 7.11), Prentice-Hall, Inc. (Figure 5.30), Princeton University Press (Tables 1.1 and 1.2), Rellim Technical Publications (Figure 5.9), Society of Economic Paleontologists and Mineralogists (Figures 5.34 and 5.35), Royal Society for the Protection of Birds (Figures 9.9, 9.11, 9.12 and 9.14), V.L. Streeter (Figures 6.19 and 6.22), UNESCO (Figure 4.19), and Water Resources Publications (Table 1.3 and Figures 4.19 and 7.22).

1 Introducing the Medium

1.1 WATER AS A FLUID

Water is a widespread, life-sustaining substance, comprising 70–90% of living materials and covering a similar percentage (71%) of the Earth's surface. Of the Earth's total moisture, however, some 97% is contained in the oceans, with less than 0.003% flowing through its rivers and streams. Although it may change from ice to liquid water to water vapour, or move from one location to another, the total amount of moisture on Earth has remained essentially constant over time.

Water is a substance with many unique chemical and physical properties. Unlike most substances which contract when frozen, water expands, allowing ice to float on the surfaces of lakes and streams. It is found as a liquid at temperatures common to most places on Earth. With its great heat capacity it can absorb or lose a large amount of energy before showing a change in temperature. As a universal solvent, it dissolves gases, nutrients and minerals. Its internal cohesion gives rise to surface tension, which allows water striders to traverse a pool's surface or even run upstream. Because of its physical properties, a quite different set of environmental conditions is presented to amoebae and fish that both live in the same waters.

Depending on the temperature, water can exist as a liquid, a gas (water vapour) or a solid (ice). Combinations such as steam–air mixtures or water with entrained air fall into a specialized category called **two-phase flows**.

The general term **fluid** describes both gases and liquids, examples being oxygen, motor oil, liquid glass and mercury. The differences between fluids and solids are not always obvious. As fluids, both liquids and gases flow readily under the slightest of forces. They do not have a definite shape, and vessels are required for containing them. **Solids** are substances which are considered to have both a definite volume and a definite shape. Thus, the line is drawn between molasses as a fluid and gelatine as a solid.

Liquids are distinguished from gases by their cohesiveness. Whether in a laboratory beaker or in a frog pond, a liquid will have a definite volume.

It will also have a free surface, which is horizontal when the fluid is at rest. **Gases**, in contrast, do not have a definite volume, and will expand to fill a container enclosing them.

The next section will introduce some basic principles of physics and the system of units used in the text. These principles are applied to the description of physical properties of water in Section 1.3.

1.2 THE PHYSICS OF FLUIDS

The quantities used in describing the properties and motion of a liquid such as water are derived from four basic quantities: mass, length, time and temperature. The magnitude of these quantities (e.g. how hot or how large) are expressed in **units**. In the International (SI) system, the **fundamental** or **base units** are given as:

- meter (m)—length,
- second (s)—time,
- kilogram (kg)—mass, and
- Kelvin (K)—temperature.

In studies of aquatic systems, absolute temperatures are not normally of interest, and for the purposes of this text, temperature will be expressed in °C (Celsius), where 273.15 K = 0°C and a change of 1°C is the same as a change of 1 K.

The metre was originally proposed as 10^{-7} of the length of the meridian through Paris (Blackman, 1969). It is now represented in terms of the wavelength of a specific type of orange light. The unit of time, the second, is now defined by an atomic standard based on caesium. The unit of mass was originally based on the mass of a certain volume of water at prescribed conditions. Thus, conveniently, a litre (0.001 m^3) of water at 4°C has a mass of about 1 kg.

Whereas these base units are all defined in reference to some standard, there are other quantities, such as velocity, for which standards are impractical. These quantities have units which are defined in terms of the base units and are thus called **derived units**. Some of the quantities associated with the area of physics known as "mechanics" and relevant to the study of water will be discussed. A summary of both fundamental and derived units, their dimensions and the associated symbols is given in Table 1.1. For tables of conversion factors and other information relevant to water resource studies, VanHaveren's (1986) handbook is a highly useful reference.

Table 1.1. Common quantities used in the description of fluids. Adapted from Vogel (1981), by permission of Princeton University Press

Quantity	Symbol	Dimensions[a]	SI units
Length, distance	$x, y, d, r, h, k, w, \delta$ H, D, W, L, R, P	L	metre (m)
Area	A	L^2	square metre (m^2)
Volume	$\forall$	L^3	cubic metre (m^3) or litre (l)
Time	t	T	second (s)
Velocity	V, v	L/T	metre per second (m/s)
Discharge	Q	L^3/T	cubic metres per second (m^3/s)
Acceleration	g	L/T^2	metre per second squared (m/s^2)
Mass	M	M	kilogram (kg); tonne (1000 kg)
Force	F	ML/T^2	newton (N or $kg \cdot m/s^2$)
Density	ρ	M/L^3	kilogram per cubic metre (kg/m^3)
Heat, work, energy	Ω, KE, PE	ML^2/T^2	joule (J or N·m)
Power	ω	ML^2/T^3	watt (W or J/s)
Pressure	p	M/LT^2	pascal (Pa or N/m^2)
Shear stress	τ	M/LT^2	pascal (Pa or N/m^2)
Dynamic viscosity	μ	M/LT	newton second per square metre ($N \cdot s/m^2$ or $kg/m \cdot s$)
Kinematic viscosity	ν	L^2/T	square metre per second (m^2/s)
Surface tension	σ	M/T^2	newtons per metre (N/m)
Temperature	T	—	degrees Celsius (°C)

[a]Dimensions are: L = length, M = mass, T = time.

Velocity

Motion is defined as a change of position. **Speed** refers to the rate at which that position changes with time; i.e. if a raft floats 500 m downstream in 5.5 min, then its average speed is about 1.5 m/s. Technically, **velocity** refers to the speed in a given direction; however, in ordinary speech, no distinction is usually made between velocity and speed.

Discharge or streamflow

Discharge, or **streamflow**, is the rate at which a volume of water flows past a point over some unit of time. In the SI system it is expressed in m^3/s. For example, if a small spring filled a 0.01 m^3 bucket in 2 s, its discharge would be 0.005 m^3/s. Discharge is normally symbolized by Q.

Acceleration

Acceleration is the rate at which velocity changes with time. An object dropped off a cliff on Earth will accelerate at 9.807 m/s^2 (this **gravitational**

acceleration (*g*) varies about 1% with position on the Earth's surface). The distance (h) covered by a dropped object (starting at zero velocity) is:

$$h = \frac{1}{2}gt^2 \tag{1.1}$$

where t is the time in seconds from when it was dropped and h is in metres.

Force

Force is described in terms of its effects. It may cause an object to change its direction of motion; to stop or start; to rise or fall. By Newton's second law of motion, force is proportional to mass multiplied by acceleration. In the SI system, the unit of force is the **newton (N)**, defined as the force necessary to accelerate one kilogram at one metre per second squared:

$$\text{Force (N)} = \text{mass (kg)} \times \text{acceleration (m/s}^2\text{)} \tag{1.2}$$

A very small ("Newton's") apple with a mass of 0.11 kg experiences a gravitational force on Earth of about one newton.

The term "weight" does not appear in the SI system, and can create confusion, particularly when converting from the Imperial to the SI system. **Mass** is an expression of the amount of matter in something, whether a brick, a balloon or a bucket of water. **Weight** is a gravitational force. If Newton's apple were taken to the moon, it would still have a mass of 0.11 kg, but its weight (the force due to gravity) would be considerably reduced. On Earth, if an American buys 2.2 lb of apples at the supermarket to make a pie, and an Australian buys 1 kg of apples at the greengrocer to make apple slices, they will both get the same amount of produce. In this case, the distinction between mass and weight does not matter. However, to a researcher studying the behaviour of fluids, the distinction is essential!

Pressure

The **pressure** at any point is the force per unit area acting upon the point. For example, a human of 70 kg standing on the top of an empty aluminium can with a surface area of 0.02 m^2 would exert a pressure of

$$\left(\frac{70 \times 9.807}{0.02}\right) \approx 34\,000 \text{ N/m}^2 \text{ or 34 kilopascals (kPa)}$$

—probably sufficient to crush it.

Shear stress and shear force

Shear stress, like pressure, is force per unit area. The difference is the direction in which the force is applied. In pressure, the force acts *perpendicular* to a surface: ☐ whereas a **shear force** acts *parallel* to it: ☐ For example, a glob of liquid soap rubbed between the hands experiences shearing forces. Shear stress is the shearing force divided by the area over which it acts. For the soap, the shearing force acts over the surface area where the soap contacts the hand. Shear stress, symbolized by τ (tau), has the same units as pressure, N/m^2.

Energy

Energy and work have the same units. **Work** is a quantity described by the application of a force over some distance, measured in the direction of the force:

$$\text{Work (N·m or J)} = \text{Force (N)} \times \text{distance (m)} \tag{1.3}$$

For example, if a force of 200 N is required to push a floating log 10 m across a pond, then the amount of work done is 2000 N·m or 2 kilojoules (kJ).

Energy is the capacity for doing work. Thus, the quantity of work which something (or someone) can do is a measure of its energy; i.e. it would take about 700 kJ for a person of average ability to swim one km. Energy is usually symbolized by Ω (omega).

Power

Power is the amount of work done per unit time:

$$\text{Power (J/s or Watts)} = \text{Work (J)/time (s)} \tag{1.4}$$

Power is usually symbolized by ω (lower case omega). For a flow of water, Q, falling over a height, h, the relevant formula for calculating power is:

$$\omega = \rho g Q h \tag{1.5}$$

where ω has units of watts, Q has units of m^3/s, h is in metres, ρ is the density of water (kg/m^3) (see Section 1.3.1) and g is the acceleration due to gravity (m/s^2). As an approximation, this can be simplified to:

$$\omega = 10Qh \tag{1.6}$$

with ω in kilowatts. Thus, if a waterfall of 10 m height is flowing at 1.0 m^3/s, the power of the falling water is 100 kW. If the flow were diverted into a small hydroelectric plant rather than over the waterfall, much of this water power could be converted to electrical power. Because of losses associated with the turbine, electrical generator and the diversion works, efficiencies of 70% are common. In this example, then, approximately 70 kW of electricity could be produced.

1.3 PHYSICAL PROPERTIES OF WATER

1.3.1 DENSITY AND RELATED MEASURES

Density

Because the formlessness of water makes mass an awkward quantity, **density**, or mass per unit volume, is typically used instead. Density is normally symbolized by ρ (rho), and in the SI system it is expressed in kilograms per cubic metre (kg/m^3).

Pressure can be assumed to have an insignificant effect on the density of water for most applications. Thus, unless one is dealing with water at great depths within the ocean, density can be considered constant with changes in pressure. However, density is affected by temperature, decreasing as the temperature increases above 4°C (i.e. tepid water floats on top of colder water). Water density reaches a maximum at 4°C under normal atmospheric pressure. As the temperature decreases below 4°C, water becomes less dense, and upon freezing, it expands. The densities of selected fluids at different temperatures are listed in Table 1.2.

Materials dissolved or suspended in water such as salt or sediment will also affect its density. Thus, fresh water will float above salt water in estuarine environments or where saline groundwater enters a stream. Density is also reduced in the frothy whitewater of rapids, under waterfalls, or in other areas where large quantities of air are entrained in the water. Swimmers have more trouble staying afloat or propelling themselves in these regions; hence, fish tend to "jump" towards their upstream destinations from less-aerated areas (Hynes, 1970).

Specific weight

Specific weight is a non-SI measure, which is commonly used in practice in the Imperial system instead of density. Usually symbolized by γ (gamma), specific weight is equal to the product of density and gravitational acceleration, ρg. Thus, in the Imperial system, where the specific weight of water (at 4°C) is 62.4 lb/ft^3, one can calculate the weight of a 10 ft^3 "bathtub-full" of water as $62.4 \times 10 = 640$ lb. This measure will not be used in this text, and is included here only because it appears so often in the literature.

Relative density

Relative density is usually defined as the ratio of the density of a given substance to that of water at 4°C. It is thus a dimensionless quantity (it has no units). For example, the relative density of quartz is about 2.68. Relative

Table 1.2. Values of some fluid properties at atmospheric pressure. Adapted from Douglas et al. (1983) and Vogel (1981), by permission of Longman Group, UK, and Princeton University Press, respectively

	(°C)	Density, ρ (kg/m^3)	Dynamic viscosity, μ ($N{\cdot}s/m^2$)	Kinematic viscosity, ν (m^2/s)
Fresh water	0[a]	999.9	1.792×10^{-3}	1.792×10^{-6}
	4	1000.0	1.568×10^{-3}	1.568×10^{-6}
	10	999.7	1.308×10^{-3}	1.308×10^{-6}
	15	999.1	1.140×10^{-3}	1.141×10^{-6}
	20	998.2	1.005×10^{-3}	1.007×10^{-6}
	25	997.1	0.894×10^{-3}	0.897×10^{-6}
	30	995.7	0.801×10^{-3}	0.804×10^{-6}
	40	992.2	0.656×10^{-3}	0.661×10^{-6}
Sea water[b]	0	1028	1.89×10^{-3}	1.84×10^{-6}
	20	1024	1.072×10^{-3}	1.047×10^{-6}
Air	0	1.293	17.09×10^{-6}	13.22×10^{-6}
	20	1.205	18.08×10^{-6}	15.00×10^{-6}
	40	1.128	19.04×10^{-6}	16.88×10^{-6}
SAE 30 oil	20	933	0.26	0.279×10^{-3}
Glycerin	20	1263	1.5	1.190×10^{-3}
Mercury	20	13 546	1.554×10^{-3}	0.115×10^{-6}

[a]Ice at 0°C has a density of 917.
[b]Sea water of salinity 35%. The salinity of sea water varies from place to place.

density is equivalent to **specific gravity**, used in the Imperial system, where specific gravity is defined as the ratio of the specific weight of a substance to that of water.

Example 1.1

Calculate: (a) the mass of a 5 l volume of 15°C fresh water and (b) the gravitational force ("weight") it experiences on Earth:

$$\text{(a)}\ (5\,\text{l})\left(\frac{.001\ \text{m}^3}{1}\right)\left(999.1\,\frac{\text{kg}}{\text{m}^3}\right) = 5.0\,\text{kg}$$

$$\text{(b)}\ 5.0\,\text{kg}\left(9.807\,\frac{\text{m}}{\text{s}^2}\right) = 49.0\,\frac{\text{kg m}}{\text{s}^2} = 49.0\,\text{N}$$

1.3.2 VISCOSITY AND THE "NO-SLIP CONDITION"

Viscosity is a property which is intuitively associated with motor oil and the relative rates with which honey and water pour out of a jar. It is related to how rapidly a fluid can be "deformed". When a hand-cranked ice cream maker is empty the handle can be turned relatively easily. If it is then filled with water, the amount of effort increases, and if the water is replaced with molasses, the handle becomes extremely difficult to turn.

Viscosity, or more precisely, **dynamic** or **absolute viscosity**, is a measure of this increasing resistance to turning. It has units of newton-seconds per square metre ($N \cdot s/m^2$), and is symbolized by μ (mu). Of interest to aquatic organisms and aquatic researchers is the fact that there is almost no liquid with viscosity lower than that of water (Purcell, 1977).

The dynamic viscosity of water is strongly temperature-dependent. Colder water is more "syrupy" than warmer water. For this reason, it takes less effort for a water-boatman to "row" across a tepid backyard pond in summer than the equivalent distance in a frigid high-country lake. It also takes more work for wind to produce waves on a water surface when the water is colder. Dynamic viscosity can be calculated directly from temperature using the Poiseulle relationship, shown in Figure 1.1 as an equation and a curve. The equation will give slightly different values than those listed in Table 1.2. It should also be noted that this relationship holds true only for fresh water. As shown in Table 1.2, salt water has a higher dynamic viscosity than fresh water at the same temperature. Vogel (1981) describes instruments for measuring the viscosity of fluids for which published values are not available.

The influence of viscosity is perhaps most significant in the region where fluids come into contact with solids. It is here that fluids experience the equivalent of friction, which develops entirely within the fluid. When a solid slides across another solid, like shoes across a carpet, friction occurs at the

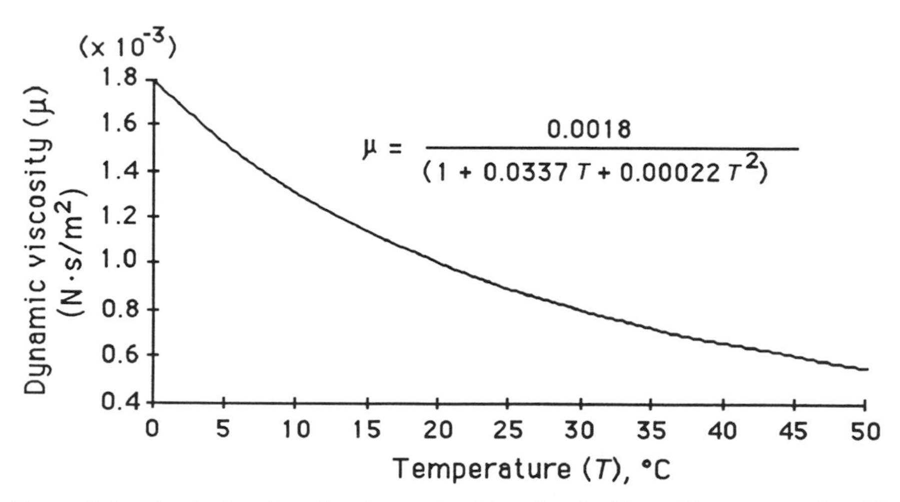

Figure 1.1. Graph showing the change in dynamic viscosity with temperature. The equation given is Poiseulle's relationship (Stelczer, 1987), from which dynamic viscosity can be calculated from temperature

interface between the two solids. When a fluid encounters a solid, however, the fluid sticks to it. There is no movement at the interface. According to this **no-slip condition**, at the point where a viscous fluid contacts a solid surface like a cobble on a streambed or a fish scale, its velocity is the same as that of the solid.

Thus, when water flows by a stationary solid object the velocity of the water is zero where it contacts the solid surface, increasing to some maximum value in the "free stream"—the region "free" of the influence of the solid boundary. Velocity profiles—curves which describe the way velocity changes with distance away from a solid—will be discussed in Chapter 6.

Kinematic viscosity, symbolized by ν (nu), is the ratio of dynamic viscosity to density:

$$\text{Kinematic viscosity } (\nu) = \frac{\text{Dynamic viscosity } (\mu)}{\text{Density } (\rho)} \tag{1.7}$$

where ν has units of m^2/s. This ratio shows up frequently in important measures such as the Reynolds number, and is another way of describing how easily fluids flow. The quantity was introduced by engineers to simplify the expression of viscosity by eliminating mass from the description (kinematic viscosity only has dimensions of length and time).

From Table 1.2 it can be seen that the kinematic viscosities of air and water are much more similar than their relative dynamic viscosities. The similarities in the behaviour of air and water make it convenient to model air currents, chimney plumes or aircraft using water tanks (after applying appropriate scaling factors).

1.3.3 SURFACE TENSION

A beetle darting across the surface of a pool, beads of dew on a waxy leaf, the curve of water spilling over a weir and the creep of water upwards from the groundwater table into the pores between fine sand grains—are all illustrations of the phenomenon, **surface tension**. Surface tension can be thought of as a stretching force per unit length (or energy per unit area) required to form a "film" or "membrane" at the air–water interface (Streeter and Wylie, 1979). It is symbolized by σ (sigma), and has units of N/m.

Surface tension of water in contact with air results from the fact that water molecules attract each other. Within a body of water, a water molecule is attracted by the molecules surrounding it on all sides, but molecules at the surface are only attracted by those beneath them. Therefore, there is a net pull downwards which puts tension on the water surface. The surface region under tension is commonly known as the **surface film**. Because this film is under tension, any change in shape which would add more surface area (and further increase the tension) is resisted. Water drops and submerged air bubbles, as examples of air–water interfaces, are almost perfectly spherical because a sphere has less surface area per unit of volume than other shapes.

The surface tension of water is temperature-dependent. It decreases as temperature increases by the relationship given in Figure 1.2.

Surface tension also affects whether a droplet will bead up or spread out on a solid surface. The angle of contact between a liquid and a solid is related not only to the *cohesion* of the water molecules (attraction for each other) but also to the *adhesion* of the liquid to the solid. If this contact angle (θ in Figure 1.3(a)) is less than 90°, the liquid is said to "wet" the solid. If the angle is greater than 90°, the liquid is "non-wetting".

Water is wetting to a clean glass surface or a bar of soap but does not wet wax (White, 1986). Non-wettable objects with a higher density than water can be supported by the surface film up to a certain point. For example, in water at 18°C, a dry sewing needle of 0.2 g will "float", whereas at 50°C, it will sink. Near sandy banks, patches of fine dry sand may likewise be supported by the water surface. Insects which dart quickly around on the water surface tend to have a waxy coating which functions as a water repellant (Vogel, 1988).

Adding a wetting agent such as detergent to the water will reduce the surface tension, making it more difficult for mosquitoes to "attach" to the surface film from the underside or for water-striding insects to walk across it. If a baby duck is placed in a tub of soapy water the water-repelling oil in its feathers dissolves, releasing air trapped within the feathers, and it

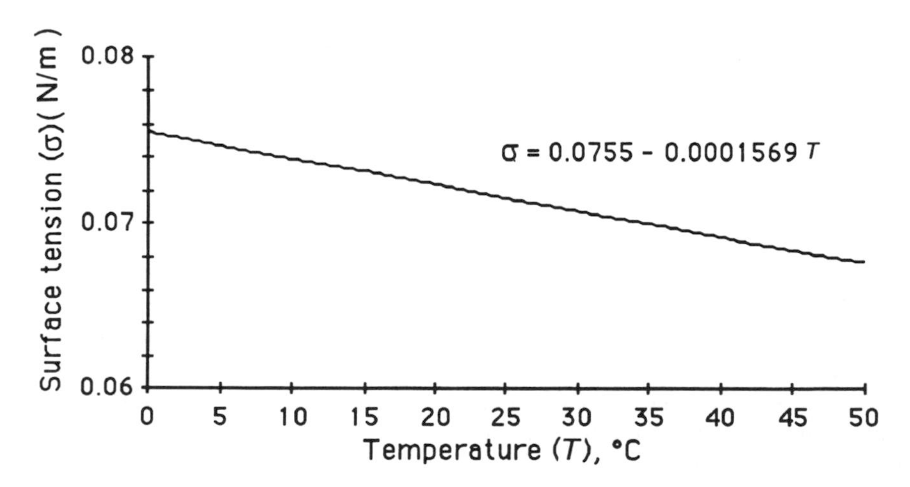

Figure 1.2. Relationship between surface tension and temperature. The equation is from Stelczer (1987)

sinks (Bolemon, 1985). Cormorants do not have water-repellent feathers, and they become saturated when they dive for fish. After struggling out of the water, they must spread their wings out to dry.

Wetting agents are also added to liquid pesticides so that they spread out and cover more surface area on plant leaves. Similarly, laundry detergents reduce surface tension, allowing water to penetrate through dry clothes more readily (Vogel, 1988).

Another important implication of surface tension is that pressure within a droplet of water in air, or within an air bubble under water, is higher than the pressure outside. The increase in pressure is given by (White, 1986):

$$\Delta p = \frac{2\sigma}{r} \tag{1.8}$$

where Δp is the increase in pressure (in N/m^2) due to surface tension and r is the radius (in metres) of the droplet or bubble. It can be seen that the pressure becomes larger as the radius gets smaller. Because of the increased pressure, the air held in small bubbles will tend to go into solution, and the bubble will shrink. Thus, very small air bubbles will quickly collapse and disappear. Vogel (1988) offers a fascinating discussion on how bubbles form at scratched surfaces in beer glasses and other biologically related implications of surface tension.

Capillarity is another phenomenon caused by surface tension. Capillarity, which causes water to rise in plant stems, soil pores in levees or streambanks and thin glass tubes, results from both adhesion and cohesion. Its height is positive (capillary rise) if liquids are wetting and negative (capillary depression) if liquids are non-wetting, as shown in Figure 1.3(b). It should also be noted that the meniscus is concave for wetting liquids and convex for non-wetting.

The formula for capillary rise (or depression), h (m), is (White, 1986):

$$h = \frac{2\sigma\cos\theta}{\rho g r} \tag{1.9}$$

where r is the radius (m) of the tube or mean radius of soil pores, ρ is the density of the water (kg/m^3) and the other symbols have been explained earlier in this section.

From Equation 1.9 it can be seen that capillarity increases as the tube or pore radius decreases. For water in glass tubes with diameters over about 12 mm, capillary action becomes negligible (Daugherty, 1961). It can also be seen that h is positive for $\theta < 90°$ (wetting liquids) and negative for $\theta > 90°$ (non-wetting). For open-water surfaces and soil pores the simplification $\theta = 0$ is usually made so that the ($\cos\theta$) term drops off (Stelczer, 1987). In soils, organic matter and certain mineral types can

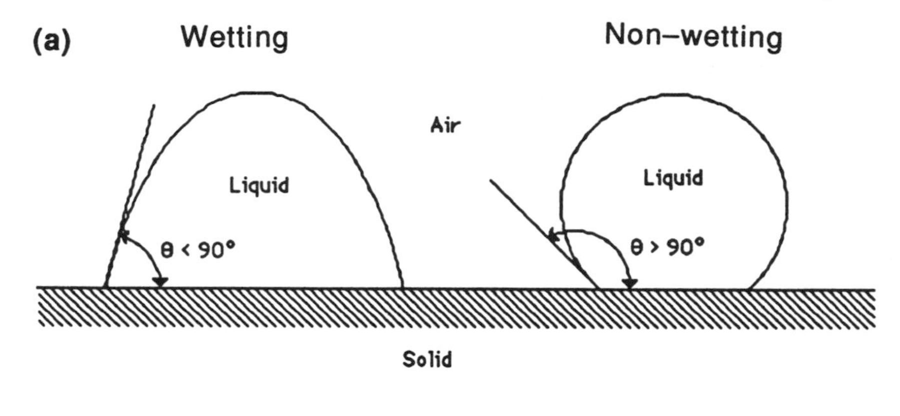

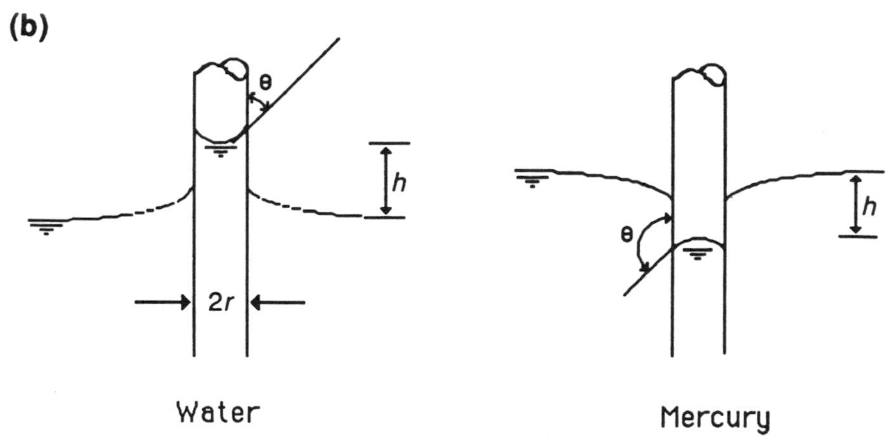

Figure 1.3. Effects of surface tension: (a) on the angle of contact, θ, in wetting and non-wetting liquids, and (b) on capillarity in circular glass tubes of radius r, where h is capillary rise or depression

increase the contact angle above 90°, in which case the soil will not wet (Kirkham, 1964). For example, after intense fire, soils can become "hydrophobic", preventing water from infiltrating (Branson et al., 1981).

1.3.4 THERMAL PROPERTIES

Temperature has an effect on most of the other properties of water such as density, viscosity and dissolved oxygen concentration. It also has an important influence on the metabolic rates of aquatic organisms.

Streams, as a rule, exist between the temperature extremes of ice floes and boiling hot springs. Pure water freezes at 0°C and boils at 100°C. The presence of dissolved solids raises the boiling point and depresses the freezing point as compared with pure water. Since aquatic organisms normally concentrate salts in different proportions to those in the surrounding solution, their "boiling" and "freezing" temperatures will be different from those of the surrounding medium, and some primitive organisms such as blue-green algae and bacteria can tolerate great extremes of temperature.

In studies of aquatic systems, temperature data are sometimes converted to **degree-days** to correlate temperature with snowmelt, plant germination times or developmental times for aquatic insects, where:

$$\text{degree-days} = nT_{avg} \tag{1.10}$$

Here, n is the number of days from a given starting date and T_{avg} the mean daily temperature above some base, usually 0°C (Linsley et al., 1975b). A day with a mean temperature of 30°C would therefore represent 30 degree-days above 0°C. An equivalent calculation can be performed to obtain the number of degree-hours in a day using average hourly temperatures.

The amount of heat a body of water absorbs depends on the amount of heat transferred to it from the air and streambanks, as well as the **thermal capacity** of the water. For water, the thermal capacity is very high in comparison with other substances, meaning that it can absorb a large amount of heat before its temperature increases substantially. Thermal capacity, T_c, has units of J/°C, and is defined as (Stelczer, 1987):

$$T_c = cM \tag{1.11}$$

where M is the mass of water (kg) and c is the specific heat of the water (J/kg·°C). **Specific heat** is the amount of heat required to raise the temperature of a unit mass of water by 1°C. As shown in Table 1.3, it is temperature-dependent, reaching a minimum at 30°C. Thus, a kilogram of water at 10°C would require 4.19 J of heat energy to raise its temperature to 11°C.

Energy is released when water freezes, a fact used by citrus fruit growers when they spray their trees with water to protect them from frost damage.

Table 1.3. Values of specific heat for water at various temperatures. From Stelczer (1987), by permission of Water Resources Publications

Water temperature (°C)	Specific heat (J/kg·°C)
ice	2.039
0	4.206
10	4.191
20	4.181
30	4.176
40	4.177
50	4.183

The **latent heat of fusion**, the energy needed to melt ice or the energy which must be taken away for it to freeze, is 335 kJ/kg for water at 0°C. At the other extreme, additional energy is required when water reaches the boiling point to get it to vaporize. Vaporization actually reduces the temperature of the remaining water. The **latent heat of vaporization** for water at 100°C is 2256 kJ/kg. These latent heat values are relatively high in the natural world, and are caused by hydrogen bonding. This bonding is also responsible for the unusual behaviour of water density near the freezing point.

1.3.5 ENTRAINED AIR AND DISSOLVED OXYGEN

Dissolved oxygen (DO) is actually a chemical property of water, but is included because it is affected by physical properties such as temperature and turbulence, and because of its biological relevance. Oxygen enters water by diffusion at the interface between air and water at the surface of a stream or at the surface of air bubbles. It can also be produced from the photosynthesis of aquatic plants.

Entrainment of air under waterfalls and in the frothy whitewater of rapids increases the amount of interface area where diffusion can occur. Most of this entrained air soon escapes, however, and it is the escape of these air bubbles which produces the roar of rapids and the murmur of meandering brooks (Newbury, 1984). The amount of air remaining in the water is determined by the gas-absorbing capacity of water, which is dependent on temperature and ambient pressure.

Under normal atmospheric pressure and a temperature of 20°C, water will contain about 2% (by volume) dissolved air (Stelczer, 1987). As temperatures rise, the gas-absorbing capacity of water decreases rapidly, reaching zero at 100°C. Although the concentration of oxygen (O_2) in the atmosphere is about 21%, oxygen is more soluble in water than nitrogen, and the dissolved air contains from 33% to 35% O_2, depending on the temperature—a fact which has no doubt played an evolutionary role in the dimensions of gills and other respiratory mechanisms.

At any particular temperature the highest possible dissolved oxygen (DO) content is termed **oxygen saturation**. Table 1.4 gives oxygen saturation values for a range of water temperatures. The amount of DO actually present in the water can be expressed as a percentage of the saturation value:

$$\text{Percentage saturation (\%sat)} = \frac{\text{Actual DO concentration}}{\text{Saturation concentration}} \times 100 \quad (1.12)$$

Concentrations are usually given in milligrams per liter (mg/l). For example, if a 20°C water sample has a DO content of 8.2 mg/l, then %sat = (8.2/9.05) × 100 = 90.6%.

Organic matter in streams is assimilated by bacteria which use dissolved oxygen for the aerobic processing of organic materials. An increase in the

Table 1.4. Oxygen saturation concentrations at normal atmospheric pressure. After Best and Ross (1977), by permission of Liverpool University Press

Water temperature (°C)	Oxygen saturation (mg/l)
0	14.2
5	12.7
10	11.3
15	10.1
20	9.05
25	8.25
30	7.83
40	6.60

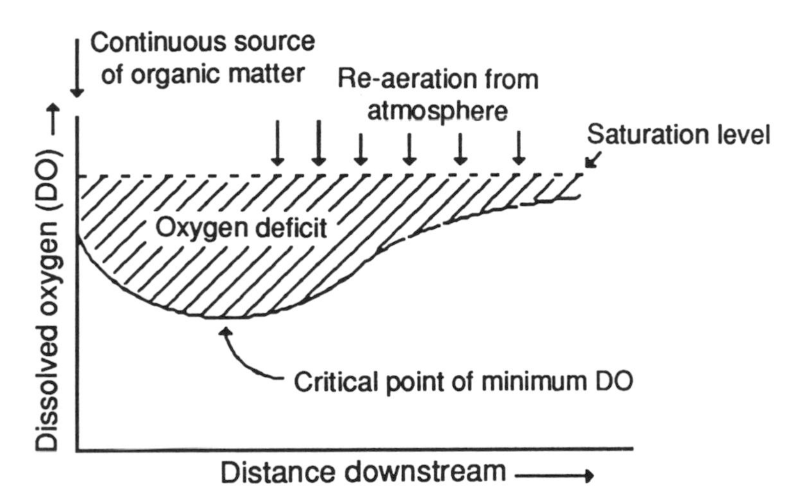

Figure 1.4. Dissolved oxygen sag curve

amount of organic matter (e.g. sewage or detritus stirred up by dredging or an overload of autumn leaves in temperate climate zones) stimulates bacterial growth. If the organic load is extremely excessive nearly all the dissolved oxygen can be used up by the bacteria, leading to anaerobic conditions. In these streams, conditions can become unfavourable to forms of aquatic life sensitive to oxygen levels (Best and Ross, 1977).

The process of deoxygenation and re-aeration of streams produces a pattern in the DO concentration known as the **dissolved oxygen sag**, first described by Streeter and Phelps in 1925 (Clark et al., 1977). A sag curve is illustrated in Figure 1.4, representing the dissolved oxygen "deficit" (amount below saturation level) as it varies with distance downstream. A light organic load and adequate aeration will only cause a slight dip in the curve with a quick recovery, whereas a heavy load and low re-aeration rate may cause DO to decrease to 0%, from which it recovers only slowly. Equations for estimating the sag curve are given by Clark et al. (1977, pp. 296–8), and require field studies to determine the degree of organic pollution and the re-aeration characteristics of the stream.

2 How to Study a Stream

2.1 FOCUSING ON PHYSICAL HABITAT

Before beginning, the definitions of the terms **river**, **stream** and **catchment** should be clarified. Dictionaries make little distinction between the words "river" and "stream", although, in general, river is associated with "larger" and stream or creek with "smaller". In colloquial terms, one is *up* a creek but sent *down* the river, implying relative sizes. This relative definition will be retained, but the word "stream" will be used as a generic term for flowing waters throughout the text. A **catchment** is the area above a specific point on a stream from which water drains towards the stream. Catchments and their characteristics will be described further in Chapter 3.

At the interface between aquatic ecology and hydrology, studies of streams fall roughly into the following categories:

(1) Description or classification of aquatic habitats based on their biota and environmental characteristics. Descriptions of the flowing environment are also needed for simulating the same conditions in laboratory flumes;
(2) Monitoring programs to determine variability in the natural environment over time or to detect some trend due to environmental deterioration or recovery (Green, 1979);
(3) Comparison of conditions at one place/time with conditions at another place/time; e.g. comparing effects of management or of some experimental treatment, either between sites or at the same site at different times (Platts et al., 1987);
(4) Development of relationships between variables, e.g. local water velocity and blackfly larvae populations, or catchment area and stream width.

A variety of factors control the abundance, distribution and productivity of stream-dwelling organisms such as competition for space, predation, chemical water quality, nutrient supplies, the presence of waterfalls or dams, and flow variability. Thus, studies of streams may involve the measurement and analysis of biological, chemical and physical parameters. The emphasis throughout this text, however, will be on physical habitat: those factors which form the "structure" within which an organism makes its home. Physical factors are generally more predictable, less variable and more easily measured than biological or chemical ones, and are thus preferable for general, consistent descriptions of streams.

Patterns of physical habitats are created along a stream and within the pools, riffles and boulder clusters within particular stream reaches. It is to these patterns that stream biota respond and adapt. An individual species will have a range of tolerance to any given factor, with some factors more critical for some species than others. Thus, if the physical factors are known, predictions can be made about the abundance and/or diversity of organisms. The accuracy of these predictions will depend on the extent to which biological or chemical factors (e.g. predation or pollution) are also affecting the stream biota.

Following is a discussion of the physical factors which are of the most ecological significance: streamflow, velocity, channel shape, substrate and temperature. Dissolved oxygen and salinity, although chemical factors, have been included since they are influenced by physical factors, they have high ecological relevance and they can be measured relatively easily. Vegetation has also been included as a related factor because of its influence on the physical nature of streams. These factors are all highly interrelated, making it difficult to distinguish the effect of one from that of another. However, this fact also makes it possible to use one measure as a "surrogate" for others if their interrelationships are known.

More information on the effects of physical factors on the distribution of biota can be obtained from texts on aquatic biology or stream ecology such as Barnes and Mann (1980), Bayly and Williams (1973), Brown (1971), DeDekker and Williams (1986), Fontaine and Bartell (1983), Goldman and Horne (1983), Hynes (1970), Maitland (1978), Moss (1988), Resh and Rosenberg (1984), Townsend (1980), Uhlmann (1979) and Welch (1935).

Streamflow

Streamflow is an important factor geomorphologically because of its relationship to the form of the stream channel; i.e. streamflow increases and channels become larger in the downstream direction. It also has ecological relevance because of the influence of streamflow patterns on organisms, both in the downstream direction and over time. Ephemeral streams (Section 4.1), for example, usually (but not always) support different species than perennial ones. Snowmelt-fed and spring-fed streams generally have more predictable and less variable flow patterns than rainfall-fed ones. Their flora and fauna will be different from those found in highly variable streams, where organisms may require flexibility in their feeding, growth and/or reproductive behaviours. Fish community patterns will also tend to be influenced more by the streamflow in variable streams, and more by biological factors such as competition, predation or food resources in stable streams (Pusey and Arthington, 1990).

Floods and droughts can have significant impacts on riverine species. Periodic scouring of banksides and inundation of floodplains regulate plant growth and nutrient input to the stream. The patterns of flooding affect the distribution of plant species both within the stream and along a gradient from the river's edge to upland areas. Prolonged flooding of wetlands is needed for waterbirds to feed, rest and reproduce. The survival of juvenile fish may also depend on the inundation of floodplains, billabongs and backwaters. Moreover, floods serve as a signal for some fish species that it is time to spawn. Floods turn over rocks, altering the configuration of the streambed and "resetting" the ecosystem by allowing a succession of organisms to recolonize the substrate.

During low flows, temperature and salinity levels may rise, and plant growth within the channel can increase. Some species may rely on low-flow periods for a part of their life history; for others, it is a time of stress. The stream may dry to a series of scattered pools connected only by subsurface flows, limiting movement and increasing predation and competition for nutrients and space within the remaining waters.

Intermittent streams, which dry up completely for periods of time, experience a greater range of physical and chemical variation (e.g. in temperature and dissolved oxygen levels), and thus support unique biological communities. These streams generally have a lower species richness as compared to perennial ones (Lake et al., 1985). For coping with temporary waters, some organisms have developed special adaptations such as dormant phases which allow them to survive. Some larvae of aquatic insects, for example, burrow deep into the streambed to find sufficient moisture, or species may have drought-resistant eggs or spores and quick regenerative powers for the time when favourable conditions return. Often, only part of a stream will go dry and affected reaches will be repopulated rapidly from remaining pools or upstream tributaries. Williams (1987) provides additional information on the ecology of intermittent streams.

Streamflow is a particularly important factor in the study of regulated waters, where modifications to the natural patterns of flow can have marked effects on the stream's flora and fauna.

Current

As stated by Hynes (1970, p. 121), "current is the most significant characteristic of running water, and it is in their adaptations to constantly flowing water that many stream animals differ from their still-water relatives". Some species have an innate demand for high water velocities, relying on them to provide a continual replenishment of nutrients and oxygen, to carry away waste products and to assist in the dispersal of the species. At a given temperature, the metabolic rates of plants and animals are generally higher in running water than in still water (Hynes, 1970). However, it takes a great deal of energy to maintain position in swift waters, and most inhabitants of these zones have special mechanisms for avoiding or withstanding the current.

On average, water velocity tends to increase in the downstream direction, even though mountain torrents give the impression of high speeds in comparison to the more sluggish-looking lowland streams. Within a particular region, however, local variations create a mosaic of patterns which support species with different preferences. The velocities actually encountered, then, are of more relevance than average velocities (Amour et al., 1983). As flow levels increase, these velocity patterns will shift, forcing organisms to find refuge in calmer backwaters, behind rocks or snags, within vegetation stands or beneath the streambed.

At a finer scale, the leaves and stems of plants or the arrangements of rocks can vary the local flow environment, creating "microhabitats" for other organisms. Moving even closer to the surfaces of these features, very small animals such as protozoans can live in a thin fluid layer of almost zero velocity. Complex communities of bacteria, fungi, protozoans and other microscopic organisms form "biofilms" on surfaces within the stream, which constantly grow and slough off under the influence of the current.

A factor related to velocity patterns is turbulence, which is important in the aeration of waters and the ability of a stream to carry sediments. This has a "buffeting" effect on organisms exposed to the current. Near the streambed, small turbulent vortices stir up the substrate and circulate foodstuffs to animals at their edges.

Current affects the distribution of sediments on the streambed through its influence on lift and drag forces (Section 6.5). Organisms subject to these same forces often show morphological adaptations such as streamlined or flattened bodies, or the presence of hooks or suckers for clinging to the substrate. Blackfly larvae (Diptera: Simuliidae), characteristic of very fast waters, attach to the substrate with hooks and have a silk "lifeline" with which to reel themselves in when they are dislodged. Stream-dwelling mollusc species have heavier, thicker shells than still-water forms, perhaps for ballast as well as protection from moving stones (Hynes, 1970). Species of fish which must negotiate strong currents tend to be streamlined, whereas those which spend almost all of their time near the streambed are more flattened from top to bottom (Townsend, 1980).

Species unable to tolerate high currents may use behavioural mechanisms to escape by burrowing into the streambed, hiding under rocks or building shelters. Even fish and eels utilize the dead-water regions behind rocks for shelter, moving in short bursts from one to the next.

The distribution of current within a stream can be considerably affected by channel modifications such as desnagging or straightening. The comparison of velocity distributions in modified and unmodified streams is valuable in stream-rehabilitation work (Section 9.4).

Water depth and width

The depth and width of a stream is related to the amount of water flowing through it. However, variations in channel form such as pools and riffles, wide meander loops and sand bars will create variation in water width and depth even where the streamflow is the same.

Water depth has an influence on water temperature, since shallow water tends to heat up and cool down more rapidly. It affects light penetration (more so in turbid waters), influencing the depth at which aquatic plants have enough energy for photosynthesis. Hydrostatic pressures (Section 6.2) also increase with depth, affecting the internal gas spaces in both plants and animals.

Depth has an influence on the distribution of benthic invertebrates, with most preferring relatively shallow depths (Wesche, 1985). Both depth and width affect the physical spacing between predator and prey species. In general, larger fish tend to prefer to live in pools and smaller fish in shallower water. Depth may become a limiting factor for fish migration when the water is too shallow for passage. Changes in water level can also affect the survival of species (for example, by stranding fish or eggs or inundating seedlings at the wrong time).

According to Pennak (1971), the width of streams determines much of the biology of stream habitats. The influence of overhanging streamside vegetation on shade and nutrient input decreases as the stream width increases. Thus, instream photosynthesis generally increases in the downstream direction.

For resting and nesting, some waterbirds need a water "barrier" of a certain width to protect them from predators. Migrating birds may require open riverine corridors for navigation and a certain width of water unobstructed by trees for landing and take-off. Mammals such as beavers also have specific requirements for width, depth and slope (Statzner et al., 1988). For terrestrial animals, width and depth will affect their ability to migrate from one side of a stream to the other.

Substrate

In a stream, "substrate" usually refers to the particles on the streambed, both organic and inorganic. Inorganic particle sizes generally decrease in the downstream direction. On a more local scale, larger particles (gravel, cobble) are associated with faster currents and smaller particles (sand, silt, clay) with slower ones. Generally, stability of the streambed decreases as the particle size decreases, but this will also depend on the mix of particle sizes and shapes. Studies of substrate composition should consider the average and range of particle sizes, the degree of packing or imbeddedness, and the irregularity or roundness of individual particles (Chapter 5).

Substrate is a major factor controlling the occurrence of benthic (bottom-dwelling) animals. A fairly sharp distinction exists between the types of fauna found on hard streambeds such as bedrock or large stones and soft ones composed of shifting sands. Slow-growing algae, for example, require stable substrates such as large boulders (Hynes, 1970). Additionally, a whole complex of microfauna can occur quite deep within streambeds. These may carry out most or all of their life histories underground in this **hyporheic** zone.

In general, lowland streams with unstable beds tend to have a lower diversity of aquatic animals. Aquatic plants, however, may prefer finer substrates, the plants then becoming substrates for other organisms. Silt substrates may support high populations of burrowing animals, particularly if the silt is rich in organic matter. Freshwater mussels, for example, mostly occur in silty or sandy beds. Clay substrates typically become compacted into "hardpan" (Pennak, 1971), supporting little except encrusting algae and snails.

The greatest number of species are usually associated with complex substrates of stone, gravels and sand. The mix of coarser particles in riffles provides the richest aquatic insect habitat, and is considered the "fish food" production zone in upland streams. With larger substrate materials, more shelter is available in the form of crevices and irregular-shaped stones, and firmer surfaces are provided for the insects to cling to. Crustaceans such as freshwater crayfish also use rock crevices for shelter.

Fish have substrate requirements in terms of shelter from the current, places for hiding from predators, and sites for depositing and incubating eggs. Salmonids, for example, dig nest-like "redds" in gravel substrates by lifting particles with a vacuum-generating sweep of their tails. Successful incubation of the deposited eggs depends on circulation of water through the gravels to supply oxygen and carry away waste products.

Streambed particles are subject to dislodgement during floods, from dredging or other human disturbances and to a much lesser extent when the activities of bottom-feeding fish or burrowing animals stir up sediments. Suspended sediments reduce light penetration and thus plant growth; they can also damage the gills of insects and fish. Larger grains in suspension have a "sandblasting" effect on organisms, and rolling stones can crush or scour away benthic plants and animals.

The composition of stream substrates can be altered by sediment influxes from upland erosion and by channel modification. Excessive siltation of gravel and cobble beds can lead to suffocation of fish eggs and aquatic insect larvae and can affect aquatic plant densities. This, in turn, can result in changes in mollusc, crustacean and fish populations. Generally, these changes tend to cause a shift towards downstream conditions (i.e. unstable beds of fine materials), effectively extending lowland river ecosystems further upstream.

Temperature

In general, water temperature increases in the downstream direction, to a point where the water reaches an equilibrium with air temperatures. Water temperature changes both seasonally and daily, but to a lesser degree than air temperature. Seasonal fluctuations tend to be more extreme in lowland streams whereas daily fluctuations may be more extreme in the smaller, upland ones, especially where they are unprotected by vegetation or other cover. In temperate or cold climates, upstream reaches may actually remain warmer in the winter than those downstream, particularly when these upper reaches are spring-fed.

Local variations in shade, wind, stream depth, water sources (e.g. hot springs) and the presence of impoundments will alter the general trends caused by geographical position. Many organisms take advantage of these local variations. For example, Wesche (1985) cites studies which have shown that some trout species select spawning sites in areas with groundwater seepage, where the warmer waters protect the eggs from freezing and reduce hatching times.

When water cools, it becomes more dense and sinks. In most stream reaches, turbulence keeps the water well mixed, but temperature stratification can occur where waters are more stagnant, such as in deep pools. One of the unique properties of water is that it is less dense as a solid. In winter, ice and snow can form an insulating blanket over streams, under which aquatic life can continue. Ice usually starts forming when the entire water mass nears 0°C, beginning with low-velocity areas near the edges of streams, along the streambed and on the underwater surfaces of plants.

The temperature of a stream is critical to aquatic organisms through its effects on their metabolic rates and thus growth and development times. With the exception of a few aquatic birds and mammals, most aquatic animals are cold-blooded; i.e. their internal temperatures closely follow that of the surrounding water. As a general rule, a rise of 1°C increases the rate of metabolism in cold-blooded aquatic animals by about 10%. Thus, these aquatic organisms will respire more and eat more in warmer waters than in colder ones. Each organism will have maximum and minimum temperatures between which they can survive, and these limits may change with each life stage. For fish, unusually high temperatures can lead to disease outbreaks, cause the inhibition of growth and cause fish to stop migrating (Platts, 1983).

Water temperature is thus an important factor in regulating the occurrence and distribution of riparian vegetation, fish, invertebrates and other organisms. Temperature also affects other properties of the water such as viscosity, sediment load and concentrations of nutrients and dissolved oxygen. It can thus be difficult to separate the direct effects of temperature on biota from the indirect effects of these other properties. Stream classification systems (Section 9.2) often include temperature as a factor.

Dissolved oxygen

Dissolved oxygen (DO) is essential for respiration in aquatic animals as well as being an important component in the cycling of organic matter within a stream. Since gas solubility generally decreases as the temperature rises, this can lead to lower DO levels during the summer.

As mentioned in Section 1.3.5, the oxygen concentration of air dissolved in water is higher than it is in the atmosphere. Some organisms use this property indirectly, such as the beetles which carry bubbles of air underwater to function as gills: oxygen diffuses into the bubble from the surrounding water as it is used (Hynes, 1970).

In the turbulent, well-mixed waters of upland streams, dissolved oxygen concentrations are usually near saturation levels. As these turbulent reaches gradually give way to more poorly mixed waters downstream, biological sources of oxygen become more important. Photosynthesis can actually supersaturate the water with oxygen during the day. Then, night-time respiration and decomposition demands for oxygen can lead to diurnal changes in DO level. Diurnal variations in DO have been used as a basis for computing primary production in streams (Bayly and Williams, 1973).

The distribution of DO influences the patterns of species found along streams. Carp, tench and bream, for example, can survive in low-oxygen waters, whereas trout require higher levels. The amount of oxygen will also affect the activity of organisms. As mentioned by Hynes (1970), the swimming speed of young salmon varies with DO concentration.

Oxygen is supplied continuously to stream organisms by the current. In fact, some swift-water invertebrates rely on the current to carry water across their gills, being unable to produce their own currents for respiration. Many of these organisms can tolerate lower DO concentrations if the velocity is sufficiently high (Hawkes, 1975).

Low-oxygen conditions can result from both natural and artificial causes. When organic matter such as sewage or detritus undergoes aerobic decomposition by bacteria, oxygen is removed from the water. Oxygen deficiencies can be found in stagnant waters at the edges of streams or when the stream is totally covered by ice or mats of water weeds. Reductions can also be caused by influxes of deoxygenated groundwater or downstream releases of anoxic water from the bottom of reservoirs (Section 9.1). These effects can create conditions unfavourable to those aquatic species sensitive to oxygen levels.

Dissolved salts

Salinity refers to the concentration of ions dissolved in water. The ions most commonly contributing to salinity are the cations sodium, magnesium and calcium, and the anions chloride, sulphate, carbonate and bicarbonate. Generally, the concentration of salts increases in the downstream direction, especially if streams originate in areas with resistant rock and flow into regions with sedimentary rocks which erode more rapidly (Townsend, 1980). Salts can also enter a stream from saline groundwater, from sea salts dissolved in the rainwater of coastal areas and from agricultural runoff. In general, water originating as groundwater tends to have a higher dissolved salt concentration than surface runoff.

Since saline water is denser than purer water, stratification can occur in slow-moving pools fed by saline groundwater or in estuarine regions. Here, the fresher water "floats" as a layer over the denser saline water. Salinity levels are generally (although not universally) inversely related to discharge levels: the highest salinities occur during low flows, with higher flows having a diluting effect.

As with other factors, tolerance of saline conditions can influence the distribution and abundance of stream biota. Most freshwater plants and animals are unable to maintain their internal ionic balances in saline waters. Water then diffuses out of cells, leading to dehydration, or excessive amounts of ions can diffuse into cells, producing toxic conditions. The concentration of individual ions is also important; for example, increases in water hardness favour some groups such as molluscs and crustaceans (Hawkes, 1975). Some organisms, particularly those which have evolved from marine species, have special mechanisms for restricting salt movement and/or excreting salt against an osmotic gradient. Hart et al. (1990a,b) provide a comprehensive review of the effects of salinity on Australian plants and animals.

Vegetation

Vegetation influences the physical habitat of streams by providing shade and altering the structure of channels. The flexibility of plants can absorb erosive forces directed against the streambanks from water currents and ice and debris flows. Overhanging vegetation can help keep streams cooler in summer and warmer in winter. Shade affects the growth of algae in streams and the distribution of other organisms which have shade or sun preferences. Dappled shade also has a camouflaging effect on fish which helps prevent predation.

Trees and shrubs are both important for bank stability, for shading the banks and streams, for providing nutrient input to the streams in the form of litter, and as wildlife habitat. Standing dead trees offer hollows for breeding and shelter for birds and mammals. When trees fall into a stream they provide cover for fish and insects, and dams of logs and other debris form pools and restrict the downstream movement of sediments (see Section 7.5.1). Shrubs, grasses and other plants growing along the sides of streams slow the current and encourage silt deposition.

Aquatic macrophytes and algae grow within the stream, especially in less-turbulent and less-shaded waters. The presence of large quantities of aquatic plants is often considered undesirable, since they retard the water flow, causing waterlogging of adjacent lands and increasing the risk of flooding. Aquatic vegetation is cleared by mechanical cutting in many countries (Dawson et al., 1982). The retarding effect of both aquatic and riparian plants depends on their density and their flexibility, and will thus vary with species and life stage.

As an aid to separating "zones" of a riparian ecosystem, Lewis and Williams (1984) suggest dividing the river cross-section into mid-channel, channel edges, banksides and banktops (Figure 2.1). Aquatic macrophytes, algae, and mosses may inhabit the **mid-channel zone** if scour and shade are not excessive. At the **channel edges**, pond-like conditions can develop in backwater areas and during low-flow periods, allowing rich plant growth including floating, unrooted plants. Rooted plants with floating leaves, reeds, rushes, liverworts, mosses and other species tolerant of inundation may also inhabit this zone. **Banksides** form a transition between the channel

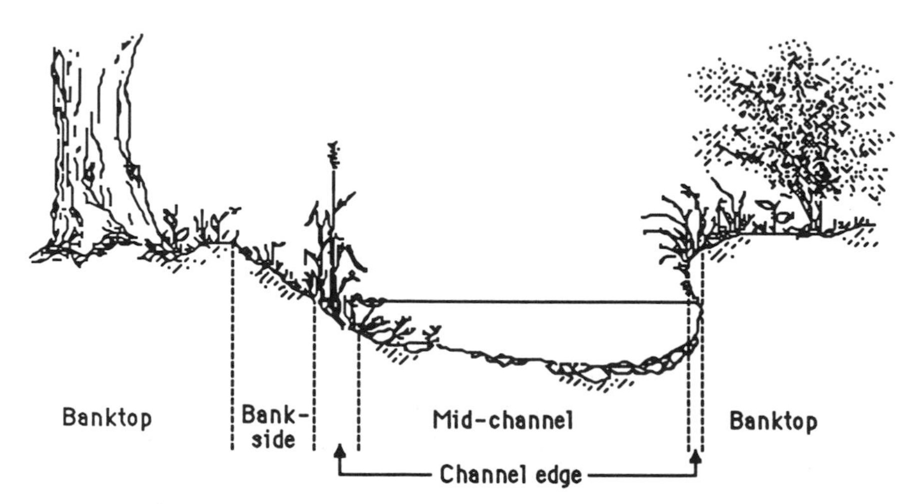

Figure 2.1. Channel cross-section showing vegetation zones as defined by Lewis and Williams (1984). Comparing this diagram with Figure 8.1 can give an indication of the frequency of inundation of the various zones

and the floodplain, with a merging of river and terrestrial plant species. Tree species may grow in this zone, and, with grasses, rushes, sedges and other herbs, they form a dense mat of roots which protects banks from erosion. Annual plants often spring up in eroded or unshaded areas, their flowers and seeds being important food sources for insects and birds. **Banktops** are primarily dry, and vegetation is usually influenced by the adjacent land use. Some plants may be characteristic of this riverside zone, such as ferns and berry bushes, even though they may also be found in upland areas.

The removal of streamside vegetation can lead to bank instability and changes in aquatic biota. Revegetation is an important component of stream-rehabilitation work, and is discussed in Section 9.4.2.

2.2. THE PLANNING PROCESS

2.2.1 GENERAL

When charting the course of a new study, the object is to choose an approach that allows one to "miss the snags", so to speak. It may be tempting to head out to a stream with a Surber sampler, a current meter and a vague notion that a statistician can work magic with the results, but a carefully thought-out plan greatly increases the chances of obtaining meaningful results. As Hamilton and Bergersen (1984, p. I-7) put it, "no amount of data juggling will make the results of a poorly planned study useful".

A well-organized study plan will maximize the amount of useful information collected while minimizing wasted efforts and worthless data. Expending up to 20% of the total study effort on the planning process should not be considered unreasonable. The time is well spent and can prevent wasted efforts in the field, even though, as Green (1979, p. 31) says, "we all find time saved in the hand to be more attractive than time saved in the bush".

In the planning process, all phases of the study should be considered, from purpose through the presentation of results. The process follows the scientific method, and can be summarized by the following steps:

(1) Define a question (or series of questions);
(2) Choose a method for answering the question(s);
(3) Collect the appropriate information;
(4) Analyse the information; and
(5) Answer the question(s).

The steps are interconnected and the process itself a delicate balancing of statistical ideals, choice of measurement techniques, the unpredictability of nature, and available resources of time, money, equipment and personnel. The stages of the planning process are presented sequentially, but an "iterative" approach should be taken in developing a final plan.

2.2.2 WHAT IS THE QUESTION?

The first question to be asked (after "is this study necessary?") might be: "What is the object?" Are high or low flows of interest? Will the study be localized or representative of a larger region? Will the results be used to classify habitat or to test a hypothesis about the effects of some treatment? Fundamental research on the habitat of freshwater mussels, for example, will have a different objective than studies on the reclamation of a dredge-mined stream.

Objectives should be defined in the initial stages of a project in order to set goalposts for the study. The objectives should be clearly stated, communicating the nature and depth of the problem, and they should be achievable within the limitations of time and budget. Excessive ambition and vagueness are common faults at this stage (Platts et al., 1987). It is better to successfully carry out a small project than to leave a larger, more complicated one unfinished.

The **study question** is an outgrowth of the objectives. This may start out in a general form, such as: "Does the length of crocodiles vary with river width?" But in the process of defining boundaries for the study, it may be necessary to make the question more specific, e.g. "Is the average length of freshwater crocodiles (*Crocodylus johnstoni*) measured during daylight hours in October in the main stem of the Wildman River different between lower and upper reaches?" It is important to clearly formulate the study question based on the extent to which generalizations (inferences) can be made from the answers. The physics of flowing waters is essentially the same anywhere on earth; crocodile lengths have a less universal relevance.

Four basic study objectives are listed at the beginning of Section 2.1. In the first two categories, the question is "What's out there?", whereas the third asks "How is this different from that?" and the fourth "How is this related to that?" On the surface, there are similarities between the different types of studies, but they each represent different objectives, sampling designs, analysis procedures and methods of presenting results.

A little preliminary or "exploratory" work can help to focus both the design and scope of the project. In this preliminary stage, maps and photographs are collected, the study area is defined and tentative sites for data collection are identified. If possible, a trip to the site or a fly-over should be made. An appropriate period of study is chosen, based on whether the study is for existing conditions or long-term trends. Data on the hydrology and meteorology should be obtained for the study area and/or for nearby or similar regions (Chapter 3). Additionally, a literature search can provide life-history data on species to be evaluated, as well as information on pertinent ecological and hydrological measurement techniques (Hamilton and Bergersen, 1984). In the course of this preliminary process, initially fuzzy boundaries should sharpen into a clear definition of the study question(s).

2.2.3 CHOOSING YOUR METHOD: THE STUDY DESIGN

"Methods" include both data-collection techniques and statistical analysis methods: what to measure, when and how to measure it, and how to analyse the results. The choice of methods may be patterned after previous studies to allow the comparison of results, or modified based on recommendations of "those who have been there before". If the results will be used in a model such as PHABSIM (Section 9.3.4), specific data or methods may be required.

When first considering the possible number of environmental variables to measure, all may appear interesting and "critical" to the study. However, the variables should be carefully chosen so that the study gives relevant information rather than just being an exercise in data collection. Data needs will depend on the stream(s) to be studied, the study objectives, the level of detail required and budget restrictions.

The length of study and sampling times will also be based on study objectives. Monitoring studies may extend over a period of years, whereas others may coincide with a particular life stage of a species such as the spawning season. Alternatively, a period of high or low flow may be chosen for study. Automated sampling, either continuous or triggered by a rise in water level, will be helpful in the sampling of flow-dependent variables.

A tentative approach should be developed for collecting field information. This should consider the practicality of techniques and their potential for achieving appropriate levels of accuracy and precision (Platts et al., 1987). Units, scale of measurement and taxonomic level of identification should be selected. Advantages and disadvantages of possible methods in terms of equipment prices, time requirements, and the expertise and availability of personnel should be balanced. The "glamour factor" associated with complex, more precise, expensive methods makes them attractive, but often a simpler method will suffice. It would be overkill, for instance, to try to measure a stream's discharge with a laser Doppler anemometer.

Statistics has affected the basic philosophy of biological sciences to such a large extent that study design, analysis and the conclusions drawn are now virtually inseparable from statistical methods. Each study will utilize statistics to a different degree. Different procedures are needed if the purpose is to measure something specific (e.g. length of a certain crocodile named "Jaws") or to make inferences from the results about a larger population (e.g. lengths of all freshwater crocodiles in Australia).

In a well-designed study the method of statistical analysis will be built in at the start (Green, 1979). The optimal design for environmental studies will differ from that of agricultural or laboratory studies because the same degree of control over variables is not possible (Armour et al., 1983). Care should be taken in the application of statistical methods. Techniques such as the analysis of variance are only powerful if sampling is structured properly and certain assumptions are met. If possible, a statistician's advice should be sought early on in the planning phase, especially for long-term, large-scale studies.

A useful way to check the study design for flaws is to write out a "mock" analysis detailing the statistical analyses that will be used, and their limitations and assumptions. Some assumptions such as random sampling are crucial, and it should be determined beforehand if and how these assumptions can be satisfied. Others, such as equality of variance or the presence of an underlying normal distribution, cannot always be checked ahead of time. In this case an alternate plan should be formulated in case the assumptions are not met. For instance, can the data be transformed? What are the consequences of not meeting the assumptions?

Also, there will be limits to the number of samples or measurements which can be taken and analysed. A well-designed study will ensure that enough samples are taken to meet the study objectives. At the same time, oversampling should be avoided since it wastes time and energy which could be spent on something else, such as another problem or playing volleyball.

A completed study design should, therefore, include a plan map showing the study sites, a listing of the variables to be measured, a choice of approximate dates for field surveys and the number of samples or measurements to be collected. More details on statistical sampling methods and methods of site selection are given in the following sections.

2.2.4 COLLECTING INFORMATION: THE VALUE OF A PILOT STUDY

As a general rule, it is a good idea to test the water before plunging into a new study. Reviewing pertinent literature will provide background information, but a pilot study in the field is advisable for checking out the proposed methodology. This "sneak preview" can provide advance estimates of the biological and physical variability which are needed for designing an efficient sampling program. Pilot studies may point out the need for adjustments in the size and number of sampling sites, the number of samples required, and the sampling design. For example, the presence of spatial patterns may make stratified sampling (Section 2.3.5) desirable. Excessive variability, created by either the methods or by the object of study, may indicate the need to select an alternate variable or develop new sampling methods (Armour et al., 1983).

Additionally, practical difficulties will often show up during pilot studies. For example, if a site is not easily accessible the sampling equipment must be portable. New species may be collected for which the taxonomy is not known. A pilot study is useful, too, for familiarizing field personnel with equipment and methods. The feasibility of the study itself may be questioned during this stage, and a trip "back to the drawing board" indicated.

A pilot study is well worth the investment. Problems can be detected and corrected at less expense in this stage than after the full-scale study is launched.

2.2.5 ANALYSING AND PRESENTING THE RESULTS

In the planning process the method for drawing conclusions from results should be outlined (Cochran and Cox, 1957). This is a good time to consider the purchase of statistical computer software (see Appendix 1). It is also a good idea to become familiarized with the software before data are collected. If these steps are taken and statistical methods integrated into the study design, the analysis should be just a matter of entering numbers into a computer and interpreting the output.

Software packages should be examined in light of requirements for "user-friendliness" and the appropriateness of statistical methods and output, whether as a graph on a computer screen or as a printed summary. Graphical displays are valuable for revealing patterns that might otherwise escape notice (Green, 1979). Graphics are also useful in the presentation of results. Clearly illustrated and labelled graphs can "say a thousand words" in a relatively small space. Three-dimensional and colour plots can be particularly effective in some cases. Methods of presenting the results should be considered during the planning phase. Publishing results in a refereed journal, for example, will require a different approach than presenting a report to a river management agency or giving evidence in court.

Thus, conclusions can be only as good as the study design, the accuracy of measurements and the appropriateness of statistical tests. The odds of obtaining legitimate conclusions can be improved with careful planning. Designing a good study is something of an art, based on professional experience and judgement. It requires knowledge about the object of the study, methods for obtaining the information, and statistical design and analysis. For large-scale, longer-term studies the co-operative judgement, experience and knowledge of a larger group of professionals is needed. However, at any scale, plans must be customized to the individual study;

there are no "cookbook methods". For further information, Green (1979) has written an excellent reference for guiding the development of environmental study designs.

2.3 STRATEGIC SAMPLING

2.3.1 POPULATION, SAMPLE AND OTHER VOCABULARY OF THE TRADE

Wardlaw (1985) points out that science rarely gives a complete and final description. Even though we may never arrive at the Absolute and Complete Truth about anything in nature, statistics can provide some useful approximations and descriptions.

Statistics can be either descriptive or inferential. **Descriptive statistics** summarize the information contained in data in terms of indices which describe central tendency, spread, "clumpedness" and so forth. The objective of **inferential statistics** is to draw conclusions about the characteristics of a large (even infinite) group from a small sampling of that group (Zar, 1974). In environmental studies, both types are commonly of interest. Data are collected, described and related, and generalizations made from the results.

Statistical analysis is a tool for extracting useful information from observations and measurements. In discussing statistical analysis methods it is helpful to develop a common language. A few of the more common terms are described as follows.

A **population** refers to the entire collection (group) of observed or measured **elements** (items) about which one wishes to draw conclusions. Populations can be finite (e.g. the number of logs in a short stream reach), or infinite (e.g. the width of a stream, measured everywhere along its length), or so large as to be considered infinite (e.g. the number of crocodiles in Australia). The qualities or characteristics of interest, such as length, width or counts, are termed **variables**. Variables which cannot be expressed numerically (for example, the quality of pools, colour or a property such as "dead or alive") are sometimes referred to as **attributes**. Attributes can be treated statistically when combined with frequency of occurrence (Sokal and Rohlf, 1969).

If the population is small enough it may be possible to measure every element in it (a **census**). For example, one could count (perhaps on one hand) the number of ecology journal citations found in a mechanical engineering journal during a certain time period. Each citation would be an element of the total population of citations. In most cases, limitations on time, money, site access or the destruction of an element during measurement (e.g. when measuring the life of a light bulb) make a "total census" impractical (Barnett, 1982). Therefore a **sample** is collected in such a way that it represents the whole population. A sampling **unit** is a collection of elements (e.g. a selected stream reach, within which measurements are taken). From information contained in the samples, **inferences** (i.e. "leaps of faith" bounded by levels of confidence) are made about the population from which the samples were drawn.

Parameters describe some fixed characteristic of the population, such as the total, mean or variance. An estimation of a population parameter is called a **statistic**, and is calculated from sample measurements. As the sample size approaches the population size, a statistic will become a better **estimate** of the parameter it is estimating (Zar, 1974). For example, a sample of crocodiles might be chosen from all the crocodiles in a region to obtain a mean crocodile length. The mean crocodile length calculated from the sample is the statistic and estimates the mean length for the region. If *all* the crocodiles in the region are measured, then the computed mean value *is* the population mean. **Efficiency** describes how well the statistic approximates the parameter being estimated (Zar, 1974).

A brief mention of **hypothesis testing** should be made, as it is a frequent application of statistics in biological research. A **null hypothesis** is a formal answer to a study question, stated in a way that is "testable and falsifiable" (Green, 1979). The study question given in Section 2.2.2 can be restated as a null hypothesis (H_0): crocodile lengths are the same in upper and lower river reaches: with the **alternative hypothesis** (H_A) lengths are different. Statistical tests of hypothesis verify or reject the null hypothesis at a certain **level of significance**.

A test designed to have significance level 0.05 will reject H_0 with a probability of 0.05 when H_0 is actually true. For the given example, this means there is only a 5 in 100 (or 1 in 20) chance that we will incorrectly conclude that crocodile lengths are different when they are actually the same. Such an error is called a **type I error**. Depending on the seriousness of making such an error, we may wish to use a test with a significance level smaller or larger than 0.05. A certain amount of loyalty to one's initial choice of significance level and null hypothesis is necessary. One should not rationalize unexpected or unfavourable results to come up with a "better" hypothesis and significance level (Green, 1979).

Some trade-off exists between a type I error and a **type II error**, which is the acceptance of a false null hypothesis (i.e. there really is a difference in crocodile lengths, but the study did not detect it). Associated with the type II error is the **power** of a test. This is the probability of rejecting the null hypothesis when it is false. Statistical power depends on (1) the variation in the data (the more variable the data, the lower the power), (2) the actual difference between experimental treatments (the larger the difference, the easier it is to detect), and (3) the number of replicates taken within each category. Commonly, the power of tests applied to field data is low because of large variability and small actual differences. In general, the more replicates, the more powerful the test. However, there is a "point of diminishing returns" once a certain level is reached, where further sampling does not substantially increase the power.

Hypothesis testing and the vocabulary of statistical sampling is covered in most statistical texts. Although focusing on aquatic insect studies, Allan (1984) is a particularly good reference on hypothesis testing in environmental work. Coehn (1988) provides information on statistical power analysis and Goldstein (1989) gives a recent review of software for calculating the power of a study design. An example of hypothesis testing is given in Section 2.5.

2.3.2 THE ERRORS OF OUR WAYS

Error can creep into a study through faulty sampling design, sloppy measurements, improperly calibrated instruments and poor sample site location. It will affect how closely the statistics from the sample data resemble the population parameters. **Error of estimation** refers to the distance by which an estimate misses the "true" population value (Platts et al., 1987). The error of estimation can be defined by a level of confidence. For example, we might state that the mean freshwater crocodile length in a given region is 1.5 m, with 95% confidence that the "true" mean length is between 0.9 and 2.1 m. This "confidence" is based on the sample variability, which reflects both natural variability in crocodile length and the investigator's ability to measure it.

Thus, both measurement and statistical errors can occur. **Measurement error** refers to the difference between a recorded measurement and its true value. **Accuracy** is the closeness of a measured (or computed) value to its true value. **Bias** is a term used when measurements are "off" due to an introduced source of error (e.g. inaccurate instrument calibration or the stretch of a measuring tape). A constant bias is insidious because it will not be revealed by any manipulation of the sample data (Cochran, 1977). **Precision** is the closeness of repeated measurements to each other. The precision of field measurements is related to the "vagueness" of the methods used. As Hamilton and Bergersen (1984, p. I-4) say, "don't measure it with a micrometer, mark it with a grease pencil, and cut it with an axe". Care in collecting data, in applying the same standards to all sites and in the recording and processing of data can greatly improve the quality of information collected.

Statistical error actually includes both measurement error and the inherent natural variability of the observed phenomenon. It refers to the difference between the parameter of interest and the value derived from sampling. Assumptions about the structure of errors are sometimes necessary in order to calculate confidence intervals, perform tests of significance and draw valid inferences from information contained in the measurements. These **parametric** statistical techniques, which assume that the population follows some statistical distribution (see Appendix 1), require errors to be (1) additive (homogeneous in terms of variance), (2) normally distributed, and (3) independent (Green, 1979). Sokal and Rohlf (1969) discuss these assumptions and the consequences of not meeting them.

Although the natural world almost never satisfies these assumptions, all is not lost. Pilot sampling or the literature may reveal a simple transformation which can reduce violations of the first two assumptions, which commonly occur together (Green, 1979). Alternatively, one can proceed without a transformation but with an awareness of the effects on statistical analysis. Distribution-free, **non-parametric** techniques can also be used, and do not require the same assumptions. These are discussed in a number of statistical texts and, if selected, should be built into the initial study design rather than used as a last-ditch effort to salvage data (Green, 1979).

Independence of errors is the only assumption for which violation is impossible to cure by transformation after the data have been collected. Random sampling, built into some level of the sampling design, is the only way to prevent this violation (Green, 1979; Sokal and Rohlf, 1969). In agricultural and laboratory settings, random sampling is more easily employed than in riverine environments, where, in practice, there is almost always a certain degree of subjectivity involved in the collection of data.

It might be said that true randomness exists only in theory, the real world being a mixture of order and randomness. Even the notion of a "random" coin-tossing experiment fails in the hands of statistician–magician Persi Diaconis, who can control the flip of a coin (Kolata, 1985). Randomization will be discussed further in Section 2.3.3.

Large errors can lead to **outliers**, values which depart significantly from the bulk of the data. These should be eliminated from an analysis only if there is knowledge about why they are abnormal (e.g. someone had been giving growth hormones to Jaws). Graphical exploration of the data can quickly reveal the presence of outliers as well as the homogeneity of error and other patterns which indicate a need for transformation. Statistical tests are also available for measuring the validity of assumptions and determining whether an outlier should be rejected. The reader is referred to statistical texts for these techniques.

2.3.3 CONSIDERATIONS IN CHOOSING A SAMPLING DESIGN

Hydrological and ecological variables follow their own patterns of ebb and flow in space and time, and sampling designs should reflect these differences. It might be of interest to measure populations of benthic invertebrates in a riffle, for example, yet a riffle may not be the best location for measuring average stream discharge. Similarly, measurements of discharge or bedload movement from a bridge may be feasible at high flows but drift-diving for fish observation during floods would be unwise. Often, some variables can be measured to give information about how to sample others more efficiently. Statzner (1981), for example, found that precision was increased when abundance of benthos was first correlated with hydrodynamic factors.

The object of sampling is to estimate population parameters from information contained in a sample. A good sample survey design will maximize the amount of information for a given cost. Replication and randomization are the key considerations in obtaining useful results.

Replication

Replicates are taken in order to quantify the variability—within a stream, between different streams, and/or over time. For monitoring change, it is best to use the same study site across time. However, if the study variable is sensitive to sampling methods (e.g. electrofishing), other sites may be necessary. The channel itself will change shape over time, in which case pool–riffle locations may change and new sites should be chosen.

The number of replicates needed will vary with the sampling design. It is extremely important to replicate at the appropriate level. For example, if we want to make inferences about a region, taking 10 samples on different streams is preferable to taking 10 at one site on one stream. In the latter case the results would only be applicable to the one site! Hurlbert (1984) provides a discussion on "pseudo-replication". This point is also illustrated in the example of Section 2.5.

Pooling of samples (combining several samples into one) can be used in some cases to reduce costs. Pooling several small samples is preferable to taking one large one. For example, several water samples can be collected over an hour or across a section, and then combined before analysing for suspended sediment. The pooled sample would be considered a single replicate.

Randomization

Randomization means that the selection of any member (e.g. sample or sample site) must not influence the selection of any other member (Zar, 1984). Efforts should be made to **randomize** at some level of a study, whether in randomly selecting a set of rivers, randomly locating a series of transects, placing samplers randomly within a transect or randomly placing a wire grid within a reach. Randomization need not take place at all levels; in fact if a recognized variation exists (e.g. pools versus riffles), a stratified design (Section 2.3.5) with randomization within strata will be more efficient.

Accessibility sampling, haphazard sampling and representative sampling are examples of non-random sampling. **Accessibility sampling** means that the samples or observations are those which are the most easily obtained (Barnett, 1982). "Easily accessible sites" close to roads are often chosen because thick overgrowth, cliffs, non-portable equipment and/or private land ownership make sampling of other sites impractical. **Haphazard sampling** refers to the use of one's own judgement to "randomly" select a sample. An example would be the common practice of tossing a wire quadrat, where sampling is not random but subject to the personal bias (and arm strength) of the tosser. **Representative sampling** involves choosing a sample that is considered to be "typical" or "representative" of the population. These are all forms of **judgemental sampling**, and are subject to investigator bias. Their use may invalidate statistical tests and "lead to estimators whose properties cannot be evaluated" (Scheaffer et al., 1979, p. 32).

However, if a researcher's intuition is sound, judgemental sampling may actually yield better estimators than random sampling, especially if a random sample includes an extremely odd condition. People constantly form associations and determine averages, and a certain degree of "natural history intuition" develops from years of exposure to the literature, the field environment and the opinions of others. According to Konijn (1973), a judgement sample may be better than a random sample if the correlation is high between what is being measured (e.g. invertebrate populations) and the characteristics used to select a sample (e.g. eye estimates of substrate size).

The experienced eye will integrate a number of characteristics; unfortunately, the "view" is inconsistent—a biologist will "see" differently from an engineer. Thus, judgement may eliminate anticipated sources of distortion but introduce others because of personal prejudices or lack of knowledge about the population (Barnett, 1982). The appearance of bias in the selection process may also mean that the data will not be valid if challenged in court.

Some compromise is needed to balance the demands of statistical perfection and the practicality of data collection. In the selection of sampling sites it might be feasible to have personnel from different disciplines (e.g. engineering, hydrology, biology) each choose "representative" reaches. A random sample could then be taken from their choices. Another alternative would be to use judgement to eliminate unacceptable sites rather than to pre-select sites. An excess number of candidate sites (which can be streams, reaches or grid plots) can be selected at random, and judgement used to guide the ultimate selection of the study site(s). Criteria can be imposed on the selection process to "weed out" atypical sites. If all the candidate sites are essentially equal with respect to the variables being studied, then perhaps it can be reasoned that an easily accessible site is as good as any other.

Judgemental sampling can be used to obtain "suggestive information" during pilot studies (Konijn, 1973). It is recommended, however, that randomness be introduced at some level of a full-scale study, tempered by judgement, so that statistical analysis techniques can be applied and valid conclusions drawn.

2.3.4 PARTITIONING THE STREAM

The selection of measurement sites begins with decisions about the extent to which results are to be generalized (e.g. catchment or continent). Is the objective to study a critical site, to sample only tributaries of a certain size or to study changes in a stream from headwaters to mouth?

Selected streams are normally divided into relatively homogeneous sections, based on topography, geology, slope, streamflow and/or biological characteristics (Bovee and Milhous, 1978). This type of division is called **stratification**, which will be discussed further in the following section. Bovee (1982) provides a checklist for establishing study sites which considers a number of hydrological, geographical and ecological characteristics at several scales.

Selection of the number and lengths of segments in a stream, reaches in a segment or transects or plots within a reach is ultimately a problem in sampling design. This must take into account the variability of the stream, the precision required, and budget and time limitations. Some basic guidelines for partitioning a stream are as follows (see Figure 2.2):

(1) *Identify and eliminate anomalous areas.* Bridge or dam sites, road crossings, large waterfalls or other atypical features should normally be eliminated from the study area unless they represent a large portion of the stream or are the object of study. An example would be "critical reaches", defined by Bovee and Milhous (1978) as those which limit the success of a particular life stage of fish, such as a location which becomes a barrier at low flows. These areas would be set aside for separate study rather than included in statistics which are used for making inferences about the segment.

(2) *Divide a stream into segments.* A segment of a stream is a section where

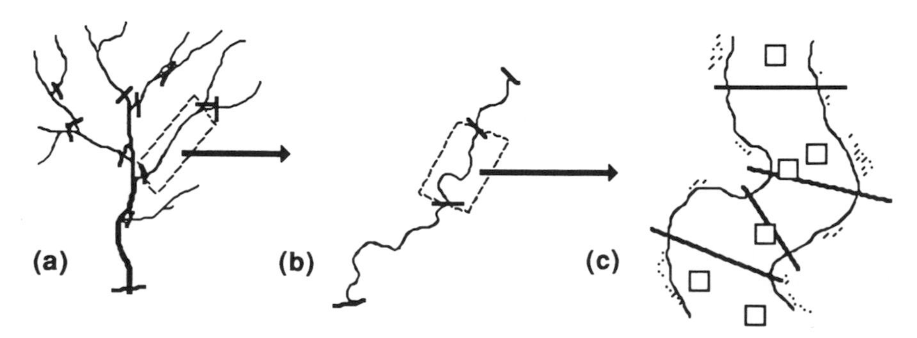

Figure 2.2. Location of sampling sites: (a) division of stream system into homogeneous segments, (b) division of a segment into reaches, and (c) location of transects (lines) and plots (boxes) within each reach

the flow and morphology are fairly uniform. Bovee (1982) recommends locating segment boundaries at locations where the average streamflow changes by more than 10%, such as at major tributaries or diversions. He also recommends placing boundaries where abrupt changes occur in slope, sediment input, bank materials and channel morphology. Where changes are gradual, boundary locations become more subjective. Stream ordering (Section 4.2.5) is another method of segmenting a stream. Based on the study design, a certain number of segments are chosen from which inferences are made about the river system(s).

(3) *Divide stream segments into reaches*. A segment can be further subdivided into reaches, some of which are sampled to make inferences about the whole segment. Reach boundaries can be located on a uniform basis (all reaches are of equal length) if the segment is homogeneous or changes smoothly from one end to the other. Alternatively, as in locating segment boundaries, divisions can be based on more localized changes in channel structure. If the stream morphology is composed of meanders, pools and riffles, or other easily discernible changes in channel shape, reaches can be divided at the point of change. For example, two cycles of riffles and pools or two meander loops constitute a workable reach length for hydraulic geometry surveys (Section 7.2.6).

A common practice in stream surveys is to select "representative" reaches which are considered to be representative of conditions in the segment. Selection is usually based on the ability of the investigator to integrate knowledge about the stream segment and choose a reach which contains the same features in the same relative proportions. To introduce randomness at this level (see discussion in last section), several candidate reaches can be located and one or a few chosen at random for further study.

(4) *Select transect, plot or point measurement sites*. A reach may be measured in its entirety (for example, in counting fish populations or the number of debris dams) or sites might be located within a reach for measuring bed material sizes, cross-sectional profiles, aquatic insect numbers, etc. A **transect** is typically a line along which measurements or samples are taken. The term **plot** is a throwback to agricultural terminology, and refers to an area within which samples are taken. **Point** measurements are made over an area determined by the size of the sampling instrument (e.g. rainfall gauge or substrate sampler).

Transects, plots or point measurement sites can be located at random, uniformly, at representative sites or within noticeable microhabitats. These may constitute subsamples at either the segment or reach level. For instance, cross-sections might be surveyed at uniformly spaced transects along an entire segment but crocodile populations sampled only within a few reaches.

Most of the steps in partitioning a stream can be done in the office if adequate streamflow records, topographical maps and recent aerial photographs are available. Critical areas, segment and reach boundaries and measurement sites should be marked on maps and/or aerial photographs. The precise location of reaches or measurement sites should be done in the field, perhaps in conjunction with pilot sampling. As much as possible, efforts should be made to avoid bias in the selection of sampling sites by introducing randomization at some level.

Prior establishment of study sites is an essential part of planning. This eliminates confusion and time lost in the field, improves the validity of results and reduces the impact of the study itself on the study area.

2.3.5 BASIC SAMPLING DESIGNS

Statistical texts normally present three basic sampling designs: (1) simple random sampling, (2) stratified random sampling, and (3) cluster sampling. Variations on these designs include two-stage sampling, systematic sampling and sequential sampling. Table 2.1 compares the various methods and their potential applications.

Statistics may be computed from formulae given in various texts such as Barnett (1982), Ching (1967), Cochran (1977), Elliot (1977), Konijn (1973) or Scheaffer et al. (1979) or calculated using statistical computer packages (see Appendix 1). Platts et al. (1987) give pertinent examples for several of the designs. Tests of significance associated with each sampling design will not be covered in this text, but information can be found in these statistical texts and in manuals for statistical computer packages.

In selecting between sampling designs, important considerations are statistical efficiency, relative cost, ease of computation, time available for sampling, distance between sites and compatibility of the sampling designs with measurement techniques. The efficiency of design is influenced by the type of variable being sampled. Bovee (1986) gives the example that aquatic macroinvertebrates tend to be more prevalent in streams than fish, so random sampling may be efficient for the former group but may not pick up a rare, highly mobile fish species. Spacing of sites is also an important consideration when sampling at one site can affect conditions at adjacent sampling locations. For example, control plots should almost always be located upstream from treatment plots.

Rules for selection of statistical design are not set in concrete. Design selection is based on prior knowledge of the stream, the variable(s) being measured, costs and logistics. By being aware of the various designs and their limitations, researchers can structure their studies more efficiently.

Simple random sampling

Simple random sampling requires that every possible sample of the same size has an equal probability of being chosen from the population. Additionally, no population member is included more than once.

Each element of the population should be identifiable as an individual (e.g. fish or stream reaches). To select elements at random from the larger population, a random number table from a statistical text such as Powell (1982), numbered slips of paper or even the last two digits of the numbers in a phone book can be used. However, in practice, selecting a random sample is more difficult than riffling through a phone book. Some of the pitfalls of so-called random sampling are discussed in Section 2.3.3.

The normal approach in random sampling is to superimpose a numbered grid system over an area, then randomly choose grid numbers. A "grid" may be one-dimensional, as in dividing a stream segment into uniform-length reaches; two-dimensional, as in gridding a gravel bar for estimating particle size; or perhaps even three-dimensional. Grid size may depend on the desired intensity of sampling and/or the area sampled by a measurement device (Bovee, 1986).

In the types of studies performed in aquatic environments, simple random sampling will be most successful in relatively small, homogeneous areas. Otherwise it may result in "samples that (1) are widely dispersed, causing considerable travel expense, and (2) leave some areas totally unsampled" (Platts et al., 1987, p. 3). The advantage of simple random sampling lies in its relative simplicity of design and ease of statistical computation.

Stratified random sampling

Stratified random sampling is more efficient than simple random sampling if there are obvious patterns (e.g. in stream form, vegetation or substrate) which can be separated into "strata". Each stratum should be more homogeneous than the whole population. Pools and riffles, for example, may be separated into different strata not only for statistical purposes but because sampling techniques are different. In addition to being more efficient in estimating population parameters, stratified random sampling provides separate estimates for each stratum (Platts et al., 1987). A summary of this sampling procedure is given by Ching (1967):

(1) Group the population into several subpopulations called strata. Each item in the population is put into one and only one stratum and no item of the population is left out. The strata are said to be "exclusive and exhaustive".
(2) A random sample of items is then drawn from each stratum; these samples form the stratified sample.

Table 2.1. Comparison of statistical sampling techniques. Based on Green (1979) and Platts et al. (1987)

Sampling design	Description	Considerations	Potential field use
Simple random	Each element of the population has an equal chance of being sampled. Random sampling is necessary to meet assumptions of independence.	If used without judgement, anomalous samples can lead to poor estimates. Also, if elements are widely scattered throughout a large area, costs can be excessive.	Relatively small, homogeneous areas with a randomly distributed population.
Stratified random	A population is divided into uniform "strata" and random samples taken from within each stratum. More samples should be taken in more variable strata. The method yields estimates for strata as well as for the population.	Variation between strata should be high and within strata low. Finding the optimal size of strata and sample size within strata requires trade-offs.	Areas with recognizable patterns of heterogeneity in channel form, habitat, etc.
Cluster sampling	A population is divided into clusters, each of which has heterogeneity similar to that of the population. A random sample of clusters is taken and all or a subsample of elements within a cluster are measured.	Clusters should be somewhat alike, but heterogeneous within each cluster. Again, trade-offs exist between number and size of clusters. Intensive sampling within a few clusters can be an economical approach.	Areas where populations are dense so that the distance between elements is small.

Two-stage sampling	This is a variation on the other methods, where a sample or cluster or stratified sample is further subsampled.	Subsamples should be well mixed (if applicable) and of sufficient quantity to characterize the sample.	When analysing or counting a large sample is cumbersome; when elements are so similar that counting all of them is inefficient; or when it is desirable to use different precision at different levels.
Systematic sampling	Samples or sampling units are spaced uniformly.	Method is simple to use, but standard error of mean cannot be evaluated. Also, periodicities in environment may coincide with spacing interval.	When gradual, regular changes in the channel occur or if uniform spacing is required for modelling.
Sequential sampling	Samples are collected until a desired level of precision is reached.	Flexible, but population distribution must be constant throughout study. Impractical if costs are high for travel to site or if there is a large time lag between sample collection and analysis.	Not recommended for ecological sampling unless population distributions are stable over time.

(3) The estimates (e.g. mean, variance or total) are calculated separately from each of the strata and combined to get estimates of the population parameters.

The stratified sample mean and variance are weighted by strata size. Weighting factors might be based on area, volume or counts. Stratified sampling is most efficient if the variation between stratum means (e.g. substrate size in riffles versus pools) is large compared to the within-strata variation (substrate size within riffles) (Barnett, 1982). The sample and strata sizes should be based on the comparative costs of sampling within and between strata and the acceptable precision.

Proportional allocation is a type of stratified sampling in which the sample size is proportional to the relative *size* or *variance* of the stratum. With this method, stream reaches with many different habitat types, for example, would be sampled more intensively than more homogeneous reaches (Bovee, 1986). **Optimum allocation**, another alternative, minimizes the amount of error for a given *cost* of a sample (Hamilton and Bergersen, 1984). Optimum and proportional allocation are discussed in several of the above-mentioned statistical texts and an example of proportional sampling is given by Bovee (1986).

Cluster sampling

Cluster sampling is similar to stratified sampling in that the population (or area) of interest is divided into units (sub-areas), from which samples are taken. In cluster sampling, each sampling unit is a cluster of items or elements (Scheaffer et al., 1979). All or a further subsample of the elements are measured. Platts et al. (1987) give the example of measuring all the trees within 2.5 ha plots. The plots are considered the "clusters" and a sample of clusters would be selected rather than a sample of individual trees.

The difference between cluster sampling and stratified sampling is that in the latter, strata should be as homogeneous as possible within each stratum but one stratum should differ as much as possible from another with respect to the characteristic being measured. Clusters, on the other hand, should be as heterogeneous as possible within the cluster and one cluster should look very much like others (Scheaffer et al., 1979). Cluster sampling is most effective when the physical distance between elements (like trees) is small and only a few clusters are needed, so that travel expenses and time required for sampling are minimized.

The steps involved in cluster sampling are given by Ching (1967):

(1) Separate population into mutually exclusive, heterogeneous subpopulations called clusters.

(2) Choose a random sample of clusters.
(3) A cluster sample may include all the items in clusters, or a sample of items can be drawn from each of the chosen clusters.
(4) Estimates are calculated separately for each cluster and then combined to give an estimate of the population parameter.

The cluster sample mean and variance are influenced by both the sampling of clusters and the sampling of items from within the selected clusters. The size and number of clusters sampled will affect the efficiency of this sampling design. As a general rule, each cluster should be representative of the diversity of the entire population (Konijn, 1973). When measurements of elements within a cluster are highly correlated a large cluster size is inefficient, and it is better to take more samples of smaller clusters (Scheaffer et al., 1979).

In general, sampling a larger number of small clusters is more efficient than sampling more items within each cluster (Scheaffer et al., 1979). However, the larger number of clusters increases both travel expenses and sampling time, so a compromise must be made. Compared to either simple random or stratified random sampling, cluster sampling offers one major advantage: the cost per element sampled is lower. The disadvantages are that the variance among elements tends to be higher and the required computations are more extensive.

Two-stage sampling

Two-stage sampling is a variation on the above methods, in which samples are again "subsampled". For example, it may be desirable to subsample a one-litre sample of water to analyse the zooplankton count or sediment concentration. In cluster sampling, subsampling is used when it is obvious that many near-identical elements are included within a cluster sample, and to sample them all would be a waste of effort (Platts et al., 1987).

Two-stage sampling is also useful when a variable is difficult or expensive to measure precisely. In this case, a "quick and dirty" method can be applied to a large number of sites (e.g. remote sensing) and a more precise method (e.g. vegetation surveys) applied to only a small number of the same sites. The imprecise method can then be calibrated against the results of the precise method by a ratio or regression method (see Armour et al., 1983).

Systematic sampling

Systematic sampling, or uniform sampling, is another method of subsampling. The first unit in a sample is randomly selected, followed by the selection of the other units at fixed (uniform) intervals (e.g. transect locations in a stream reach). The design is particularly relevant if the data are to be input into a model requiring a fixed spacing interval. Bovee (1986) outlines a variation on this method, "the systematic random walk", in which the distance between sampling locations is uniform but the bearing from one location to the next is randomly selected.

Uniform spacing is most effective when gradual, regular changes occur (Hamilton and Bergersen, 1984). However, this design is inefficient if the sampling interval coincides with some periodicity in the stream geometry or aquatic populations, much like sampling only the valleys between evenly spaced mountains (or only the tops). Also, there is no valid way to estimate the standard error of the sample mean (Elliot, 1977). The method is probably not applicable to most habitat studies.

Sequential sampling

In sequential sampling, described by Green (1979), statistics are initially calculated from preliminary samples (such as those taken during a pilot survey). Sampling is then continued until a desired level of precision is reached. For the design to be effective, the population distribution must remain the same during the study (perhaps unrealistic in ecological work). It is also impractical if most of the cost is in travel to the site rather than the cost per sample at the site.

2.3.6 SAMPLE SIZE

The number of samples taken is dependent upon the number of replicates needed to compensate for within-site sampling variation (Armour et al., 1983). After obtaining an indication of the expected variability (e.g. from pilot studies or from the literature), the number of samples required can be estimated, based on the chosen level of precision. If there is much variation—either natural or due to measurement error—the number of samples may seem unrealistic. Statzner (1981) cites several examples to illustrate that hundreds of samples were necessary for estimating abundance of benthos, even when considering only a small, uniform section of a stream.

For random sampling, assuming that the data follow a normal distribution (see Appendix 1), the number of samples can be estimated as:

$$n = c\left(\frac{s}{\bar{x}}\right)^2 \tag{2.1}$$

where

n = sample size,
s = sample standard deviation,
$\bar{x}$ = sample mean,

and c is a factor. Based on assumptions of an infinite population, a reasonably large sample, and 95% confidence limits on the error of estimation of the mean, c can be approximated by:

$$c = \frac{4}{\epsilon^2} \qquad (2.2)$$

where ϵ is the percentage error of the mean, expressed as a decimal (modified from Armour et al., 1983). The term $s/\bar{x}$ is also called the coefficient of variation (C_v; see Appendix 1). The factor c would vary from 4 for an estimate of the mean within ±100% to 1600 at ±5%. If the standard deviation and mean are approximately equal (i.e. $C_v = 1$), then c is equal to the sample size; thus 1600 samples are needed to obtain an error of ±5%. Obviously, if the variation is higher, even more samples will be needed for the same level of error.

There is actually another variable in Equation 2.2, the level of confidence. For confidence limits of approximately 68% (±1 standard error), c would be approximately $1/\epsilon^2$. Thus, the required sample size would be smaller but we would be less certain about our estimate of the mean.

Realistic goals should be set regarding the precision of estimates which can be achieved with a reasonable number of samples. "Acceptable" may mean a boundary on the "true value" of ±40% or more. Since the power of a test is also affected by sample size (Section 2.3.1), one should also be realistic about the magnitude of effects which can be detected. In sampling benthic invertebrates, Allen (1984) recommends 10 to 20 replicates for modest precision and at least 50 for high precision.

There will be a trade-off between precision, budget and the number of samples taken at different levels (e.g. the number of reaches versus the intensity within each reach). A common approach is to first determine the maximum sample size based on time, personnel and budget and then decide whether this will yield an acceptable level of precision (Armour et al., 1983). In general, it is preferable to:

- Sample the study area by taking many small samples rather than a few large ones;
- Have more main plots at the expense of less within-plot sampling;
- Draw a larger sample from a stratum which is larger in size and/or more variable;
- Balance the expense of taking samples on-site with travel expenses.

2.4 KNOW YOUR LIMITATIONS

Platts et al. (1987) and Armour et al. (1983) discuss several "confounding factors" which can affect the results of a study: institutional, political, biological, statistical and those relating to equipment and personnel. To these can be added effects imposed by weather, flow conditions or site accessibility.

Institutional and political factors may lead to a study being terminated before its conclusion because of budget cuts, or the unpopularity of potential results or of the study itself. If a study extends across a long period of time it is important that the management of the study area remains under constant administration. Land-use changes during the period of study (such as logging of the study area) can seriously affect results if undesired and unforeseen.

Budget also affects the number and expertise of personnel. Although sometimes unavoidable, especially if seasonal workers are hired, changes in personnel can have potentially drastic effects on measurements. The quality of sampling by the same personnel may also vary over time (e.g. they may be affected by certain times of the day or year or they may become more efficient with experience) (Armour et al., 1983). Administrators and investigators both need to ensure a continued commitment of time, personnel and money until the study is finished. If it seems that a study will go over budget further funding should be sought or a scaled-down plan developed early in the planning process.

Biological and physical factors will affect both the natural variation of the stream and its biota. If the objective is to detect the effect of some treatment, such as a management practice or a point pollution source, excessive natural variation may mask treatment effects. Another confounding factor is that there may be response thresholds rather than a linear response of biota to a treatment. Clumped distributions of biota, fishing pressure or migration of a species out of an area may cause problems with sampling. Additionally, variability which causes "noise" in one study may be the information of interest in another (Green, 1979).

Statistical confounding factors are common. Some variables may be assumed independent when they are not. The sampling size or number of replicates may not be sufficient to give adequate power to a test. Rounding errors or errors in recording or tabulating information can also affect statistics. Thus, methods and sampling procedures should be efficient and consistent throughout the study. To reduce bias, instruments should be calibrated as needed and personnel made aware of the effects of different conditions on the precision and accuracy of readings.

Additionally, the inferences from statistics must be tempered by sanity. A significant relationship is not proof of causality; similarly, results from a small area should not be extended beyond reasonable boundaries. If a significant relationship is found between the frequency of frog croaking and the number of eddies along a stream reach it does not mean that one caused the other or that the relationship is a universal one.

Weather and flow conditions and the accessibility of a site affect both stream biology and field workers. More remote sites may be more representative, but will perhaps receive less intensive or careful attention by workers who have had to "bush bash" to get to them.

Although directed towards stream gauging, AWRC (1984, p. 11) nicely sums up the difficulties in working in a stream environment:

> Gauging streams under high flow conditions is expensive, dangerous, uncomfortable and requires skilled and dedicated staff. Cuts in funding usually result in less resources being available for high flow gaugings, loss of morale of staff trying to do a professional job and an eventual unsatisfactory set of records which rely heavily on subjective and tenuous extrapolations of rating curves.

2.5 EXAMPLES OF HOW AND HOW NOT TO CONDUCT A STUDY (M. Keough)

Three different ways of answering the same question will be presented. The study question is:

> "Do streams with trout have lower densities of blackflies (Diptera: Simuliidae) than streams lacking trout?"

Blackflies live in fast-flowing riffle areas, where they feed on micro-algae growing on rock surfaces. They are absent from slow-moving sections of streams. All the sampling will therefore be done from riffles. Rocks will be removed from the stream and all the blackflies counted. These stream insects are large enough to see with the naked eye.

The null hypothesis (H_0): There is no significant difference in blackfly abundance between the two types of stream.

The statistical method: Analysis of variance. A significance level of 0.05 is selected for testing H_0.

Data to be collected: The number of blackflies on the surfaces of rocks, each having an area of approximately 0.009 m^2.

In each case it will be assumed that time and money allow a minimum of 100 rocks to be sampled, and that travelling time between streams or riffles is insignificant compared to the time needed to collect and examine the rocks. The data will be analysed using the computer program SYSTAT (Wilkinson, 1986).

Sampling program 1

Two streams are chosen, one with brown trout and one which has lacked these fish for as long as records have been kept. In each stream, a riffle area is selected and 50 rocks sampled from it.

The data are analysed using a one-way analysis of variance. We find that the mean number of blackflies on rocks from the trout stream is 20.94, with a standard deviation (s) of 7.49, while the estimates from the trout-free stream are 24.36 and 6.00, respectively.

The following table is generated by SYSTAT:

Analysis of variance

Source	Sum-of-squares	DF	Mean-square	F-ratio	p
Fish	292.949	1	292.949	**6.703**	**0.011**
Error	4282.793	98	43.702		

From the analysis (F-ratio and its associated probability, in boldtype) we conclude that our two samples differ (i.e. that the fish have some effect).

The problem with this sampling method is that only two streams were sampled, one with fish and one without. Further, only one riffle from each stream was sampled. What would happen if we had sampled a different riffle from each of these streams? Would we have obtained the same result? Would we have achieved the same result if we had used a different pair of streams? We know that there is variation between the streams, but with only one stream of each fish type we cannot know whether the results reflect a true effect of fish or whether the two streams would have differed regardless of the presence of fish. *There is no way to identify the effect of fish from this design!*

Sampling program 2

A better approach would be to sample at least two streams of each type. Suppose, then, that four streams had been chosen from a map: two with trout and two without. The 100 rocks are again sampled, 25 from each stream. The sampling now has two levels: rocks within streams and streams within each fish category.

The mean values for rocks in each stream are 22.49 and 19.39 for the two trout streams and 23.17 and 25.55 for the streams without trout. (Note: the same data have been used in all examples, so if the data were pooled from the streams in each category, the same mean and standard deviation would be obtained as in the first example.) The mean for the two trout streams is 20.94 (s = 2.19), and 24.36 (s = 1.68) for the trout-free streams.

The analysis in this case must take account of these sources of variation and so a nested analysis of variance is used. The SYSTAT output is as follows:

Analysis of variance

Source	Sum-of-squares	DF	Mean-square	*F*-ratio	*p*
Fish	292.949	1	292.949	**3.060**	**0.222**
Streams within fish categories	191.488	2	95.744	2.247	0.111
Error	4091.305	96	42.618		

In this case we can see from the analysis that there is no evidence (the *F*-ratio and probability in boldtype) for an effect of fish.

The problem here is not in the analysis but in the sampling design. Practically, much effort has been spent counting blackflies on 25 rocks in each stream, only to use an average value for that stream in our test about trout. Unfortunately, we then have a statistical test with very low power: the *F*-ratio has one degree of freedom for its numerator and only two for the denominator.

Did we need to look at 25 rocks in each stream? Are there better ways to do the sampling? As might be guessed, one method is to use more streams and fewer rocks per stream.

Sampling program 3

Instead, suppose 10 streams with trout and 10 without had been chosen (preferably randomly), and only five rocks from each stream sampled. The mean for rocks in the 10 streams with trout are: 21.14, 21.37, 24.34, 22.42, 23.18, 15.45, 19.30, 22.05, 17.86, 22.23, while for the streams lacking trout, the values are: 25.70, 26.02, 24.72, 16.84, 22.57, 23.22, 26.46, 27.93, 25.29, 24.87. For the two stream categories, the means are 20.94 (s = 2.67) and 24.36 (s = 3.06), respectively.

This time the analysis shows a significant effect of fish:

Analysis of variance

Source	Sum-of-squares	DF	Mean-square	*F*-ratio	*p*
Fish	292.949	1	292.949	**7.118**	**0.016**
Streams within fish categories	740.858	18	41.159	0.930	0.547
Error	3541.935	80	44.274		

What is different about this example? The key change is that more effort has been devoted to examining the variation between streams and less to documenting variation among individual rocks. The statistical test is more powerful because the *F*-ratio used to test whether fish have an effect now has 18 degrees of freedom for its denominator.

In this example, with fewer rocks per stream, higher standard deviations are obtained than in the second example. Statistical power can be calculated for each case. Assuming that trout really do decrease the number of blackflies per rock by five (the means are 20.9 and 24.4), we can then ask if we are likely to detect such a difference if we sample, say, two, five, 10 or 500 streams. If we use the observed standard deviations we find that the chance of detecting an effect of this magnitude with only two streams of each type is only 30%, but when the number of streams is 10, we are 94% certain of detecting this fish effect.

This example shows the benefits of prior planning of sampling effort. In the first two cases, efforts were largely wasted. In the first sampling program no conclusions could be drawn from the sampling because we did not account for the fact that trout vary at the level of whole streams—some streams have fish and others do not. If we had done a mock analysis we would have seen that there are actually two levels of variation—rocks within streams and streams within fish categories. In this first design there was replication at the lower level but no replication at the level of whole streams.

In the second design streams were properly designated as the sampling units, but the test of hypothesis about trout was not very powerful because most of the effort was spent on looking at rocks. Again, if we had done a mock analysis we would have seen that the power of the statistical test depended on the number of streams, not the number of rocks.

The third design demonstrates the benefit of advance planning of field surveys. More information was obtained by concentrating sampling efforts at the proper level. This was done by minimizing the number of rocks from each stream and maximizing the number of streams.

This example is a very simple one, but the task for which these sampling programs are designed is fairly common. All three designs are, at face value, quite plausible but one is clearly better than the other two. The design flaws illustrated are, unfortunately, also common. The term "pseudoreplication" has been applied to this error by Hurlbert (1984).

Field sampling programs like this cannot determine whether trout cause changes in blackfly density but can only test whether the blackfly density is higher (or lower) in streams with trout than in those without them. If we need to show causality, we would need to experimentally manipulate the abundance of fish. In such an experiment, we would still need to guard against errors in the sampling program.

One cannot emphasize too strongly the importance of careful thought and planning of the statistics *before* going out and collecting the data. The three sampling programs described above all require approximately the same field and laboratory effort, but the first two represent poor allocation of time and money.

3 Potential Sources of Data (How to Avoid Reinventing the Weir)

3.1 DATA TYPES

A researcher typically starts out on the path to discovery by first retracing someone else's footsteps. An investigative phase can provide valuable background information on the study stream and its species, it can reveal existing sources of data, and it prevents the unwitting replication of someone else's work.

Finding the information, however, can be a time-consuming task. Computer-assisted literature searches are of immeasurable help in sifting through the profusion of written "fallout" dispersed by the information explosion. However, the search for data may require more detective work, as much of it is not catalogued or readily obtainable.

Maps and data summaries are usually published in a form compatible with the requirements of the majority of users. The quality, quantity and availability of data will vary from one place to another and from time to time, depending mainly on who has collected them and their budgets and priorities. It is best to approach any form of data with a healthy sense of scepticism and a realistic attitude about how much information can be drawn from it.

Although the concepts of study design introduced in Chapter 2 relate to scientific investigations in general, the remainder of the text will emphasize the analysis of the physical properties of flowing water. Relationships between hydrological and biological variables will be discussed, but methods of collecting and sorting through biological samples will not be covered.

Physical data usually take the form of (1) a time series or (2) spatial data. **Time-series data** are collected at regular or sporadic intervals, and may either be instantaneous values or an accumulated or average quantity measured across some time period. Hydrological and hydraulic data, sediment data, water quality data, and climatic data can be included in this category. **Spatial data** are observations made across a line, an area or a space, and include maps, photographs and other remotely sensed images. When temporal and spatial data represent different view of the same variable they can form a powerful combination of site analysis. For example, maps from different dates can be compared to detect river channel changes over time, and annual rainfall averages from different gauges can be combined to create regional maps.

3.2 PHYSICAL DATA SOURCES, FORMAT, AND QUALITY

3.2.1 WHERE TO LOOK FOR DATA AND WHAT YOU ARE LIKELY TO FIND

When looking for data, one thing to bear in mind is that the agencies responsible for data collection are not necessarily the same as those disseminating data. National data centres and principal data-collection agencies will be primary sources of data, and should be the starting point of a search for information. Other potential sources include water authorities at national, state and local levels, environmental protection agencies, departments of agriculture or lands, electricity departments involved in hydroelectric generation, research centres and universities. Even highway or railway departments or consulting firms may collect data for specific projects such as design of bridges, culverts, or drainage systems. Increasingly, data are being supplied on a commercial basis from private firms. An example is the US West Optical Publishing Company in Denver, Colorado, which supplies climatic, streamflow, and water quality data from the USA and Canada on floppy disk or compact (laser) disk, along with software for analysing the data.

A shrewd investigator will turn over as many rocks as possible to uncover data hidden in unlikely places such as newspapers, diaries of pioneers, files of state and local historical societies, and interviews with local residents.

The availability of data will depend on the restrictions on information dispersal imposed nationally or by an individual agency. Free access may not be allowed, especially while the data are still in a provisional state. Research organizations may be reluctant to give away their data until they have extracted the information they need for research publications.

The density of data-collection sites will also vary. Streamflow gauging networks, for example, may average one gauge per 1–2000 km^2 in Europe and North America but only one per 10 000 km^2 or more in tropical countries (Budyko, 1980). Rain gauge networks tend to be more dense and water quality stations much less so.

Each establishment will have its own way of collecting, processing, storing and presenting data. A researcher may thus be faced with the problem of assimilating information acquired by a variety of methods and recording techniques, especially if the study area extends across state or national boundaries. For example, the units of measurement may vary. Care should be taken in determining the units of tabulated values, as they may even change over time at the same station. Methods of recording data range from a pencilled number written down by an observer to the automatic digital recording and telemetry of sensor data obtained at a remote site. Records will vary in quality from the original "raw data" to the finalized form in which records have been refined, polished, and missing values estimated and annotated.

Corrected and summarized data will typically be available in published report form such as the Water Supply Papers of the US Geological Survey (USGS) or volumes of *British Rainfall* (replaced in 1968 by *Monthly and Annual Totals of Rainfall for the United Kingdom*), as computer printouts or on computer storage media (Shaw, 1988). A drawback of processed data is the long time lag (two years is common) between collection and release of the data. Raw data may be available but will usually contain blemishes which must be repaired by the user, and will need to be processed into a usable form. Older, unprocessed records on streamflow, climate and morphologic and hydraulic characteristics of rivers may also be mothballed in dark closets of various state, federal or private and university institutions. The reduction of unprocessed data is a specialized and time-consuming task, but worthwhile if they are the only data available for a site. Practical references for streamflow data processing include Beven and Callen (1979), Brakensiek et al. (1979), and Woolhiser and Saxton (1965).

The investigator should make more than superficial inquiries about data sources, as newer personnel in particular may be unaware of the existence of archived data. When collecting data one should be sure to obtain notes on station operation, its location, instrument type(s), historical changes and estimates of data quality.

In conventional systems of the past, data were stored as handwritten observations, replaced later by computer printouts. Written forms are still convenient for quick reference, and field records are especially valuable for obtaining information about a site. Microfilming has become an alternative to voluminous piles of paper, requiring about 1/300th of the storage space needed for original paper records (Starosolsky, 1987).

Computer analysis of data, however, has been greatly facilitated by the storage of data on magnetic tape or disk. Computerized data storage and retrieval systems, like the WATSTORE system of the USGS, allow access to records via telephone line. Data formats may differ from one agency to another. Some institutions may be able to provide data in alternate formats more suited to individual needs. For example, "flags" on data values (e.g. to indicate that precipitation fell as snow or a runoff value was obtained from an extrapolated rating curve) may be kept or suppressed.

As standards rise for both research and design work, and as competition for water increases, more pressure will be placed on data-collection agencies to provide higher-quality information in a shorter period of time—collected more frequently and at more locations. Advances in instrument and computer technology will continue to increase the efficiency, accuracy and consistency of data collection and processing as new instrumentation and procedures are developed and standardized. The following is a description of some of the types of physical data which may be available.

Hydrological and hydraulic data

Streamflow data are normally obtained at either natural sections or gauging structures (e.g. weirs). Typically, the water level (stage) is recorded and related to streamflow (discharge) by a stage–discharge relationship, called a rating curve (Section 5.8). Readings may be taken on an intermittent basis (e.g. weekly readings of a staff gauge) or as a continuous record.

Data are usually available on a daily basis, with the streamflow values expressed as a total daily volume or an average daily discharge. Instantaneous peak discharges and daily maximum and minimum flows are often available. Monthly and annual summaries are also commonly published for established stations. In the USA the cubic foot per second (ft^3/s or "cusec") is the basic discharge unit; in Australia, megalitres per day (Ml/d) is commonly used; however, in the SI system, the cubic metre per second (m^3/s or "cumec") is the basic unit for discharge. Monthly or yearly totals may also be expressed in volume units such as million m^3.

Information on the regulation of dams and irrigation diversion works will be available from agencies in charge of the structures. Current operation procedures should be verified by direct correspondence.

Paleohydrological data may be of interest for obtaining long-term trends in climate or improving estimates of flood frequencies (Chow, 1964b). Data from tree rings, fossil pollen, clay varves, fluctuations in levels of closed lakes, or glacial movement may be available commercially or through research agencies. A reference edited by Gregory (1983) provides information on the interpretation of these data.

Channel characteristics

Data on channel characteristics will most likely be obtained from agencies involved in detailed studies on water surface profiles, local scour, velocity and direction of flow, or channel changes, particularly in the vicinity of bridges, dikes, dams, or in channelized or dredged reaches. Size distributions of bed and bank material may also be available.

Water data agencies are usually able to supply cross-sectional profiles at gauged sites. During and after high flows, river-regulation agencies may collect data on streambed roughness, magnitude of scour, and changes in channel form. Observations of flooding made by either professionals or local residents, including high-water marks, lateral distribution of flow, stages achieved by ice dams, or the amount of sediment or debris in the river, are valuable in evaluating the response of a channel to high flows.

Climatic data

For the purposes of this text the only climatic data which will be emphasized are those of precipitation and temperature. A researcher carrying out environmental studies may also wish to collect data on evaporation, wind, solar radiation, dewpoint, humidity, and other climatic variables.

Precipitation data are normally available as a daily reading of water depth in a rain gauge, published in monthly or yearly summaries. Information may also be available about the amount of snow which fell over a day. A "history" of each rain gauge station should be obtained in order to evaluate inconsistencies in the data. If the gauge is read daily by an observer, the actual time of observation should be noted, as should missing data or rainfall amounts which represent an accumulation of several days' precipitation. Near-continuous records are sometimes available, either in digital form (for example, from a tipping bucket gauge) or as data processed from a pen trace on a chart. Agencies will sometimes process the data to derive maximum depths of rainfall per unit of time, or minimum duration of certain rainfall depths. Regional maps are commonly published from the results, such as the one shown in Figure 3.7.

Temperature data are usually recorded as daily maxima and minima. Summaries of monthly and yearly averages are also commonly available. Both precipitation and temperature data can usually be obtained from meteorology agencies.

Water quality

Water quality data may be available from national or state agencies (typically, environmental branches), municipal water and sewage-treatment facilities and industrial plants sited along rivers. Some water-quality sampling stations on main rivers in the USA and the UK now have many years of record, but it would be more typical to find short, discontinuous records. Because of the cost and number of possible parameters to measure, the length of record and type of data available are highly inconsistent. The quality of laboratory analyses probably does not vary as much as the type of analyses done and the frequency of sampling.

Water-quality data may include electrical conductivity, pH, and concentrations of heavy metals, ions (e.g. ammonium, chloride or sulphate), organics such as pesticides, and dissolved oxygen. Daily records of turbidity, salinity and temperature may be available if the site is automatically monitored, but it will be more common for samples to be collected manually at fixed intervals of time (for example, once or twice per month) without regard to the magnitude of stream discharge. If the aim is to determine seasonal variability, data should be reviewed to see if the range of discharges is adequately sampled (AWRC, 1984).

Sediment

Sediment data provide useful information for analysing sediment influxes and throughput, identifying potential sources and evaluating problems of deposition of sediments in reservoirs, channels and irrigation diversions. Decisions about the management of land resources can also be assisted if the effects of land-use options on sediment production are known. Sediment data are collected at least as widely and as often as those on water quality, and may be obtained from some of the same sources. Data may also be found in engineering reports and in flood control and other water resources investigation reports.

Sediment data may be given as an instantaneous concentration or as a yield (daily, monthly or yearly). Yield is usually calculated by combining concentrations with flow data to obtain the total amount of sediment transported past a station over a given time period (Section 7.5). Occasionally, sediment data will be analysed by particle size or separated into categories of bedload, suspended load and wash load (Section 7.1).

The value of published data will vary. If "daily" sediment samples are only taken at a designated time once per day, the peak events will often be missed; thus the "representativeness" of one instantaneous value is questionable. This problem is not as great in larger streams which are more consistent over a day. In all cases, however, one should investigate the method of data collection before using sediment data.

3.2.2 DATA FORMATS FOR AQUAPAK PROGRAMS

AQUAPAK, a computer package which can be used in conjunction with this text, contains a number of programs for analysing streamflow (see Appendix 2). The same techniques can be applied to the analysis of other variables such as rainfall, sediment, or water quality. The help program 0HELP gives information on the standard data formats. Basically, the data files consist of a continuous series of daily or monthly values, identified by date. Data can be entered by hand to create a data file using the data entry program INPUT. A separate word processing or other file editing software package must be utilized to view or edit the entered data.

If commercial or agency-supplied data are obtained they will need to be re-formatted by the user. A re-formatting program should be developed, taking into account the presence of leap years, treatment of missing values, rearrangement of the date to yyyymmdd (year, month, day) form, skipping of blank or text lines in the supplied data, and compilation of daily totals if the data are based on a shorter time interval.

Since hand entering of data is laborious and high levels of detail are not needed in many cases, many of the programs may be run using monthly as well as daily data. Daily streamflow data should have units of m^3/s and monthly data should be expressed as volumes, in million m^3. Program CONVERT will convert from different units (e.g. cubic feet per second or acre-feet) to these standardized units. From the input data (daily or monthly), SUMUP is used to create files containing annual and/or monthly totals. It is recommended that the data be error-checked (see following section) either by plotting the data or visually scanning the data file with a word processing package.

3.2.3 DATA QUALITY

The record of data is, in reality, only a sample of the total population of values which have occurred and which may be expected to occur. It is thus subject to statistical errors resulting from the fact that it is only a sample of a larger population, as well as measurement errors (see Section 2.3.2). Wide variations may be found in the incidence of missing or inaccurate records and the frequency with which records are taken at each station.

For statistical analyses to provide projections into the future based on past properties of hydrologic data, data must have tolerable measurement errors, be of sufficient length to give a good representation of the total population and be "homogeneous", meaning that the data are from the same population. The following is a discussion of measurement error, representativeness and homogeneity, including methods of error detection and correction.

Measurement errors

Measurement quality depends on the precision and accuracy of instrumentation, site characteristics and the conscientiousness of the observer (Shaw, 1988). In a high-quality monitoring program, personnel will calibrate and maintain measuring devices and apply corrections to the data to adjust for problems such as slow clocks, leaky rain gauges or the growth of mosses or algae at stream-gauging sites. The checking of records and processing of data is best carried out by the actual observer who is most likely to know about the causes of errors, how to detect them and how to fix them.

Unfortunately, it is often difficult to assess the quality of data because information on errors is not always collected or recorded (or admitted). Once hydrological data have been stored in a computer or published in a report, it is a common tendency to accept these figures as "accurate". The greatest danger in using the data is not in the actual quality of the data collected but in ignorance about what are good or bad data. Digital data especially give a false sense of validity; a pressure transducer may provide precision readings of stream levels but these readings will be of marginal value if the transducer is bouncing on the streambed. If possible, an investigator should take a visit to a measurement site to learn about the quirks of the particular site and instrument. At the least, enough inquiring should be done to develop a sense of how much attention is given to detection and correction of error and to establish the degree of "faith" one can have about the data.

A quick visual scan of the data time series to detect gross errors should be an initial step in data analysis. Program PLOTFLOW in AQUAPAK will plot a time series of daily streamflow. When scanning data, a quick look at yearly or half-yearly data series using automatic scaling should be done to detect order-of-magnitude jumps in the data (e.g. from a misplaced decimal point), long periods of missing records, and other erroneous readings. Figure 3.1 gives an illustration of some of these errors.

The data should then be examined at a smaller time interval (e.g. 30 days), and the maximum vertical axis (y-axis) limit set to a lower value so that truncations of low-flow values, short-term missing values, and erratic data can be detected. When erroneous data are discovered one can then reject that part of the record altogether or attempt to reconstruct it. Methods for filling in data are covered in Section 3.2.4.

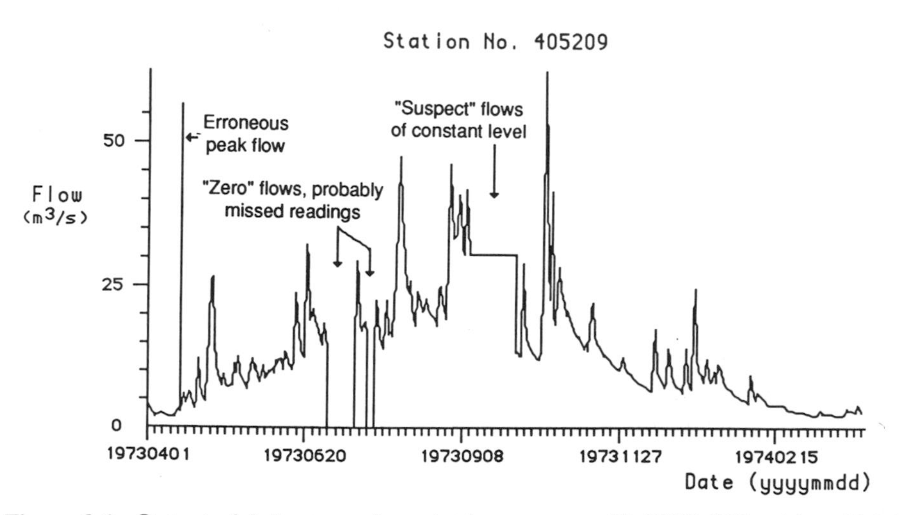

Figure 3.1. Output of daily streamflow plotting program, PLOTFLOW, with artificial errors to show problems which might occur in a data record

Representativeness

The degree to which a record is representative of the total population is difficult to determine, since, geologically, records have only been collected for a relatively short period of time. The longest record of river stage in the world is a mere 1050 years for the Nile at the Roda gauge (Leopold, 1959). For most rainfall and runoff records, 100 years would be an exceptionally long record and 20 years more typical.

Representativeness may be affected by the inherent properties of a stream such as its "flashiness" (see Chapter 4) or sampling deficiencies. Paleohydrology records can be used to get an idea of long-term trends, but should be used with caution since they depend on subjective interpretation.

Data should cover a representative range of values rather than being exclusively from unusually wet or dry periods—unless, of course, this is the record of interest. Records of only a few years in length are not likely to be representative of the long-term variability at a site (Yevjevich, 1972). One has little control, however, over the "sample" provided by nature.

The number of years of record required for statistical analysis is related to sampling design. If a normal distribution (see Appendix 1) can be assumed, Equation 2.1 can be used to approximate the number of years of data needed to calculate a mean value of annual runoff or rainfall within a certain margin of error. Section 8.2 gives information on the length of record needed for flood-frequency analysis. For habitat studies, Bovee (1982) recommends a minimum of 10 years of record, encompassing at least one high and one low sequence.

Homogeneity

A "homogeneous" record is one which is drawn from the same population or statistical distribution (Bovee, 1982), an analogy being homogenized milk versus its non-homogeneous form where cream and skim milk have separated. In hydrology, non-homogeneity can result from changes in the hydrological environment, which can occur slowly or abruptly as a result of various human or natural activities (Yevjevich, 1972). It can also be due to changes in the type or location of instrumentation.

Although for most purposes a long record is preferred to a short one, changes in the physical conditions of the stream and its catchment, or in methods of data collection, are more likely to occur during a longer period (Searcy and Hardison, 1960). Causes of non-homogeneity in streamflow records include (Bovee, 1982; Linsley et al., 1975b; Yevjevich, 1972):

(1) Movement of a gauge;
(2) Change in observer;
(3) Change in data recording method (e.g. a staff gauge manually read on a daily basis may be replaced by a recording gauge);

(4) Change in channel configuration at the gauging site;
(5) Installation of a dam, irrigation works, trans-basin diversion works, levees; or the pumping of large quantities of groundwater upstream (and, in some cases, downstream) of the gauging site;
(6) Sudden changes in hydrologic parameters from catastrophic natural events (e.g. wildfires, landslides, earthquakes or large floods) or land-use changes (e.g. urbanization or deforestation).

For rainfall records the first three causes of non-homogeneity listed above are also applicable. Additionally, the erection of a building or fence or the growth of vegetation around a rain gauge can affect the "catch".

The **double-mass curve** is a useful method for detecting non-homogeneities in a record. Searcy and Hardison (1960) provide a definitive reference on the technique. The curve is based on the concept that a graph of the cumulative data of one variable versus the cumulative data of another is a straight line as long as the relation between the variables is a fixed ratio (Searcy and Hardison, 1960). Data from streamflow gauges, for example, can be compared with those from other streamflow gauges from the same general area to detect non-climatic changes in flow regime. A double-mass program (DMASS) is included in AQUAPAK, and a sample graph is shown in Figure 3.2. Streamflow is converted to millimetres by dividing by catchment area to make records from different sites comparable.

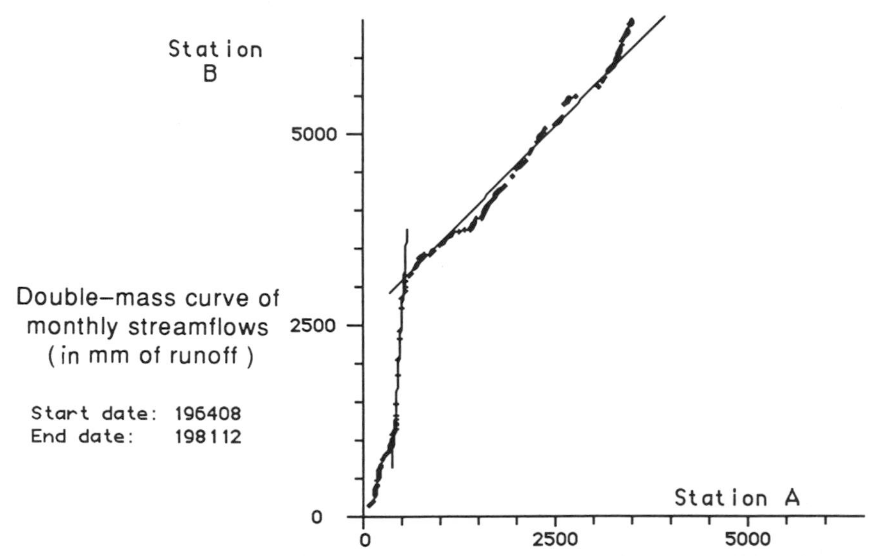

Figure 3.2. Output from double-mass curve program, DMASS, showing cumulative monthly streamflow data from Station B (suspect station) plotted against data from Station A (reference station). In this illustration, lines have been fitted through the data by eye to highlight the break in slope

When only two stations are plotted against each other, as illustrated in Figure 3.2, it is not possible to determine which station is inconsistent. It is preferable to average values for a group of surrounding stations since the average is less affected by an inconsistency at any one station. This average becomes the "base station" against which individual stations can be compared.

Streamflow should not be plotted directly against rainfall, as the relationship between them is rarely linear. A more effective approach is to first develop a relationship between streamflow (mm) and precipitation (mm). Estimates of streamflow derived from this relationship are used for comparison with the suspect station (Linsley et al., 1975). The regression program REGRESS and program TRANSFRM in AQUAPAK can be used to develop and apply a regression equation. This procedure is also discussed by Searcy and Hardison (1960).

When inconsistencies are noted, further investigation into their cause is warranted. The inherent variability in data causes spurious breaks in the curve, and these should be ignored. Major breaks in slope, such as the one illustrated in Figure 3.2, indicate that the data are non-homogeneous. If breaks are discovered, historical records on the gauge and/or the catchment area should be carefully reviewed to find causes corresponding with the time and direction of the changes. At this point one can either eliminate all but the part of the record which is of interest (e.g. that representing "present" or "natural" conditions) or adjust the record to make it homogeneous.

Methods of adjustment differ for rainfall, runoff and sediment, and Searcy and Hardison (1960) should be consulted for details. Rainfall records can be brought back into alignment by using a ratio of the slopes of the two segments, since the slope is the constant of proportionality between the stations.

With streamflow records the double-mass curve should be used only for detection of inconsistencies. Adjustment of the record should be made using techniques which take into account the cause of the inconsistency. This may be a simple matter if, for example, a gauge is moved downstream and an adequate adjustment can be made by using a ratio of the new and former catchment areas, or it may be a complex process if reservoirs, diversions and levees have modified the flows. If only coarse monthly estimates are needed, attempts can be made to add diversion amounts back into a flow record and adjust for evaporation differences (e.g. from reservoir surfaces rather than vegetation). If daily "natural" flow estimates are required for a stream now highly regulated by reservoirs the only solution may be to reconstruct a flow series from rainfall and catchment characteristics using simulation methods (Linsley et al., 1975b).

3.2.4 HOW TO FILL IN A STREAMFLOW RECORD

Missing data is a widespread problem in hydrology. Gauges can be damaged during floods, power supplies may run low, pens may run out of ink, inlet pipes to gauging stations can clog with sediment, or an observer might miss several readings. Various techniques are available for filling in missing data, and some of these can also be used to extend a record back in time for statistical analyses. Ideally, the gaps in data will be filled by the data-collection agency and identified as estimated values.

Short gaps can sometimes be filled from straight-line interpolation (graphical or numerical) between correctly recorded discharges. A formula for linear interpolation is given in Figure 3.3.

Longer periods of missing records or those which include rainfall or snowmelt events should be reconstructed by establishing a regression relationship with a nearby gauge or group of gauges. Stations should be located in a "hydrologically homogeneous region" (see Chapter 8), and preferably on the same stream. It is advisable to develop regression equations using data from time periods during which conditions were similar to those for the period of missing data (e.g. storm type, water level, condition of the channel). For short periods of daily record, linear regression may be sufficient. Searcy (1960) advises the use of exponential (log-log) regression to normalize the data and linearize the relationship. For higher flows he demonstrates that this relationship between two basins is generally parallel to the **equal-yield line**, a line of 1:1 slope drawn through the origin. Estimates of monthly or yearly values will typically have less error than daily ones, and estimates for more stable regimes will be better than those for flashy streams. Regression analysis is covered in Appendix 1, and a simple regression program (REGRESS), is contained in AQUAPAK.

Errors will always remain in data, no matter how carefully they are scrutinized and corrected. Measurement errors are normally assumed to be random and normally distributed for the purpose of computing statistics. However, systematic errors which show up as false trends in data can also

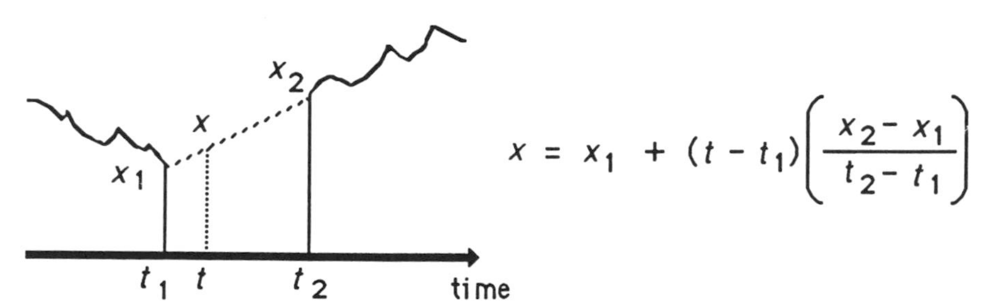

Figure 3.3. Linear interpolation by graphical and numerical methods. The points x_1 and x_2 represent values of a variable (e.g. water level) at times t_1 and t_2, respectively. The unknown value x is interpolated for time t

occur—for example from the accumulation of bed materials at a gauging site or the growth of vegetation around a rain gauge (Yevjevich, 1972). A decrease in the magnitude of errors from improvements in measurement techniques may be desirable; however it also means that a constant variance cannot be assumed.

These cautions are not meant to discourage potential investigators from using data or performing conventional statistical analyses. A little wariness is advisable to ensure that conclusions are based on changes in the parameter of interest and not the quirks of an instrument or the biorhythms of the observer.

3.3 MAPS

3.3.1 WHAT IS A MAP AND WHY WOULD YOU NEED ONE?

When working with a riverine system a bird's eye view of the situation can provide information about the stream in relationship to its surroundings. A map is a graphic depiction of an area in two-dimensional form: its roads, soils, land use, vegetation, topography, geology, etc. Maps relate information such as distance, direction, shape, position, and relative size, according to a given scale and method of projection. They give a "filtered" version of the real world, in which information is presented symbolically, as aerial coverage or as "isolines" of equal value (e.g. contour lines on a topographic map). Each country will have its own type of mapping system, criteria and standards for mapping, and degree of strictness in following the criteria. In some, maps are considered classified documents, although they are readily available in most Western countries. As with time-series data, maps have their limitations, and field checks may be necessary for detailed studies.

In hydrology, maps are useful for assessing land characteristics which influence runoff patterns and erosion rates, and as tools for predicting these variables when data are unavailable. In conjunction with field surveys, maps are helpful for locating the extent of a floodplain, the position of a river channel, and points of diversions for irrigation and hydropower or intakes for municipal and industrial water. They are also useful for calculating channel slopes and other measures associated with channel morphology (Section 4.2). Enterprising individuals have taken old maps developed before construction of a dam to develop "fishing maps" which describe the contours of the inundated land. Finally, but certainly not least important, maps provide a means of orienting oneself in the field to obtain the geographic position of data-collection sites and to avoid getting lost.

Maps are basically divided into two types: **topographic** and **thematic**. A topographic ("topo") map describes a site in terms of its horizontal and vertical position. The leg muscles of anyone who has walked the tracks and trails of the world no doubt have a "feel" for what closely spaced contour lines on a topo map mean. "Thematic" literally means theme; thus, thematic maps follow a special theme such as population, natural resources, or economics. They are an excellent source of standardized, cheap, readily obtainable and easily captured land data. Thematic maps are usually based on topographic maps, although topography itself may be simplified or completely left off. The classification of "themes" is simple in concept; for example if one soil or vegetation type is more common in an area than any other, then that area is classified by the prevailing type. However, there are rarely sharp demarcations between types. As with any map type, the larger the ground surface represented, the greater the generalization.

Maps have specific purposes and applications. Bell and Vorst (1981) point out a need for an international scheme of mapping for hydrology. UNESCO (1970) has developed an international legend for hydrogeological maps to provide this consistency. Until such time as "hydrological" maps become available, however, the stream hydrologist may use a number of different map types, some of which are described as follows.

Topographical maps portray the relief of the land, its water features, developments, roads and tracks. An example is given in Figure 3.4 and an aerial photograph of the same region is shown in Figure 3.5. Relief is sometimes supplemented with hill shading, which makes recognition of landform features much easier. Topo maps are essential for outlining catchment areas and evaluating drainage patterns. "Nicks" in contour lines can give clues to the presence of watercourses even if they are not marked as blue lines on the map (see Section 4.2).

Topographical data are now available in digital form, as horizontal and vertical co-ordinates on a grid system termed a Digital Elevation Model (DEM) or Digital Terrain Model (DTM). Computer software packages have been developed for computing aspect or slope from the digital data. National cartographic or survey agencies should be contacted for information on these products.

Orthophoto maps are a relatively new product in which topographic maps are plotted over a black and white photomosaic base, which has been corrected (ortho = "correct") to true geometric precision. The advantage of this product is that it is as accurate as a map for obtaining measurements but identification of ground objects is enhanced because of the rich detail of the photographic image. At present, orthophoto maps are typically available only for populated areas, but they have great potential for use in hydrology.

Geological maps indicate geophysical features such as the age and distribution of rock formations and glacial and river deposits. Cross-sectional diagrams showing subsurface composition are often included, from which patterns of groundwater movement can be inferred (Wilson, 1969). Since geology is a major factor in stream formation, geological maps are important

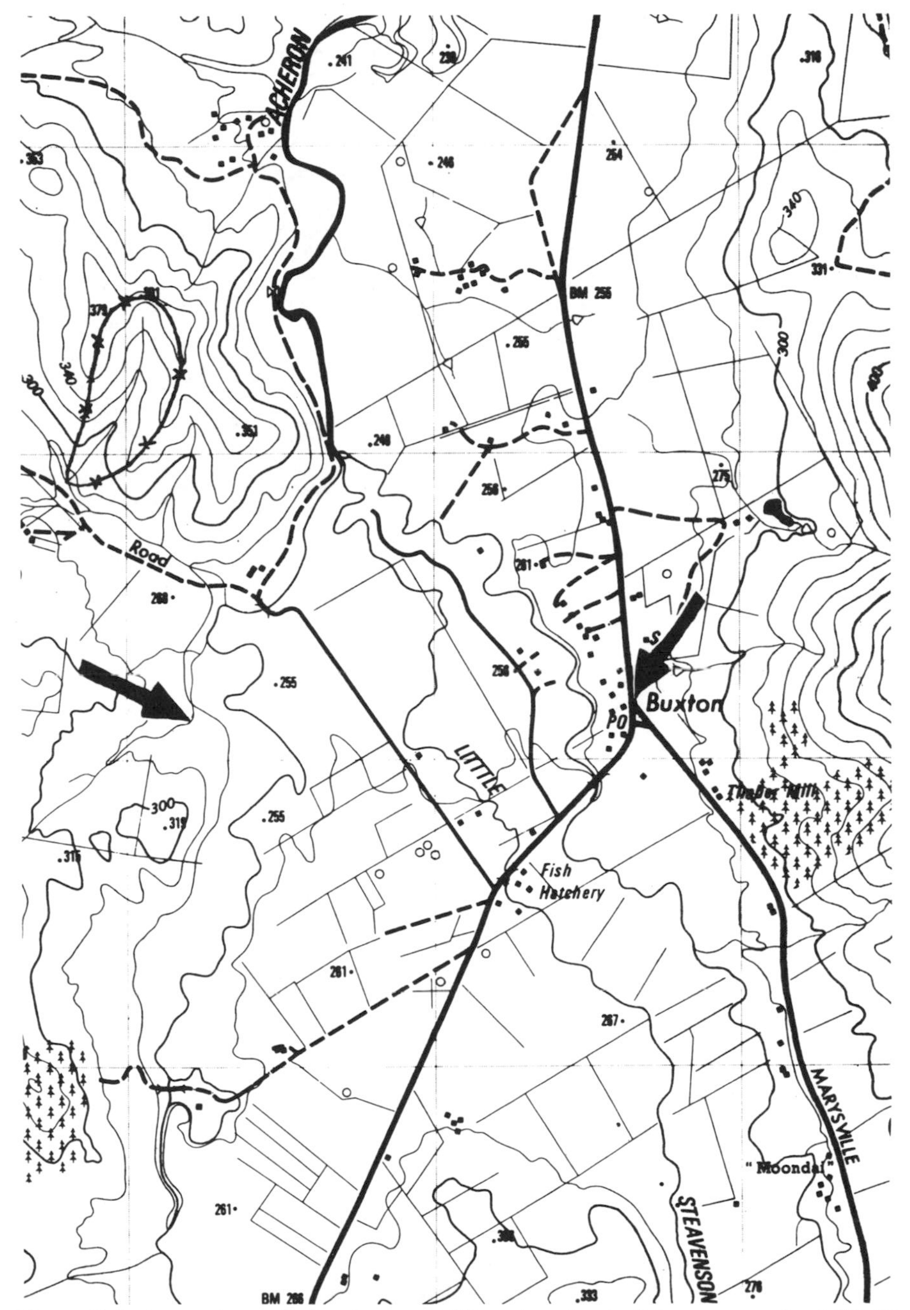

Figure 3.4. A section of the Buxton (Victoria, Australia) map, scale 1:25 000. Contour intervals are 20 m and grid lines are 1000 m apart. The map was compiled from 1972 aerial photography. The ground distance between the arrows is the same as in Figure 3.5 (see discussion of Equation 3.2). A catchment boundary is indicated by the line x---x (see Section 3.2.1). Black and white reproduction of colour original, © 1978, Crown (State of Victoria) copyright, reproduced by permission; not to be sold or copied without the written permission of the Surveyor-General

Figure 3.5. A section of a black and white aerial photograph, taken over Buxton, Victoria, on 5 February 1984. Height of the aircraft was 16900 ft (5150 m) above sea level and focal length of the camera was 152.57 mm. The ground distance between the arrows is the same as in Figure 3.4 (see discussion of Equations 3.1 and 3.2). © Crown (State of Victoria) copyright, reproduced by permission; not to be sold or copied without the written permission of the Surveyor-General

to hydrologists for interpreting the evolution of channel patterns and stream characteristics. Channel geometry, roughness, and the source and size of streambed materials may be associated with geological features shown on maps. A black and white reproduction of a segment of a colour-coded geological map is shown in Figure 3.6. The light regions near the centre with drainage-like patterns depict alluvial deposits.

Morphological maps symbolically indicate breaks of slope in the landscape—the convex/concave boundaries. They thus give an indication of the form and steepness of landscape, which is useful in locating areas which are too steep for transport or agriculture. **Geomorphological mapping** relates information on surface form, giving the location of features formed by fluvial, glacial, fluvioglacial and aeolian (wind) processes (Gerrard, 1981).

Soils maps may be helpful in determining erosion potential of lands. The ability of a soil to absorb water will also affect the amount entering streams,

Figure 3.6. A section of the Warburton (Victoria, Australia) geological map, scale 1:250 000; 1977 edition. Buxton is just below centre in the area shown. Black and white reproduction of colour original, by permission Director of Geological Survey of Victoria

and is somewhat dependent on the soil type (CSU, 1977). Soil classifications differ, however, and some may relate more to crop-growing potential than to hydrology.

Vegetation maps often depict "theoretical" or "potential" natural vegetation, although in extensive areas of the world the natural vegetation has been removed or replaced. Global maps in particular should be used with care, as concepts of major plant associations and "potential" vegetation types may not be relevant on a more local scale. Regional maps are more likely to show conditions existing at the time the map was made.

Climate maps are useful for assessing patterns of temperature, rainfall, evaporation and other related variables. These patterns can be classified (see Section 9.2) to develop maps showing different climate zones. Climate maps are of interest in making generalizations about the behaviour of streams and erosion rates and about an area's potential for supporting vegetation or people. As well as general climate, maps may also be available for individual climatic elements such as rainfall and evaporation. Rainfall maps may depict average depths, frequencies and/or intensities. Figure 3.7 shows an example of a rainfall intensity map developed specifically for flood analysis.

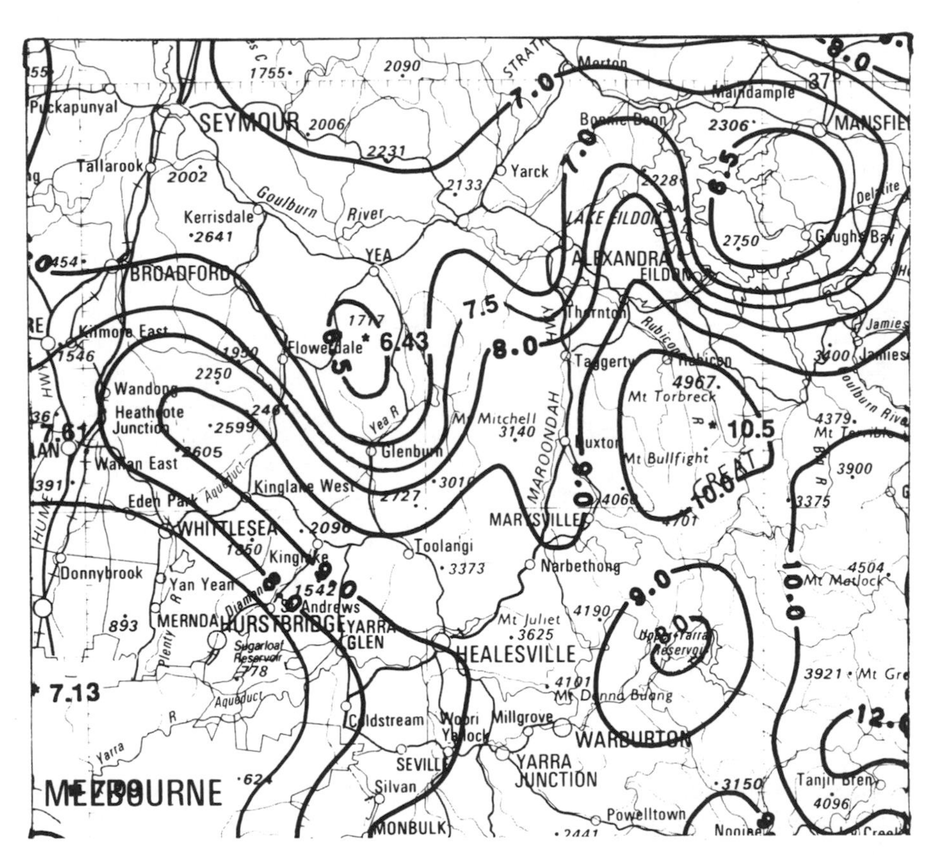

Figure 3.7. A section of a 1:1000000 map showing design rainfall isopleths of 12 h duration and 50-year recurrence interval. Isopleths are lines of equal rainfall intensity in mm/h. Buxton is located to the right of centre. Black and white reproduction of original, © 1987, Institution of Engineers, Australia; reproduced by permission

Runoff maps usually display the mean annual runoff for a region and are useful for obtaining a general picture of geographical variations in runoff (Linsley et al., 1975a). Depending on the amount of detail included in the maps, they may prove useful for extrapolating data from one region to regions of similar runoff, for general classification studies and for study site selection.

Custom mapping may be appropriate if available maps do not contain sufficient detail for a particular analysis. Finding the headwaters of streams or obtaining accurate contours of a streambed or eroded gully may require the addition of detail to commercial maps or the drafting of topo maps for very small areas. Special aerial photography and photogrammetric mapping can be commissioned for this purpose at varying degrees of cost and accuracy (see Section 3.4.2). Also, the aerial photography from which published maps are made can usually be purchased from the mapping agency.

3.3.2 MAP INTERPRETATION

Marginal information

Some of the most valuable information on a map is contained in its margins. This normally includes the name, scale and date of the map; the units of measurement; grid and contour intervals; an explanation of symbols; and arrows giving true, grid and magnetic north. Dates of the original map and revisions are particularly important, especially if the map is assumed to represent current conditions. Symbols will differ with the map type, country of origin, and scale, and should be examined before making interpretations. For example, different symbols may be used to indicate road or river type, or the locations of mining claims or boat docks. International symbols for many hydrological features have been standardized, such as permafrost, groundwater contours, karst formations, the disappearance point of a stream, and springs. Attributes such as vegetation or soil type will often be noted on a map by a specific colour or label, with a key provided as part of the legend.

Location and orientation

Latitude and longitude provide a co-ordinate system for pinpointing any location on the Earth's surface. Both latitude and longitude are measured in degrees: 360 degrees in a circle, 60 minutes to the degree, 60 seconds to the minute (e.g. 37°19′30″). Of historical interest, the system of using a base of 60 originated with the Babylonians, who discovered that the number 12 had more divisors (2, 3, 4, 6, and 12) than 10 (2 and 5). The number 60 combines the advantages of both base 10 and base 12 systems, and is used throughout most of the world as a basis for measuring time and degrees.

"Orienting" oneself by locating a position with respect to the points of the compass literally means finding the east or "Orient". Co-ordinates are usually indicated at corners of maps and by ticks along margins. Besides latitude and longitude, maps may also be gridded in kilometres or quadrants. These lines will either intersect at right angles or converge towards the top or bottom of a map, depending on the type of projection used in making the map and the distance away from the Equator; i.e. this effect will be much more apparent in Alaska than in Tonga.

Latitude is the angular distance measured north and south of the Earth's midriff. The Equator is 0°, the North Pole 90°N, and the South Pole 90°S. A line connecting all points of the same latitude is referred to as a **parallel**. North (as in "true North") differs from the direction a compass points as the needle orients itself within the Earth's magnetic field. Magnetic north drifts slightly east each year. The amount of correction (declination) and its annual rate of variation is normally indicated in the marginal information on topographic maps. If the map is more than about 10 years old the declination should be adjusted by the factor: (annual variation) × (number of years since map was made).

Longitude is the angular distance measured west and east of the prime meridian, where **meridians** are lines connecting equal longitudes. The prime meridian passes through the Royal Observatory at Greenwich, just east of London. This location, chosen in 1884, had more of a political than a scientific basis, influenced mainly by the prominence of British map makers and the power of Great Britain at that time (McKnight, 1990). Longitude is measured east or west of a plane through the prime meridian, to 180° in either direction. The **international date line**—where today changes to tomorrow or yesterday—generally follows the 180th meridian, with a few detours around island groups so that places like Fiji and Tonga remain on the same side of the line.

Scale

The scale of a map is the relationship between the distance measured on a map and the corresponding horizontal distance on the ground. It is usually given in the margin area of a map, as a graphic scale and/or as a fractional scale. The **graphic scale** is a line marked off in graduated distances, and remains correct even if the map is reproduced in a larger or smaller size. A **fractional scale** compares the map distance with the ground distance as a ratio (e.g. 1:10 000) or fraction (e.g. 1/10 000). A ratio of 1:10 000, read "one to ten-thousand", means that a measure of one unit (whether millimeters, inches or spans) on the map represents 10 000 of the same units in the real world. The use of metric units considerably simplifies conversions.

Scale may not be constant over a map because of the projection and method of mapping, but can normally be assumed constant if the map covers a relatively small area. It is helpful when reading maps to know the ground distance represented by 10 mm or some other convenient unit, so quick estimates of distance can be made.

Large scale and **small scale** are comparative terms. Just as a "large" cup of coffee at one fast-food restaurant may be classified as "small" at another, the difference between large and small scales is also dependent on the viewer. The distinction can be confusing, because "large" and "small" refer to the ratio or fraction rather than the amount of Earth's surface covered by a map. For example, 1/100 is a larger fraction and thus a larger scale than 1/100 000. The International Cartographic Association (ICA, 1984) uses the following classification:

Large scale: greater than 1:25 000,
Medium scale: 1:25 000 to 1:250 000,
Small scale: 1:250 000 to 1:2 500 000, and
Very small scale (e.g. maps in atlases): less than 1:2 500 000.

Each scale of map is suitable for different purposes; i.e. there is no "best" scale. A comparison of Figures 3.4, 3.6 and 3.7 will illustrate the effects of scale, with the larger-scale maps showing more detail than the smaller-scale ones. For hydrological work a medium- or small-scale map is useful for identifying the location of a project within a riverine system, obtaining information on the stream network and regulating structures, and calculating parameters such as sinuosity. Larger-scale maps of the actual vicinity are needed to obtain detailed characteristics of the reaches under consideration.

Elevation

Elevation is given on most maps in terms of distance above a certain level, usually a standardized mean sea level or a national height datum such as the Australian Height Datum (AHD). Heights of significant mountains, benchmark points and other features are often shown as spot elevations.

On topographical maps the elevation is represented by **contour lines**: lines joining points of equal elevation. The **contour interval**, or vertical distance between contours, may be altered to suit the terrain or may be fixed for a given map series. This interval is normally specified in the marginal information on maps, and elevations are marked every few contour lines. Contour lines may contain error when aerial photography is used to develop the maps, especially where vegetation obscures the ground surface. Jennings (1980) cites an example of a map made from surveys taken between the two world wars. During later surveys the nature of the errors in the map made it clear that much of the mapping had been done from ridge crests without ever entering the valleys!

3.3.3 REVISION, ACCURACY AND STANDARDIZATION OF MAPS

When using maps it is important to realize that they have their limitations in terms of accuracy and representation of the landscape. Map users cannot expect mapping agencies to produce map series for specialized needs, such as large-scale maps with close contours for drainage analysis. On the other hand, one should not be timid about questioning map accuracy. "As all who have made maps know, as finished products they have a definiteness about them which somehow conveys an accuracy even greater than that of the printed word and as with the latter it is not always fully justified" (Jennings, 1967, p.80).

Representation of natural features on maps has become more accurate over time as the amount of quantitative information extracted from them has increased. The depiction of streams on maps, however, is fraught with problems of interpretation. In limestone terrain, parts of streams may be completely underground at low stage and above ground at high stage. Even when stream patterns are more typical, the classification of a stream as perennial or intermittent is a difficult one (see Section 4.1). Map makers may choose to avoid the distinction altogether (Drummond, 1974).

Erosion and deposition normally causes slow change over historical time, but other landforms such as watercourses, glacier snouts, sand dunes and coastal spits change rapidly (Jennings 1967). New settlements, roads and areas of timber harvest also cause the landscape to change quickly, and repeated surveys and map revisions are necessary to keep up with the changes.

The use of aerial photography and computer-supported photogrammetric systems has greatly increased the accuracy of topographical maps, as well as enabling more efficient revision. A danger in using maps of different editions, however, is that the revised map may represent only the changes in man-made features and not in the surrounding environment. Before maps are used for quantitative analysis, information on mapping criteria and standards should be obtained from the agency. Field surveys and aerial photography can also help to verify map accuracy. Mostly, one should be aware of how far a map can be "stretched" to provide information—whether getting an idea about how hard a hill will be to climb or calculating the bifurcation ratios of river systems.

3.4 PHOTOGRAPHS AND OTHER REMOTELY SENSED DATA

3.4.1 WHAT IS REMOTE SENSING?

Remote sensing (a term coined in 1960) is the "measurement or acquisition of information by a recording device that is not in physical contact with the object under study" (McKnight, 1990). This definition includes more than just imagery, such as the remote acquisition of data on snow water content or wildlife movements. In the environmental sciences, however, remote sensing is defined more precisely as the recording of images of the environment using electromagnetic radiation sensors, and their interpretation (Carter, 1986; Curran, 1985). It includes conventional aerial photography and aerial photograph interpretation.

The amount of information which can be gained about a region is greatly enhanced by the acquisition of imagery. Imagery is attractive, too, where field survey costs are high or where areas are physically inaccessible. River patterns are dynamic, and aerial survey and satellite image analysis methods have contributed to the ability of hydrologists to monitor catchment changes. Environmental and resource surveys, crop conditions, snow cover, extent of urban area development or timber harvest, and the assessment of the extent of bushfires, flooding or drought are some of the practical uses of remotely sensed images.

In contrast to maps, where interpretations have been made by a map maker, images can be interpreted directly. The amount of information on an image may at first seem overwhelming because of this lack of prior interpretation, but with experience and repeated association with field conditions, qualities can be quickly inferred from imagery.

Maps, photographs and remotely sensed images each provide their own brand of information. High-altitude space imagery will cover the entire face of the planet, whereas large-scale colour or black-and-white photographs are more appropriate for detailed terrain studies. Knowledge of each product and its advantages and limitations can help in the selection of the best type of imagery for particular applications.

3.4.2 PHOTOGRAPHS

Compared to a map, a photograph is much closer to reality. Photographs can be used *before* a field trip to familiarize oneself with an area, to select suitable study sites, or to identify major landforms or drainages. They can be used *while in the field* to locate and orient oneself, to check the identification of features previously observed on photographs and to discover what lies beyond the immediate field of view. Additionally, they can be used *after* a field trip to refresh one's memory or to extend results from one site to similar areas nearby (Chapman et al., 1985).

Photographs provide a unique record of the past, and are especially valuable when it is difficult to revisit a site. They can be re-examined—sometimes many years later—when techniques (and time) for extracting more information become available. Baseline photographs can be compared with monitoring photographs taken five to ten years later to evaluate changes, either natural or due to management efforts (Platts et al., 1987).

The earliest aerial photographs were taken from hot-air balloons in the mid-1800s (McKnight, 1990). Conventional aerial photography remains a primary technique when it is necessary to resolve the detail of ground conditions. Features which are not marked on maps, such as sinkholes and the fine patterns of upper stream drainages, may be observable on aerial photographs. Aerial photography is versatile because of the wide range of scales available and the large combinations of films and spectral filters which can be used. Limitations of aerial photography are those of cost and quality, which are dependent on the type of film, camera and aircraft, the distance above ground and the suitability of flying and sensing conditions.

For small riparian area studies, Platts et al. (1987) suggest less expensive methods of acquiring photographs with a 35 mm camera and a small helicopter. However, for obtaining precise measurements of distance or aerial coverage, higher-quality images are needed. The timing of aerial photography is critical, not only for optimizing lighting or avoiding cloud cover but also for obtaining the best resolution of the features of interest. Soil-tone changes, for example, are more distinct in winter and spring; crop patterns in summer (Barrett and Curtis, 1982). Aerial photographs are useful in the analysis of river reaches: the sizes and locations of sandbars, river-control structures, time changes in the channel form, sediment concentrations, and seasonal variations in vegetation along banks or changes in floodplain vegetation following inundation (CSU, 1977). Hooke and Kain (1982) provide a reference for using historical information to study changes in the physical environment.

Although "remote" implies a great distance, photographs taken from hand-held cameras with one's feet firmly planted on terra firma (or slightly higher, from a ladder or "cherry picker" boom truck) are of great value in environmental studies. For stream studies, 35 mm colour print photos should be taken laterally, upstream and downstream at a site (Platts et al., 1987) for monitoring purposes or to provide a historical record. Photographs of substrate are useful for direct analysis of particle size, roughness and percentage coverage (Section 5.3).

Photographs are either taken vertically (meaning that the optical axis is perpendicular to the ground) or obliquely (at an angle). Oblique photographs give a more familiar point of view, but because of perspective effects, measurements are more difficult. **Photogrammetry** is the science of obtaining measurements from photographs. These, however, are subject to error because of distortions in the image caused by the lens system (e.g. at the edges of photographs), from the relief of the terrain, or from the aircraft banking in flight. For point-to-point comparison with maps, an image must be adjusted ("rectified") to ground distances with the use of previously placed markers or established features such as highway intersections or small water bodies. Agencies developing maps from aerial photographs of hilly terrain must allow for the fact that closer objects appear larger on a photograph. Computerized rectification is commonly used to create corrected "orthophotos" of high precision.

Vertical as well as horizontal measurements can be calculated from photographs if stereo pairs—two photographs of overlapping areas taken at slightly different angles—are available. Viewing these with a stereoscope yields a three-dimensional mental image. Using the same process, stereo-plotters are used commercially for tracing contours. Methods of using close-range photographs for describing and measuring surface microtopography have been compared by Grayson et al. (1988), and additional information on measuring heights from stereo pairs is given by Curran (1985) and Lillesand and Kiefer (1987).

Scale

The scale of a photograph has the same definition as for maps and, as with maps, it is helpful to work out the distance represented by 10 mm or some other unit length. Approximate scale can be calculated for aerial photographs by:

$$\text{Scale factor} = \frac{\text{Focal length of camera}}{\text{Height of camera (or aircraft) above surface}} \tag{3.1}$$

The camera height and the focal length must both be expressed in the same units. This scale applies to the negative and contact print, and does not apply if the photograph has been enlarged or reduced.

Information on the intended aircraft elevation above sea level (not the ground surface) and camera focal length, the date and time when the photograph was taken, the general location, roll and exposure number, and name of the agency responsible is normally printed in the margin of commercial aerial photographs. The aerial photograph shown in Figure 3.5 was taken at an elevation of 16 900 ft (5150 m) above sea level, and the focal length of the camera was 152.57 mm. Equation 3.1 presents a problem at this point, since the elevation of the ground surface is needed to compute the height of the camera, yet the ground surface is not a consistent elevation. Using an average elevation of 275 m for the photograph the scale comes out 1:31 950, which is somewhat smaller than that of the topographical map shown in Figure 3.4.

A more practical method of obtaining scale is to measure a distance between two landmarks on a photograph and compare it to the equivalent distance obtained from a map. The scale of the photograph is then obtained by:

$$\text{Scale factor} = \frac{\text{Distance measured on photo}}{\text{Distance measured on map}} \times \text{scale of the map (as a fraction)} \tag{3.2}$$

Distances must have the same units. Using measurements taken between the arrows shown in Figures 3.4 and 3.5, a scale of (46 mm/58 mm) × (1/25 000) = 1/31 520 is obtained. This method of calculating scale is preferable for obtaining more accurate measurements, and is essential if the photograph has been enlarged or reduced.

Platts et al. (1987) recommend a scale of 1:2000 as "acceptable and achievable" for photographing riparian areas on small streams. Channel patterns are best obtained from smaller-scale photos such as that in Figure 3.5.

Film types

The four main types of film are: panchromatic black and white, colour, and both black and white and false colour infra-red (IR). Panchromatic film has been typically used in the past, and remains the least expensive of the four types. Black and white IR is also relatively inexpensive. When a red filter is used, black and white IR film cuts through haze more effectively than panchromatic. It is also sensitive to soil moisture and provides high contrast for the tracing of watercourses. Colour film allows the discrimination of a larger number of hues and is more easily interpreted since the human eye is not used to viewing the world in black and white.

False-colour IR film was introduced as the "camouflage detection film" because it could discriminate between living vegetation and the withering vegetation used to hide objects (McKnight, 1990). For the same reason, both black and white and colour IR are useful for determining the health of plants. Astroturf, as an example of unhealthy vegetation, is blue on a false-colour IR image, but healthy grass appears red—even though the two types of "grass" would look the same with panchromatic film (McKnight, 1990). False-colour IR film has also proved to be suitable for monitoring sediment transport and for identifying sediment sources in streams, being very sensitive to the concentration of suspended material. Tones change from a very dark blue for clear water to a very light blue for waters with high sediment concentrations. These tones should be calibrated against field samples.

3.4.3 REMOTE SENSING IMAGERY

Remotely sensed images can be divided into two types: data collected by (1) passive systems which sense natural radiation and (2) active sensing systems in which electromagnetic radiation is emitted and the reflected signal detected by a sensor. Passive systems detect radiation emitted within a specific wavelength range such as visual or infrared, and are most often associated with satellites such as Landsat. Active systems include sonar (sound navigation ranging) and radar (radio detection and ranging). The use of microwaves has also been investigated and found to be sensitive to subsurface soil moisture (McKnight, 1990), although cost has limited its use.

Passive systems: Landsat

Remotely sensed data are largely used for mapping land uses and for monitoring changes in characteristics of the Earth's surface (Curran, 1985). As compared to aerial photographs, satellite data tend to be less expensive per unit area covered (Carter, 1986). As with photographs, satellite data are used for environmental and resource surveys at up to a global scale. A beautifully written book by Kotwicki (1986) gives an example of how remote sensing can be applied in hydrology. Landsat imagery was used to document the drying of Lake Eyre following flooding in January 1984. Lake Eyre is a large inland lake in Australia which is dry for long periods of time and highly saline when it fills.

The science of remote sensing was rapidly propelled into the future with the launching of the first Landsat satellites in the 1970s. Since then, millions of images have been obtained for the evaluation of Earth's resources. The Landsat satellites have all carried two sensors: a multispectral scanning system (MSS) and either a Return Beam Vidicon (RBV) panchromatic television camera (which was largely unsuccessful) or, more recently, a Thematic Mapper (TM) scanner (Curran, 1985). Sensors record reflected light of various wavelengths as "grey levels", transmitted from the satellite to ground stations as digital numbers. Computers process the data to re-create a false colour composite image. Landsat satellites now repeat coverage of the same area every 16 days. The potential amount of data is mind-boggling, and not all of the possible images are collected or processed. The development of new products and software has grown rapidly in response to the need for efficient methods of filtering, error-checking, classifying and interpreting the digital data.

MSS data consist of four images of a scene taken in green, red and two near-IR wavebands. Sensors detect 64 levels of radiance, which are recorded as grey level values for each pixel, where a "pixel" or *pic*ture *el*ement is the unit of resolution of an image. Ground resolution is 79 m, although the pixel size is 56 m by 79 m, and is actually trapezoidal in shape as a result of the sensor viewing angle and Earth rotation. Each image covers a ground area of 185 km by 185 km (Barrett and Curtis, 1982).

Thematic Mappers, carried on Landsat satellites starting with Landsat 4 in 1982, record 256 radiance levels in seven wavebands: blue/green, green, red, near-IR, near-middle IR, middle-IR and thermal IR (Curran, 1985). The pixel size is 30 m by 30 m for six of the seven sensors; the thermal sensor has a lower resolution. Each image covers the same area as MSS data.

Different properties of a surface can be ascertained from each wavelength or from combinations of wavelengths. For example, blue/green, green and near-IR are good for detecting vegetation reflectance, and near-middle-IR and thermal IR are moisture sensitive (Curran, 1985). Thermal IR scanners, developed in the 1940s for "nocturnal snooping" by the military (Curran, 1985), have special applications related to temperature differences: location of frost hollows, estimation of plant moisture stress, location of thermal water pollution or hot springs, and the location and extent of vulcanism or above- and below-ground wildfires (Curran, 1985; McKnight, 1990).

Active sensing systems: SLAR

SLAR (or "Sideways-Looking Airborne Radar") senses terrain to the side of an aircraft's path by sending out long wavelengths (up to radio wavelengths) and recording the returned pulses (Curran, 1985). Radar imagery is particularly useful for terrain analysis (McKnight, 1990). It yields images with long "shadows" which enhance microtopography and relief and is also sensitive to surface soil moisture. The major advantage of radar is that it has better than 99% cloud-penetration ability, so is useful where cloud cover restricts the use of conventional photography or satellite data (Barrett and Curtis, 1982).

The relationships between image tone and characteristics of the land surface are different from those of conventional photography, and SLAR imagery has not yet gained wide acceptance. Although the cost is relatively high, SLAR imagery has potential in hydrological studies for detecting drainage patterns and mapping land cover, soil moisture and vegetation.

3.4.4 SOURCES OF IMAGERY

Recent issues of remote sensing guidebooks such as Carter (1986) and Cracknell and Hayes (1988) are excellent references for sources of aerial photographs and satellite sensor data, software for processing the data, and addresses of consulting, mapping and research establishments. Other good references on remote sensing are the companion volumes written by Curran (1985) and Lo (1986), texts by Lillesand and Kiefer (1987) and Thomas et al. (1987), and the "bible" of remote sensing published by the American Society of Photogrammetry and Remote Sensing (Colwell, 1983). The best way to obtain current information in this rapidly changing field is to make direct inquiries to the establishments involved with the collection and processing of imagery, and refer to recent issues of major journals and symposia on remote sensing.

There is a very large amount of imagery available, although some governments place restrictions on the use of coverage within their territory. Landsat is operated in international public domain, and data are available free of access or copyright controls (Barrett and Curtis, 1982; Carter, 1986). Data are also available from earth-sensing satellites of other countries such as the SPOT (Système Probatoire D l'Observation de la Terre) satellite, developed by the French space centre CNES. SPOT collects panchromatic data (with 10 m spatial resolution) and MSS data (20–25 m resolution), its orbit repeating every 26 days. Frequency of coverage depends on the latitude and camera angle. The cameras can be set for oblique viewing, and can thus provide more frequent coverage of the same area as well as stereoscopic imagery.

Satellite data may be acquired as a false-colour "photograph" or as digital data. Digital data are normally available on tape or disk; however, complete coverage of an area over several periods of time can be quite costly. A number of computer-assisted techniques for mainframe and microcomputers are now available for rectifying an image, contrast stretching, colour and edge enhancement, filtering and other types of image processing and analysis.

In using remotely sensed data the same cautions are advised as for maps and hydrological data. One should know how the data were collected, whether from the same or different sensors or satellites; be aware of possible errors from noise in the signal, variations in attitude, altitude and velocity of the satellite, or prevailing environmental conditions; and only push the information as far as its quality will allow.

3.4.5 INTERPRETATION, CLASSIFICATION AND "GROUND TRUTHING" OF IMAGERY

Remote sensing cannot be practised without knowledge of field conditions or the "ground truth". Otherwise the patterns on a photograph or numbers on a magnetic tape will have little meaning. Although satellite data are increasingly treated by semi-automatic computer analysis, most operational uses of remote sensing still involve manual procedures.

Interpretation of products in photographic format is more art than science, based on the ability of an interpreter to integrate the patterns on an image either consciously or subconsciously. It is a deductive process to go from qualitative information such as tone and/or color hue, location, shape, size, pattern, shadow, texture, depth (from stereoscopic views) to the identification of relief features, vegetation types or cows. Brightness and texture may be more accurately quantified with the assistance of a computer, but the eye is generally better at interpretation.

The conditions of resources as well as their identification can be assessed by inference from attributes observed on an image. For example, soil moisture conditions may be derived from the tone of vegetation. Experienced photo-interpreters work from familiar features to that which is unknown, utilizing all other evidence available for the study area: topo and thematic maps, photographs or data taken at ground level, or a visit to the site itself.

The darker patches of vegetation in Figure 3.5, for example, are pine tree plantations which contrast with the native eucalyptus forest. These plantations are also designated on the map of Figure 3.4. Photographic keys are sometimes developed by experienced photo-interpreters to communicate the photographic characteristics of vegetation, landforms, forest sites or soil conditions to other users (Barrett and Curtis, 1982). The text by Lillesand and Kiefer (1987) has numerous examples of landform interpretation.

Since remote sensing is generally less expensive than field surveys, the trend to replace "contact" methods of assessment with remote sensing will no doubt continue. By correlating remote sensing with field data, measurements such as vegetation densities or shallow water-table levels can be extrapolated to similar areas using remote sensing alone. In the same manner, remote sensing can be used to monitor changes in sites over time once a connection has been made between image patterns and what they represent on the ground. Although changes can be detected from imagery, the cause of the change must be identified or inferred by the user (Platts et al., 1987).

Geographic information systems (GIS) provide a relatively new tool for classification. The concept is the same as flipping backwards and forwards between several maps and photographs in the hope that one's brain will combine the information. Unfortunately, when more than two or three maps are visually compared, the mind tends to overload (Curran, 1985). A computer-based GIS can digitally superimpose spatial data from thematic maps, digital terrain models and imagery to yield new maps. For example, climate, topography and soils information might be combined to delineate areas with high erosion potential. An additional benefit of the digital terrain data is that GIS "maps" developed from these data can be given a three-dimensional appearance by changing the viewing angle and adding shadows.

Collecting ground truth data for correlation with imagery becomes a "question of how much ground checking will be required to produce a result comparable in objective accuracy to a ground survey" (Barrett and Curtis, 1982, p. 120). Sampling schemes should be based on the statistical sampling designs given in Section 2.3. A stratified sampling design is preferred, with samples taken within each identifiable "class" of surface features (e.g. substrate or vegetation type). In practice, sampling techniques may range from the careful location of randomly selected sites on an image which are then studied in detail in the field to casual verification from the window of a car. For extrapolation of ground truth data to other areas of the same "class", more refined sampling techniques are required so that results are statistically valid.

4 Getting to Know Your Stream

4.1 GENERAL CHARACTER

4.1.1 PRELIMINARY INTRODUCTION

The process of getting to know a stream is not unlike that of a doctor learning about a patient and his or her health. Preliminary observations, standard questions and routine measurements of temperature and dimensions arc combined with remote imagery, tests on various samples and deeper probing into the specific region of interest. A conscientious doctor will then look beyond the charts, images and the test results to obtain a sense of what causes the patient's health to be what it is. In the same manner, ecologists often use indices of catchment characteristics and streamflow patterns to obtain an indication of a stream's physical condition and suitability for various organisms.

This chapter covers the preliminary stage in which first impressions of a stream system are obtained from initial visits and from existing sources of information such as maps, aerial photographs and hydrological and climatic records. In this section, methods of general, qualitative description are given. Section 4.2 gives techniques for describing catchments and stream networks and Sections 4.3 and 4.4 cover methods of describing streamflow patterns.

4.1.2 PUTTING THE STREAM CHANNEL AND ITS CATCHMENT INTO CONTEXT

Most texts on hydrology begin with a picture of the hydrological cycle such as the one shown in Figure 4.1 and then go into an explanation of each component. In fact, much of the effort in hydrology relates to the modelling of water movement both above and below ground. In studying a stream and its biota it is important to maintain a perspective on the water's origins. Snowmelt-derived streams, for example, will have a different hydrological and biological character than the temporary streams of arid regions.

Only a small part of the water flowing in streams is a result of precipitation landing directly on the stream. Most of the water is derived from surface and subsurface runoff which results from rain or snowmelt on upland areas. The region from which water drains into a stream is termed the **drainage basin** or **catchment area**. The boundary or "rim" of the drainage basin is called the **drainage divide**, and follows the highest points between two drainage basins. The somewhat ambiguous term **watershed** is variously used to describe either the drainage basin or the drainage divide. Langbein and Iseri (1960) provide a compact and useful reference for definitions of these and other hydrological terms.

It is important to be aware of the hydrological, geological, morphological and vegetational setting of a stream. Climate is a major factor controlling streamflow patterns and the shaping of landforms and vegetation communities. It provides the energy and water necessary to drive catchment ecosystems. Geology influences the shape of drainage patterns, bed materials and water chemistry. Catchment soils are the weathering products of rock materials, which influence upland erosion potentials, water-infiltration rates and vegetation types. Vegetation is a source of biological production; in turn, it affects channel bank stability and upslope resistance to erosion, water loss through evapotranspiration and rate of runoff.

Adjustment of a stream to its climate and geology takes place continuously, causing changes in slope, rate of sediment transport and channel configuration. Associations of stream organisms are established in harmony with this dynamic nature of the channel's physical conditions (Cummins, 1986; Vannote et al., 1980).

From either a biological or hydrological viewpoint, the characteristics of a stream are dependent on the downstream transfer of water, sediment, nutrients and organic debris (Petts and Foster, 1985). Progressive changes in temperature, stream width, depth, channel pattern, velocity, sediment load and instream biota occur from headwaters to mouth. Vannote et al. (1980) hypothesize that the continuous gradient of physical conditions within a stream system results in a predictable structuring (a "continuum") of biological communities.

Recent literature by Bencala (1984), Bencala et al. (1984), Fortner and White (1988), Triska et al. (1989a,b), and White et al. (1987) has investigated the subsurface character of streams. Their work brings out the point that the interstitial zones in streambeds are important in the storage of dissolved gases and nutrients, and that for ecological purposes, the stream "boundary" may lie deep within the streambed. Triska et al. (1989a,b) define this boundary as the interface between groundwater and channel water, which may be located using piezometers (see Section 5.5.5) or tracer injections. Water may enter the streambed, travel for some distance underground and then re-enter the stream. These recharge, underflow and discharge processes are dependent on the proximity of the water table to the channel bed

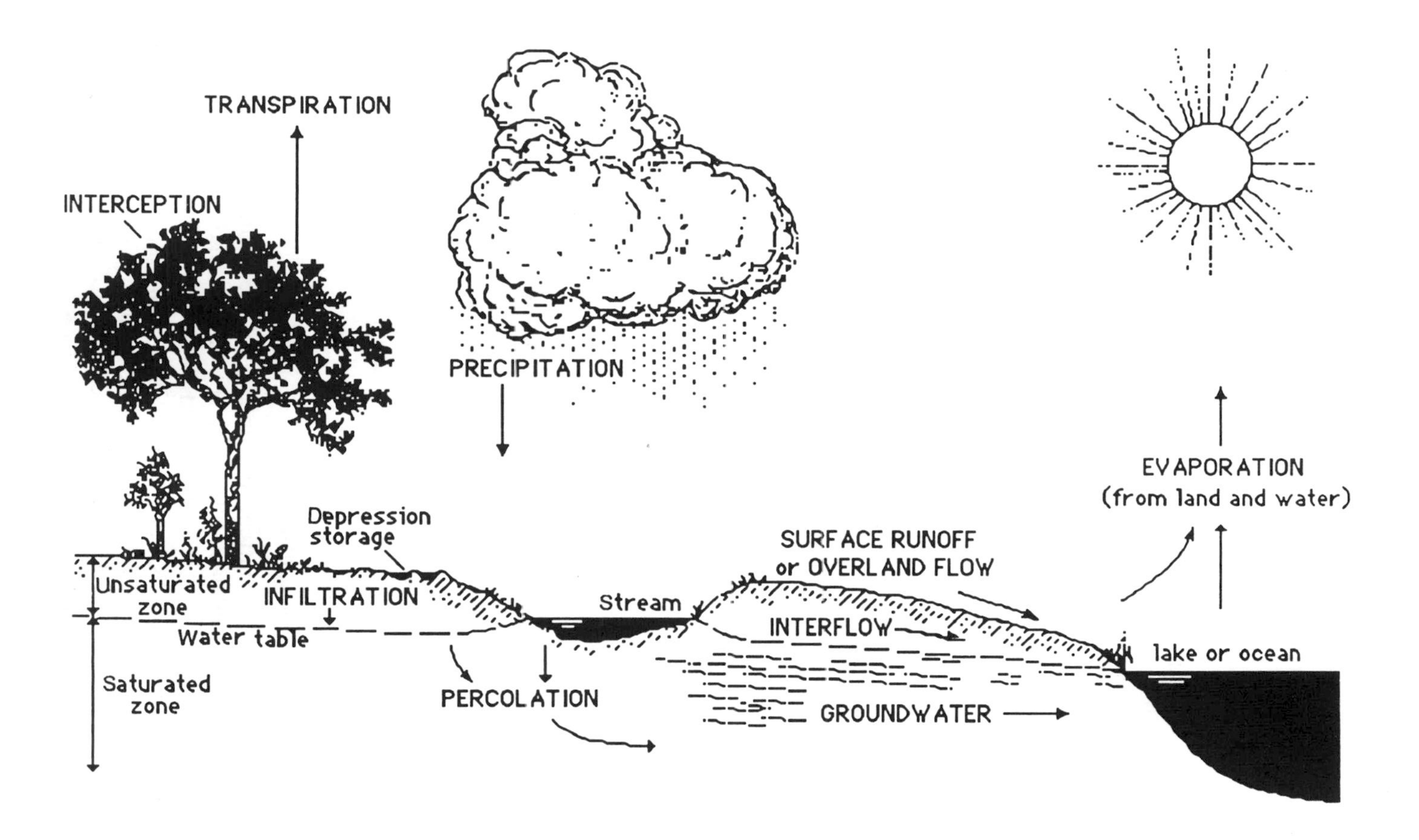

Figure 4.1. The hydrological cycle

surface, streamflow level, bed permeability and topography. Patterns of flow movement can thus affect the distribution of hyporheic organisms (see Section 2.1), rooted aquatic plants, and spawning areas for fish.

Amoros et al. (1987) have developed a methodology for considering fluvial hydrosystems as interactive ecosystems over four dimensions: (1) the upstream–downstream progression; (2) the interconnections between the main stream, side arms, floodplain and marshes; (3) the vertical interchange between regions above (epigean) and below (hypogean) the channel bed surface; and (4) the changes in a river's dynamics and ecosystems over time. The method was applied to the Rhone River, France, to develop predictive scenarios for the impact of engineering works on channel morphology and ecology. Examples of some of the geomorphic patterns and associated biological functions used as spatial units in the study are given in Figure 4.2. The geomorphological terms used in the figure are described more fully in Chapter 7.

Thus, there is no reason to draw the boundaries of a stream at the water's edge (Cummins, 1985). It is preferable to consider the catchment area as the basic ecosystem unit (Lotspeich, 1980; Moss, 1988), especially if it is extended in the vertical direction to include subsurface processes and considered over time.

4.1.3 INITIAL ASSESSMENTS OF THE STATE OF A STREAM

Natural and human modification of the stream or upland areas can have profound effects on the state of a stream. Before attempting to develop "cures" for adversely impacted streams, the current state of a stream and its ecology must first be ascertained and compared with what is considered "healthy".

In "getting to know" another, first impressions are noted and then traits are evaluated through a dialogue consisting of questions which are basically "Are you this or that?" By "asking questions" of a stream, one can diagnose its condition, put its measurements into perspective against those of other streams and classify it according to its individual characteristics. In the evaluation process, some questions which might be asked include the following.

Perennial, intermittent or ephemeral?

The terms perennial, intermittent and ephemeral are related to the terms influent and effluent. **Influent** ("losing") streams are those in which the stream feeds the groundwater as compared to **effluent** ("gaining") streams in which the stream receives water from it (see Figure 4.3). A stream may change from effluent to influent along its length depending on the geological formations crossed. Even large streams may disappear completely, reappearing as springs many miles away.

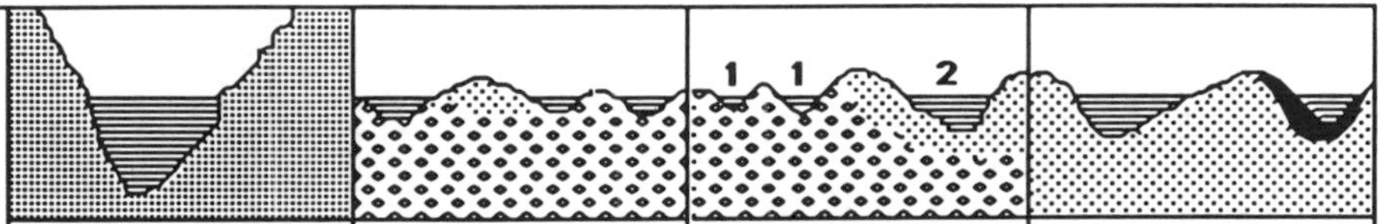

GEOMORPHIC PATTERNS	Gorge	Braided	Braided -1. and anastomosed-2	Meanders	
River bed	Very unstable	Unstable	1- unstable 2- rather stable	Rather stable	
Lateral wandering	No	Fast	1- fast 2- no	Slow	
Biotic connection to the alluvial aquifer	No	Very low	1- very low 2- medium	Medium	Aggradation of the plain
		High	*	High	Deepening
Habitat diversity of the plain	Low	Medium	Very high	High	
Development stages of the ecosystems occurring in the plain	Juvenile	Juvenile to medium	Juvenile 1-2 to mature 2	Juvenile to mature	
Expected biomass production of the plain	Low	Medium	High	High	

* Anastomosed pattern occurs only with aggradation

Figure 4.2. Associations of geomorphic patterns and their ecological implications. Redrawn from Amoros et al. (1987), by permission of John Wiley & Sons Ltd

Boundaries between the definitions of perennial, intermittent and ephemeral streams are vague. They apply to the general nature of a stream's water flow under average conditions. **Perennial** streams are those which essentially flow year-round. Perennial streams are primarily effluent, and consist of baseflow (see Section 4.3) during dry periods. Most large streams and streams in humid regions will be perennial, although a continuous low flow may be maintained in small streams if the channel is well shaded and a source of subsurface water exists. **Intermittent** streams are those which flow for only certain times of the year, when they receive water from springs or runoff. They are thus either influent or effluent, depending on the season. During dry years they may cease to flow entirely or they may be reduced to a series of separate pools. **Ephemeral** streams are influent, having channels which are above the water table at all times. They carry water only during and immediately after rain. Most of the streams in desert regions are intermittent or ephemeral. Some of these channels are dry for years at a time, but are subject to flash flooding during high-intensity storms.

Determining the permanence of flow is more than a matter of distinguishing a solid line from a dashed line on a map, as definitions used by map makers will vary. If the stream is gauged, records will provide insight into conditions at the gauge site. Local residents familiar with the stream's behaviour may be able to give a fair estimate of how often a stream reach dries up, and in what manner. Vegetational clues may also help to determine the boundaries between temporary and permanent waters.

Bedrock-controlled or alluvial?

Alluvium is a general term for stream-deposited debris. Streams can be separated into the two major groupings, bedrock and alluvial, based on whether the channel form is predominantly controlled by geology or streamflow, respectively. In bedrock-controlled channels the flow is confined within rock outcrops and the channel morphology determined by the relative strength and weakness of the bed material. Alluvial channels, in comparison, are free to adjust their dimensions, shape and gradient, and bed and bank materials are composed of materials transported by the river under the present flow conditions (Schumm, 1977).

Headwater, middle-order or lowland stream?

The general downstream trends of energy input, water quality and physical conditions (see Section 2.1) lead to a longitudinal succession of fish, benthic invertebrates, plants and other organisms. These general changes can be divided into three zones: headwater, middle-order and lowland. This classification is similar to one of the oldest and most famous three-zone classifications of Davis (1899), who used the categories "youth", "maturity"

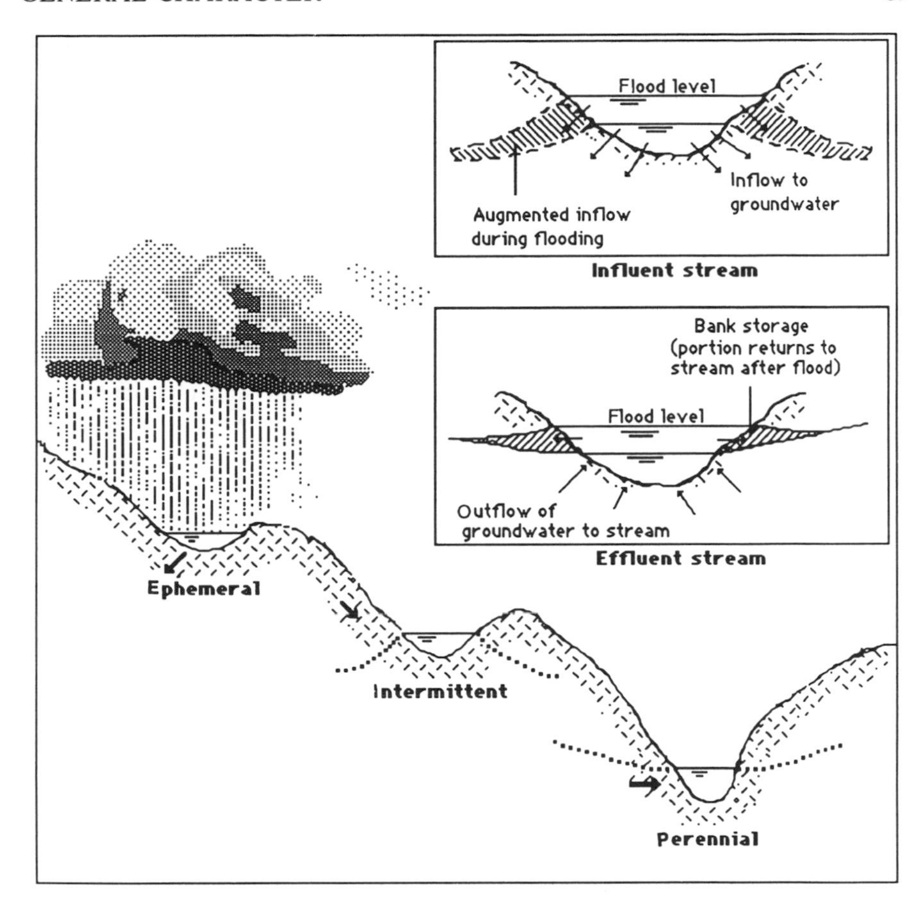

Figure 4.3. Descriptions of flow permanence. Water table level is shown as a dotted line in the main figure. Inserted diagrams illustrate how water enters the banks as the stream swells during a flood; in effluent streams, much of this bank storage returns to the stream (see also Section 4.3). Inserted diagrams are redrawn from Wilson (1969), by permission of Macmillan

and "old age". Variations on this classification have been developed for both ecological and geomorphological purposes, and are presented with other classification schemes in Section 9.2. Although distinctive differences exist between upland, middle-order and lowland streams, wide variations in physical and biological characteristics can occur within each division and over time. General characteristics of the three major types are summarized as follows (see also Figure 9.3):

(1) *Headwater zone*: Typically, the upper, "young" reaches of a stream are incised into V-shaped valleys, with steep slopes and few, short tributaries. The channel bed material consists of coarse gravels, boulders and rock

outcrops. Water temperatures are relatively cool and stable. Narrow widths mean that riparian vegetation can shade more of the stream and there is proportionately more material entering the stream as leaves and logs. Shading and scour from coarse sediments restrict the growth of algae and other plants. Thus, organic matter from outside the stream supplies most of the food for fungi, bacteria, macroinvertebrates and others, which in turn become food for higher organisms such as small fish. Habitat diversity may be low because of the restricted temperature range and low input or production of nutrients.

(2) *Middle-order zone*: Stream slopes lessen in the middle, "mature" reaches of a river. These reaches transport sediment from bank erosion and from upstream supplies, and have highly variable physical characteristics. Floodplain development begins and bank cutting replaces downward cutting (CSU, 1977). Channels are wider and aquatic plants contribute organic nutrients and oxygen to the stream, augmenting supplies transported from upstream. The coarse substratum, diversity in channel form, diversity of nutrient sources, variable discharge and wider range of temperatures favour a diverse fauna since the range of conditions encompass the optimum conditions for a large number of species (Petts and Foster, 1985). Vannote et al. (1980) relate this to a convergence of organisms which evolved downstream from terrestrial origins (e.g. insects) and upstream from marine origins (e.g. crustaceans).

(3) *Lowland zone*: In the lowland, "old-age" zone, bed materials are composed of fine sediments, discharges are relatively stable, and temperature fluctuations are buffered by the large volume of water. Valleys are very broad, deeply filled with alluvium, and marked with the evidence of frequent channel changes: meander scars, oxbow lakes (billabongs) and swamps. Deposition of sediment occurs through this zone and to the terminus of the river, where sediment may deposit out on an alluvial plain, delta or in an estuary. Natural levees may border the stream, and because of deposition, a stream can actually flow at a level higher than the floodplain. Increased turbidity and depth may restrict the growth of aquatic plants, and the macroinvertebrate populations tend to be dominated by those which collect fine particles of organic matter received from upstream. Overall biotic diversity may be low, although fish species diversity may increase with the presence of larger fish which feed on smaller ones.

"Stable, aggrading or degrading?"

A **stable** channel is one that does not exhibit progressive changes in slope, shape or dimensions, although short-term variations may occur during floods (Schumm, 1977). **Degradation** refers to the downcutting of a stream into its bed materials and **aggradation** is the accumulation of bed materials (see Section 7.2.2). Bedrock streams do not degrade very quickly, but alluvial channels can change rapidly in response to alterations of flow, sediment supply or base level.

A river is a delicately balanced system, and any alterations will cause the stream to attempt to re-establish an equilibrium condition (CSU, 1977). Disequilibrium results from changes in sediment or runoff conditions (e.g. from diversion of a major amount of a stream's discharge or from modifications to upland conditions). Stream stability is also important biologically, as mentioned in Section 2.1. When a stream aggrades, fine sediments can impede the exchange of water, organic matter and organisms between bed surface and hyporheic habitats, whereas during degradation, fine sediments are washed out, enhancing exchanges (see Figure 4.2) (Amoros et al., 1987). The concept of equilibrium is discussed further in Chapter 7.

Evaluation of channel stability either from aerial photographs or in the field is relatively difficult, although partly buried fences or other permanent landmarks may indicate changes. Changes in bed cross-section and elevation is best done by comparing old channel surveys to recent ones (Bovee, 1982).

An associated question is: "*narrowing or widening*?" Trees falling into a river from both banks may indicate widening; however, narrowing occurs more slowly and may only be detected by increases in vegetation on islands or a trend from a braided to meandering condition (see Section 7.3). The stability and direction and rate of change of a channel is important when evaluating habitat potential, designing stream rehabilitation works and determining minimum instream flow recommendations (see Section 9.3).

"Regulated or natural?"

Dams and diversion canals can be easily located from maps, aerial photographs or ground surveys, but their effects are much more difficult to evaluate. Thus, "regulated or natural" is not an either/or question but a matter of degree. Effects depend on whether regulated releases are returned to the river further downstream and on the operating policy of reservoir storages (McMahon, 1986).

Cadwallader (1986) has indicated that regulation may change the seasonal distribution of flow (the regime), reduce the incidence and severity of flooding and decrease long-term average flows downstream. Regulation can also have a marked influence on low-flow behaviour by increasing the duration and frequency of low-flow extremes (including periods of zero flow). In addition, reservoirs can influence sediment movement, stream temperatures and water quality. Changes in flow regime and sediment supply can, in turn, lead to changes in downstream channel dimensions.

Once a river is regulated, it is likely to remain so. Thus, a goal of river management should be to determine the habitat value of modified streams and the degree to which regulation structures and their operation can be modified to optimize it. The effects of dams and reservoirs on stream habitat and the provision of reservoir releases for habitat improvement are discussed in Section 9.3.

"Channelized or non-channelized?"

The clearing and straightening of a stream to improve water conveyance is termed channelization. This usually increases channel slope and thus water velocity and sediment transport capacity, which causes scour within the modified section and sediment deposition in flatter sections downstream. Higher peak flows may also occur downstream of the channelized section where the gradient lessens. Vegetation and shading is usually reduced in channelized sections, leading to an increase in stream temperature and a lowered supply of organic nutrients for biota. Water tables in the region of channelization may also be affected. Effects of channelization are further discussed in Section 9.1.

"What is the condition of the upland catchment?"

Streamflow and ecology are both affected by catchment conditions. It should be kept in mind that the catchment and the stream system are integrated; thus, a change in one part of the system will be felt elsewhere (Morisawa, 1985). Changes in stream discharge and sediment loading caused by modification of a catchment are reflected in variations in the rate of sediment transport, channel shape and stream pattern (Bovee, 1982). Responses to a change may be immediate, delayed or dependent upon a critical factor reaching some threshold level.

Vegetation removal, as from fire or logging or conversion from forest to pasture, can change the natural drainage system and the rate at which water and sediment runs off the land surface. Road construction can be a major point source of sediment. Urbanization has a more drastic effect on hydrology—roads, parking lots and houses prevent infiltration, increasing total runoff and the magnitude of peak flows (Morisawa, 1985).

Regulation, channelization and catchment conditions were all considered by Macmillan (1986, 1987) in the development of five categories for evaluating the "naturalness" of a catchment, and these are presented in Table 4.1. Leopold and O'Brien Marchand (1968) also describe a system for ranking the quality of river landscapes.

4.1.4 AN EXAMPLE OF A QUALITATIVE SURVEY: STATE OF THE STREAMS, VICTORIA, AUSTRALIA

The State of the Streams Survey was a major collaborative effort of several water resource agencies to conduct an inventory of all the rivers in Victoria, Australia (Mitchell, 1990). The 1986 study identified the current status of the streams; additionally, it provided baseline data for future monitoring work.

A workshop was conducted by experienced river managers to orient all participants on details of evaluation standards and to delineate candidate representative reaches (see Section 2.3.4) on maps of each stream. Land type was classified as either "forest" or, if cleared, as a landscape type such as "flat volcanic plain" or "sedimentary steep hills". Land types were marked on 1:250 000 maps and the lengths of major and minor streams were measured within each type.

Sample reaches were selected based on local knowledge, one for each 25 km of major stream length or 250 km of minor stream length. Reach length was equivalent to five bends or 25 times the stream width, depending on the stream configuration. Each river was inspected from the air to see if selected reaches were sufficiently representative of the channel configuration, vegetation and engineering works in each stream segment, and 35 mm aerial photographs were taken of each sample reach. As a result, adjustments were made to the locations and number of sample reaches before an intensive ground survey was undertaken.

Table 4.1. Criteria for evaluating "naturalness" of a catchment. Modified from Macmillan (1986)

Pristine: The entire catchment system represents an unmodified ecosystem which can act as a baseline reference area.

Slightly modified: Catchment processes are largely intact. The flow regime has been modified to only a minor extent, and the only input of pollution is sediment. There are no barriers to the movement of instream biota.

Moderately modified: Catchment processes, hydrology and instream biota have been noticeably altered. There may be direct manipulation of the flow regime by impoundment, and sediment input may have altered the stream substrate. Levels of biostimulants may be elevated but not other toxic inputs.

Heavily modified: Catchment processes, riparian and instream biota have been substantially modified. The flow regime may be highly manipulated, sediment input may be substantial and levels of biostimulants are substantially elevated. Toxic substances may be present at significant levels.

Severely degraded: Major modification of the stream has taken place, leading to severe degradation of riparian and instream biota. Examples would be streams grossly affected by elevated levels of salinity, heavy metal pollution and/or enclosure within a concrete channel.

Hydrographers of the Rural Water Commission, already familiar with conditions in their home regions, were assigned the task of surveying the selected streams in their own districts. At each site, 35 mm colour slides were taken of (1) the upstream view, (2) the downstream view, (3) an oblique view, from a distance back from the bank of about three times the channel width, (4) special features, and (5) each bank type. An identifying code was printed on a card, photographed as the first frame in each film, and slides were checked against recorded data soon after processing. The photographs proved to be particularly useful records for verifying field estimates, as well as providing a data source for other uses such as landscape assessment. Visual estimates were made of:

- Channel characteristics: valley type, floodplain features, channel pattern and dimensions, levee banks, water depth and velocity, aquatic habitat;
- Bank characteristics: shape, soil type, location, vegetation and bare ground, bank damage and stability;
- Bed and bar characteristics: underwater vegetation, canopy cover and vegetation on dry beds or bars, size composition, origin, angularity, shape and consolidation of bed materials, organic debris, bed aggradation and degradation;
- Characteristics of the stream verge (a strip of vegetation between the top of the stream bank and surrounding land use); and
- Occurrence of any bed or bank works in the sample reach.

Samples of the field-evaluation criteria have been reproduced in Figure 4.4.

The study was almost wholly of a qualitative nature. Distances were measured by pacing and stream criteria estimated by eye. As a result, sampling was relatively fast, taking only a few hours per site to accumulate a wealth of observations. Statistical limitations, resulting from assessor bias and the use of judgmental sampling to extrapolate from sample reach to stream segment, were not considered to be a problem at the chosen level of analysis. Criteria found to be highly inconsistent between observers, such as bank stability and bed aggradation/degradation, were left out of the final evaluation.

The aim of the survey—to obtain an approximation of stream stability and general environmental health and to assist in identifying areas for remedial measures—was satisfied. It had the additional benefit of being a rewarding experience for the personnel involved as well as "opening their eyes" to features which they had not noticed in the past.

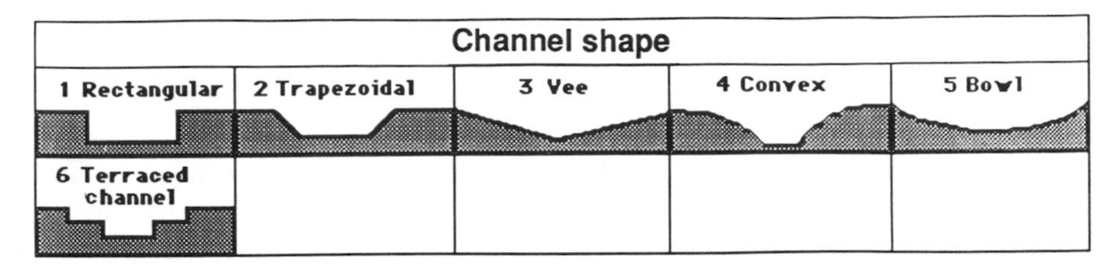

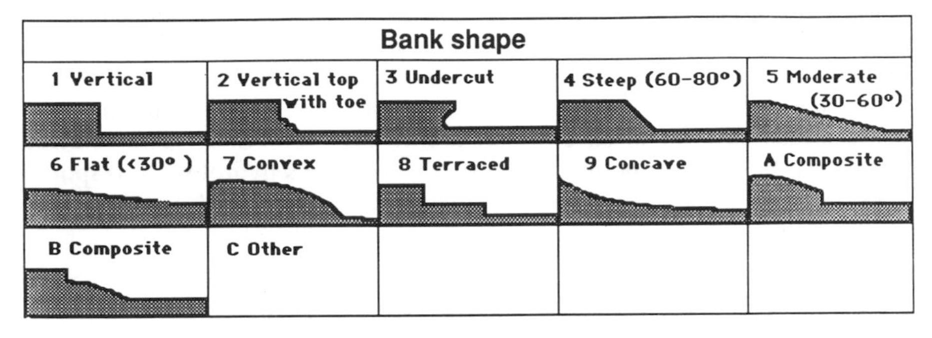

Consolidation of bed material				
1 Excellent Array of sizes; tightly packed, overlapping; difficult to dislodge with foot.	**2 Good** Array of sizes; overlapping; some can be dislodged by foot	**3 Moderate**	**4 Fair** Poor grading; some packing; can be dislodged by foot.	**5 Poor** No packing; loose array; easily moved by foot.

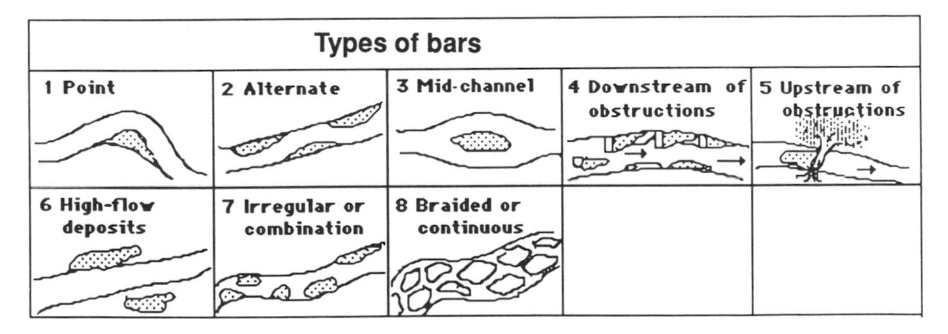

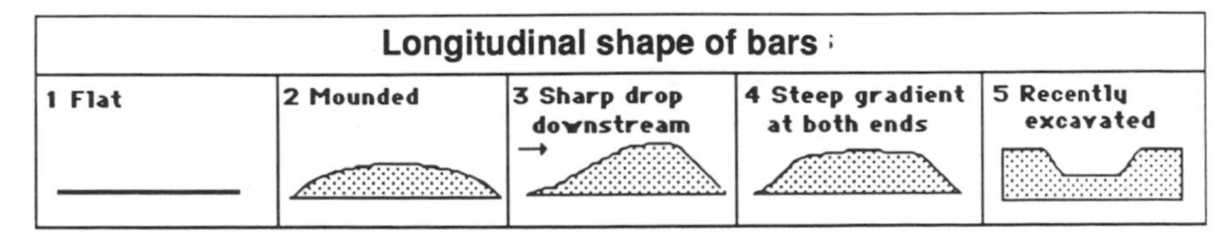

Figure 4.4. Examples of field criteria used in the State of the Streams Survey, Victoria, Australia. Reproduced by permission of the Department of Water Resources, Victoria

4.2 CATCHMENT CHARACTERISTICS

4.2.1 GENERAL

A number of factors affect the way water and sediment move from upland areas to the stream and from there to its terminus. Many geomorphic descriptors have been developed which are related to catchment hydrology. The general term **morphometry** is applied to the measurement of shape and pattern. Measures given in this section can be relatively easily obtained from maps.

Because of interrelationships between factors, one (usually the one most easily measured) can often serve as a surrogate for others. The selected factors can be used in the prediction of a catchment's hydrological response to rainfall and for distinguishing one catchment from another for comparative or classification purposes.

4.2.2 DELIMITING AND MEASURING THE CATCHMENT AREA

Catchment area is one of the more important descriptors of a basin since it influences the water yield and the number and size of streams. It includes all the upstream land and water surface area which drains to a specific location on the stream. We can speak of catchment areas for a whole stream system, or catchment areas for a particular point on a stream (e.g. at a gauging station or study site). The area is delimited by the topographic divide, a theoretical line which passes through the highest points between the stream system and those neighbouring it.

Catchment boundaries are located by using the contour lines on a topographical map. These can be supplemented with stereo pairs of aerial photographs. Boundaries are drawn by following the ridge tops which appear on topo maps as downhill-pointing V-shaped crenulations. The boundary should be perpendicular to the contour lines it intersects. The tops of mountains, often marked as dots on a map, and the location of roads which follow ridges are other clues. In areas of little relief, it will be difficult to locate boundaries precisely. A line has been drawn around the catchment of a small stream in the upper left-hand section of Figure 3.4.

Catchment area can be measured directly from the marked maps using a planimeter or digitizer with appropriate software. Other methods include superimposing a grid of squares or a dot grid over the map and then counting the number (and fractions) of squares or the number of dots which fall within the catchment area (Figure 4.5). Any of the methods must be calibrated to the map scale. It is a good idea to check the calibration by first measuring a section of known area (e.g. bounded by grid lines).

Another technique suggested by Gregory and Walling (1973) involves tracing the catchment onto a high-grade paper and then carefully cutting it

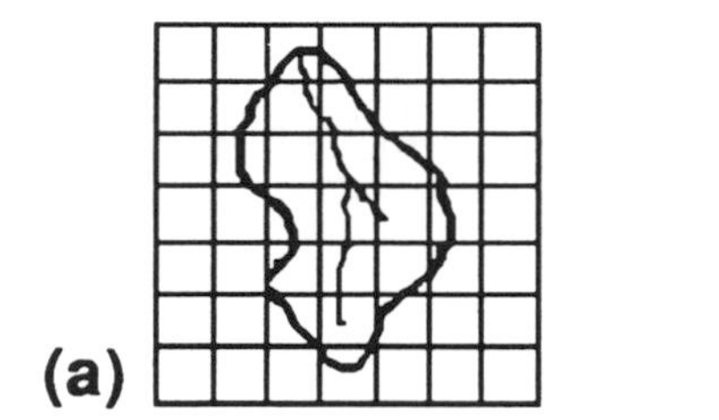

Figure 4.5. Methods of measuring catchment area using (a) square grid and (b) dot grid

out using a sharp cutting tool. A square of known map area is also cut from the same paper. Both pieces are weighed on an accurate balance. The catchment area is then obtained from

$$\text{Catchment area} = \frac{\text{Catchment "cutout" weight}}{\text{Weight of square "cutout"}} \times \text{area represented by square} \tag{4.1}$$

where the catchment area and square map area have the same units. For example, if a catchment "cutout" weighed 3.5 g and a square cutout representing 16 km^2 weighed 10.5 g, the catchment area would be $(3.5/10.5) \times 16 = 5.3$ km^2.

The topographic divide may not represent the true area from which water in a stream is derived. Subsurface water may move from one drainage basin to another, as illustrated in Figure 4.6. The distinction between the topographic divide, which determines the direction of surface drainage, and the phreatic divide, which determines the direction of subsurface drainage, is an important consideration in hydrological studies.

For the purposes of this text, the topographical definition will be used. The stream network and catchment area for the Acheron River, Victoria, which is used as an example throughout the text, is shown in Figure 4.7.

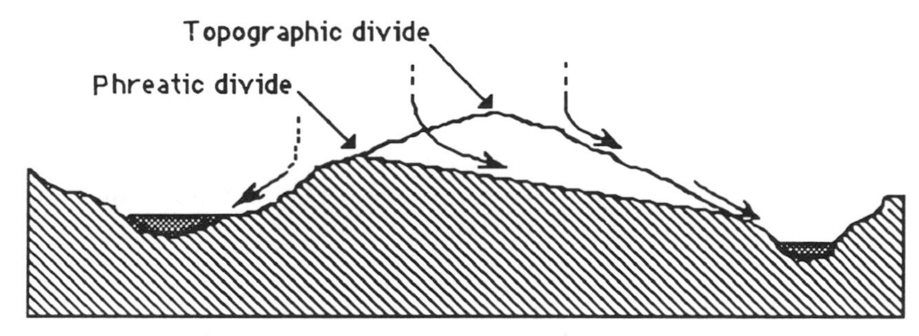

Figure 4.6. Description of topographic and phreatic divides. Redrawn from Gregory and Walling (1973), by permission of Hodder and Stoughton Ltd

In addition to measuring the whole catchment area, the area covered by a vegetation or soil type or other feature designated on a thematic map (Section 3.3) may also be of interest. Measurements are made using the same methods.

4.2.3 STREAM LENGTH

Stream length will influence the amount of stream habitat area in a catchment. The lengths of streams also affect the travel time of water in a drainage system and availability of sediment for transport. Stream length is most often obtained from measurements of the blue lines on topographic maps. The actual length of channel containing surface water changes constantly; thus the map measure should only be considered a standardized index.

Measurements of stream length may simply be made by following the line with a rule or scale graduated in map units, a digitizer, or a pair of dividers. Map wheels (opisometers) which measure distance as they are rolled across a map are also helpful for measuring distances, although they are difficult to use when a stream line is highly tortuous.

Differences in map scales and accuracy will lead to variations in measured stream lengths. In fact, one can pose the philosophical question: What is the true length of a stream? Is it the distance measured down the middle of the stream channel or down its deepest path (the **thalweg** distance)? Does it extend to the drainage divide? Should it represent the path of a mass of water or a particle of organic matter as it passes around boulders and travels across and through the streambed?

Gan et al. (1989) argue that the length of an intricately sinuous line tends to increase as it is measured more accurately. They suggest the use of a **fractal stream length** as a standardized measure. Using maps of a single scale, the stream is stepped off repeatedly with dividers set at different spacings. In this manner, a stream length (L) is obtained for each spacing (X). Using simple regression, the data are fitted with a log–log equation of the form:

$$L = aX^b \tag{4.2}$$

where a and b are regression constants. A log–log plot of stream length against step size for the Acheron River is shown in Figure 4.8 with the fitted regression line. The fractal stream length (L') is simply equal to the constant a:

$$L' = a \tag{4.3}$$

which is 55.0 km for the example illustrated. The fractal dimension (f) is given by:

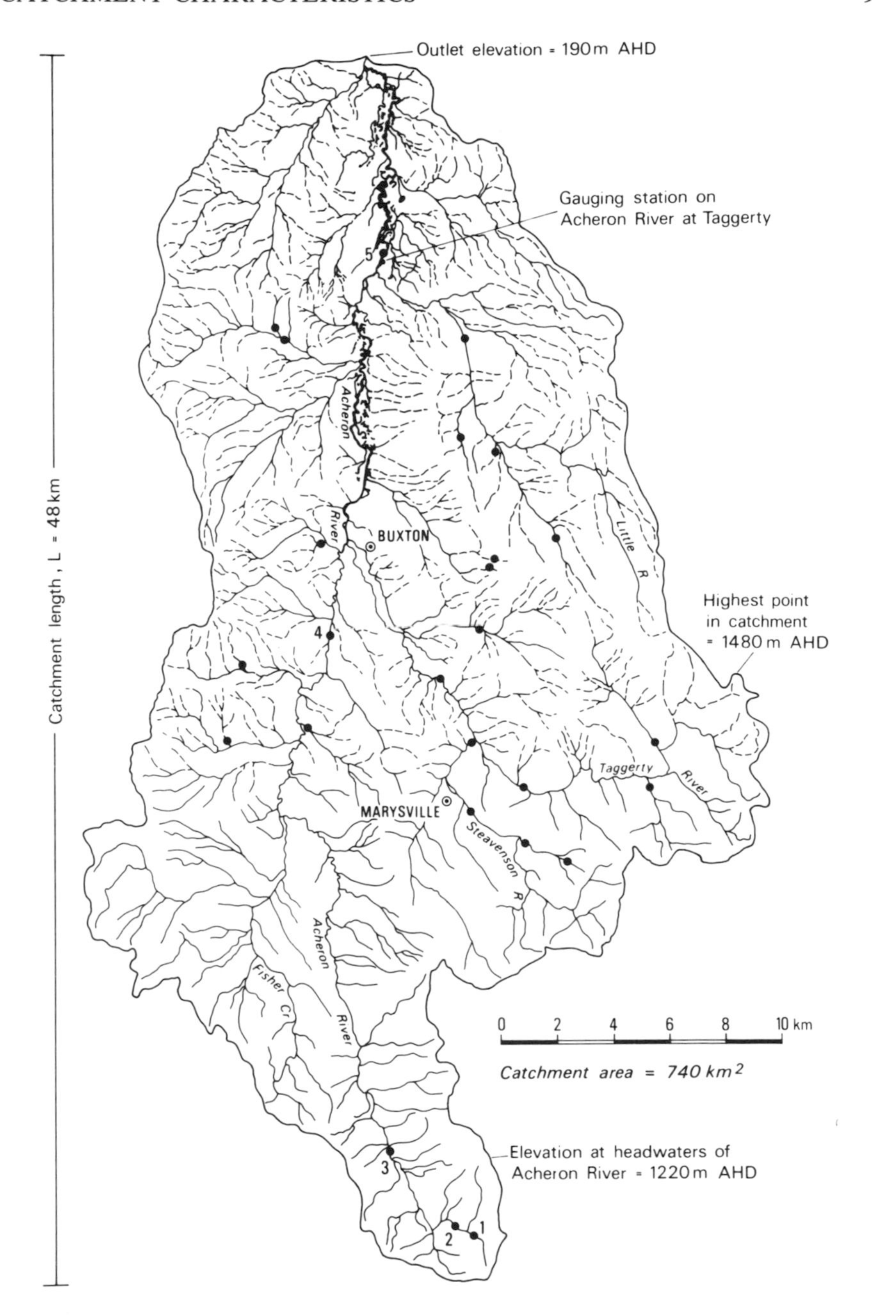

Figure 4.7. Catchment for the Acheron River, Victoria, Australia, traced from 1:100 000 maps. Both map and scale have been reduced. Dots designate stream sampling sites, discussed in other chapters. The catchment area for the gauging station at Taggerty is 619 km^2

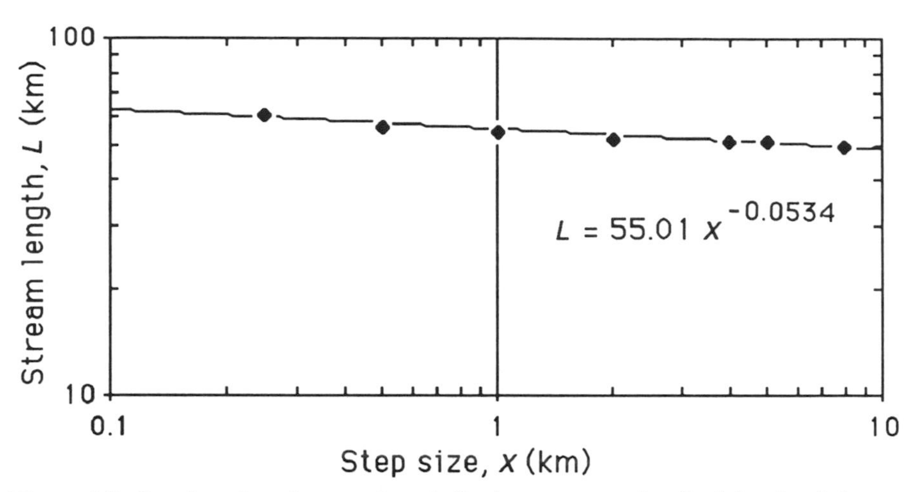

Figure 4.8. Log-log plot of stream length (km) versus step size (km) for the Acheron River, as measured from headwaters to mouth on 1:100 000 scale maps. A power equation and fitted line obtained by simple regression are shown

$$f = 1 - b \tag{4.4}$$

If an average value of f is computed for a basin, the fractal length (L') of any stream in the basin can be calculated as:

$$L' = X^f N \tag{4.5}$$

where X is again the step length and N is the number of steps needed to "walk" the length of a stream.

For example, if the fractal dimension of 1.0534 obtained for the Acheron River is applied to a segment measured as 60 steps of size 0.1 km, then $L' = (0.1)^{1.0534} \times 60 = 5.3$ km, as compared to a conventional measure of 6.0 km (60×0.1).

The advantage in using the fractal stream length is that it is a robust measure, independent of both map scale and finiteness of measurement. In contrast, the conventional measure of stream length tends to increase as the step size (in map units) decreases (Gan et al., 1989).

4.2.4 STREAM PATTERNS

Catchments can be described according to their stream channel patterns, as viewed from maps or from the air. All drainage patterns "are tree-like, but different patterns resemble the branchings of different kinds of trees" (Horton, 1945, p. 300). Each stream has its own individual characteristics, based on the particular topographical and geological obstacles encountered as it seeks the "path of least resistance" in its journey towards the sea.

Stream patterns may develop randomly on uniform soils, or in response to weaknesses in the underlying geology. Some of the basic drainage patterns are (McKnight, 1990; Morisawa, 1985):

Dendritic: found in areas of relatively uniform geological structure;
Trellis: usually develops on alternating bands of hard and soft strata;
Pinnate: forms in very fine-grained surfaces;
Rectangular: common in areas with right-angled faults and/or joints, such as some types of granitic bedrock;
Radial: forms where streams flow outward from a dome or volcanic cone;
Centripetal: results from a basin structure where streams converge centrally;
Annular: develops around a dome or basin where concentric bands of hard and soft rock have been exposed;
Parallel: occurs in areas of pronounced localized slope;
Distributary: refers to diversion of channels (for example, in deltas or alluvial fans).

Examples of the drainage types are shown in Figure 4.9. These are only some of the more common and easily recognizable regional patterns of drainage. The patterns in the Acheron drainage (Figure 4.7) are mainly dendritic.

The development of branching networks is a fascinating topic in itself. A variety of phenomena which distribute or collect matter or energy exhibit branching patterns: root systems, tree branches, veins in leaves or in animals; lightning strikes, fern-like precipitations of manganese in rocks, and highway and telephone systems. Knighton (1984) provides a discussion on the evolution of drainage networks on land surfaces, and Jarvis and Woldenberg (1984) have compiled a number of papers on river networks.

One could pose the question: given a certain area, what is the most efficient method of draining water from its surface? As an illustration, we can consider a number of strategies for "draining" water from all points in Figure 4.10(a). To study the relative efficiency we will look at the total "channel" length and the average length from each point to the delivery point. Figure 4.10(b) shows a spiral "snail-shell" configuration, Figure 4.10(c) a "starburst" pattern and Figure 4.10(d) a branching system. It can be seen that the pattern of Figure 4.10(c) provides the maximum possible drainage efficiency because each point has a direct path to the outlet. This is reflected in the low average channel length. However, this is the "big budget" alternative, with its large total channel length. In the branching system (Figure 4.10(d)) the average length is only slightly higher yet the total channel length is greatly reduced. Thus, the branching system requires less channel length to maintain close to the maximum drainage efficiency (Newbury, 1989).

In natural streams it is typical for major tributary branches to drain about

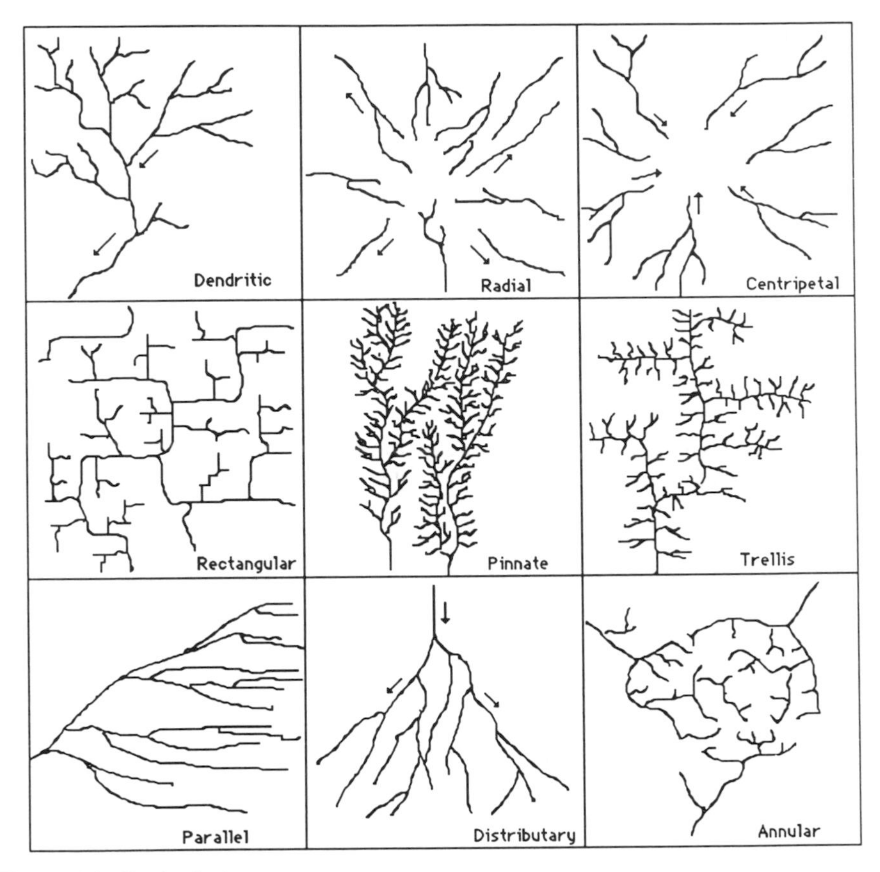

Figure 4.9. Basic drainage patterns

half the total catchment area, with the remaining area draining directly into the main stem of the stream. For the Acheron River system (Figure 4.7), 52.6% of the area is drained by the main tributaries, the Little River, Steavenson-Taggerty, and Fisher, leaving 47.4% for the main stem. The implication of this for flood control or water supply is that dams built on these major tributaries would catch water from only about one half of the basin area, in comparison to one dam built at the mouth which would control runoff from the whole basin (Leopold and Langbein, 1960).

4.2.5 STREAM ORDERS

Stream ordering is a widely applied method for classifying streams. Its use in classification is based on the premise that the order number has some relationship to the size of the contributing area, to channel dimensions and to stream discharge (Strahler, 1964). This premise is often criticized—for

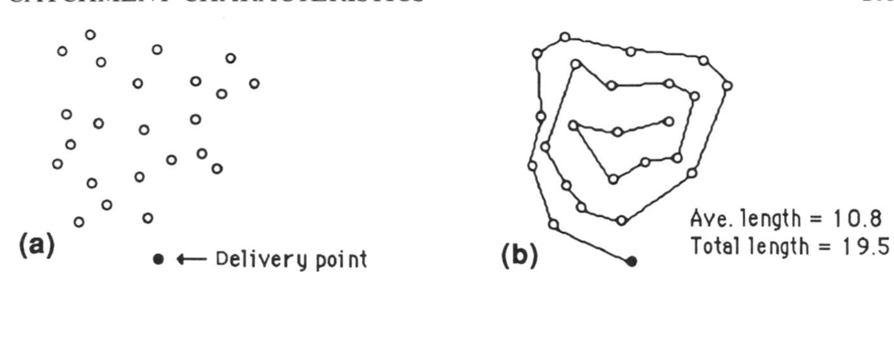

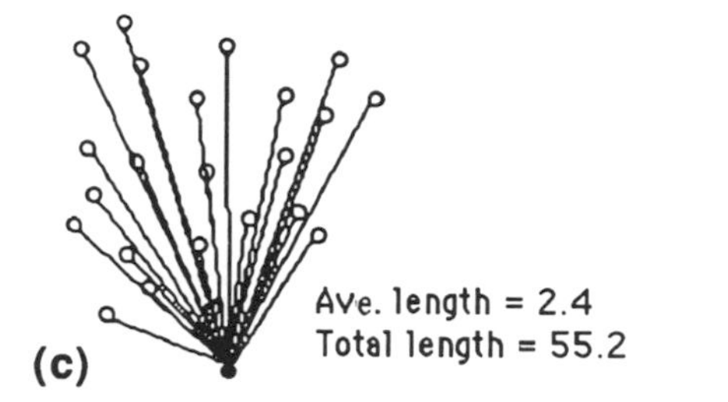

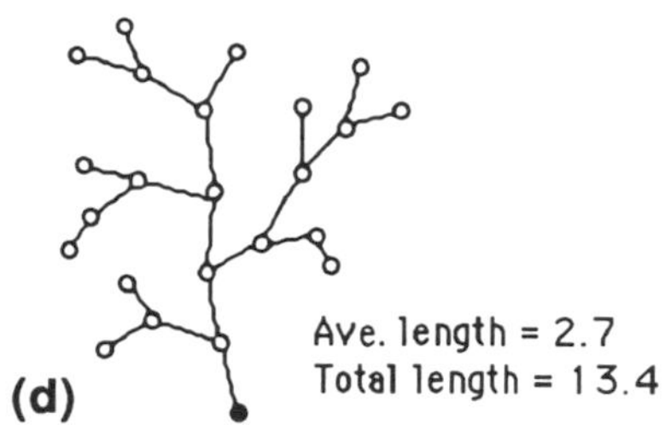

Figure 4.10. A theoretical comparison of drainage pattern efficiencies: (a) points to be drained through the delivery point, (b) a spiral drainage pattern, (c) a "starburst" configuration, and (d) a branching pattern. Average length = average distance from each point to the delivery point and total length = sum of all channel segment lengths (both in arbitrary units). See text for discussion

example, by Blyth and Rodda (1973), who emphasize that the network of flowing water is dynamic rather than fixed on a map surface. However, because of its simplicity, stream ordering provides a rapid "first approach" method of stream classification and a convenient means of stratification for sampling designs (Section 2.3).

Stream-ordering methods

Specifically, stream orders provide a means of ranking the relative sizes of streams within a drainage basin. In most stream-ordering methods, smaller tributaries are given a lower order than major streams and tributaries. Catchments can also be classified by order number (for example, the area draining a second-order stream would be labelled a second-order basin). Figure 4.11 shows the different ordering systems which will be discussed.

Horton (1932) first introduced the concept of stream order in the United States, after reversing the European practice of giving main streams an order of one and the fingertip tributaries the highest order. In Horton's method, first-order streams are those tributaries which have no branches. Streams receiving only first-order streams are of second order, third-order streams receive only first- and second-order streams, and so on. The "parent"

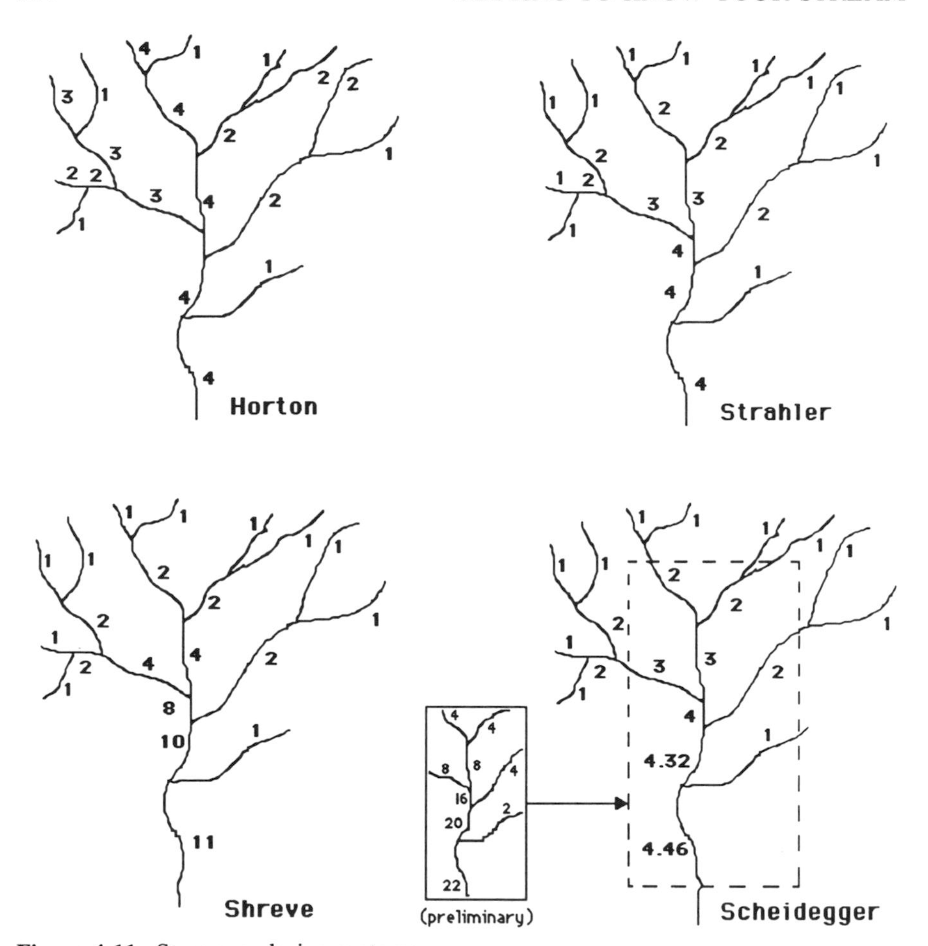

Figure 4.11. Stream-ordering systems

stream is denoted by the same number all the way back to its headwaters. At the furthest upstream junction it may be difficult to determine which is the parent stream and which is the tributary. In this case, the tributary is the one which joins at the largest angle, or, if angles are the same, it is the shorter of the two.

Strahler (1952) modified Horton's method in a small footnote in a paper on area–altitude analysis. Strahler's method has been widely accepted as less subjective than Horton's, and is commonly used by stream biologists. In this system, all the small, exterior streams are designated as first order, those which "carry wet weather streams and are normally dry" (Strahler, 1952, p. 1120). A second-order stream is formed by the junction of any two first-order streams; third-order by the junction of any two second-order streams. Here, only one stream segment has the highest order number, rather than the whole parent stream.

A limitation of Strahler's method is that a large number of minor tributaries may intersect a larger-order stream, adding substantially to its discharge but not to its stream order. To overcome this drawback, Shreve (1967) proposed a system whereby the order of a stream is the sum of the orders of the upstream tributaries. Thus, the joining of a second order and a first order produces a third order. This system may produce a classification which is more descriptive of the total network and average streamflow volumes.

Another alternative is given by Scheidegger (1965). This method also takes into account the effect of tributary inflow, but the order of the main stream increases less rapidly than in the Shreve system. In Scheidegger's method, first-order streams are first assigned the number 2. Then, as in the Shreve system, numbers for other segments are obtained by adding the orders of upstream segments. Stream order is then calculated by taking base-2 logarithms of the numbers (e.g. $\log_2(2) = 1$). Figure 4.11 shows the preliminary magnitudes and final orders for this method.

Calculation of base-2 logarithms can be awkward because most calculators provide only base 10 or natural logarithms. However, this is easily remedied by using the formula:

$$\log_a n = \frac{\log_b n}{\log_b a} \tag{4.6}$$

which, for this method, becomes:

$$\log_2 n = \frac{\log_{10} n}{\log_{10} 2} = \frac{\log_{10} n}{0.30103} \tag{4.7}$$

As an example,

$$\log_2 22 = \frac{1.342}{0.30103} = 4.46$$

Stream ordering, although a relatively quick and simple means of classification, has some major drawbacks. These include the variability in mapping standards, the problem of deciding which map scale is appropriate, and inconsistent definitions of first-order streams (Hughes and Omernik, 1983). A first-order stream interpreted from a map may turn out to be as much as third or fourth order when interpreted in the field. To get around the limitations of map accuracy, some have used contour crenulations (V-shaped contours) and the location of the headwater divide to extend the stream patterns marked on maps (Drummond, 1974; Mark, 1983). In a fisheries study, Platts (1979) defined first-order streams as being the first recognizable drainage on 1:31 680 scale maps, whereas Lotspeich (1980, p. 582) gives a biological definition of a first-order stream as one "with sufficient continuous flow to support an aquatic biota at all seasons".

The popular choice of Strahler's method is based on practicality. The Shreve and Scheidegger methods are computationally more difficult, and an ordering will need to be completely reworked if any streams are missed the first time through. It is essential to state the map scale and method used when citing a stream order.

Relationships between stream orders and other measures

There are a number of relationships which can be demonstrated using stream orders. Several "laws" of drainage basin geometry have been developed which state that as stream orders increase: (a) the number of streams decreases, (b) the average stream length increases, (c) catchment area decreases, and (d) average slope decreases, and these relationships are geometric (Selby, 1985).

A stream ordering using Strahler's method was done for the entire stream system shown in Figure 4.7, including both intermittent (dotted lines) and perennial (solid lines) streams. Stream lengths were measured from 1:100 000 maps with a ruler and the relationships that were derived are shown in Figure 4.12.

Figure 4.12(a) demonstrates that there are more small-order streams than large-order ones, and that the relationship approximates a geometric progression. It can also be seen in Figure 4.12(b) that stream length increases with stream order. However, for the stream and mapping system used, the data do not follow the expected semi-logarithmic trend, perhaps due to a lack of resolution at the level of order-one streams or to the nature of the drainage system.

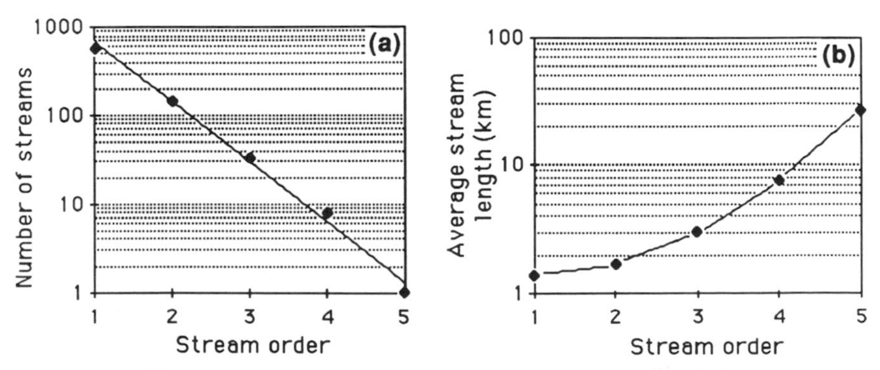

Figure 4.12. Semi-log plots of (a) the number of streams and (b) average stream length versus stream order (by Strahler's method) for the Acheron River system (see Figure 4.7). A regression line has been fitted to the data in graph (a); in graph (b) the lines simply connect the data points

These simple, regular relationships provide a quantitative description of drainage development in a basin. Sampling designs should take these relationships into account; i.e. if the object is to describe all streams in a basin, then more small streams should be sampled than large ones. The relationships for a basin can also be extrapolated from large streams to their smaller tributaries. For example, the number or average lengths of first- and second-order streams might be estimated for large basins where the sheer number of streams make measurements impractical (Leopold et al., 1964).

Bifurcation ratio

Horton (1945) introduced the term bifurcation ratio (R_b), where bifurcation means dividing in two. The data plotted in Figure 4.12(a) can be described by this index, given as:

$$R_b = \frac{\text{Number of stream segments of given order}}{\text{Number of stream segments of next highest order}} \tag{4.8}$$

The average R_b for the Acheron River system is approximately 4.0. This compares to a US average of about 3.5 (Leopold et al., 1964). Bifurcation ratios normally range between 2 and 5 and tend to be larger for more elongated basins (Beaumont, 1975). As a matter of interest, the average R_b for trees is about 3.2, for lightning strikes, 3.5, and for blood vessels, 3.4, perhaps implying that ratios in this range approach a natural optimum (Newbury, 1989; Stephens, 1974).

4.2.6 MISCELLANEOUS MORPHOMETRIC MEASURES

Drainage density

Drainage basins with high drainage density are characterized by a finely divided network of streams with short lengths and steep slopes. In contrast, a basin with low drainage density is less strongly textured. Stream lengths are longer, the valley sides flatter and the streams further apart. A comparison of high and low drainage density patterns is shown in Figure 4.13.

Drainage density (R_D) is calculated by dividing the total stream length for the basin (ΣL) by the catchment area (A):

$$R_D = \frac{\Sigma L}{A} \tag{4.9}$$

Drainage density has units of 1/length; thus its value will vary with the units used. It is preferable to express it in units of length/unit area. For example, the Acheron basin (Figure 4.7) has a drainage density of 1209 km/740 km^2 = 1.63 km of channel per square kilometre.

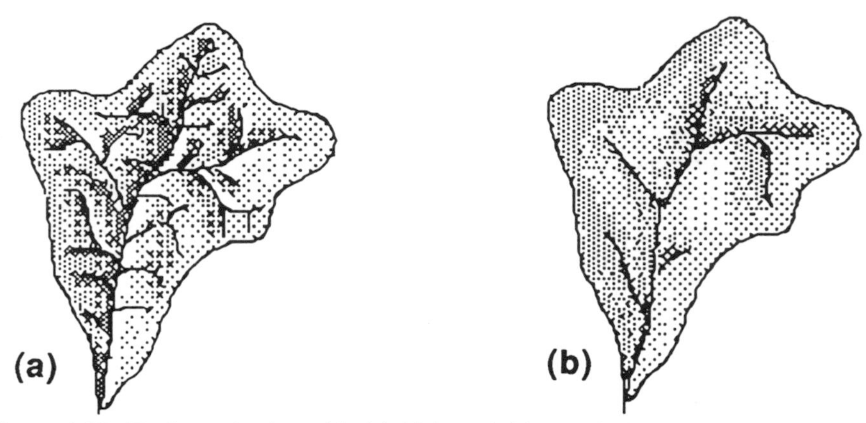

Figure 4.13. Drainage basins with (a) high and (b) low drainage density

Drainage density thus represents the amount of channel required to drain one unit of catchment area. The inverse, as the amount of drainage area needed to maintain one unit of channel length, was termed the **constant of channel maintenance** by Schumm (1956).

In relation to drainage basin processes, Gregory and Walling (1973) state that drainage density is perhaps the most useful single index. The density of a stream network reflects the climate patterns, geology, soils and vegetation cover of a catchment. Drainage density is highest in semi-arid areas where surface runoff from intense thunderstorms erodes sparsely vegetated slopes. High sediment yields reflect a more highly developed channel system (Knighton, 1984); thus the relationship between drainage density and precipitation is similar to that proposed for sediment yield (see Figure 7.19).

Relief ratio

Schumm (1956) gives a simple expression for describing relief, the relief ratio (R_r):

$$R_r = \frac{h}{L} \tag{4.10}$$

where L is the maximum length of the basin (along a line basically parallel to the main stream; see Figure 4.7) and h is the difference in elevation between the mouth of the basin and the highest point on the drainage divide. Units of h and L should be equal so as to make R_r dimensionless. For the Acheron River basin, R_r = (1.480 − 0.190) km/48 km = 0.0269.

Drainage density and slope of the upland areas are both influenced by the basin relief. In a study on small basins in the western United States Hadley and Schumm (1961) demonstrated that annual sediment yields (Section 7.5) increase exponentially with the relief ratio.

Mean stream slope

The average channel slope is one of the factors controlling water velocity. Mean channel slope (S_c) is simply given by:

$$S_c = \frac{(\text{Elevation at source} - \text{elevation at mouth})}{\text{Length of stream}} \tag{4.11}$$

Using the fractal stream length for the Acheron River, this value is: (1.220 − 0.200)/55 = 0.0185 or 1.85% or 1.06° (see Section 5.2.4 for slope conversions).

Mean catchment slope

The slope of the catchment will influence surface runoff rates, and is related to drainage density and basin relief. Van Haveren (1986) gives several definitions of mean basin slope (S_b), the simplest method of computation being:

$$\frac{S_b = (\text{Elevation at } 0.85L) - (\text{elevation at } 0.10L)}{0.75L} \tag{4.12}$$

where L is again the maximum length of the basin, as defined for Equation 4.10, and measurements are taken along this line ($0.10L$ near the lower part of the catchment, $0.85L$ towards the upper end). For the Acheron River basin, S_b = (1.080 km − 0.200 km)/0.75 (48 km) = 0.0244 or 2.44% or 1.40°.

Another method of computing average catchment slope is to superimpose a grid with approximately 100 points over the catchment. The slope at each point is tabulated and an average computed. This is similar to the methods described for aspect and hypsometric curves (this section).

The longitudinal profile

The longitudinal profile of a stream describes the way in which the stream's elevation changes over distance, as shown in Figure 4.14. The x-axis represents the distance along a stream, as measured from some outfall point such as a stream junction, a lake, or an ocean.

For many streams, the longitudinal profile shows a characteristic concave shape, with slope decreasing from the upper "eroding" reaches to the lower "depositional" ones. This shape is associated with both an increase in discharge and a decrease in sediment size in the downstream direction (Schumm, 1977). In catchments of high rainfall the concave shape is more pronounced, whereas in rivers such as the Nile, which derive their flow from headwater regions and do not increase in discharge downstream, a concave profile does not develop (Petts and Foster, 1985). In some semi-

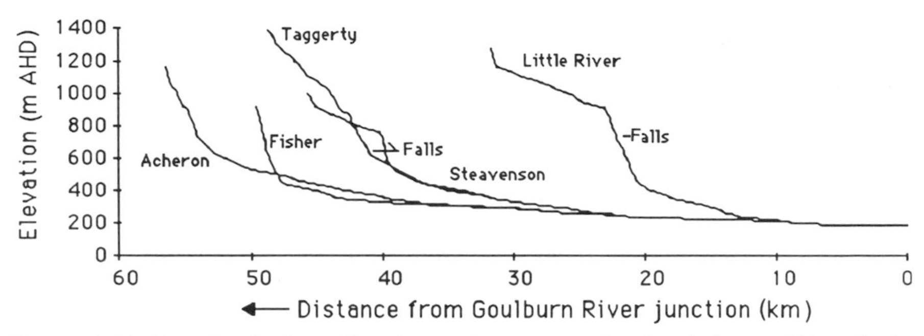

Figure 4.14. Longitudinal profiles for major streams in the Acheron River basin, Victoria, Australia (see map, Figure 4.7). AHD = Australian Height Datum (Section 3.3)

arid streams the average discharge may actually decrease in the downstream direction due to evaporation and transmission losses, and in these streams the general shape of the profile can be convex. There have been a large number of attempts to quantify the shape of the longitudinal profile, including the use of exponential-decay equations (Morisawa, 1985).

As well as general trends, the profile is also modified by local topography, bedrock features, changes in bed material, etc. In general, the profile will tend to be steeper on harder rock types and flatter when the streambed is less resistant to erosion. Abrupt changes in slope called **knickpoints** may occur at stream junctions or where the geology changes. For example, the waterfalls shown in Figure 4.14 are knickpoints. These sites are good candidates for locating boundaries of study segments (Section 2.3.4).

Hypsometric (area–elevation) curves

A curve can also be developed for describing the distribution of catchment area with elevation, called a hypsometric or area–elevation curve. These are useful for describing hydrologic variables which vary with altitude; for example, rainfall or snow cover (Beaumont, 1975).

A hypsometric curve is derived by measuring the area of contour "belts" from a topographical map (the amount of basin area between two contours). The cumulative area above (or below) a given elevation is then plotted against the elevation. It may be convenient to express areas as percentages of the total catchment area in order to compare curves between different catchments. Elevation can also be expressed as a relative height from 0 at the stream outlet to 1 at the highest point on the drainage divide.

Linsley et al. (1975a) suggest an alternate method of deriving hypsometric curves which reduces the amount of effort required. A grid is superimposed on a topographic map to obtain at least 100 points within the catchment. The curve is then plotted using the number of points falling within each elevation range, expressed as cumulative percentages. The technique is similar to the development of cumulative frequency curves (see Appendix 1).

Drainage basin shape

Drainage basin shape is difficult to express unambiguously (Gregory and Walling, 1973) and a number of quantitative indices have been suggested by various authors. Selby (1985), for example, lists seven different measures.

Horton (1932) gives a simple **form ratio** (R_f) for describing basin shape:

$$R_f = \frac{A}{L^2} \tag{4.13}$$

where A is catchment area and L is the length of the basin, as defined for Equation 4.10. Units should be chosen to make the ratio dimensionless. For the Acheron River basin, $R_f = 740 \text{ km}^2/(48 \text{ km})^2 = 0.32$.

Morisawa (1958) determined that the elongation ratio (R_e) given by Schumm (1956) had the best correlation with hydrology:

$$R_e = \frac{D_c}{L} \tag{4.14}$$

where L is the same as for Equation 4.10 and D_c is the diameter of a circle with the same area as that of the basin. Working backwards from an area of 740 km^2, the diameter of a circle with the same area is $\sqrt{[(4 \times \text{area})/\pi]} = \sqrt{[(4 \times 740)/\pi]} = 30.7$ km. Inserting this value into Equation 4.14 gives a value of $R_e = 0.64$ for the Acheron River basin.

Aspect

The aspect of a hillslope is the direction it faces (e.g. south-west). Aspect influences vegetation type, precipitation patterns, snowmelt and wind exposure. To determine catchment aspect a uniform or random grid is superimposed over a map of the catchment and the aspect at each grid point is noted. Aspect is measured as the bearing (in the downhill direction) of a line drawn perpendicular to the contour lines at the grid point. The number of points within each aspect category (e.g. 330° to 30°) is tabulated to estimate the percentage of catchment area within each aspect interval. The distribution of aspect in a basin is usually plotted as a polar or "rose" diagram, with the aspect shown as an angle (0–360°, with zero representing north), and percentage area as distance from the origin (Linsley et al., 1975a).

4.3 STREAMFLOW HYDROGRAPHS

4.3.1 DEFINITIONS

A hydrograph is a graph of water discharge or depth against time. The term hydrograph can refer to the pattern of streamflow which occurs over a season or over a year, e.g. the pattern of daily flows shown in Figure 3.1. In this section, however, we will consider hydrographs which result from a single runoff event, either as a result of snowmelt, rainfall, or both. These are sometimes referred to as "flood" or "storm" hydrographs. For flood analysis, it is preferable to plot these using near-instantaneous readings; however, average daily values will be used here.

Engineers need to know the response of a catchment to rainfall in order to design structures such as overflow spillways on dams, flood-protection works, highway culverts and bridges. Also, if modifications to a channel are planned, it is necessary to predict their effects on flood hydrographs. **Rainfall–runoff** analysis is an important and highly researched aspect of engineering hydrology, and readers are referred to hydrology texts such as Chow (1964), Chow et al. (1988), Linsley et al. (1975a,b) and Shaw (1988) for a more detailed coverage of the subject. Stream biota are also affected by how quickly a stream rises and falls. Thus, hydrograph characteristics are useful in the classification of streams for biological purposes (Hawkes, 1975) (see Section 9.2).

Figure 4.15 shows a flood hydrograph for the Acheron River, developed from daily streamflow data. The section of the hydrograph where the flow is increasing is called the **rising limb** and the section where the graph falls off the **falling limb** or **recession curve**. Some catchments will produce narrow, peaked hydrographs which rise and fall quickly. Streams which characteristically have this type of hydrograph are termed **flashy**. In contrast, **sluggish** streams are those which have wide, rounded hydrographs, with the runoff spread over a longer time period. In general, streams from smaller catchments will be more flashy than those from larger ones, and hydrographs from sudden, intense thunderstorms will be more flashy than those from snowmelt events or low-intensity rainfall.

When precipitation falls on a catchment or when snow melts a certain amount of time elapses before the stream level begins to rise. The water may be intercepted by vegetation, trapped by depressions in the land surface, absorbed by the soil, or it may evaporate. Any excess water will make its way to the stream as overland flow (surface runoff) or subsurface interflow (see Figure 4.1). Overland flow will reach the stream more quickly than interflow, and it may take many days for all the upslope water to reach the stream. A measure of the catchment response time, the **time of concentration** or **lag time**, has a number of definitions, usually based on the elapsed time between the rainfall event and the runoff peak.

When the channel fills, groundwater levels near the stream are temporarily

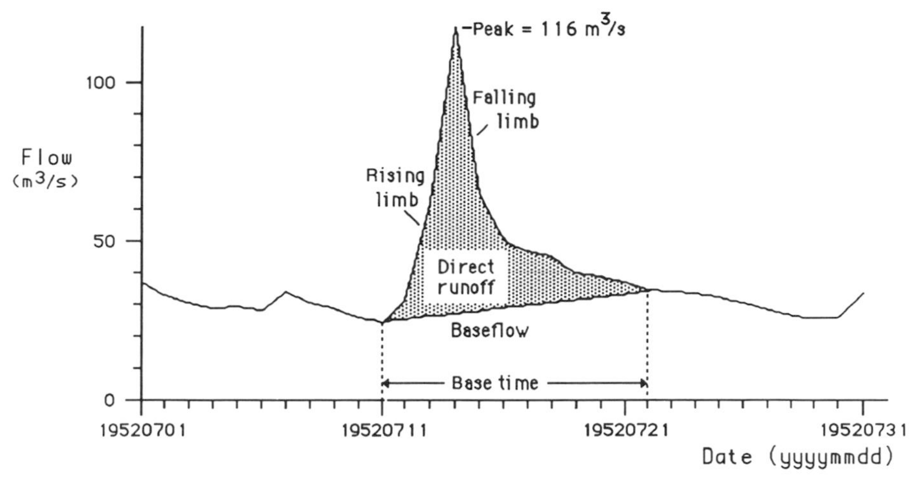

Figure 4.15. Flood hydrograph for the Acheron River at Taggerty (see map, Figure 4.7), July 1952, developed using daily data. Hydrograph components are labelled

raised. The portion of the water entering the bank is referred to as **bank storage**. When the hydrograph drops off, some of this stored water again re-enters the stream under effluent conditions (see Figure 4.3). Bank storage is important hydrologically because of its effects on the shape of the hydrograph. It may also have ecological relevance, since dissolved salts and nutrients can be leached from the soil and introduced to the stream in the stored waters.

4.3.2 HYDROGRAPH SEPARATION

A hydrograph can be separated into two main components:

(1) *Direct runoff or quickflow*—the volume of water produced from the rainfall or snowmelt event;
(2) *Baseflow*—the volume of water representing the groundwater contribution.

The amount of direct runoff is of interest since, for rainfall events, this will best correspond with the amount and rate of precipitation falling on the catchment. Direct runoff is sometimes divided into surface runoff (the water flowing over the land surface) and interflow (the water which moves through the soil) (see Figure 4.1). The relative contributions of direct runoff and baseflow will vary between events.

Methods of separating hydrographs into components are commonly based more on characteristics of the hydrograph shape than the actual origin of the streamflow. Thus, the division is somewhat arbitrary. Choice of a method is less important, however, than the consistent use of one method. Linsley et al. (1975b) suggest using a straight line which tapers up slightly from the point of rise on the hydrograph to a point on the other side (as in Figure 4.15). The angle of the line is somewhat a matter of choice, based on what "looks right". More objective methods exist, as discussed by Nathan and McMahon (1990b). Methods which use water quality data as a basis for separating the components are presented by Gregory and Walling (1973).

The **base time** is the base width of the direct runoff portion of the hydrograph. A simple measure of hydrograph "peakedness" is the peak to base time ratio (Gregory and Walling, 1973). For the hydrograph of Figure 4.15 this value is 116/11 = 10.5 m^3/s/day. Again, the ratio will vary with the units used, so the units should be consistent for comparative purposes.

Another ratio which may be of interest is the ratio of direct runoff to baseflow (with both expressed as volumes). This ratio will be higher for flash floods in arid lands and lower for events from gradual snowmelt.

4.3.3 FACTORS INFLUENCING THE HYDROGRAPH SHAPE

The rate at which runoff moves towards the stream is dependent on the drainage efficiency of the hillslopes. Drainage efficiency is influenced by the slope and length of the upland surface, its microtopography (i.e. existence of a drainage pattern and depressions), the permeability and moisture content of the soil, subsurface geology and vegetation cover. The hydrograph shape will be affected by these factors as well as catchment shape, drainage density, channel characteristics and storm patterns.

The drainage density affects the time taken by water to reach a given point on the stream (the time of concentration). A catchment with a well-developed drainage system will have a shorter time of concentration than one with many marshy areas, lakes, reservoirs and other surface depressions (Petts and Foster, 1985). Drainage efficiency is also related to the bifurcation ratio. Stream systems with low bifurcation ratios tend to produce flood hydrographs with marked peaks, while those with high ratios produce lower peaks which are spread over longer time periods (Beaumont, 1975). In addition, the condition and shape of the channel itself, whether wide and shallow or narrow or deep, and the presence of vegetation in and along the channel will also have an effect on hydrograph shape.

As shown in Figure 4.16(a), a stream with a steeper longitudinal profile will show a more rapid response and will produce higher peak discharges than one which is not as steep. The effect of basin shape on the shape of the hydrograph is shown in Figure 4.16(b). Shorter, wider catchments will produce a faster stream rise and fall than longer, narrower ones due to the shorter travel times. Basin shape factors, however, may not correlate well with hydrograph shape because they do not reflect the orientation of the catchment.

Catchment shape also affects the hydrograph shape when a rainstorm moves across the catchment from one end to the other. If the storm moves downstream, the peak flows from all streams tend to "compress", arriving at the outlet at nearly the same time. If the storm moves upstream, the water in the lower sections of the catchment runs off long before the water in the upper catchment arrives, "stretching out" the hydrograph. This effect is more pronounced in elongated catchments.

The hydrograph will move downstream as a "wave" of increasing then decreasing discharge. In small headwater streams the hydrograph responds quickly, and then drops back to baseflow levels even before the flow peaks at downstream sites. The hydrograph lengthens and becomes rounder as it progresses downstream, collecting the volumes of water contributed at varying rates by tributaries. This "damping" of the hydrograph shape is called **attenuation** (Shaw, 1988). The behaviour of a hydrograph can be numerically followed downstream using **flood-routing** techniques. These enable predictions of hydrograph shape to be made for ungauged sites. Flood-routing methods will not be covered here but are given in most applied hydrology texts.

Vegetation cover and land use will also have an effect on hydrograph shape. Since vegetation affects infiltration rates, its removal can cause direct runoff to increase and hydrographs to become more peaked. It is well known that runoff rates increase after wildfires until vegetation becomes re-established. Grassland and agricultural land may exhibit a larger range of

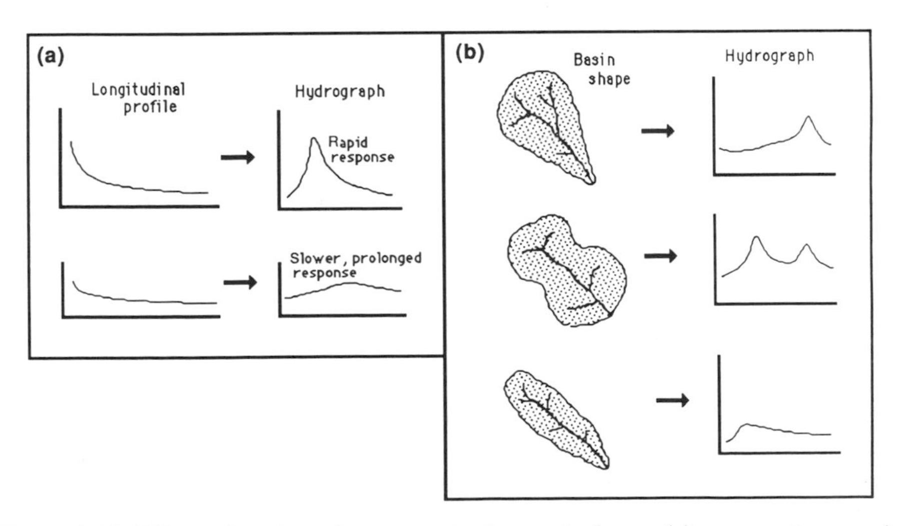

Figure 4.16. Effect of various factors on hydrograph shape: (a) stream slope and (b) catchment shape. Modified from Gregory and Walling (1973), based on (a) Schumm (1954) and (b) DeWiest (1965) and Strahler (1964), by permission of Hodder and Stoughton Ltd

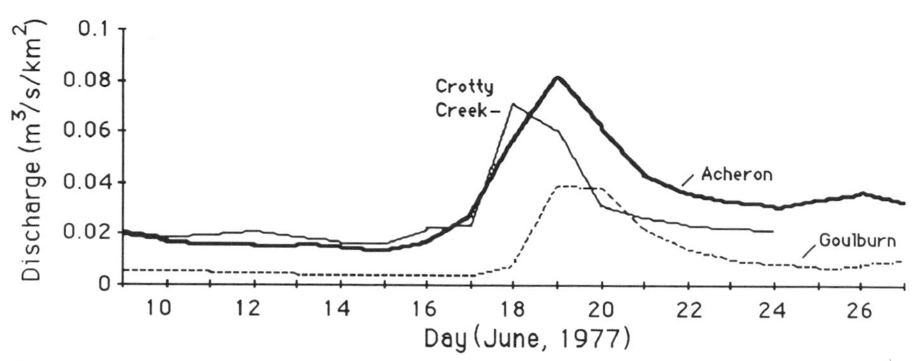

Figure 4.17. Downstream progression of hydrograph shapes. Crotty Creek is a small tributary of the Acheron River, and the Acheron flows into the Goulburn. Discharge values are mean daily flows per unit area, where catchment areas for the three basins (at the gauging sites) are 1.2 km^2, 619 km^2 and 8601 km^2, respectively

flows and an earlier, more rapid hydrograph rise than woodlands (Gregory and Walling, 1973). Branson et al. (1975) give a number of examples of studies which illustrate the effect on runoff of converting from one vegetation type to another. Urbanization leads to increased total runoff, higher peak discharges, more frequent flooding and shorter times of concentration (Morisawa, 1985).

It should also be borne in mind that only part of the basin may produce runoff at any given time. The maximum flood discharge per unit area typically decreases with basin size because of the fact that storms have a limited extent. In Figure 4.17 a hydrograph from the Acheron River is presented with those from upstream and downstream sites, with all discharge values divided by catchment area. In this case, the hydrograph from the furthest upstream site (Crotty Creek) does not show a larger maximum per unit area, possibly because precipitation did not occur as intensely in that catchment as in other parts of the Acheron basin. Complex storm patterns and variations in catchment characteristics may also lead to multi-peaked hydrographs which bear little similarity to textbook examples.

4.3.4 RECESSION CURVE ANALYSIS

The rate of rise of the hydrograph, and thus the shape of the rising limb, is influenced mainly by the character of the event (e.g. snowmelt or rainfall). In contrast, the shape of the recession curve is based on groundwater flow patterns and tends to be fairly consistent for a particular location on the stream. Thus, if a flood peak has passed through, the time required for the flood waters to recede can be predicted, given that no additional runoff occurs. Linsley et al. (1975b) give a simple exponential decay equation for describing the shape of the recession curve:

$$q_t = q_0 K_r^t \tag{4.15}$$

where q_t is the discharge at t time units (e.g. 1 day, 2 days, 30 days) after q_0 and K_r is a recession constant which is less than 1.0. This equation will plot as a straight line on semi-log paper, with q_t on the logarithmic axis and t on the linear one. Since q_0 is the starting value, it is also the y-intercept value. Usually, K_r changes gradually, approaching 1.0 as the flow levels off. Thus, Linsley et al. (1975b) recommend the use of one curve to represent baseflow and another to represent the recession from direct runoff.

For perennial and intermittent streams, this technique can also be used to describe the seasonal fall of groundwater levels (Todd, 1959; WMO, 1983). For example, it may be of interest to analyse the general downward trend following the snowmelt period to predict the time when a stream will reach a certain level (e.g. zero flow).

If a number of hydrographs are analysed, an average or **master recession curve** can be developed. It is preferable to develop separate curves for snowmelt and rainfall events. Recession analysis is treated in more detail in engineering hydrology texts. Nathan and McMahon (1990b) also present and discuss several techniques.

4.4 HOW DOES THIS STREAM MEASURE UP?

4.4.1 GENERAL

When two personalities come into contact for the first time a certain amount of "sizing up" takes place. Similarly, a great deal of information about the characteristics of a stream can be obtained from some very basic analyses of streamflow data. Annual, seasonal and daily patterns of streamflow determine many of the physical and biological properties of streams. A description of this "hydrological habitat" is necessary for interpreting changes in communities of stream organisms (Hughes and James, 1989). Statistical measures can reveal differences between catchments or between locations on the same stream, and changes due to natural trends or land-use modifications.

Preliminary analyses of streamflow data might include the calculation of total volumes, averages, maxima and minima, the degree of variability, seasonal distributions and/or trends over time. Statistical definitions and formulas are given in Appendix 1. In AQUAPAK, program STATS will compute totals and basic statistics and HISTO will plot histograms. Commercial statistical software packages can provide other means of presenting and analysing data. For example, Bren et al. (1988) used box-and-whisker plots to show the monthly statistical distribution of the fraction of red gum forest flooded for pre- and post-dam conditions on the Murray River, Australia.

This section will only present descriptive statistical methods for characterizing the runoff patterns of a stream. For descriptive purposes, no assumption about the underlying population distribution is necessary for computing statistics (e.g. mean, coefficient of variation). However, for making inferences from samples to a population, and for putting confidence limits on estimates of population parameters, the distribution of error must be known (see Section 2.3).

In these preliminary computations it will also be assumed that the streamflow data are independent. In reality, rainy days tend to occur together and dry days together. In the same manner, days of high flow tend to follow other high-flow days, and low flows follow other low flows. This tendency is called **persistence**, and the implication is that more data are actually required for estimating the mean than if the data were truly random (Leopold, 1959). One measure of persistence is the autocorrelation coefficient, which is described in Appendix 1.

Because of variability, length of the data series will affect the statistics. For comparative purposes, then, it is best to use the same time period for analysis.

4.4.2 ANNUAL STATISTICS

Before computing annual statistics, one should first decide what is meant by "annual". Results can depend on which starting month is chosen (McMahon and Mein, 1986). The calendar year is a fairly standardized annual measure, and most precipitation data will be published as calendar-year summaries. However, in most hydrological studies it is preferable to use another interval, called a **water year** or **hydrological year**. This is defined such that the flood season is not split between consecutive years. Especially in regions where runoff originates from winter snowmelt, the cycle of snow accumulation and melt should be contained in one interval.

In the United States and Britain the usual water year runs from 1 October to 30 September. Water year 1990, for example, would end on 30 September 1990. McMahon and Mein (1986) reviewed several methods for determining the start of a water year, and concluded that the most appropriate starting month for a hydrological year is the one with the lowest mean monthly flow. Options are available in some of the programs in AQUAPAK for specifying the starting month and SUMUP will compute either calendar or water year totals.

Mean annual flow (either as a total volume or an average discharge) gives an indication of the size of a catchment, its climate and the "typical" amount of water delivered from it. Mean annual flow, as an average discharge in m^3/s, can be computed by averaging daily data from complete years. It can also be estimated (in m^3/s) by dividing the average annual volume (in million

m^3) by 31.5576. Table 4.2 lists discharges of selected rivers of the world, so that individual catchments can be placed in context with other rivers.

Mean annual runoff (as a depth) represents the difference between annual evaporation and precipitation. It is useful for obtaining gross estimates of the water resources of a catchment. Mean annual runoff (mm) is obtained by dividing the mean annual flow volume (million m^3) by the catchment area (km^2) and multiplying by 1000. For the Acheron River, this figure is $(337/619) \times 1000 = 544$ mm. Figure 4.19(a) shows the distribution of mean annual runoff by continent. As a general rule, mean annual runoff decreases as basin area increases; however, there is a great deal of variability in this relationship worldwide (McMahon, 1982). Because of problems when comparing stream orders (Section 4.2.5) between different regions, Hughes and Omernik (1983) recommend the use of mean annual flow per unit area and catchment area as an alternative to stream ordering for classifying stream and catchment size.

Plots of annual totals (Figure 4.18) show their range and distribution over time. The pattern is determined by climatic conditions, soil moisture and changes in land use. A line illustrating the 5-year moving average has been superimposed on the graph of annual totals. This average is calculated by averaging the value for the given year with those from the two previous years and the two following years. This has the effect of smoothing the year-to-year fluctuations to reveal longer-term trends. For the Acheron River it can be seen that the 1970s were generally wetter than the 1980s.

Table 4.2. Selected rivers of the world, as ranked by average discharge measured at the river mouth. Data sources: Holeman (1968), and Knighton (1984); converted to SI units

River	Catchment area (10^3 km^2)	Discharge (10^3 m^3/s)
Amazon, Brazil	6130	181.0
Congo, Congo	4010	39.6
Orinoco, Venezuela	950	22.7
Yangtze, China	1940	21.8
Brahmaputra, East Pakistan	666	20.0
Mississippi, USA	3220	17.8
Yenisei, USSR	2470	17.4
Mekong, Thailand	795	15.0
Parana, Argentina	2310	14.9
St Lawrence, Canada	1290	14.2
Ganges, East Pakistan	956	14.1
Danube, USSR	816	6.2
Nile, Egypt	2980	2.8
Murray-Darling, Australia	1060	0.7
Acheron, Australia	0.619	0.01

Since streams of high variability experience a rapid turnover of organisms, one statistic of particular interest as an index of hydrological variability is the annual coefficient of variation (C_v). A high C_v may be indicative of high disturbance and low predictability (Lake et al., 1985). In studies on streams in Tasmania, Australia, Davies (1988) concluded that interannual flow variability was likely to be a principal factor limiting trout abundance in streams. In general, streams in arid or semi-arid regions and in areas affected by tropical cyclones are characterized by high variability (AWRC, 1984). In these areas, the runoff tends to be "all or nothing".

Based on the work of McMahon (1982), Figure 4.19(b) shows a plot of the C_v of annual flows against the mean annual runoff (MAR). It can be seen that variability generally decreases with an increase in the yearly runoff. However, the streams of arid zones show a much higher variability at all levels of yearly runoff. McMahon et al. (1987) demonstrate that the annual coefficient of variation of runoff for Australia and Southern Africa is nearly twice that of the other continents. The paper by McMahon (1982) provides useful information on the hydrological characteristics of many of the world's rivers.

In intermittent and ephemeral streams it may also be of interest to calculate the amount of time each year that the stream is "dry", where "dry" refers to the lack of surface flows. Water and subterranean life may still persist below the channel bed. The length of the aquatic phase will vary with geographical location and local hydrology. Long-term patterns of

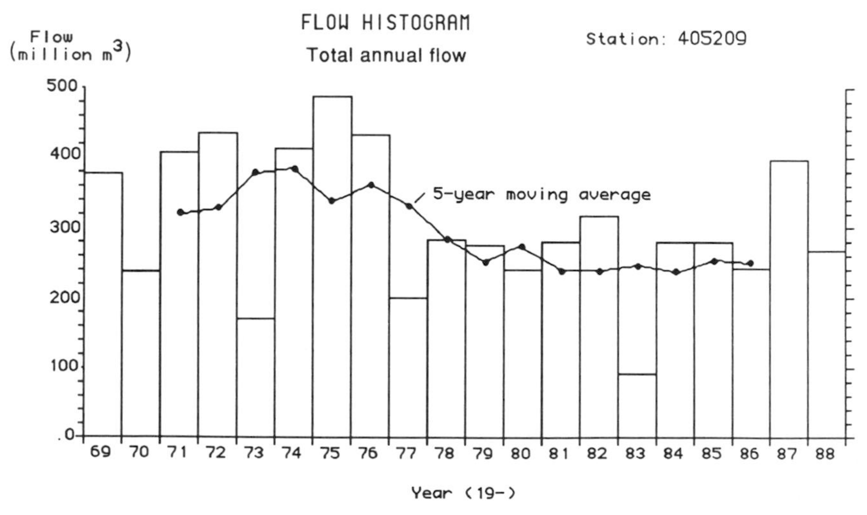

Figure 4.18. Histogram of annual streamflow totals from the Acheron River at Taggerty, as output by program HISTO in AQUAPAK. A 5-year moving average has been added which smooths year-to-year fluctuations

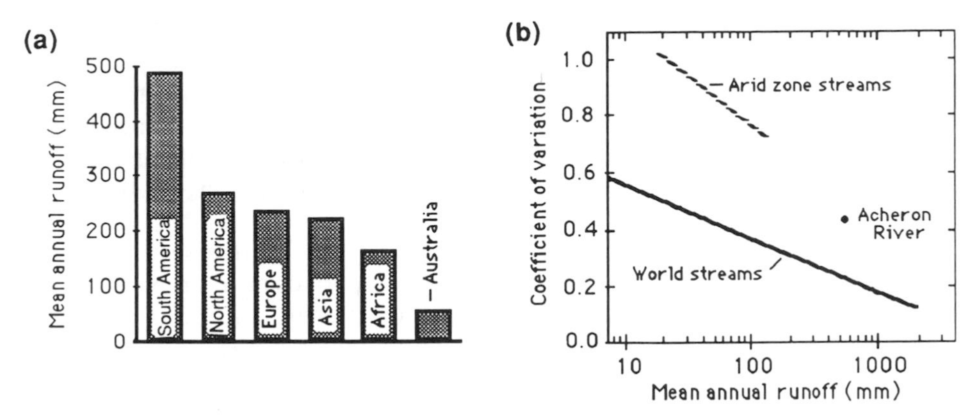

Figure 4.19. (a) Mean annual runoff for world streams, by continent (redrawn from McMahon and Mein, 1986, by permission Water Resources Publications), and (b) relationship between coefficient of variation of annual flows and mean annual runoff for world rivers and arid zone streams (redrawn from McMahon, 1982, by permission of UNESCO)

drought impose constraints of an evolutionary nature on the stream biota (Resh et al., 1988). Program STATS will compute the number of zero values in a data series.

4.4.3 MONTHLY STATISTICS

STATS and HISTO in AQUAPAK can be used for computing statistical measures for each month and for plotting monthly averages and coefficients of variation. These are of interest in studying seasonal variations in discharge, which are controlled by climatic patterns and channel and catchment characteristics. Some regions, for example, will show a strong snowmelt contribution in spring. In studies of rivers in Victoria, Australia, Hughes and James (1989) found that streams in high-rainfall areas had less variable monthly flows than rivers in dry regions.

A simple index can be derived from monthly statistics for describing average low-flow (baseflow) conditions in perennial streams:

$$\text{Baseflow index} = \frac{\text{Lowest mean monthly flow}}{\text{Mean annual flow}} \times 100 \qquad (4.16)$$

where the flows should be expressed as average discharges, in m^3/s. An index value near one would indicate that the flow remains fairly constant over the year, whereas a value of zero would be indicative of an intermittent or ephemeral stream (Hamilton and Bergersen, 1984).

The **regime** of a river refers to its seasonal pattern of flow over the year. River regimes have been identified by Hynes (1970) as having an important influence on instream biota, along with average and extreme water temperatures. The classification of river regimes has a wide range of potential applications, including the description of natural regime type in regions of extensive river regulation, the identification of potential water availability or shortage in particular seasons, the extrapolation of hydrological predictions within like regions and the classification of stream organisms or riparian vegetation by regime type.

The effect of regulation on regime patterns can be demonstrated by comparing regimes from pre- and post-regulation periods or reservoir inflows with outflows. For example the regulation of streams for irrigation may cause the regime to turn "upside down" due to winter storage of water and the augmentation of low flows in summer.

Using mean monthly flows from a global data set of 969 stream-gauging stations from 66 countries, Haines et al. (1988) developed 15 regime classes for perennial streams, as illustrated in Figure 4.20. Month one is the first summer month (December for the Southern Hemisphere, June for the Northern). Average monthly flows are expressed as a percentage of the average annual flow.

Cluster analysis (Section 8.6) was used to group similar patterns of flow. Algorithms in the form of a "decision tree" were developed for applying the global classification. These have been included in the monthly histogram program, HISTO. The decision tree is much like a dichotomous key for classifying plants or animals. At each "branch" criteria are imposed to assign an individual to one category or another. The first "branch" separates out patterns belonging to the uniform flow regime (Group 1 in Figure 4.20). From there, patterns are separated into summer- and winter-dominant regimes and then into individual classes.

The limitations of this technique as a method of stream classification are that the absolute volume of flow is not considered in the classification—only its pattern. Also, the method does not address the regularity of the regime type from year to year. However, it does provide a method of classifying streams based solely on streamflow data. Other hydrological classification methods are discussed in Section 9.2.

4.4.4 DAILY STATISTICS

As mentioned in the discussion of annual statistics, daily data can be analysed to obtain the mean daily discharge for a stream. Statistics on daily data can be computed using the same program, STATS. The daily C_v, like the annual C_v, also characterizes hydrological variability. Horwitz (1978), for example, demonstrated a relationship between fish community structure and the C_v of daily discharges.

An average annual hydrograph can also be developed using daily data. Points on the hydrograph represent the average value for each day, computed

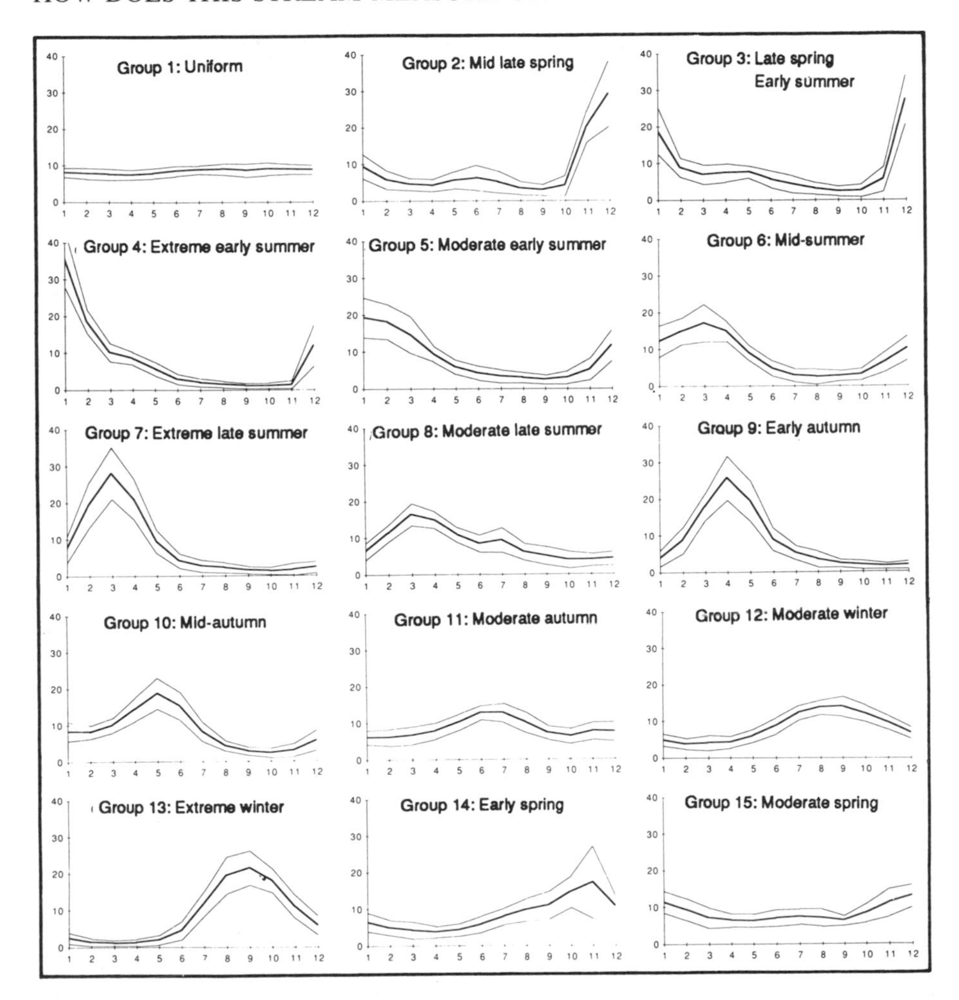

Figure 4.20. River regime patterns for the global classification of Haines et al. (1988). Average flows are expressed as percentages of the mean annual flow, and are shown with bands of plus and minus one standard deviation. Month 1 in the classification is the first month of summer. Reprinted from Haines et al. (1988), by permission of Butterworth-Heinemann

as an average over all complete years of interest. It is simply another view of the runoff regime (Section 4.4.3) at a finer scale of resolution. To give some indication of the amount of variability, maxima and minima for each day and/or the standard deviation can also be plotted. The average hydrograph is probably most useful for describing runoff patterns which are fairly consistent from year to year.

4.4.5 A METHOD FOR DESCRIBING HYDROLOGICAL PREDICTABILITY: COLWELL'S INDICES

The predictability of flow patterns from year to year, from one season to another and even from day to night is ecologically important for a number of reasons. For example, flow predictability is thought to influence the evolution of behavioural mechanisms and the timing of life-history stages in stream biota (Resh et al., 1988). If a seasonal pattern is repeated exactly the same way each year it would be totally predictable. Then, knowing the time of year, we could exactly predict the state of the phenomenon (e.g. a certain fish species will spawn for exactly 10 days from 12 May to 21 May).

Resh et al. (1988) state that a quantitative technique for comparing the predictability of flow patterns among streams must address both the frequency and intensity of flows as well as the contribution of seasonal phenomena to the annual runoff pattern. Colwell (1974) presents a method based on information theory which satisfies these requirements.

Colwell defines three simple measures for describing fluctuations in physical and biological phenomena over time: predictability, constancy and contingency. **Predictability** is a measure of the relative certainty of knowing a state at a particular time. It is the sum of two components, constancy and contingency, where all measures have a range of 0 to 1. **Constancy** is a measure of the degree to which the state stays the same (e.g. the year-round leaf drop of eucalyptus trees would have higher constancy than the seasonal leaf drop of maples). **Contingency** is a measure describing how closely the different states correspond to different time periods (e.g. for many trees, fruiting occurs in late summer, leaf drop in autumn and bud formation in spring).

For streamflow data, Colwell's method can yield indices that are biologically meaningful, since periodicities in flow have an important influence on riverine ecosystems. For instance, they can be used to characterize flow seasonality, patterns of drought or occurrence of peak flows. "Predictable, seasonal" streams can be separated from, say, "unpredictable, unseasonal streams" for classification purposes, or streamflow indices calculated from "before" and "after" time periods can indicate the effects of river regulation or catchment changes.

To apply the technique, a matrix is constructed with the states of the phenomena represented by rows and the time periods within some cycle represented by columns. An example is shown in Table 4.3, where months form the time periods and the "state" is designated by different streamflow intervals (classes). Entries in the matrix are the number of months in which the streamflow falls within the given interval. Constancy reaches a maximum (value of 1) when only one row has values (i.e. there is only one state). Contingency is zero when all columns are identical, and it reaches a maximum when there is only one non-zero entry in each column and each row. As stated previously, predictability is simply the sum of the other two components. These indices are shown in Table 4.3 for the Acheron River data.

Both the length of record and the way in which continuous records (such as streamflow data) are partitioned into "state" classes can affect the computed values. Gan et al. (1990a) investigated the use of Colwell's indices for analysing periodicity in monthly rainfall and streamflow data. They found a tendency for predictability and contingency to be biased towards high values for short periods of record. From data-generation methods they concluded that 40 years of data are needed to stabilize these measures. The authors also commented that the lack of a consistent classification of states for continuous hydrological records was a major shortcoming of the method. They adopted seven classes for their study: $< 0.5\bar{Q}$, $0.5\bar{Q}$–$1.0\bar{Q}$, $1.0\bar{Q}$–$1.5\bar{Q}$, . . ., $> 3\bar{Q}$, where $\bar{Q}$ is the mean monthly flow. These are the default classes used in the AQUAPAK program and illustrated in Table 4.3.

Bunn and Boughton (1990) suggest that a log-2 scale (i.e. 0.5, 1, 2, 4, 8, . . .) be used rather than a linear one, on the premise that a larger change in streamflow constitutes a disturbance in larger rivers whereas only a small change would be needed to produce an equivalent effect in smaller streams (program TRANSFRM in AQUAPAK can be used to produce a file of log-2 transformed values). They argue that trade-offs are necessary in selecting both the number of states and the length of record. Too few categories can produce high constancy, the extreme example being one category which would lead to a constancy of 1.0. In contrast, too many categories can lead to low predictability but high seasonality. In regard to the amount of data, they recommend a record length of 10–15 years, stating

Table 4.3. Colwell's indices: sample output from COLWELL program in AQUAPAK using monthly streamflow data from the Acheron River at Taggerty, Victoria (see Figure 4.7). Matrix entries are the number of months in which the flow falls within the given interval

Class	Jan.	Feb.	March	April	May	June	July	Aug.	Sept.	Oct.	Nov.	Dec.
< 13.629	26	37	38	37	24	12	5	1	1	2	6	12
to 27.257	13	4	4	4	10	19	10	5	4	7	16	21
to 40.886	1	1	0	1	4	5	11	8	10	15	11	6
to 54.514	2	0	0	0	2	1	4	7	12	1	1	1
to 68.143	0	0	0	0	1	3	7	8	7	12	4	0
to 81.771	0	0	0	0	1	0	1	7	3	3	1	2
> 81.771	0	0	0	0	0	2	4	6	5	2	3	0

Number of years of data: 42
PRED = 0.38 CONST = 0.17 CONTING = 0.21

that the patterns in recent time are more important ecologically. Further comparative studies are needed to develop consistent methods which provide indices of the greatest ecological relevance.

5 How to Have a Field Day and Still Collect Some Useful Information

5.1 VENTURING INTO THE FIELD

Velocity, water depth, substrate, discharge, sediment concentrations and channel configurations all interact to form the hydrological and hydraulic environment of stream-dwelling organisms. The subject of this chapter is measurement of these variables in the field. There are many possible approaches to the collection of physical habitat data and some of the more common techniques are presented here. For further information, Dackombe and Gardiner (1983) is a compact guide to standard geomorphological field techniques and publications of the US Geological Survey provide time-tested methods of stream data collection. Other useful references for the field evaluation of streams have been written by Platts et al. (1983) and Hamilton and Bergersen (1984).

Before the first visit to the field, decisions should be made about which variables are to be measured, and in what order. This helps in the choice of equipment and methods, estimation of the number of people required and their duties, and development of field data forms. It is a good idea to run through the entire field procedure as a "thought exercise" to smooth out any rough spots.

Field data forms (preferably printed on waterproof paper) should allow efficient and systematic recording of measurements and observations. Each form should have spaces for the date, time, site name and personnel names (e.g. see Figure 5.11). If the information is to be computer processed, some thought should also be given to format and coding requirements. Dataloggers, the "tech-age" answer to the field notebook, can be programmed to accept data which can be directly downloaded onto a computer. A pencil and notebook should always be carried for jotting down random notes and sketches.

Equipment should be checked and calibrated thoroughly and batteries charged as needed before venturing into the field. It is a good idea to put together a small tool kit with some spare parts, pliers, a few screwdrivers, a small adjustable wrench, a roll of duct tape ("100 mile-an-hour" tape), and some bailing wire for minor repairs and adjustments. Repair kits, waterproof gear, camping equipment and first-aid supplies are other considerations. Checklists are invaluable for remembering all the bits and pieces required.

Field personnel should be told which observations are critical and why, trained in data entry and field measurement procedures, and instructed in the use, care and repair of instruments. Safety instruction should form an integral part of training sessions, since there are many hazards in working on or in streams, drowning being only an obvious example. It is a well-known rule of thumb that the depth (in metres) times the velocity (in m/s) should not exceed 1.0 for safe wading. Life preservers should be included on the equipment list if conditions approaching this level are anticipated. A rope or cable can also be strung across swifter streams for workers to hold on to. As a minimum, field crews should have two people and preferably three to four, with at least one trained in first aid. The location and phone numbers of emergency services should be noted, and if study sites are highly remote, communication equipment should be considered.

Field personnel should be taught to be keen observers—to look beyond the depth, breadth and speed of a stream to note other factors appropriate to the study. Each person will filter out different information from their surroundings, biased by the preconditioning of past learning. Particularly if measurements have a qualitative nature, field workers and researchers should both be able to make the same interpretation; e.g. what does "degraded" or "undisturbed" mean? For example, logs in a stream may be viewed as good habitat by biologists, whereas to another, they make the stream look "messy". In a book on nature study, Pepi (1985, p. 54) gives the following illustration of preferential perception:

> A New York banker was giving a tour of the city to an upstate entomologist. As they walked along a crowded street, the entomologist tapped the banker on the shoulder, pointed to a potted shrub, and said, "Listen, there's a field cricket singing in there." The banker was amazed that the entomologist had heard the cricket's song above the city's roar and complimented the scientist on his "good ears". A short time later, the banker stopped at a crack in the sidewalk and picked up a dime.

Perception can be improved through the practice of asking questions. Why is it this way? Is it affected by the season, the weather, the time of day? Is it a nice day, and what do you mean by that? What if? What else? Why not? Researchers should constantly be on the lookout for unidentified hypotheses hurtling through the spaces between known facts.

Training and re-training of field personnel is thus essential to make sure that all share the same "vision" in terms of what is observed and the goals of the project. Avenues of communication should also be constructed prior to the study to guide the smooth flow of information between everyone involved. As mentioned in Section 2.2, planning should be done with care to ensure that the study runs smoothly and yields information of quality.

5.2 SURVEYING: A BRIEF INTRODUCTION

5.2.1 GENERAL

The objective of any surveying technique is to establish the horizontal and/or vertical location of a given point. This location can be referenced to map co-ordinates and a national height datum (see Section 3.3) or used directly, as for constructing channel cross-sections or measuring the heights of waterfalls. Whereas measurements taken from maps and aerial photographs will suffice for studies of larger scope and lower detail, surveying becomes more essential as the study shrinks from continental to microhabitat scale.

Some surveying techniques will be more appropriate than others. For example, a measuring tape and a meter rule may be accurate enough for cross-sectional profiles of smaller streams, whereas more precise methods are needed for measuring water surface slopes. Larger rivers, too, will require more sophisticated equipment simply because of the distances covered. The techniques employed will depend on the purpose of the work, the accuracy required, the equipment available and how easily it is transported, time constraints, and ultimately the sensitivity of the hip-pocket nerve.

Some of the more common methods of measuring horizontal and vertical distances and slope will be presented in Sections 5.2.2, 5.2.3 and 5.2.4, respectively. In Section 5.2.5, techniques which yield more than one of these measures are given. Specialized literature exists on surveying techniques, which can be consulted for further information. Surveying references include Brinker and Wolf (1977), Higgins (1965), and Uren and Price (1984).

5.2.2 HORIZONTAL DISTANCE

Pacing

With practice and care in taking consistent steps, pacing can be surprisingly accurate. Two natural steps constitutes one pace; i.e. the number of paces equals the number of times the right foot touches the ground. One's pace length will naturally shorten on steeper slopes, whether traversing uphill or downhill. Thus, the pace must be calibrated by stepping off a known distance such as 100 m and re-calibrated on different terrains.

Rangefinder

A rangefinder is a relatively inexpensive device for indirectly measuring distances. This instrument has a focusing mechanism, whereby two images of a "target" object are brought together. The distance between the rangefinder and the target is then read directly from the instrument. Accuracies and prices will vary. Rangefinders require calibration with objects at known distances prior to use. It is advisable that readings be taken only by the person who has done the calibration. It is also best to take several measurements each time and average them.

Measurements using a tape ("chaining")

The direct measurement of distance using a tape is referred to as "chaining", because a chain made of 100 links was historically used for distance measurement. Chaining is now more commonly accomplished by stretching a tape horizontally between two points. Tapes may be made of steel, fibreglass, cloth, or plastic, and various lengths are available. Metal tapes will normally give the most accurate measurements; however, other materials will be lighter for the backpacking researcher.

For most environmental field work, high-precision surveying is not necessary. Some sag of the tape or other deviation from horizontal is acceptable if the error is less than about 5%. It may be preferable to use tachometry under conditions where the error is greater, such as when measuring across a wide gorge or gully or when measuring horizontal distances on steep slopes.

Two or three people are normally required for chaining—one on each end of the tape and one to record measurements. If the distance to be measured is longer than the tape, some technique must be used to keep the measurements on line. This can be done by lining up ranging poles "planted" at each station or by using a compass bearing and a landmark.

If an obstacle is encountered it may be necessary to use an offset technique of some kind to make a precisely engineered detour, as illustrated in Figure 5.1. A few steps are taken at a right angle to the original path (by using some type of optical square or by using the tape itself and properties of the 3-4-5 triangle), and the same distance is measured back after passing the obstacle.

5.2.3 VERTICAL DISTANCE

Measurement with a tape or rule

A graduated tape or rule can also be used for vertical measurements. The simplest example of this is the use of a metre rule or surveying staff to

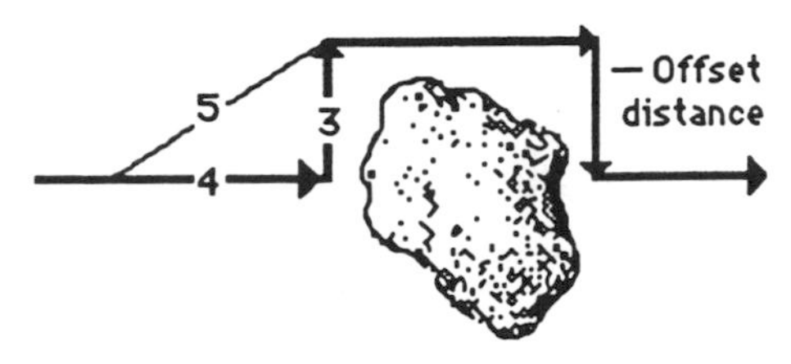

Figure 5.1. Offset technique for avoiding obstacles when chaining

measure from the streambed to the water surface to obtain water depth. If needed, a simple spirit level can be used to make sure that the measurement is taken vertically. Water depths are taken in the measurement of stream cross-sections (Section 5.4) and streamflow (Sections 5.7 and 5.8). A tape weighted on one end can also be used to take vertical measurements over longer distances—for example, to measure water depth in a well (Section 5.5.5).

Levelling

Levelling is a procedure for determining the relative heights of a number of points. The procedure involves the use of a device which allows one to sight along a horizontal (0°) line. In this section the levelling device is assumed to be a hand-held telescopic level, an Abney level or a clinometer. Levelling with a dumpy level is described in Section 5.2.5. Magnification is often needed for taking readings over distances longer than about 30 m, particularly in shaded locations.

The eye height can be kept constant by resting the level against a staff or rule (e.g. a 2-m carpenter's rule) or by using a pole with a mechanism to lock the level into place. A sighting is taken through the level to a survey staff to obtain a reading. At longer distances where the numbers on the staff become illegible, the person with the staff can hold a finger horizontally in front of the staff and raise or lower the finger until the level height is reached. In this case, the staff bearer would record the reading. Measurements taken at one location are compared to those taken at other locations, or to the "eye height" of the level, to determine the difference in elevation. This is shown in Figure 5.2(a), and the relevant calculations are given in the caption.

A simple arrangement for levelling consisting of a spirit level attached to a graduated staff by a sliding bracket is shown in Figure 5.2(b). This apparatus can be used to quickly obtain water and channel bed elevations by sighting to some fixed point such as a mark on a boulder or a nail in a tree. This technique was used by Bren (pers. comm., 1988) to rapidly survey the water level at 100 cross-sections in one day on a fifth-order stream in Oregon, USA.

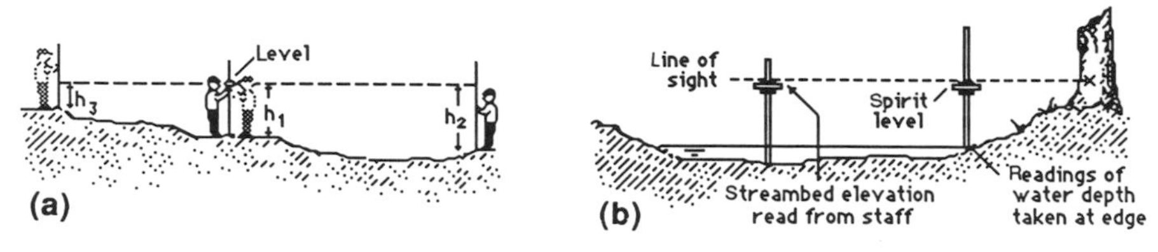

Figure 5.2. Levelling: (a) vertical distance can be measured as the difference between the level height and the height at another point (e.g. $h_2 - h_1$) or the difference between elevations at two points (e.g. $h_2 - h_3$); (b) a simple method for measuring channel and water elevations

Indirect methods for measuring height

Vertical heights of features such as trees or waterfalls may require indirect methods. The properties of right triangles can be used to obtain vertical distances by measuring one side and one angle of the triangle (see Figure 5.3). Vertical angles can be measured either upward or downward from the horizontal with a clinometer or Abney level. Commercial versions are relatively inexpensive, and easy to use and carry. To obtain a vertical angle, the reading is normally taken by looking through a viewing window with one eye while lining up the clinometer with the other. An Abney level is operated in a similar manner, although the angle must be set on a protractor-like device.

To measure the height of a feature, a specific distance, L (e.g. 50 m), is first measured out from its base. One then sights through the clinometer or Abney level to the base of the feature and then to the top, recording each vertical angle. The vertical height (H_v) is then calculated as:

$$H_v = L(\tan \theta_1 + \tan \theta_2) \tag{5.1}$$

where θ_1 and θ_2 are the upper and lower angles as shown in Figure 5.3 and L and H_v are in metres.

For example, if the top reading was 12.2° above horizontal, the lower reading 2.3° below horizontal and the measured distance, L, 50 m, then the feature would be 50(0.256) = 12.8 m tall. If the person is totally above or below the object, a little more trigonometric juggling is needed to come up with an appropriate formula, and this is left to the reader (or see Dackombe and Gardiner, 1983, pp. 25–7).

5.2.4 SLOPE

The **slope** or **gradient** of a stream, road or hillside is the amount of vertical drop per unit of horizontal distance. Measurement of slope, then, involves either a direct measurement of the slope angle or measurements of both

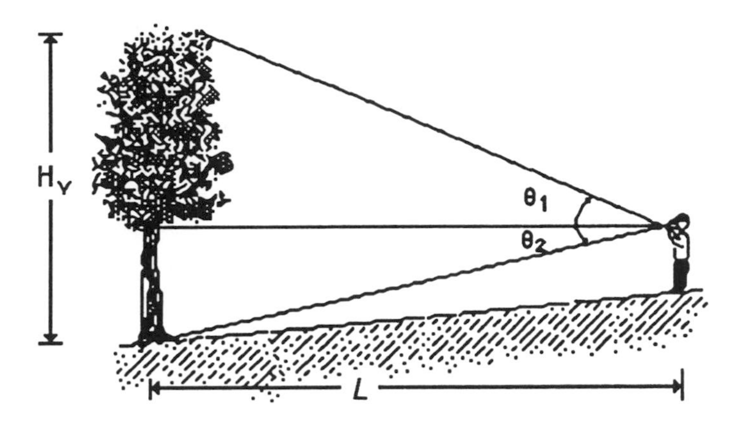

Figure 5.3. Measuring the height of a tree using vertical angles (see Equation 5.1)

horizontal and vertical distance over some length. Thus, combinations of methods in the preceding two sections can be used as well as those presented here and in Section 5.2.5.

Slope can be expressed as a fall per unit distance (e.g. metres per kilometre). If the distances are measured with the same units, the slope can be expressed as a ratio (e.g. 1 in 10 or 1:10), a decimal (e.g. 0.10) or a percentage (e.g. 10%). To convert from one form to the other is easily done with a scientific calculator:

$$\text{Percentage slope} = \text{slope (in decimal form)} \times 100 \quad (5.2)$$

$$\text{Slope (in decimal form)} = \tan \text{(slope in degrees)} \quad (5.3)$$

$$\text{Slope (in degrees)} = \tan^{-1} \text{(slope in decimal form)} \quad (5.4)$$

For example, a 45° slope would be a "1 in 1" slope, and a 0.009 ("9 in 1000") slope would be 0.9% or $\tan^{-1}$ (0.009) = 0.52°.

Measuring slope with a clinometer or Abney level

Slope can be read to about one degree with a clinometer or Abney level (described in Section 5.2.3). To obtain a direct reading of slope, the instrument can be roughly aligned with a hillslope by eye. For local slopes, such as on streambanks, the clinometer can be placed on a board which is set on the ground (see Figure 5.12). This method is not accurate enough for measuring water surface slopes in streams since these are normally very close to zero.

For more accurate measurements of slope, two people are required: one to use the clinometer and one to hold a staff or a pole marked at the clinometer "eye height". Measuring the slope is then a simple matter of looking through the clinometer with one eye and viewing the staff with the other, and raising or lowering the clinometer until the cross-hair lines up with the mark on the staff (see Figure 5.4(a)).

By including a measurement along the ground between the two people (L), horizontal and vertical distances can also be determined, as shown in Figure 5.4(b). The vertical angle (θ) measured with the clinometer is used to calculate distance as follows:

$$\text{Horizontal distance} = (L)(\cos\theta), \text{ and} \tag{5.5}$$

$$\text{Vertical distance} = (L)(\sin\theta) \tag{5.6}$$

where θ is the slope angle in degrees and distances are in metres.

Hydrostatic levelling

Hydrostatic levelling, a principle familiar to most carpenters, is an inexpensive, light and yet accurate method of slope measurement. A manometer for levelling can be simply made from two metre rules and a long length of hose marked at convenient intervals with indelible ink (Figure 5.5). Twenty metres of 10 mm i.d. (inside diameter) clear tubing is a practical size. The hose is filled with water, with care taken to exclude all air bubbles.

At the site to be measured the hose is extended over the length of the slope. It may be curled up as necessary, as this will not affect the readings. After the water level stabilizes in the manometer a measurement is taken at the bottom of the meniscus at each end and the difference in readings determines the vertical drop. An adequate measure of the horizontal distance between the two rules can be made by pulling the calibrated hose taut and using it as a tape measure. Slope is then calculated by the formula given in the caption for Figure 5.5.

In comparing slopes measured with a manometer to measurements with a surveyor's level and stadia rod, LaPerriere and Martin (1986) reported agreement within 2%. An added advantage of this equipment is that it can be used to measure slope around bends, meaning that neither a straight

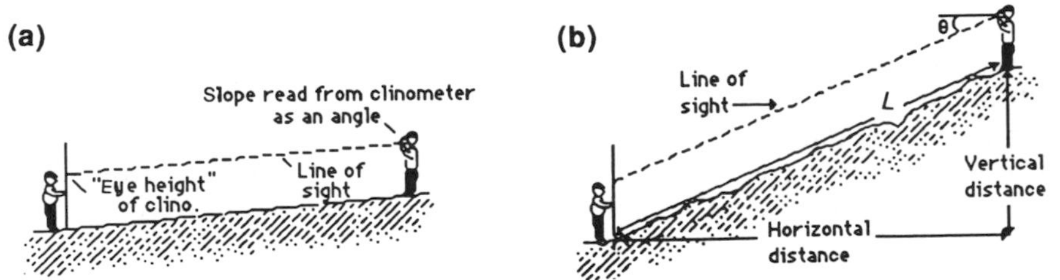

Figure 5.4. Using a clinometer to (a) measure slope and (b) measure slope and horizontal and vertical distance. L is the distance measured along the ground and θ is the angle read from the clinometer

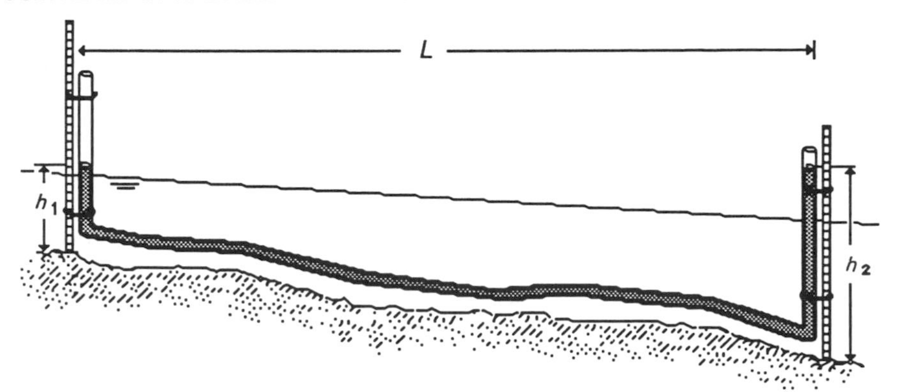

Figure 5.5. Hydrostatic levelling. Slope of channel bed is $(h_2 - h_1)/L$, with all measurements in metres

stretch nor a clear line of view are required. In streams, the slope of both the water surface and the streambed can be measured at the same time.

Estimation of slope

Slopes can also be estimated within about plus or minus 10° or better by eye or by "feel". It is a good idea to "calibrate" oneself each time an estimate is to be made and to recognize that overestimation is common. There are few natural slopes of 90°. Main roads seldom exceed a slope of 1 in 7 (8°), the maximum gradient for a four-wheel drive vehicle is about 1 in 2 (27°), and a 35° slope is about where it becomes difficult to stand upright on a hill.

5.2.5 THE FULL CONTINGENT OF CO-ORDINATES, INCLUDING METHODS OF MAPPING

In previous sections, methods were presented for measuring vertical and horizontal distances and slope. This section presents methods for fixing the measurement points in space. For many applications an estimate of position using landmarks and a map or aerial photograph is sufficient. In other cases, such as sketching a scaled map of a stream reach, a compass or plane table may be used to locate one's position more accurately. When even higher accuracy is needed (for example, when re-surveying a stream cross-section at a historically surveyed site) more sophisticated surveying techniques will be appropriate.

Tachometry using level or theodolite

Tachometry is a traditional surveying method of optical distance measurement. In tachometry, a sighting instrument is used to obtain readings from a staff. From the readings obtained, the horizontal distance to the staff, the difference in surface elevations and the location in reference to other points can be derived.

The instruments most commonly employed in tachometry are dumpy levels or theodolites. A dumpy level is basically a telescope fixed in the horizontal plane. The advantage of a theodolite over a level is that it can be tilted vertically, making it more versatile. This is especially true in irregular terrain or deep cross-sections where a dumpy level would need to be moved frequently to cover the range of elevations.

The level or theodolite should be positioned to give the clearest possible view of the area of interest. The instrument is preferably located near the middle of the measured distances (e.g. stream reach or hillslope) to reduce any bias introduced by the instrument or the way in which it was set up. The instrument is set on a tripod and levelled by adjusting the tripod leg heights and tilting screws on the instrument platform (tribrach). On hillslopes, the tripod should be set up with one leg uphill and two downhill to improve stability. In other situations, aligning one leg in the main direction of surveying will keep the operator from having to straddle the tripod.

Before taking measurements, the elevation of the instrument should be determined with reference to a real or assumed benchmark elevation. The zero azimuth point for the horizontal angle is then set with reference to some origin (e.g. a tree or benchmark) or to a compass direction (e.g. north). The horizontal angle is usually read on a scale viewed through an eyepiece.

With either a level or theodolite, the staff bearer holds the staff vertically at the point to be surveyed and the instrument is focused on the staff. As shown in Figure 5.6, the view is of a centre cross-hair bracketed by two stadia cross-hairs. All three readings and the vertical and horizontal angles to the staff should be recorded. The stadia reading is the distance between the stadia cross-hairs, as read from the staff (or twice the distance between one stadia hair reading and the centre line). Thus in Figure 5.6 the vertical reading is 2.70 m and the stadia reading is $2.76 - 2.64 = 0.12$ m. All readings should be recorded in a systematic manner. Special booking forms are available for this purpose from equipment suppliers.

With a **level**, all the sights are horizontal and the relevant measures are shown in Figure 5.7(a). The vertical distance is simply the staff reading at the middle cross-hair. For some surveys a vertical reading is taken at a benchmark in order to relate all other measurements to that elevation. In cases where only a difference in elevation is needed (as for calculating slope) this is not necessary.

The horizontal distance, L, is obtained from

$$L = 100s \tag{5.7}$$

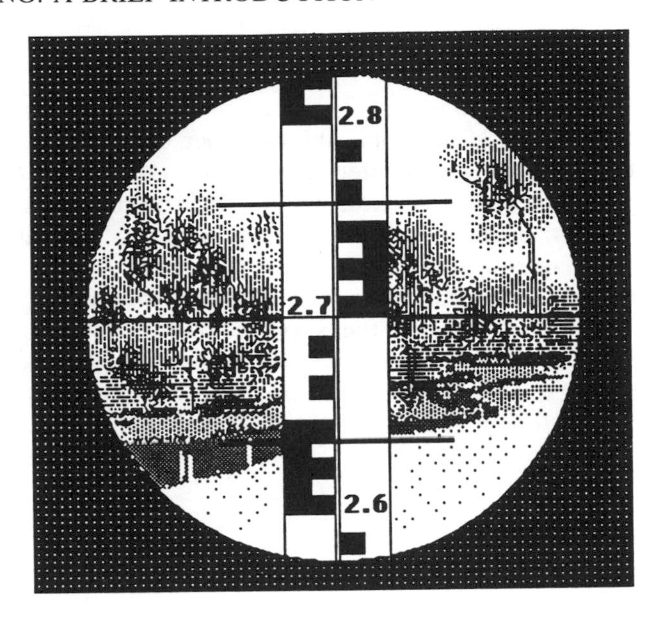

Figure 5.6. View through the theodolite or level, showing metric levelling staff plus centre and stadia cross-hairs

where s is the stadia reading and both s and L are in metres. The multiplying constant, 100, should be verified from the instrument reference manual or by calibration over a known distance.

With a **theodolite**, the calculations become more complex. The zenith angle (θ_z), or angle of inclination (θ_v), shown in Figure 5.7(b), is obtained from a scale on the instrument. A quick tilt of the instrument will reveal whether it is measuring θ_z or θ_v. To simplify computations, the instrument can be set to horizontal (in which case it becomes a level) and the vertical adjustment used only when needed. The horizontal distance, L, is calculated as:

$$L = 100s(\sin^2 \theta_z) \tag{5.8}$$

or

$$L = 100s(\cos^2 \theta_v) \tag{5.9}$$

where 100 is the multiplying constant of the instrument and should be verified. All other terms are defined in Figure 5.7. When the vertical angle is horizontal, the equations reduce to Equation 5.7. The vertical distance, H_v, is calculated as:

$$H_v = 100s\ (\sin \theta \cos \theta) \tag{5.10}$$

or, equivalently,

$$H_v = 100s\left(\frac{1}{2}\right)(\sin 2\theta) = 50s\,(\sin 2\theta) \qquad (5.11)$$

Equations 5.10 and 5.11 are applicable whether θ is θ_z or θ_v. H_v will be negative if the theodolite is pointed downwards (below horizontal) and positive if pointed upwards. The calculations may be programmed using a scientific calculator or read from a "stadia slide rule", available from surveying equipment suppliers.

The horizontal and vertical distances and the horizontal angle establish the x, y and z co-ordinates of the surveyed point with respect to the instrument position. These can then be referenced to a reference origin or to map co-ordinates. These calculations are based on simple arithmetic and trigonometry and are left to the reader.

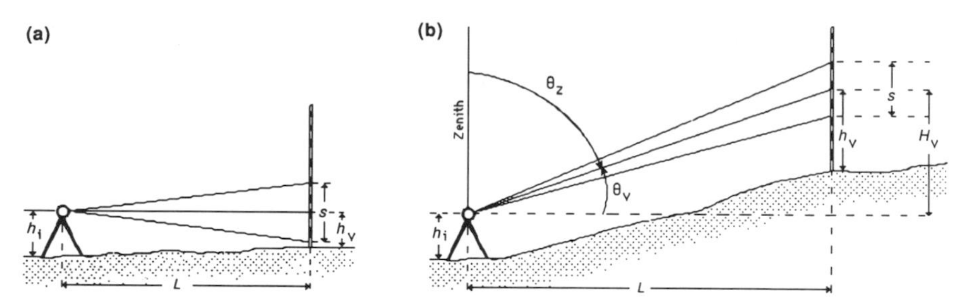

Figure 5.7. Measures used in tachometry: (a) with a level and (b) with a theodolite. Shown are the horizontal distance (L), vertical distance (H_v), zenith angle (θ_z), angle of inclination (θ_v), stadia reading (s), vertical reading (h_v), and instrument height (h_i). All distances are in metres and angles in degrees

Example 5.1

If the stadia reading is as shown in Figure 5.6 the measured zenith angle (θ_z) is 80° (meaning that the angle of inclination, θ_v, is 10°), find the horizontal and vertical distances with respect to the instrument.

Horizontal

L = 100(0.12 m) $\sin^2(80°)$
= 100(0.12 m)$(0.9848)^2$
= 11.64 m

Vertical

H_v = 100(0.12 m)(1/2) sin 2(80°)
= (6 m)(sin(160°))
= 2.05 m; so elevation of the ground is 2.05 − 2.7 = −0.65 (0.65 m below "eye level" of instrument)

Plane table mapping

In plane table mapping a scaled map is constructed on-site as the readings are taken. The plane table consists of a drawing board that may be levelled on top of a tripod, and a sighting instrument or alidade. A piece of sturdy paper is taped to the drawing board for constructing a map. There are two main types of alidade used: a simple "peep sight" type with two folding vertical sights attached to either end of a straight line, and a "telescopic" alidade, which is similar to a theodolite but is attached to a flat base that acts as a rule on the plane table (Figure 5.8). With the simple alidade the position of objects on the map must be determined either by using a tape or by a method of intersection which will be described. No vertical elevations may be determined with the simple alidade, although these are often obtained with a normal dumpy level set up near the plane table. Both horizontal and vertical measurements may be obtained with the telescopic alidade using the same methods described for the theodolite.

To map on the plane table, a convenient location is selected to one side of the site and the table is levelled. If a **peep sight alidade** is used, a measured baseline (e.g. 100 m) is established along the longest axis of the sight to a second position for the plane table. The baseline is plotted on the map, its length determining the map scale. The alidade is then set on the baseline and the plane table is rotated horizontally until it can be locked with the sight on the position at the other end of the baseline. With the table now levelled and locked, points to be mapped are sighted in the alidade and a line is drawn along that bearing from the initial position (station) at one end of the baseline (see Figure 5.14(b)). The radiating lines from this position are labelled. When all the points have been located, the plane table is moved to the other end of the baseline and levelled. With the alidade along the baseline, the table is then rotated and locked with the initial station sighted. The same mapping points are now sighted again. The location of a point may now be found from the intersection of the alidade bearing and the bearing line drawn from the first station. More baselines may be added to the map to cover larger areas. Depending on the care taken, the simple alidade can produce a scaled map with unsophisticated equipment. An even simpler version of this technique can also be used to develop a scaled map using two compasses, where the maps would be drawn on graph paper with a rule and a protractor.

If a **telescopic alidade** is used, points may be located along the bearing line of the alidade from a single position by using the stadia method for distances and elevations described for theodolite surveys. Because a scale ruler is attached to the alidade, the point can simply be plotted on the drawing paper at the appropriate distance. If the elevation of the plane table is established from a real or assumed benchmark, the elevation of the points may be noted as well. Contours of equal elevation may then be

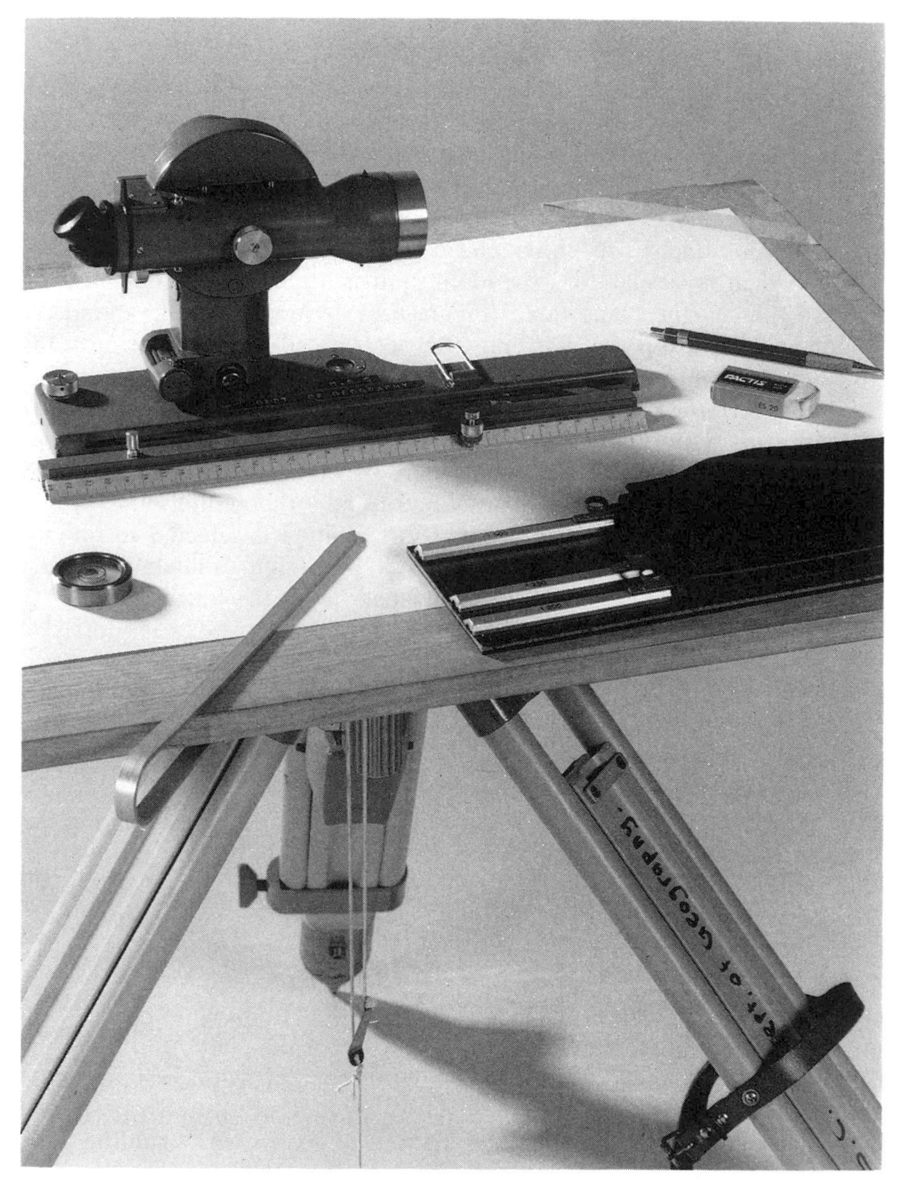

Figure 5.8. A plane table, showing tripod, drawing board, alidade, and scale ruler. Photo: P. Cardiyaletti

sketched in the field with additional points added as necessary. If the plane table must be moved to cover the entire site, at least two and preferably three points of reference should be established which can be sighted from the next station. A new piece of drawing paper is used at the next station, and the two maps are later joined together by overlaying the reference points.

In general, if good weather prevails, a plane table survey requires two or three times the field time required for other surveys. However, a higher level of detail is possible with more on-site observations of features such as soil types, pools, boulders, plant species and possum nests, and the mapping does not require further plotting and drafting. More information on plane tabling is given by Higgins (1965) and Low (1952).

Electronic Distance Measurement (EDM)

Electronic distance measurement (EDM) is more precise but also more costly than a theodolite. The instruments, called EDMs or total stations, measure the distance from a station to a target by the time or phase difference between a transmitted and reflected beam of radiation. Either microwaves, visible light, laser or infra-red light can be emitted. The target is normally a retroprism or reflective surface of some kind. As with theodolites and levels, calibration and careful levelling of the instrument is required prior to use.

Most EDM instruments include a microprocessor for calculating distance and height automatically. In some models the data can be stored on a datalogger which is later downloaded onto a computer for processing. One of the major advantages of using an EDM or total station is that it can be used over long distances (up to several kilometres). This considerably simplifies surveys of large rivers since the equipment does not need to be moved as frequently. It should be noted, however, that obstructions of any type can result in a weak signal. Heat waves can also cause interference; thus early morning or evening surveys are preferable on hot days.

The major disadvantage of these instruments is the price. They also require some prior training and experience in both care and use. Charged batteries and thus some logistical forethought are needed. They are also more difficult to pack into the field when compared with a clinometer, compass and tape.

Traversing: moving from one station to another

With a level, theodolite, plane table or total station, the instrument will often need to be moved because of obstructions such as hills, stream bends or vegetation or due to the limitations of the instrument. When the instrument is moved, each new location must be referenced to the previous one. This procedure is called **traversing**. To ensure that a few crucial measurements are not left out, a little redundancy and overlap is a good idea. For an effective survey the following precautions should be taken:

- At each position a marker should be placed in the ground directly below the instrument. The instrument height should be measured; with an EDM, the target height is measured as well.

- A sighting should be made to a reference point—if possible, one of known elevation such as a national survey marker. This can be used as an origin to which all points are referenced.
- Staff positions used for relocating the station are called **turning points**. Before moving from one position to the next, a **foresight** to the turning point should be taken. After the instrument is set up at the new location, a **backsight** is taken to the same turning point (see also Figure 5.14(b)).
- The last position should always be tied back to the starting point by taking a few measurements on the way back, to complete **closure**. This allows an estimate to be made of the error.

5.3 METHODS OF MEASURING AREAL EXTENT

5.3.1 GENERAL

The areal extent of various attributes such as different soil types, vegetation densities, fish cover or substrate composition can be measured by several sampling techniques. These measurements can be averaged to find percentage cover for individual categories or used to "ground truth" aerial photographs (Section 3.4), from which areas are measured. The methods which follow are general-purpose, and can be adapted for application in a variety of situations.

A stratified sampling design (Section 2.3.5) will be appropriate for most estimates of areal cover, with samples taken within a relatively homogeneous area. As mentioned in Section 2.3, randomness should be introduced at the proper level of sampling. For example, if an investigator wishes to sample specific areas, such as those meeting the spawning requirements of a particular fish species, it is much more effective to have a fisheries biologist identify spawning sites from which a random sample is taken than to sample randomly across a whole reach (Hamilton and Bergersen, 1984).

5.3.2 VISUAL ESTIMATION OF PERCENTAGE COVER

In stream hydrology, visual estimates of substrate sizes, vegetative canopy, snow or ice cover or amount of woody debris are often made. For example, one might look up through a tree canopy and estimate the percentage of the sky which is blocked from view. "Calibration" of one's eyes is crucial. This can be accomplished using visual charts such as the one shown in Figure 5.9.

It is important to first define the area within which an estimate is to be made; e.g. the area inside a grid cell, a hoop of known diameter thrown on the ground, or an area "as far as the eye can see" from a helicopter or

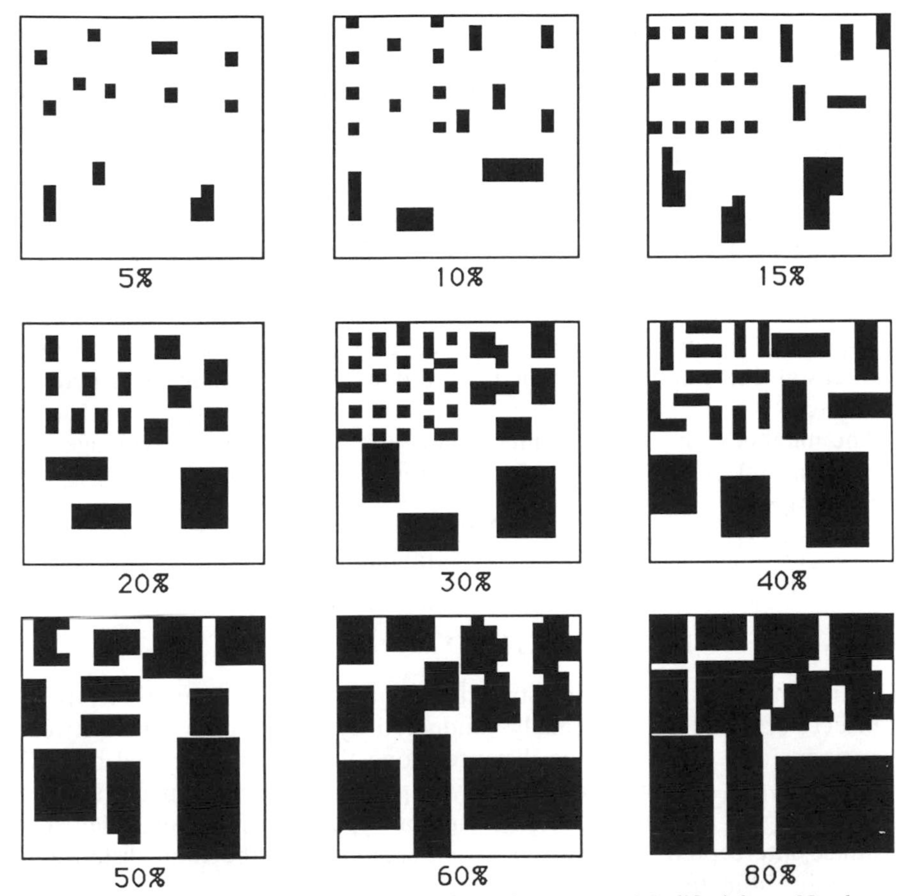

Figure 5.9. Chart for visual estimation of areal coverage. Modified from Northcote (1979) by permission of Rellim Technical Publications

car window. An investigator can become fairly proficient at visually estimating areal cover, reducing some of the subjectivity of this technique. It should also be realized that proficiency will change with experience, and periodic "re-calibration" is a good idea.

5.3.3 POINT INTERCEPT METHOD

The point intercept method is used for surveying along a transect. At set points along a horizontal line the object or feature on which the point falls—or "intercepts"—is noted. One method of doing this is to lay out a tape and use a constant interval between points. The "point" can be established simply by looking vertically downward over the tape or, more precisely, by vertically lowering a thin, sharpened metal rod from the edge of the tape, or by using a cross-hair telescope aimed downwards. A less rigorous technique consists of walking across a bar, bank or reach and stopping every few paces to note what is intercepted by the toe of one boot (a mark or notch can be added for more precision). In taking measurements, one must guard carefully against "selective drift" (e.g. to intercept more aquatic plants and fewer beer cans).

A point frequency frame, illustrated in Figure 5.10(a), is a more formal arrangement. Pins are dropped and the "hits" recorded. The height of each pin can also be measured to give an indication of surface topography. If vegetation is being measured, the pin may "hit" several layers as it is dropped. In estimates of percentage cover only the first hit is needed, whereas all hits might be recorded if the density of the plant itself is to be estimated.

The number of hits in a certain category is calculated as a percentage and used as an estimate of the relative amount of area covered:

$$\text{Percentage cover for category} = \frac{\text{Number of hits in category}}{\text{Total number of hits}} \times 100 \qquad (5.12)$$

5.3.4 LINE INTERCEPT METHOD

The line intercept method, illustrated in Figure 5.10(b), is another type of transect survey. This is normally conducted by extending a tape between two points and measuring the distance along the tape intercepted by each category type. Examples of category types are: different plant species, overhanging or eroded banks, different soil types, and pools or riffles. The distance intercepted is expressed as a percentage of the total distance to obtain an estimate of cover:

$$\text{Percentage cover for category} = \frac{\text{Distance intercepted}}{\text{Total distance}} \times 100 \qquad (5.13)$$

5.3.5 GRIDS

Grids can be used to define regions of known area for estimating areal cover. Superimposing a grid on an area before taking a photograph, for example, allows one to later analyse the photograph for percentage cover. This can be done by using a dot grid or planimeter (see Figure 4.5), by counting the number of grid nodes intercepted by different features, or by using digital image processing.

As shown in Figure 5.10(c), a 1 m square grid, sectioned into 200 mm squares, might be used to establish the percentage of different bed material sizes. Within each "cell" the average or dominant size would be measured or estimated. Alternatively, the areal coverage of each substrate category (e.g. boulder, gravel, sand; see Section 5.10) can be estimated within each cell. The areal coverage of algal mats or other aquatic plant species can be estimated in the same manner.

The average area of cover for a category is calculated either as the number of cells of that category divided by the total number or as the average of the cover in all cells:

$$\text{Percentage cover for category} = \frac{\Sigma(\text{area of cover in each cell})}{\text{Total area of grid}} \times 100 \tag{5.14}$$

5.4 SURVEYING STREAMS

5.4.1 GENERAL

To describe the physical characteristics of a stream reach, a basic survey should include a measurement of channel slope, several cross-section profiles representative of the channel form, a description of bed materials and a sketch of the reach. The reach selected should be long enough to include a full meander amplitude with two sets of pools and riffles, generally twelve to fifteen times the bankfull width (see also Section 2.3.4). Field data forms for recording stream survey data are given in Figure 5.11. A space is also provided for discharge measurements, which will be described in Section 5.7. Newbury (1984) further describes stream survey procedures for biological studies.

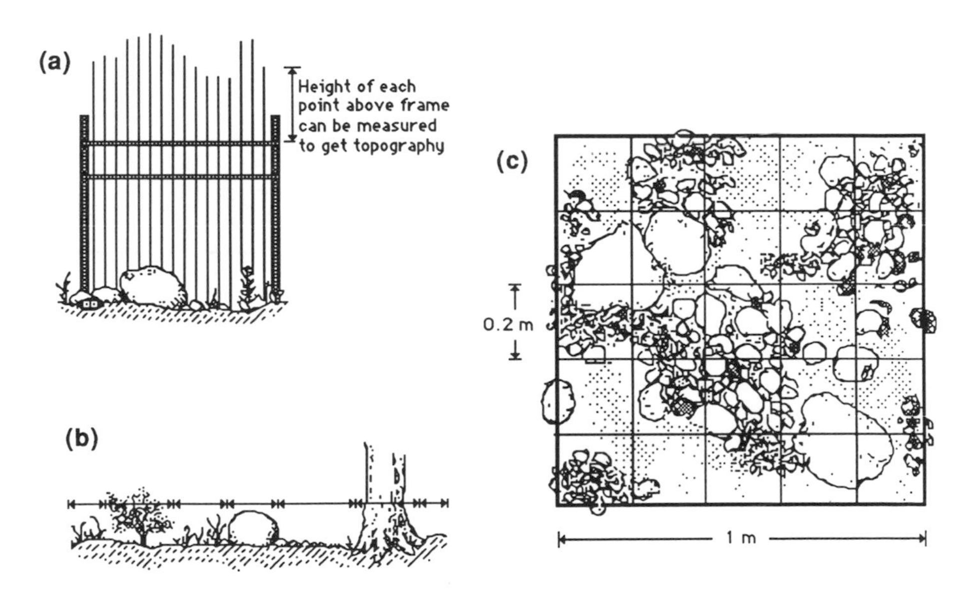

Figure 5.10. Methods of estimating areal cover: (a) a point frequency frame, (b) the line intercept method, and (c) a grid for estimating coverage of substrate types

5.4.2 CROSS-SECTIONAL PROFILES

Once a representative reach has been chosen, three to five sites for measuring cross-sections within the reach are selected. Fewer are needed in uniform reaches, more in complex channels. Sites can be located at random, spaced uniformly or selected as representative of a smaller area of the reach. For example, two sites in pools and two sites in riffles might be selected, with each site reflecting average conditions in terms of width, depth and bed topography. The number and spacing of cross-sections is a problem in statistical sampling and study design (Section 2.3). Figure 5.12 gives some of the terminology for describing channel dimensions at a cross-section.

A cross-sectional profile can be obtained with a dumpy level or theodolite, or in smaller streams, by using a measuring tape and metre rule or survey staff. If the stream has water in it, the water surface provides a horizontal

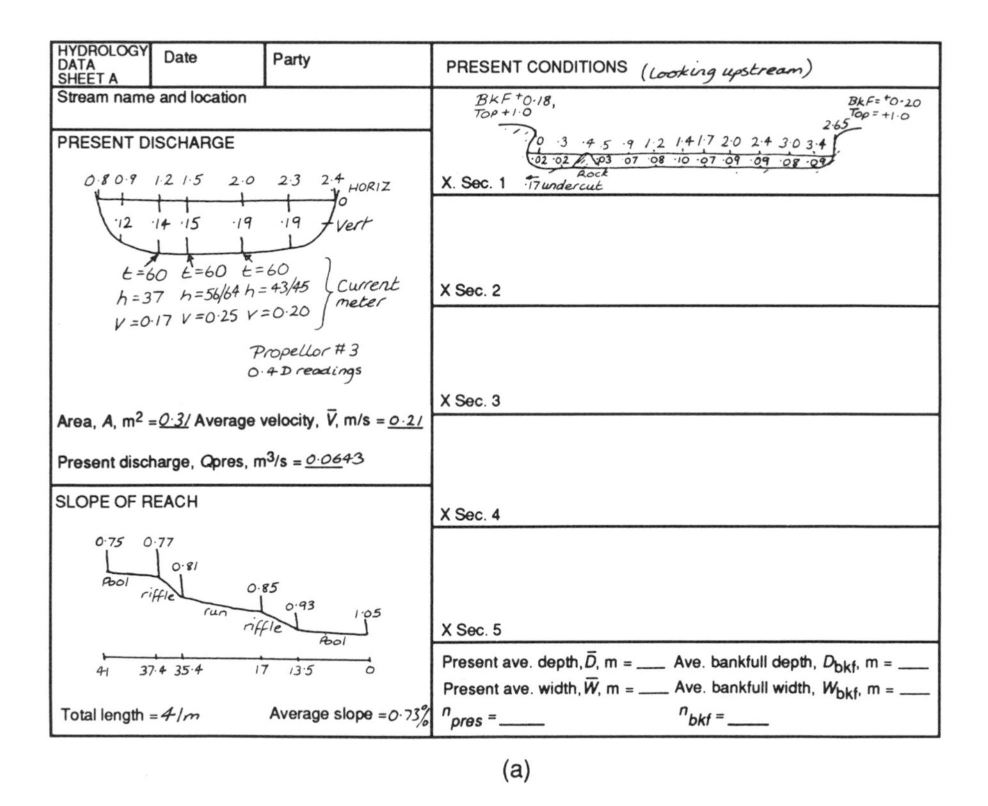
HYDROLOGY DATA SHEET A | Date | Party

Stream name and location

PRESENT DISCHARGE

0·8 0·9 1·2 1·5 2·0 2·3 2·4 HORIZ
0
·12 ·14 ·15 ·19 ·19 Vert

t=60 t=60 t=60
h=37 h=56/64 h=43/45
V=0·17 V=0·25 V=0·20
Current meter

Propellor #3
0·4D readings

Area, A, m^2 = 0·31 Average velocity, $\bar{V}$, m/s = 0·21

Present discharge, Q_{pres}, m^3/s = 0·0643

SLOPE OF REACH

0·75 0·77 0·81 0·85 0·93 1·05
Pool riffle run riffle Pool
41 37·4 35·4 17 13·5 0

Total length = 41m Average slope = 0·73%

PRESENT CONDITIONS (Looking upstream)

X. Sec. 1
BkF +0·18, Top +1·0
BkF = +0·20, Top = +1·0
2·65
0 ·3 ·4 ·5 ·9 1·2 1·4 1·7 2·0 2·4 3·0 3·4
·02 ·02 ·03 ·07 ·08 ·10 ·07 ·09 ·09 ·08 ·09
Rock
·17 undercut

X Sec. 2

X Sec. 3

X Sec. 4

X Sec. 5

Present ave. depth, $\bar{D}$, m = ___ Ave. bankfull depth, D_{bkf}, m = ___

Present ave. width, $\bar{W}$, m = ___ Ave. bankfull width, W_{bkf}, m = ___

n_{pres} = ___ n_{bkf} = ___

(a)

Figure 5.11. Partly filled forms for collection of physical data at stream survey sites. These forms may be freely copied. Modified from Newbury and Gaboury (1988), by permission of *Canadian Water Resour. J.*

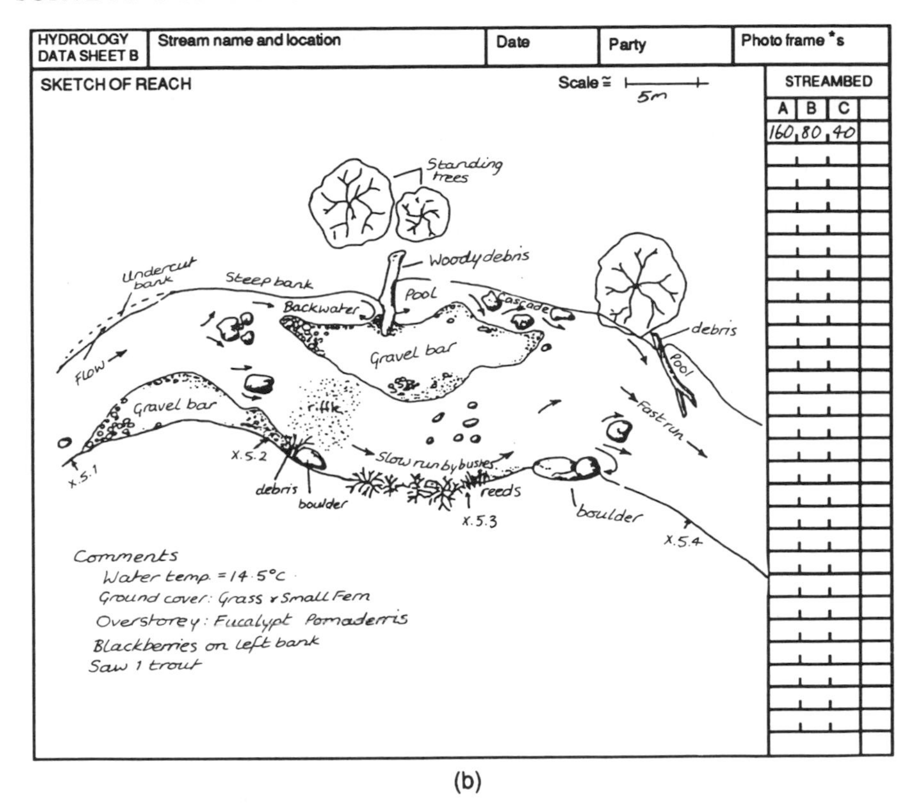

(b)

Figure 5.11. (Continued)

surface from which to take vertical measurements. If it is dry, vertical measurements can be taken in reference to a pole or tape held horizontally over the streambed or by using a hand-held level.

For those who prefer "streamside" surveys, Molloy and Struble (1988)

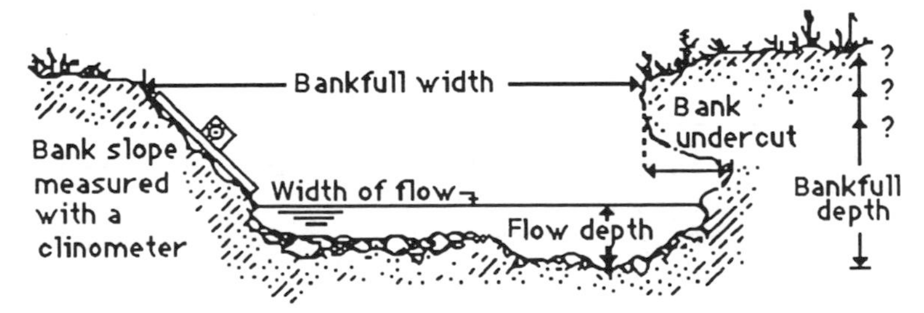

Figure 5.12. Field measurement of a stream cross-section. The definition of bankfull depth is based on field interpretation or geomorphological definitions (see Section 7.2.5)

describe a method of tossing a marked nylon string to the opposite bank to define horizontal distance, and lowering another marked, weighted string from a pole to measure water depth. This method would be appropriate for small, deep stream reaches of low velocity.

With any type of equipment, the procedure of surveying cross-sections is basically the same: vertical measurements are taken at several points along the horizontal line. The horizontal distance to the measurement point and the vertical distance to the streambed and water depth are recorded. Measurements should be taken at each break in slope along the bed, as shown in Figure 5.13. The depth of water at each edge should also be recorded. If the study involves monitoring of channel changes the survey should be continued past the edges of the active channel and out to permanent markers.

Program XSECT in AQUAPAK will plot cross-sectional profiles using present water depths as the vertical heights. The program will calculate area, wetted perimeter, hydraulic radius and a hydraulic depth for each section and as an average for the reach (see Section 6.6.1 for a description of these terms).

Bank slope and bank overhang can also be recorded. Both are indicators of bank stability, and bank overhangs are particularly important as shelter areas for fish. **Bank slope** is best measured using a staff and clinometer (Platts et al., 1987). The clinometer is held against the staff which is set against the bank, and the angle is read directly from the clinometer. **Bank overhang** is measured with a staff or metre rule from the farthest point of undercut to the most distant point of overhang. At surveyed sites on the Acheron River (Figure 4.7), overhangs of 0.30 m were quite common.

Bankfull width and **depth** may actually be more important than the area presently occupied by the stream (see Figure 5.12) and these measures provide a more standardized description of channel dimensions. At the edges of each cross-section, measurements should be made from the water surface to the bankfull level (or, if the channel is dry, these levels are surveyed). The bankfull width should also be noted at each cross-section. In the field, the bankfull elevation is identified by scour lines, vegetation

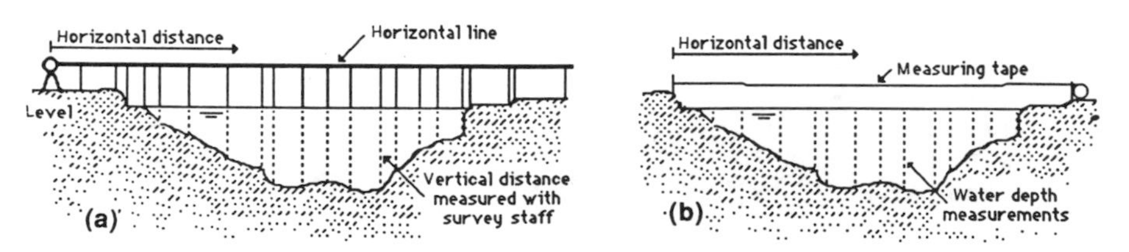

Figure 5.13. Surveying stream cross-sections: (a) with a level and staff and (b) with a measuring tape and rule, using the water surface as a horizontal line. Measurements are taken at "breaks" in the slope of the channel bed and at the edges of the stream

limits, changes between bed and bank materials, the presence of flood-deposited silt or abrupt changes in slope. This involves subjective judgement; however, training and experience will lead to remarkably consistent interpretations. A discussion of bankfull definitions is included in Section 7.2.5.

If a flood has recently passed through a reach, it may be of interest to estimate the peak discharge from **high water marks** or **trashlines**. These are indicated by tell-tale signs such as flattened vegetation, truncation of lichen growth, debris (driftwood, small seeds, strands of dried grass, plastic bags, etc.), stain lines on walls and/or coarser bedload sediments deposited at a well-defined elevation along the bank. Flood marks should be interpreted with caution; for example, water will "pile up" on the upstream face of walls or trees (see Section 6.6.4) and marks can wash away in rains. The slope–area method for estimating discharge from high water marks is given in Section 5.8.2.

5.4.3 CHANNEL SLOPE AND THALWEG PROFILE

When measuring channel slope it is important to differentiate between the slope of the streambed and that of the water surface, as the two are not necessarily the same. Also, depending on the detail of the study, the localized slope of a water surface over a rock or a pool might be measured rather than the average slope over a reach. Slope measurement is covered in Section 5.2.4.

The average slope of the water surface or channel bed can be surveyed by taking measurements at the upstream and downstream ends of a reach. If the channel has pool–riffle or pool–step sequences, slopes can also be measured separately in each type of channel form. A sample of how slope is measured in a stream reach with pool–riffle topography is shown on the data sheet in Figure 5.11(a).

The **thalweg** profile, or the path of the deepest thread of water, can be surveyed by having the staff bearer follow this path downstream. Similar to the procedure for cross-sectional profiles, measurements of channel bed and water surface elevation should be taken at distinct breaks in both channel bed and direction of the flow path, noting the location of "knickpoints" created by logs, boulders or dams. The data can be plotted as a longitudinal profile (Section 4.2.6), showing both the water elevation and the streambed elevation.

5.4.4 BED SURFACE MATERIALS

On the second field data sheet shown in Figure 5.11(b) spaces are provided for recording measurements of rock sizes to estimate channel bed resistance. About 25 to 30 rocks may be sampled by walking upstream and measuring the larger rocks at each pace or two (the distance is adjusted appropriate to the reach length) (Newbury, 1984). Measurements are normally made along three axes, as described in Section 5.10.4. Again, the number of rocks sampled is a problem in sampling design, with more rocks needed where sizes are more variable.

The measurements can be used to describe the surface streambed materials in terms of particle shape (Section 5.10.6), to estimate streambed roughness coefficients (Section 6.6.5) and to estimate whether these surface materials will move at different flow levels (Section 7.4.3).

5.4.5 MAPPING THE STREAM REACH

Maps of a stream reach can provide a useful store of information from which changes in stream morphology can be documented and the extent of various habitat types determined. Detailed survey information is helpful in the mapping of specific habitats such as larger boulders or areas of spawning gravels. Maps can be constructed in the field by freehand sketching or plane table mapping (Section 5.2.5), or they can be created in the office by plotting survey data on graph paper or as a computer-generated plot. At an "overview" scale, copies of aerial photographs and/or topographic maps can be taken into the field and details such as the location of sampling sites or unmapped features can be added in.

A freehand sketch can provide a wealth of information about a stream reach and its surroundings. Just the process of drawing will force people to become better observers, and notes made in the field are often found to be valuable when sifting through the data back in the office. The drawing scale should be adjusted to the map size and the length of the reach. For example, for a 300 mm standard sheet, a 1 m wide stream could be mapped at a scale of 1:60 and a 10 m wide stream at a scale of 1:600, with 50 mm for margins. A sketch might include the location of riffles and pools, large woody debris, bedrock outcrops, gravel bars, abandoned channels, undercut or trampled banks and visual assessments of vegetation type (aquatic and riparian), substrate size and the character of surrounding slopes. The scale, orientation, date, and names of field personnel should be included on the map. An example sketch is shown in Figure 5.11(b).

Another method of mapping is to use a longitudinal transect. A tape is extended down the axis of the stream, as shown in Figure 5.14(a), and the distance from the transect line to the bank, thalweg or other features is measured from it at right angles. The right angle can be fixed by carrying a plastic drafting triangle, by using the properties of the 3-4-5 triangle (see Figure 5.1) or simply estimated, depending on the precision required. A compass reading should be taken along the direction of the transect tape each time it is moved so the readings can be tied together.

With a level, theodolite or EDM, the distances and angles are measured

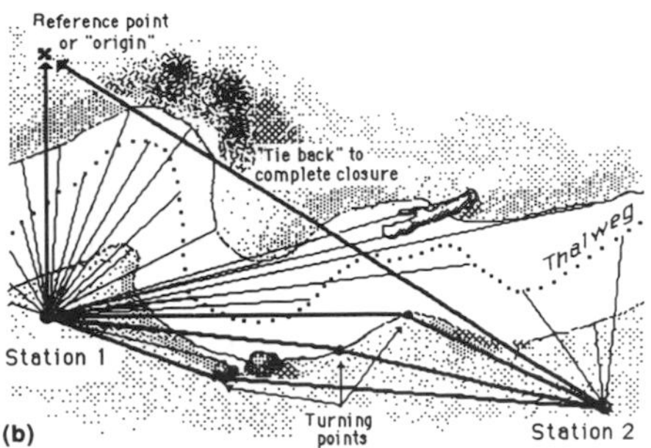

Figure 5.14. Mapping the stream using: (a) longitudinal transect or (b) a level, theodolite, total station or plane table

to various points in the stream reach, as illustrated in Figure 5.14(b). Each set of measurements is recorded with a label (e.g. "beginning of snag", "end of pool"). These data, together with a freehand sketch of the reach, are later used to construct the scaled map. The same procedure is used with a plane table, although the map is constructed on-site (see Section 5.2.5).

When mapping a feature it is important to take enough measurements to sufficiently describe its position. If a large number of horizontal and vertical measurements are taken in a stream reach a "topo map" of channel topography can be developed by drawing contour lines between points of equal elevation. This information can also be depicted by plotting "isobaths", lines of equal water depth which correspond to a particular stage (see Figure 9.4). In this manner the distribution of shallow and deep areas can be illustrated.

5.5 MEASUREMENT OF WATER LEVEL OR STAGE

5.5.1 GENERAL

Whereas cross-section measurements will yield the current water level and estimates of bankfull conditions, water level measurements are needed over a period of time in order to more fully describe the hydrologic environment. This monitoring can be extended outward and/or downward from the stream channel to include measurement of groundwater levels.

The water level in reference to some arbitrary datum is generally referred to as the **stage**. Periodic measurements of stage in streams are converted to streamflow values with the use of a stage–discharge relationship (Section 5.8.1). If a stage–discharge relationship is to be developed the gauge must be carefully situated at a control section (Section 5.6).

5.5.2 THE STAFF GAUGE

A **staff gauge** is simply a graduated staff which is installed vertically at a relatively stable site on the streambank. Several staff gauges may be installed, progressing up the bank at fixed intervals of elevation, as shown in Figure 5.15. The gauge should be located where wave action and turbulence is minimized, and where it will be protected from damage by boats, vandals and flood-borne debris. It may be fixed to a wall, bridge pier, post, etc., as long as part of the scale is immersed at the lowest expected water level. A study of the gauge location should account for future erosion and deposition. Locations near bends in the stream, in pools or steep reaches are particularly vulnerable. Maintenance of staff gauges involves cleaning, replacement and checking the datum of the gauge scale.

Staff gauges must be read manually. At flood stages, binoculars may be helpful. Daily readings of stage form the basis for many stream-gauging records, older records in particular. They can also provide useful information before permanent stream gauging locations are established.

5.5.3 MAXIMUM STAGE RECORDERS OR CREST GAUGES

An estimate of the maximum height reached by a stream between visits to the site can be obtained with a **maximum stage recorder** or **crest gauge**. Somewhat like a max-min thermometer, some type of substance is floated upwards when the stage rises, and then remains in place when the water level drops again. Cork grains, coffee grounds, and plastic or styrofoam chips are inexpensive substances for this purpose.

A length of transparent plastic pipe can be used to contain the floating material. Depths are marked on the pipe, corresponding to the same datum as the staff gauge. The pipe is set vertically in the streambank and perforated at the base to allow water to enter. To keep debris out of the pipe, the holes can be covered with a screen. The top of the pipe is also capped, but not sealed airtight. When a reading is taken, the height of the floating material in the pipe is recorded and the gauge "reset" by washing the material back down inside the pipe. Shaw (1988) also describes a standard crest gauge made of a 55 mm steel tube, with a removable graduated rod inside.

5.5.4 AUTOMATIC RECORDERS

Permanent gauging stations are normally equipped with **water-level recorders** which allow the continuous recording of stage. A **stilling well**, connected to the stream by one or more pipes, is normally used to "still" wave action and thus provide a more stable reading of stage. In most installations a float sits on the water surface and its vertical movements are recorded digitally or by an ink trace on a chart. At some stations digital readings are telemetered to a base station for near-instantaneous access.

Figure 5.15. A series of staff gauges for measuring water level over a range of elevations. Photo: N. Gordon

Bubbler gauges are used at some stations instead of floats and stilling wells. In these gauges, nitrogen gas is bubbled slowly out of a tube, and the pressure required to force out the gas is related to the water depth. Pressure transducers can also be used to measure hydrostatic pressure, which is directly related to water depth (Section 6.2).

Continuous readings from a stage recorder can provide useful insight into daily and seasonal fluctuations. To obtain good-quality data the gauge should be carefully located (see Section 5.6), checked frequently (once every week or two) and records processed in a timely manner and compared with other gauging records during events. Brakensiek et al. (1979) provide more information on recording devices and their installation, calibration and maintenance.

5.5.5 DEPTH TO THE WATER TABLE: PIEZOMETRY

Piezometers are useful for monitoring the profile of the water table near a stream to establish whether it is influent (feeding the groundwater) or effluent (being fed by groundwater). A piezometer is simply a piece of pipe, driven vertically into the ground to form a "well". Water levels in the piezometer(s) can be measured and related to the water level in the stream to determine the direction of water movement.

The length of pipe should cover the range of water table levels expected. One to four metres of 13–25 mm i.d. (inside diameter) PVC pipe is a workable size. A hole just slightly larger than the pipe diameter is first drilled with an auger or by driving a metal pipe into the soil with a fence post-type driver. The piezometer is placed such that the lower end is below the lowest expected water level. A cavity of 70–100 mm in length should be left below the pipe. To keep the pipe from plugging with soil, a wooden rod can be inserted inside the pipe when it is installed, and then removed. The cavity should then be flushed out by lowering a plastic tube to the bottom of the pipe and gently siphoning water into the piezometer. Loosened soil will be carried out of the top of the pipe when it overflows.

The response rate of the piezometer is then tested by observing the rate at which the water level drops. In sands and gravels the rate will be so great that no overflow will occur during flushing; in clays, the drop will just be perceptible. Piezometers should be capped (not air-tight) to keep dust, leaves and insects out, and flushed and re-tested periodically.

The piezometer water level is allowed to come to equilibrium with the groundwater level before measurements are taken. Depth from the top of the pipe to the water surface can be measured by lowering a chalked tape into the piezometer (the chalk washes off to indicate the water level). Reeve (1986) also gives instructions for making a "bell sounder" (also called a "fox whistle"), a weight with a hollowed base which makes a sound upon contact with the water. Another alternative given by Reeve (1986) is to lower a length of plastic tubing, marked at length increments, until it encounters the water. Contact can be discerned by blowing into the tube and listening for the sound of bubbles.

The heights of the stream water surface and the tops of all piezometers should be tied back to a reference datum by surveying (see Section 5.2). If two piezometers are installed, one close to the stream and another some distance away, the relative heights of the water table will indicate whether the stream is influent or effluent. In an influent stream, shown in Figure 5.16, the water level will be highest near the stream. The converse would be true if the stream were effluent.

5.6 DISCHARGE MEASUREMENT: CHOOSING A CONTROL SECTION

To provide the best possible accuracy in discharge measurements, careful selection of a gauging site is crucial. This is particularly true if a staff or recording gauge is installed for monitoring streamflows over time. Ideally, the beds and banks of the cross-section should be stable to ensure a constant relationship between water depth and streamflow. The site should also provide good sensitivity in terms of the response of stage readings to changes in discharge.

The selection of a site requires the identification of a channel section with particular physical characteristics called the **control section** (see also Section 6.6). This regulates and stabilizes the flow so that for a given stage the discharge will always be the same.

Controls can either be artificial (e.g. weirs) or natural. A **section control** exists at a cross-section which constricts the channel, or where a downward break in slope occurs (for example, at the brink of a rock ledge, a boulder-covered riffle or a weir). A **channel control** exists when the friction along a section of the stream controls the stage–discharge relationship. Controls may govern this relationship throughout the entire range of stage, or a compound control may exist where the control changes. For example, a section control at low stages might change to a channel control at higher

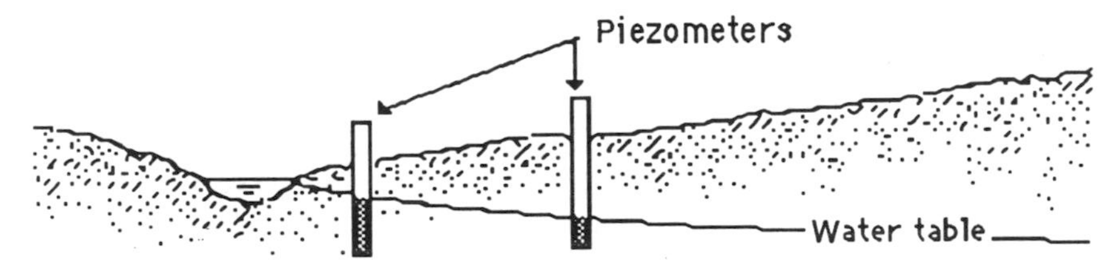

Figure 5.16. Piezometers installed next to an influent stream

ones when the constriction, weir or break in slope is "drowned out". Thus, the flow profile is affected by the control section, yet the flow determines whether a section is controlling or not (Henderson, 1966).

The gauging site is situated upstream of the control section, in a relatively uniform reach. The reach should be straight and of fairly uniform slope, bed material and cross-section for a distance upstream of about four to five times the channel width and downstream for about twice the channel width. Improvements can be made to the site by building rock dikes to cut off dead-water sections or by rearranging rocks, plants and debris to make the reach more uniform. The site should not be located just above a confluence, since varying discharges in the tributary may influence gauge readings through backwater effects. At semi-permanent installations the stream channel should also be deep enough to contain the larger flood flows.

Since accuracy will vary, the control section and gauging site should be chosen so that the rating curve is sensitive and stable over the region of most interest: high, medium or low flows. More information on control sections for gauging sites can be found in Corbett (1962).

5.7 DIRECT MEASUREMENT OF DISCHARGE

5.7.1 GENERAL

Stream discharge is the rate at which a volume of water passes through a cross-section per unit of time. In the SI system it is usually expressed in units of cubic metres per second (m^3/s), although very small flows might be recorded in litres per second (l/s). It should be remembered that measured discharge includes not only the water but any solids dissolved or suspended in it. Especially during flood flows, this may affect discharge totals by a significant amount. The selection of good measurement sites requires careful planning and evaluation.

Methods of directly measuring streamflow generally fall into four categories: (1) volumetric measurement; (2) employing some measure of average stream velocity and cross-sectional area; (3) dilution gauging using a salt or dye; and (4) the use of artificial controls such as weirs, with known stage–discharge relationships. The velocity–area method using current meters will be employed in most cases and floats can be used for coarse estimates, as during floods. Volumetric measurements are most appropriate for small flows, dilution gauging for turbulent flows and artificial structures for more permanent gauging sites. The procedures are given extensive treatment in manuals on stream gauging developed by BS (BS, 1973), USGS (Corbett, 1962), and WMO (1980). Herschy (1978, 1985) also provides a summary of streamflow measurement methods.

5.7.2 VOLUMETRIC MEASUREMENT

Volumetric measurement is the most accurate method of measuring very small flows such as those from a spring. It involves measuring the time taken to fill a container of known volume. Volumetric measurements are usually made where the flow is concentrated (for example, at a weir, overfall or pipe). If these are unavailable it may be possible to build a small temporary dam with a "trough" for collecting the streamflow. In this case the water level behind the dam should be allowed to stabilize before a sample volume of water is collected.

Discharge (Q) is simply calculated as:

$$Q = \frac{V\!\!\!/}{t} \tag{5.15}$$

where: Q = discharge in m^3/s (or l/s), $V\!\!\!/$ = volume in m^3 (or litres) and t = time (s).

The container should be calibrated by putting graduated marks on the side so that readings can be readily taken in the field. Alternatively, the container can be weighed in the field and the weight of the water (after correcting for container weight) is related to volume $V\!\!\!/$ by dividing it by density (see Table 1.2):

$$V\!\!\!/(\text{m}^3) = \frac{\text{Weight of water (kg)}}{\text{Density of water (kg/m}^3\text{)}} \tag{5.16}$$

Several measurements should be taken to check for consistency of both the streamflow and the measurer.

5.7.3 VELOCITY–AREA METHOD

This method requires measurement of the area of a stream cross-section and the average stream velocity. Discharge is then calculated as:

$$Q = VA, \tag{5.17}$$

where Q = discharge (m^3/s), V = average velocity (m/s), and A = cross-sectional area of the water (m^2). Area is calculated from cross-section measurements. Velocity measurements may be made by observing the rate of travel of a float or dye, or with current meters.

Floats

Water velocity can be measured by observing the time required for a floating object to traverse a known distance downstream. This method is appropriate for coarse estimates of discharge, particularly during floods, and requires little time or equipment. The "float" may be a specially designed surface or subsurface float, a chunk of ice, drifting branches or logs, coffee stir sticks, half-filled bottles or oranges. Even leaves can be used in emergencies, although these are apt to be affected by wind. Ideally, the float should move with the same velocity as the water at the stream surface. Objects such as oranges or orange peels or ice cubes which are only slightly buoyant are preferred as they are less sensitive to air disturbances. A fishing line can be attached for the retrieval of floats or they can be caught in a net.

A suitable, straight reach with a minimum amount of turbulence is chosen, and an interval is selected, measured and marked on the bank at each end by pegs, trees or rocks. These marks should be far enough apart to allow a travel time of at least 20 s. The interval should also overlap one or more surveyed cross-sections to determine the cross-sectional area.

A float is introduced a short distance upstream of the reach so it can reach the speed of the water before passing the first mark. A stopwatch is used to measure the time of travel between the marked sections. In larger streams (greater than 10 m width) the cross-section should be divided into three or more subsections. Floats are introduced at the midpoint of each section. Several runs should be made to obtain an average.

The surface velocity (V_{surf}) is calculated for each section as:

$$V_{surf} = \frac{L}{t} \tag{5.18}$$

where L = measured reach length (m), t = travel time (s) and V_{surf} is in m/s. Since the surface velocity is higher than the mean velocity (see Chapter 6), a correction coefficient (k) must be applied:

$$\bar{V} = kV_{surf} \tag{5.19}$$

The correction coefficient generally ranges from 0.80 for rough beds to 0.90 for smooth artificial channels. However, in mountain streams, Jarrett (1988) calculated a value as low as 0.67. The commonly used coefficient is 0.85.

Discharge is calculated from Equation 5.17 or, if more than one subsection is used, from Equation 5.20. Under favourable conditions and with repeated observations, float measurement may be accurate to within an error margin of 10%. In non-uniform sections or where wind is excessive, measurements may be in error by 25% or more.

Dye

As in the float method, coloured dye can be poured into the water and its movement timed. Fluorescein dye (see Section 5.7.4) is useful for this purpose. The major disadvantage with this technique is the inaccuracy in identifying the middle of the mass of dyed water. However, as compared to the float method, a measure of the average stream velocity is obtained rather than the surface velocity. Hence, a correction factor is not necessary.

Current meters

Current meters may be used in streams of almost all sizes. In larger streams with swift waters and debris flows these meters are larger with heavy attachments. There are three types of current meters commonly used in the measurement of streamflow: (1) propeller, (2) cup and (3) electromagnetic. A listing of current meters, suppliers and prices is provided by Hamilton and Bergersen (1984).

The first two types of meter work somewhat like a wind gauge, where the fluid moving past a vane causes it to rotate and the velocity is calculated from the rate of rotation. The **propeller-type meter** has a horizontal axis rotor. Propellors come in various sizes for operation over different ranges of velocity. These meters generally disturb the flow less and are less likely to become entangled with debris than the **cup-type meters**. These have a vertical axis rotor and are typically more sensitive at lower velocities. However, they will also rotate when buffeted by non-horizontal currents, leading to error in turbulent sections. A recent development mentioned by Jarrett (1988) is a meter with solid cups which should reduce this error.

With the rotary-type meters the number of revolutions is counted either by some type of recording apparatus or by listening to clicks heard through headphones. For low speeds (and as a "back-up system") a mark can be made so that revolutions can be visually observed. Counts are made over some time interval, which is measured with a stopwatch, and the rotor speed is calculated as the number of revolutions per second. Current meters are normally supplied with a rating curve or table giving the relationship between rotor speed and flow velocity. Modern instruments may have direct readouts which display velocity. Maintenance includes keeping the meters cleaned and lubricated and, when necessary, recalibrated.

Electromagnetic meters (Figure 5.17) work on the principle that a conducting fluid moving through a magnetic field will induce a voltage (White, 1986). Since salty water is a high-conductivity fluid, the meters are commonly used in oceanography. More sensitive instruments are needed for measurements in fresh water, and this type of meter may be impractical for extremely pure waters. These meters provide a direct reading of velocity. They are also very durable and can be used in situations where rotary meters cannot be operated such as within clumps of vegetation. The bulb portion should be kept smooth and polished as nicks can cause flow separation (Section 6.5.4).

Current meters should be recalibrated on a routine basis or whenever poor performance is suspected. Commercially, this is accomplished by using

Figure 5.17. Electromagnetic current meter, showing readout display unit and base of top-setting wading rod. Photo: N. Gordon

a towing tank, in which the meter is attached to a carriage and towed at a set speed. This procedure can be roughly imitated by towing the meter through a pool of still water. A standard length would be measured off alongside the pool and the person holding the meter would walk or run over the length while being timed. This procedure would be repeated at varying travel speeds to check the calibration.

Current meters can be operated from a bridge, boat, cableway, ice cover or by wading. The type of current meter chosen should be appropriate for the size and velocity of the stream. Most standard meters require a minimum flow depth of 0.15 m and velocities over 0.10 m/s. A fairly straight reach should be chosen in which flow depth, width, velocity and slope are relatively uniform (see also Section 5.6). Sites should be avoided which have extreme turbulence, upstream obstructions, eddies, dead-water zones, divided channels or regions where the flow path is noticeably curved. A reach which is slightly contracting in the downstream direction is preferable to one which expands outwards. Reaches can be "improved" by moving rocks to make conditions more uniform and to eliminate backwater areas at channel edges.

After the cross-section has been selected, a measuring tape or **tagline** is strung across the section perpendicular to the flow. The stream is divided into subsections within which velocity is measured along **verticals**, shown in Figure 5.18. Each subsection should have roughly the same amount of flow;

thus verticals will be spaced more closely if the water is faster and deeper. Additional verticals are added where sudden changes in depth or velocity occur. The standard procedure for hydrographers is to use a minimum of 20 verticals (Brakensiek et al., 1979; Goudie, 1981; WMO, 1980). This is an excessive number for most surveys for biological purposes, especially on smaller streams. A practical guideline is to use about one vertical per metre of channel width, with more if the section is irregular and less if it is uniform.

In the measurement of streamflow it is the **mean velocity** within each vertical which is of interest. Under the assumption that water velocity varies logarithmically from zero at the streambed to a maximum near the stream surface (Chapter 6), the mean occurs at about four-tenths of the depth ($0.4D$), as measured upwards from the streambed. Most texts refer to this as "six-tenths depth", in which case it is measured downwards from the water surface. However, depth is more easily measured upwards from the streambed and the meter is set more easily by measuring up from the base, and this will be the convention adopted here.

For measurements in small, uniform streams or where ice cover is present, or when time is limited, the single measurement at $0.4D$ is used. Because velocity profiles are often not exactly logarithmic, it is preferable to take two measurements at $0.2D$ and $0.8D$. As a rule of thumb, the $0.8D$ velocity should be greater than the $0.2D$ velocity, but less than twice as great. Three measurements should be taken if the velocity profile is distorted by overhanging vegetation contacting the water or by large submerged objects. Even more measurements can be taken for more precision or for developing a velocity profile. If for some reason it is difficult to lower the meter into the water a surface velocity measurement can be made and adjusted as for the float method. These depth settings and equations for calculating mean velocity at a vertical arc summarized in Table 5.1.

For most stream survey work the meter is attached to a graduated wading rod, with a sliding mechanism which allows for depth adjustment. A **top-setting wading rod** can be purchased which greatly facilitates setting the current meter at $0.4D$.

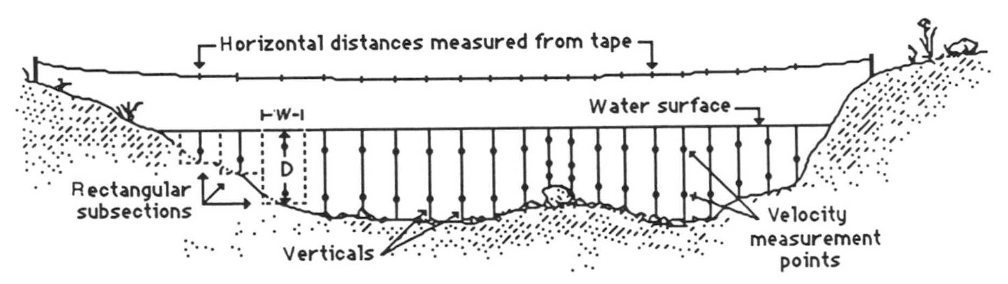

Figure 5.18. Definition of terms used in computing discharge from current meter measurements (see text). Note variable spacing of verticals

Table 5.1. Applicability and averaging procedures for determining mean velocity at a vertical. D is the vertical distance between the water surface and the streambed, and is measured upwards from the streambed. Adapted from Goudie (1981) and WMO (1980)

Number of points in vertical	Depth of measurement (from streambed)	Application	Equation for mean velocity in vertical $\bar{v}$
1	$0.4D$	When D is small (< 0.5 m) or when a measurement must be made quickly	$\bar{v} = v_{0.4}$
2	$0.2D$ and $0.8D$	This method is preferable to the single $0.4D$ method if the size of the meter allows both measurements (normally where $D > 0.5$ m)	$\bar{v} = 0.5(v_{0.2} + v_{0.8})$
3	$0.2D$, $0.4D$ and $0.8D$	Where irregularities distort the velocity profile and the stream depth is sufficient	$\bar{v} = 0.25(v_{0.2} + v_{0.8} + 2v_{0.4})$
1	Just below water surface (≈ 0.6 m, or lower to avoid turbulence)	In swift streams or in high flows when it is difficult to lower meter into the water	$\bar{v} = kv_{\text{surface}}$, where k depends on shape of velocity profile, normally taken as 0.85. k is best determined by correlating surface and $0.4D$ velocities
Many	A range of depths, incl. $0.2D$, $0.4D$ and $0.8D$	When high precision or the shape of the velocity profile is of interest	$\bar{v}$ is determined by integrating the area bounded by the velocity profile, and dividing by D

To take readings the meter is set to the appropriate depth and then placed in the stream, with the rod held vertically and the meter facing upstream. The observer stands as far to the side of the meter as possible and slightly downstream. The meter should be allowed to stabilize before readings are taken. If clicks are counted by ear, a stopwatch is started with the first click, counted as "zero", not "one". For rotary meters, measurement time is typically between 30 and 60 s. Enough time should be allowed to count 40 "clicks". For high velocities, as little as 10 s may be adequate.

At each vertical, the horizontal distance, the water depth and the current meter reading(s) should be recorded. To define the cross-section additional horizontal distances and water depths should be recorded at the water's edge, the deepest point and any other points where the topography changes (see entry under "Present Discharge" in Figure 5.11(a)). If the angle of the flow across the tape differs appreciably from 90° the angle should be recorded and the velocity reading corrected by multiplying by the cosine of the angle.

To calculate discharge, meter readings are first converted to velocities, if necessary, and averages for each vertical are calculated using the equations in Table 5.1. The mean velocities and areas for each subsection are multiplied together and then summed to obtain the total discharge. One way to do this is to assume that the subsections are rectangular, defined by the depth at each vertical and divisions halfway between each vertical, as shown in Figure 5.18. The "lost" discharge in the triangular areas at the edges is usually assumed negligible. Discharge (Q) is then calculated as:

$$Q = w_1 D_1 \bar{v}_1 + w_2 D_2 \bar{v}_2 + \cdots + w_n D_n \bar{v}_n \quad (5.20)$$

where w is the width in metres, D is the depth of the vertical in metres, $\bar{v}$ is the average velocity at each vertical (m/s) and Q is in m^3/s. This method simplifies hand calculations. Program QCALC in AQUAPAK will calculate discharge by the velocity–area method. However, it uses irregular areas which are bounded by the cross-sectional profile rather than rectangular areas. The boundary between subsections is halfway between the verticals.

Because of the variation in velocity within a stream, the accuracy of discharge measurement depends largely on the number of points at which velocity and depth readings are taken. Turbulence will also create "noise" in the readings, although it is assumed that this error is random and will not cause bias with the large number of measurements taken. Most gauging authorities cite accuracies of plus or minus 5–10%. In turbulent mountain streams, however, Bren (pers. comm., 1988) estimated an error of about 25% in current meter readings. During rapidly rising flood stages the error will be even higher. WMO (1980) gives recommendations for measurement of quickly changing flows such as flash floods.

5.7.4 DILUTION GAUGING METHODS

Dilution gauging methods involve introducing a chemical "tracer" substance such as a salt or dye into the stream and then monitoring changes in its concentration at some point downstream. These methods are especially useful in highly turbulent rock-strewn streams which provide rapid, complete mixing of the tracer and make conventional current metering difficult. Several such methods have been developed and are reviewed by Gregory and Walling (1973). Church (1974) and Finlayson (1979) also provide details on several of the techniques.

In addition to providing a valuable means of measuring total discharge in turbulent mountain streams and braided rivers, chemical gauging techniques also lend themselves to specialized discharge measurements. These include locating "dead-water zones" where there is little transfer of flow with the main stream, monitoring water flow through the hyporheos (Section 2.1) and tracing downstream movement of nutrients or heavy metals. Zellweger et al. (1989), for example, compared discharge measurements made by tracer-dilution and current meter methods in a small gravel-bed stream. They concluded that a significant portion of the total stream discharge moved as underflow through the streambed. Triska et al. (1989a,b) also gives examples of ecological applications.

The two most frequently used methods of dilution gauging are **slug-injection** and **constant rate-injection**. The types of tracers will be discussed first, followed by procedures for these two techniques. With either method the selected reach is preferably straight, with well-developed turbulence, and of sufficient length to allow complete mixing and dispersion of the tracer fluid throughout the stream water. This length can be surprisingly long. It is recommended that a dye such as fluorescein, a red powder which produces a green colour when dissolved in slightly alkaline water, be used to visually assess mixing.

Tracers

A tracer is a substance which can be followed in its downstream course by monitoring some characteristic such as its concentration. There are three main types of tracers used: chemical, fluorescent and radioactive. Radioactive tracers require specialized training and will not be covered here. Additionally, it may be possible to use natural physical properties such as temperature, salinity, ion concentrations and sediment to study the origin of stream water (e.g. as baseflow) or the mixing of two merging streams. For hydrological purposes the tracer should preferably be (Church, 1974):

(1) Highly soluble in water at stream temperatures;
(2) Stable in the presence of light, sediment or other substances found in natural waters;
(3) Easily detected at low concentrations;
(4) Absent from the stream itself or present in concentrations which do not interfere with measurements;
(5) Non-toxic to stream biota and without permanent effect on water quality; and

(6) Relatively inexpensive.

Chemical tracers are commonly electrolytes or solutions of specific ions (e.g. chloride). For small-scale field studies in streams of low salinity, common table salt (NaCl) is an inexpensive tracer. The salt should be dissolved in water and the electrical conductivity of the solution measured before introducing it into the stream. About 1 litre of 20% solution is needed for each cubic metre of discharge. If salt is added to a container of water until no more will dissolve, the concentration is approximately 20% (Finlayson, 1979). A well-cleaned large plastic rubbish bin is a suitable container for mixing up a solution of known volume.

Dilution gauging has been successfully done on larger rivers using Rhodamine WT, a dye which can be detected by fluorometric methods at very low concentrations. Another compound, Rhodamine B, is a known carcinogen and should not be used for water tracing (Smart, 1984). Wilson et al. (1984) is a source of more information on fluorometric procedures.

Detection of the tracer can be done in the field, as with an EC meter, a specific ion electrode or a fluorometer. Alternatively, samples can be collected and later analysed in the laboratory.

Slug-injection method

In this method a solution of known volume and concentration is added to the stream in one "slug" or "gulp". A cloud of marked fluid is produced which continuously disperses as it moves downstream. Concentration of the tracer is monitored downstream as the "wave" of marked fluid passes by. The "background" concentrations of the stream should always be measured before the tracer is introduced.

The discharge is calculated from integration (calculating the area under the curve) of the "concentration hydrograph", as shown in Figure 5.19. The equation for computing discharge is (Shaw, 1988):

$$Q = 1000 \frac{\forall c_t}{\int_{t_1}^{t_2} (c - c_0)\mathrm{d}t} \tag{5.21}$$

where: $\forall$ = known volume of tracer (l), c_t = concentration of tracer in introduced solution, c_0 = background concentration of stream (may be negligible), c = changing concentration of tracer measured downstream, Q = discharge (m^3/s) and t_1 and t_2 are the initial and final times of measurement (in seconds). The units of concentration must be consistent, but the actual units are not important since they cancel each other in the equation. In practice, the area under the curve (the integration term) is measured by numerical methods (e.g. adding up sub-areas for each time step) or by drawing out the curve on graph paper and planimetering the total area.

Constant-injection method

In the constant-injection method a solution of known concentration is fed into the stream at a constant rate. Downstream, the concentration of the tracer will rise and then stabilize at a constant value which can be related to discharge. The discharge is calculated from the equation (after Shaw, 1988):

$$Q = 1000 \frac{(c_t - c_1)}{(c_1 - c_0)} Q_t \tag{5.22}$$

where c_1 is the final, constant concentration of the tracer in the stream (see Figure 5.19), c_0 and c_t are as defined as for Equation 5.21, Q is in m^3/s and Q_t is the injection rate of the tracer in litres per second (l/s).

The tracer is normally injected at a constant rate using commercially available pumps of high flowrate accuracy. Smith and Stopp (1978) also give instructions for a simple apparatus which siphons a tracer solution into the stream. To maintain a constant injection rate the operator must continuously adjust the height of the outflow end of the siphon tube to maintain a constant elevation drop between it and the fluid level in the container.

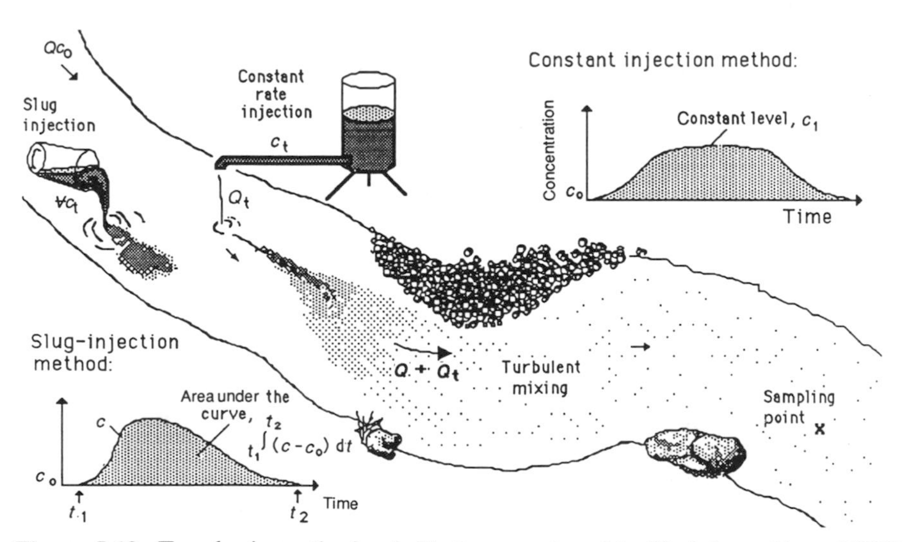

Figure 5.19. Two basic methods of dilution gauging. Modified from Shaw (1988), by permission of Van Nostrand Reinhold International

Researchers may also be able to purchase constant-rate injection tanks which make this continuous adjustment unnecessary.

5.7.5 ARTIFICIAL STRUCTURES

Pre-calibrated devices such as flumes and weirs are available for measuring discharge under a wide range of conditions. Their use is recommended when a semi-permanent station is to be established for recording streamflows on a continuous basis on smaller streams. The cost of construction may prohibit their use on large ones. Existing structures such as highway culverts, concrete sills at road crossings or bridges may also be used for routine discharge measurements if a rating curve can be developed. Small, portable devices may also be useful for the measurement of small flows when current metering is impractical.

When selecting the type and size of the device it is important to consider the magnitude of peak flows, the amount of debris and sediment carried, icing conditions, straightness of the stream reach, and soil and bed sediment conditions. Cutoff walls extending downwards and laterally from the structure into the earth are required at most sites to prevent flow from bypassing the measurement device. A full description is beyond the scope of this book, and the interested reader is referred to Ackers et al. (1978) and agency manuals on stream gauging.

Triangular or V-notch thin plate weirs are probably the most appropriate for work in small streams, as the V-shape allows very small flows to be measured accurately. The notch is machined with a sharp upstream edge so that water springs clear of the notch edge as it passes through (see Figure 5.20(a)). The angle of the "V" is normally 90°, although weir designs of other angles have been developed (see Brater and King, 1976).

With V-notch weirs the "V" should be situated some distance above the streambed. Water will pond behind the weir, which reduces and smooths the approach velocity. A control section (Section 5.6) exists at the notch, meaning that it controls the upstream depth. Measurements of this depth, as a vertical distance from the bottom of the "V", are taken upstream (Figure 5.20(a)). Alternatively, a staff gauge can also be installed on the upstream face of the weir, away from any "drawdown" effect from the notch.

Brakensiek et al. (1979) cite an equation derived from experimental work on 90° V-notch weirs which relates upstream depth or head (H) to discharge (Q) as follows:

$$Q = 1.342H^{2.48} \tag{5.23}$$

where Q is in cumecs and H is in metres. The equation is accurate to within 1% when H is between 0.06 and 0.6 m. Volumetric measurement using the weir as an overfall can be used for even smaller flows (Section 5.7.2). Care

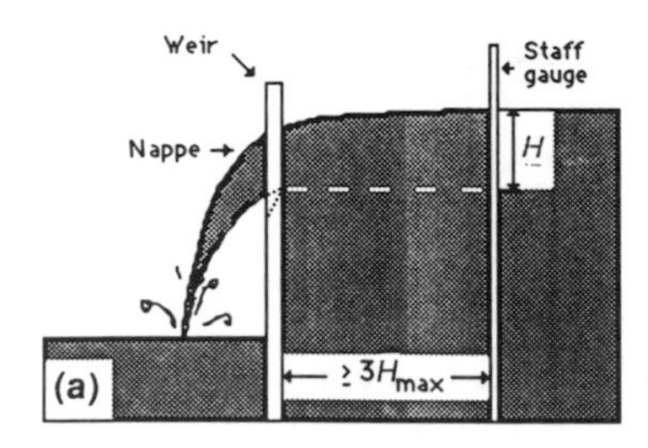

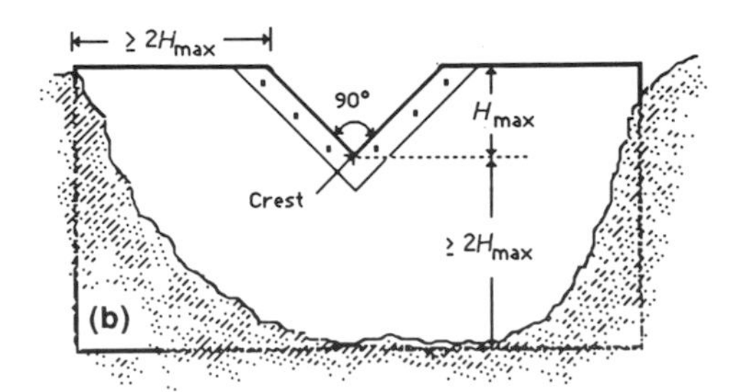

Figure 5.20. Dimensions associated with V-notch weirs: (a) side view and (b) front view. H_{max} is the maximum expected depth through the notch and H is the depth of water above the bottom of the "V", as measured in the upstream pond

should be taken to keep the notch clear of debris so that depths are not affected. V-notch weirs are not appropriate for measuring flows with high sediment or debris loads.

The weir can be constructed out of metal, plywood, perspex or concrete, but the notch should be carefully machined using stainless steel. A guide for sizing the notch is that the depth of the "V" should be smaller than 20% of the width of the approach channel (Dackombe and Gardiner, 1983). Other recommended proportions are shown in Figure 5.20. For best results, the approaching flow should have a Froude number (see Section 6.6) below about 0.5. As a semi-permanent installation, a V-notch weir may have undesirable ecological side-effects such as alterations to the site from sedimentation or scour and blockage of the channel to the passage of fish, and these should be considered by the installer.

5.8 STAGE–DISCHARGE RELATIONSHIPS AND OTHER INDIRECT METHODS OF DERIVING DISCHARGE

5.8.1 STAGE–DISCHARGE RELATIONSHIPS

If a site is visited regularly and simultaneous discharge and stream depth observations are made over a large range of discharges a **stage–discharge relationship** can be developed. Then discharges can be determined simply by monitoring the water level. This is the technique used to obtain most of the published data from gauging stations.

The stage–discharge relationship is expressed as a rating curve, a rating table or a rating equation. If the control section is a flume or weir (Section 5.7.5), a standardized equation will be available. In natural streams a relationship must be developed by plotting measured stage against discharge (Figure 5.21). The accepted procedure of plotting the dependent variable (discharge) on the *X*-axis and the independent variable (stage) on the *Y*-axis is somewhat unconventional, but makes the plot follow the direction of the rising and falling stages.

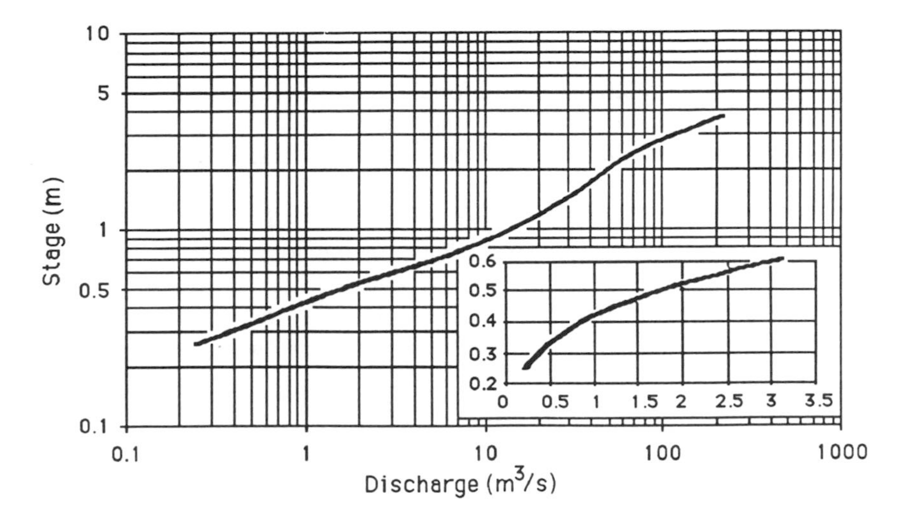

Figure 5.21. Rating curve for the Acheron River at Taggerty. Inset shows lower part of the rating, plotted using linear axes. Data supplied by the Rural Water Commission, Victoria, Australia

At gauging stations, curves are developed from a large number of data points, usually collected on the falling limb of a flood hydrograph. For environmental studies this precision may not be necessary, although it is still important to obtain data for ratings at high flows. Measurement errors are likely to be greatest in this range, due to high water velocities, rough, turbulent water surfaces, shifting bed conditions and bank erosion and deposition. From data on several streams in the Western USA, Bovee and Milhous (1978) demonstrated that developing a stage–discharge relationship using only three points produced more reliable results than Manning's equation (Equation 5.25), when extrapolated within the range 40–250% of the calibration flow. They also showed that little improvement was gained by adding more than three points.

If the control section performs adequately and the cross-sectional profile is fairly regular, the resulting relation on linear axes will be smooth and parabolic, as follows:

and an equation can be easily fitted to the curve. However, if abrupt changes in the profile exist or if the control section becomes submerged at some stage, the resulting curve may look more like this:

In the latter case, it is better to express the stage–discharge relation as a curve or a rating table from which values can be interpolated. A plot on arithmetic co-ordinate paper is preferable for this application. It is also better for plotting the lower part of the rating curve (inset of Figure 5.21) for improved accuracy and because the zero flow stage cannot be plotted on log–log paper.

The use of logarithmic graph paper is preferred in most cases since parabolic curves will plot as straight lines on log–log paper. This makes it easier to objectively extrapolate the curve to points beyond the observed range of stage heights, if needed, and to identify changes in slope which may be due to changes in control. It also allows the fitting of a rating equation of the form:

$$Q = a(h - z)^b \tag{5.24}$$

where Q = discharge (m^3/s), h = gauge height of the water surface (m), z = gauge height of "zero flow" (m) and a and b are coefficients. This equation can be fitted to the data by simple regression methods (Appendix 1). It should be noted that the regression equation is fitted with $(h-z)$ as the independent variable and Q as the dependent variable, even though the rating curve is plotted with the axes reversed.

The value of z (stage height at zero flow) must be derived by trial and error. The "true" value of z is assumed to be the value which makes the stage–discharge relationship plot as a straight line on logarithmic paper. It is thus an "effective" zero flow rather than the stage at which the channel goes dry. Successive values of z can be assumed and the relation plotted. If the assumed value of z is too small the rating curve will be concave upward; if it is too large the rating curve will be concave downward. The assumed value of z is increased or decreased until a value is found which results in a straight-line plot (WMO, 1980).

If changes to the gauging section or the downstream control section occur the stage–discharge relationship will also change. For example, if the width of the section increases, the coefficient a will rise, causing the rating to shift to the right. If the section scours, z decreases and the depth for a given gauge height increases, causing the relation to curve downwards on log–log paper. Temporary adjustments to the rating curve may also be needed to account for vegetal or ice growth or other seasonal effects. WMO (1980) also point out that changes can occur from summer holidaymakers who pile rocks in the control section to create a deeper pool for swimming. These conditions of **shifting control** are particularly common in sand-bed streams.

Ideally, new stage and discharge measurements over a range of flow levels are taken to re-rate the section after a change occurs. More commonly, however, only one or two measurements are obtained. These measurements can be plotted as points alongside the original rating on arithmetic graph paper and the whole curve "shifted" left or right so it passes through these points. Corbett (1962) and WMO (1980) both offer a more complete description of shifting control problems and adjustment of rating curves.

Leopold et al. (1964) proposed the use of a **dimensionless rating curve** after studies of river data suggested that rating curves from various stations were similar in form. Depth and discharge are both scaled by bankfull values to make the numbers dimensionless. This approach may be useful for deriving regional stage–discharge relationships which can be used at ungauged sites (see also Section 8.5).

5.8.2 SLOPE–AREA METHOD OF ESTIMATING DISCHARGE

At ungauged sites, peak discharges are often estimated "after the fact" from indirect methods based on the height of water marks left behind by floods. The same techniques can be used to estimate discharges for other water levels, e.g. bankfull depth or the depth at which the flow just covers the streambed in the main channel. Indirect methods can also be used to extend the upper part of a stage–discharge relationship. Herschy (1985) is a source of general information on these techniques.

The **slope–area method** is a commonly used technique for indirect estimation of discharge, and Dalrymple and Benson (1967) provide a comprehensive description. Typically, Manning's equation is used for the calculation of discharge:

$$Q = \frac{1}{n} A R^{2/3} S^{1/2} \tag{5.25}$$

where Q = discharge (m^3/s), n = "Manning's n", A = cross-sectional area of the flow (m^2), R = hydraulic radius (m), and S = slope. These terms are explained further in Section 6.6.5, along with a discussion of the equation and its limitations.

For the method to be valid a reach must be carefully selected such that uniform flow conditions are approximated. This means that the width and depth of flow, the water velocity, the streambed materials and the channel slope remain constant over a straight reach, and further, that the channel slope and water slope are parallel. A straight, fairly homogeneous reach should be selected with a length at least five times the mean width (Dackombe and Gardiner, 1983).

The first step is to identify the water level of interest. Methods of defining bankfull depths and high water marks are discussed in Section 5.4. These levels can be flagged with fluorescent survey ribbon to define the water surface slope. Surveys should be made of at least three cross-sectional profiles (taken at right angles to the flow direction) and of the average bed and water surface slope. The inclusion of more cross-sections and greater spacing generally minimizes some of the errors associated with this method (Jarrett, 1987).

Surveyed information is used to calculate the values of A and R in Equation 5.25. The slope, S, is actually the slope of the energy line (Section 6.6.4). However, in practice, the energy slope is commonly assumed to be parallel to the water surface slope and the bed slope, which can be used instead. The more closely the reach approximates uniform conditions, the better the results. In highly turbulent sections and in steep streams, particularly those with a pool-step structure (Section 7.3.2), this assumption may not be valid. Jarrett (1985), Jarrett and Petsch (1985) and WMO (1980) provide information on adjusting the value of S for non-uniform reaches. The last variable is Manning's n, which is basically a composite factor which accounts for the effects of many forms of flow resistance, as will be discussed. Program MANNING in AQUAPAK will compute both velocity and discharge from Manning's equation. Program XSECT can be used to compute R and A from measurements at cross-sections.

One of the greatest difficulties in applying this method is the accurate estimation of Manning's n. In general, n increases as turbulence and flow retardance effects increase. In a reach where the slope is uniform and the roughness of the bed and banks is similar (e.g. an artificial channel), Manning's n can usually be assumed to be a constant. However, in natural streams, it will often vary with flow depth, generally decreasing as the heights of protuberances such as snags or rocks become submerged by the flow. For example, Bovee and Milhous (1978) give data from a gravel-bed stream, Oak Creek, Oregon, demonstrating a variation in Manning's n from 0.05 for flows greater than 0.84 m^3/s to 0.35 for flows less than 0.03 m^3/s. This trend may reverse, however, in streams where the channel bed is smooth and the banks are densely vegetated. Dawson and Charlton (1988) provide a source of information on vegetation-related resistance. For overbank (flood) flows, n values are normally evaluated separately for the main channel and floodplain. Discharge in these compound cross-sections is computed by summing the discharge in sub-areas, as discussed in Section 6.6.5.

Estimates of Manning's n can be made by choosing a value from a table such as Table 5.2 or by making a visual comparison with photographic keys, such as those provided by Barnes (1967) and Chow (1959). Barnes' book, produced by the US Geological Survey, contains colour pictures and n values which have been verified at near-bankfull discharges from gauged readings.

Another method given by Cowan (1956) provides a means of estimating Manning's n by considering the individual effects of various roughness components. Cowan's method is suitable for small- to mid-size channels of hydraulic radius less than 5 m. Manning's n is calculated from:

$$n = (n_0 + n_1 + n_2 + n_3 + n_4)m_5 \tag{5.26}$$

where the subfactors n_0 through n_4 and m_5 separately account for the effects of various influences on n, as given in Table 5.3. Selected factors should apply to the entire study reach, not only to the cross-sections. The basic value, n_0, should represent the smoothest reach of that bed material type and then adjustments are made using the other subfactors (Jarrett, 1985). Under extremely rough conditions (e.g. steep mountain streams or during flood conditions or very low stages) even larger adjustments may be needed.

If the roughness of the banks is very different from the streambed (e.g. dense shrub growth on banks) a composite n for the channel can be derived using a weighting method (Jarrett, 1985):

$$n = \frac{P_1 n_1 + P_2 n_2 + \cdots + P_m n_m}{P} \tag{5.27}$$

where P_1, P_2, etc., are the amount of wetted perimeter with n values n_1, n_2, etc. and P is the total wetted perimeter for the section.

When using Cowan's method care should be taken to avoid double-counting the contribution of one type of roughness under more than one component. This approach is still very subjective, although it at least forces one to observe and consider the various factors affecting Manning's n at different water levels.

Table 5.2. Manning's n values for small, natural streams (top width at flood stage <30 m). From Chow (1959), by permission of McGraw-Hill

Description of channel	Minimum	Normal	Maximum
Lowland streams:			
(a) Clean, straight, no deep pools	0.025	0.030	0.033
(b) Same as (a), but more stones and weeds	0.030	0.035	0.040
(c) Clean, winding, some pools and shoals	0.033	0.040	0.045
(d) Same as (c), but some weeds and stones	0.035	0.045	0.050
(e) Same as (c), at lower stages, with less effective slopes and sections	0.040	0.048	0.055
(f) Same as (d), but more stones	0.045	0.050	0.060
(g) Sluggish reaches, weedy, deep pools	0.050	0.070	0.080
(h) Very weedy reaches, deep pools or floodways with heavy stand of timber and underbrush	0.075	0.100	0.150
Mountain streams (no vegetation in channel, banks steep, trees and brush on banks submerged at high stages):			
(a) Streambed consists of gravel, cobbles and few boulders	0.030	0.040	0.050
(b) Bed is cobbles with large boulders	0.040	0.050	0.070

Table 5.3. Component values for Cowan's method of estimating Manning's n. Sinuosity is defined in Section 7.3. Adapted from Cowan (1956) and Jarrett (1985)

Basic n value, n_0:		*Surface irregularity*, n_1:	
Earth	0.020	Smooth	0.000
Rock	0.025	Minor (slightly eroded or scoured)	0.005
Fine gravel	0.024	Moderate (moderate slumping)	0.010
Coarse gravel	0.028	Severe (badly slumped, eroded banks, or jagged rock surfaces)	0.020
Cobble	0.030–0.050		
Boulder	0.040–0.070		
Variation in cross-section shape causing turbulence, n_2:		*Effect of obstructions (debris deposits, roots, boulders)*, n_3:	
Change occurs gradually	0.000	Negligible (few scattered obstructions)	0.000
Occasional changes from large to small or side-to-side shifting of flow	0.005	Minor (obstructions isolated, 15% of area)	0.010–0.015
Frequent changes	0.010–0.015	Appreciable (interaction between obstacles which cover 15–50% of area)	0.020–0.030
Vegetation, n_4:		Severe (obstructions cover >50%; or cause turbulence over most of area)	0.040–0.060
None or no effect	0.000	*Meandering (multiplier)*, m_5:	
Supple seedlings or dense grass/weeds	0.005–0.010	Minor (sinuosity 1.0–1.2)	1.00
Brushy growths, no growth in streambed; grass height of flow	0.010–0.025	Appreciable (sinuosity 1.2–1.5)	1.15
Young trees intergrown with weeds; grass twice depth of flow	0.025–0.050	Severe (sinuosity >1.5)	1.30
Brushy growth on banks, dense growth in stream; trees intergrown with weeds; full foliage	0.050–0.100		

Example 5.2

(a) Calculate Manning's n using Cowan's method and the following estimates of component values: $n_0 = 0.028$, $n_1 = 0.005$, $n_2 = 0.010$, $n_3 = 0.020$, $n_4 = 0.045$, $m_5 = 1.05$, and (b) calculate the discharge if $A = 5.2$ m^2, $R = 0.8$ m, and $S = 0.03$.

(a) $n = (0.028 + 0.005 + 0.010 + 0.020 + 0.045)(1.05) = 0.113$

(b) $Q = \dfrac{1}{0.113}(5.2)(0.8)^{2/3}(0.03)^{1/2} = 6.87\ \text{m}^3/\text{s}$

A "calibration" value of Manning's n for present conditions can be computed by working backwards from measured values of discharge and channel dimension. A routine in program MANNING is provided for this purpose. This value is then adjusted up or down as appropriate to estimate discharge at other levels, based on field observations of how roughness changes with water depth. This can be done based on judgement and experience from evaluating Manning's n at different discharges (e.g. see Bovee and Milhous, 1978) or by applying Cowan's method to make the adjustments. Based on a review of verified channel roughness data, Jarrett (1985) found that in uniform channels the base n value, n_0, does not vary with depth of flow over the range:

$$5 < \frac{R}{d_{50}} < 276$$

where R is the hydraulic radius, as before, and d_{50} is the median particle diameter of the streambed materials (Section 5.10). A value less than 5 for the ratio R/d_{50} represents mountain streams with large bed materials and a value greater than 276 sand-bed streams which may have significant variations in n due to bedforms (Section 7.3.4).

The slope–area method using Manning's equation has been verified on low-gradient streams of relatively tranquil flow with good results (Jarrett, 1987). For good hydraulic conditions, peak discharges can be estimated with an error of 10–25%, but under poor conditions errors of 50% or more may occur. A few measurements of channel geometry and a guess at Manning's n from Table 5.2 or a picture book will be adequate for rough estimates. For more accurate predictions, one might want to consider recruiting someone experienced in selecting channel sections and estimating Manning's n—if not for the length of a research study, at least for a day or two of training. "Getting acquainted" with streams on which Manning's n has been verified is also helpful.

In an attempt to sidestep the issue of estimating Manning's n altogether, Riggs (1976) gives a simplified slope–area method which does not require estimation of a roughness coefficient. His method is based on observations that slope and roughness tend to be related in natural channels and that hydraulic radius is closely related to cross-sectional area. From an analysis of the data from Barnes' (1967) book of verified Manning's n values at bankfull conditions, Riggs developed the following equation, given in SI units as:

$$\log Q = 0.191 + 1.33 \log A + 0.05 \log S - 0.056(\log S)^2 \qquad (5.28)$$

where Q is in m^3/s, A is in m^2 and S is dimensionless. The standard error of the equation is about 20%. It should be noted that it only applies to conditions similar to those for which the equation was derived.

Jarrett (1987) reviews many of the problems of estimating peak discharge with the slope–area method in mountain rivers. He states that misapplication of the method in higher-gradient mountain streams (slopes over about 0.01) has tended to overestimate discharge, leading to erroneous "record-breaking" flood values. Jarrett recommends checking the average Froude number (Section 6.6), which will normally be below 1.0 in streams even during peak flows. A higher value may indicate that sources of energy loss have not been properly identified.

5.9 SUBSTRATES AND SEDIMENTS: SAMPLING AND MONITORING METHODS

5.9.1 GENERAL

Information on the size and distribution of inorganic particles, whether in motion or forming part of the channel bed or banks, is often needed in studies of both the ecology and hydrology of streams. Particles can be described in terms of shape, size, mineralogy, colour, concentration in the water column and orientation and degree of compaction in the bed. Samples of bank materials might be collected for determining soil texture, moisture content, the percentage of roots and other organic matter, and other factors related to bank stability. Samples of bed materials can provide information on surface roughness and benthic habitat type. Cores of deposited sediments in lakes and stream backwaters have been used to evaluate historical changes in the inputs of heavy metals and other substances. Additionally, a researcher may wish to sift through samples of either bed and/or bank materials to count the number of invertebrates either in the whole sample volume or at different depths within the substrate. Sources of sediment, distribution of sizes transported and estimates of sediment yield (Section 7.5) can also be evaluated by sampling sediments in motion.

A biological standard method of analysis of substrate has yet to be developed (Bovee, 1982); thus the methods presented are those commonly accepted for soil and sedimentological work. The sample size should be at least 100 g in finer clays, with up to hundreds of kilograms needed for coarser-grained sediments (Brakensiek et al., 1979; Gee and Brauder, 1986). Methods of sampling range from simply scooping up a handful or shovelful of material to more sophisticated coring techniques.

5.9.2 BANK MATERIAL SAMPLING (SOIL SAMPLING)

Soil samples are taken with conventional tools such as shovels or augers (spiral, tube or bucket)—either hand or power driven. The representativeness of an augered sample will diminish as the size or number of stones in the sample increases. To obtain measurements of characteristics such as texture or moisture content or frost depth, samples should be taken at several locations on the bank and/or where the soil type noticeably changes. The depth of sample taken is typically from 10 to 30 cm, but this will depend on the type of analysis.

If soil samples are to be analysed for water content they should be immediately placed in a leakproof plastic bag or soil can for transport to the laboratory. Cores of soil can also be segmented into different layers or horizons for separate analysis. Containers should be labelled with the date, site and depth of sample.

5.9.3 BED MATERIAL SAMPLING

Manual methods and bed material samplers

For most stream studies, samples of bed materials will most likely be collected **by hand**. Where the water current is sufficiently low so that finer particles are not swept away a **shovel** can be used. A wire grid or hoop can be placed on the bed surface prior to digging to define the area sampled (see also Section 5.3).

Cores of bed materials can be isolated by several means. **Piston-type bed material hand samplers** can be purchased which are much like tube-type augers but have a piston which is retracted as the sampler is forced into the streambed. This creates a suction which holds the sample in place. A **hollow cylinder** can also be driven into the bed materials to isolate a core. After digging down to expose one side, a thin metal plate is slid under the cylinder, assisted by a hammer. The cylinder is then pulled out with the plate held firmly against its base. Hamilton and Bergersen (1984) and Platts et al. (1983) describe a more sophisticated version of this method using a stainless steel McNeil–Ahnell hollow-core sampler.

Bed material samplers have been designed for the collection of samples from the channel bed in large rivers. These are usually quite heavy and are operated from a crane on a boat or bridge. Typically, they have some type of scoop or sampling bucket which may be dragged along the channel bed or triggered on contact with the streambed to trap a sample of the surface bed material.

Freeze coring

One of the better methods for sampling coarser bed materials as well as finer sediments is **freeze coring**. As illustrated in Figure 5.22, a hollow probe is driven into the streambed and is subsequently filled with a cryogenic medium such as liquid nitrogen or liquid carbon dioxide (CO_2). Liquid CO_2 generally yields smaller samples, although it is less expensive. After allowing a prescribed period of time for freezing, the probe and the adhered core of frozen sediment is extracted with a hoist or jack.

The sample is removed from the probe by chipping with a rock hammer or by thawing. One can expedite the thawing process by running a blowtorch back and forth over the sample. The vertical stratification can also be analysed by first laying the frozen core over a segmented box. Since benthic

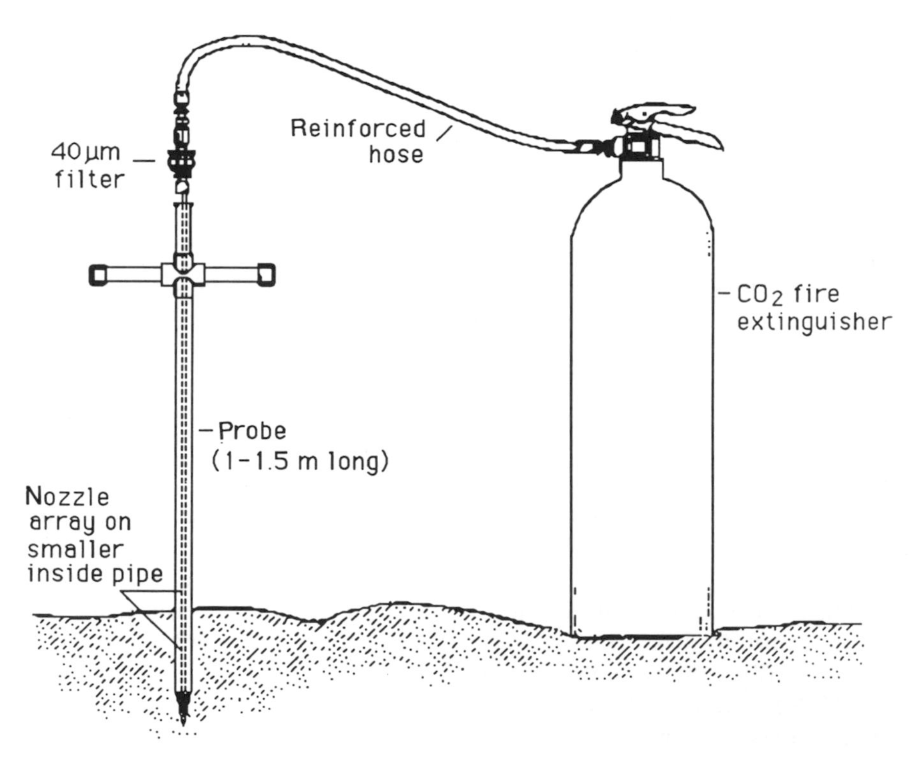

Figure 5.22. Schematic of freeze corer, showing probe and liquid CO_2 source. Modified from Hamilton and Bergersen (1984)

organisms will also be collected in the frozen sample, freeze coring thus provides a way of observing their stratification within the streambed.

When using this technique, Marchant (pers. comm., 1990) found that it was necessary to enclose the sampled area with a section of a metal drum to minimize the amount of bed material swept away by the current. He found the technique very effective for sampling particles up to the size of large rocks and for studying the numbers of invertebrates within the streambed.

Platts and Penton (1980) provide information on the construction and operation of freeze corers. Lotspeich and Reid (1980) and Everest et al. (1980) describe an improved version with three evenly spaced prongs built into a template. The size of the sample taken using this method, and thus representativeness, is increased substantially. However, the three-pronged sampler is more difficult to drive into the substrate since the chance of hitting a large rock is tripled, and it is more difficult to extract the larger sample.

5.9.4 SAMPLING SUSPENDED SEDIMENTS

When sampling streams for suspended sediment it is important to obtain a sample which accurately reflects the stream's sediment load. Since the majority of sediment transport occurs during high flows it is essential that samples be taken during these periods when developing long-term averages. Samples must be taken in such a way that concentration represents an average for the section, as sediment concentration will vary with depth and across the section depending on the particle size (see Section 7.1.4). Sediment concentrations will also vary from one instant to the next as "pulses" of sediment pass through a cross-section, followed by gaps of lower concentration. In data collected by Horowitz et al. (1989) on the Cowlitz River, Washington (affected by the Mount St Helen eruption) samples collected at 20-min intervals showed an average deviation from the mean of 4%, with individual samples varying as much as 20% from the mean. Variability was found to be higher towards the centre of the stream than towards the edges.

It should thus be recognized that sampled concentrations will be extremely variable within a section and from one moment to the next, particularly during high flows. Therefore, numerous samples should be taken near the peak discharges to establish the error margin. Fewer are needed during periods of low or stable flow. Sampling schemes usually represent a compromise between precision and the time and cost of sampling.

Sampling sites will generally be the same as those used for stream-discharge measurements (Section 5.7.3). At stream-gauging sites samples should be taken upstream of stilling ponds which act as sediment traps. It is also important to obtain samples which correctly integrate the distribution of moving sediment within the channel. This can be accomplished by taking samples at several locations across a transect, as will be described. Alternatively, the mixing effects in turbulent sections created by artificial or natural structures can be exploited so that only one relatively homogeneous sample is needed.

If only the fraction smaller than 0.0625 (silts and clays) is important a grab sample taken near the water surface in the centre of the stream is considered to be sufficient, since this fraction is assumed to be evenly distributed at a section (Section 7.1.4). During high stages it may be necessary to take a surface sample because of difficulty in lowering the sampler. For streams in India, Singhall et al. (1977) related mean sediment concentration (c_{avg}) to surface concentration (c_{surf}) by

$$c_{avg} = 2.353\, c_{surf} \tag{5.29}$$

The sample volume is adjusted according to the estimated concentration of sediment. Recommended minimum volumes for analysis of concentration by drying and weighing are given in Table 5.4. Filtration techniques are more sensitive, only requiring about one-tenth of the listed volumes. If samples are to be analysed for particle size much larger volumes are needed.

The site, date and time of collection (or numbering sequence in an automatic sampler) should be recorded on the sample container with either a waterproof marker or a soft lead pencil and waterproof label tape. For sediment sampling the type of container is not particularly critical as long as it is leakproof. If the sample is to be analysed for chemical water quality as well, then the guidelines in Section 5.11 for sample containers should be followed.

Suspended sediment samplers

If the stream is sufficiently small and/or well-mixed a **grab-sample** of part or all of the flow can be collected with a cup or bucket. Suspended sediment samplers are, however, preferable. There are two basic types: depth-integrating and point-integrating.

Table 5.4. Recommended sample volume for analysis of suspended sediment concentration by drying and weighing. After WMO (1981)

Expected concentration of suspended sediment (ppm)	Required sample volume (l)
>100	1
50–100	2
20–30	5
<20	10

Depth-integrating samplers are commonly used in upland streams. These samplers, as described by Brakensiek et al. (1979), are available in a range of sizes. They are normally made of cast aluminium or bronze and are streamlined, with tail fins to orient the sampler so that its intake nozzle points into the oncoming flow. Nozzles are available in a range of sizes to accommodate different particle diameters. Originally developed in the 1940s, the samplers were designed to hold glass pint milk bottles, which were cheap and readily available at the time. Plastic bottles can be used in some of the samplers, and teflon nozzles can be installed to minimize contact of the sample with metal when water quality samples are collected.

The samplers are designed such that the velocity at the entrance of the nozzle is the same as the local stream velocity. They generally will only sample to within about 90 mm of the streambed; the unsampled zone is sometimes considered to be the region of bedload transport (Section 7.1.4). Depth-integrating samplers are designed to extract a sample continuously as the sampler is lowered at a constant speed (the "transit rate") from the water surface to the streambed and back. This gives a velocity-weighted sample over the vertical since more water will enter the nozzle at depths where the velocity is greater.

In wadeable streams, the hand-held US DH-48 sampler, illustrated in Figure 5.23(a), is the standard tool of the trade. Finlayson (1981) also gives a design for a home-made depth-integrating sampler, shown in Figure 5.23(b), which uses a carbonated beverage bottle. The bottle opening is fitted with a two-holed stopper, the lower opening has a straight metal or rigid plastic tube through which the sample enters the bottle and a flexible hose for the escape of air is attached to the other. A handle and tail fin are also attached. Samples are transferred to storage bottles after collection.

When taking a sample with the US DH-48 or equivalent a person should stand off to one side and slightly downstream of the transect to avoid

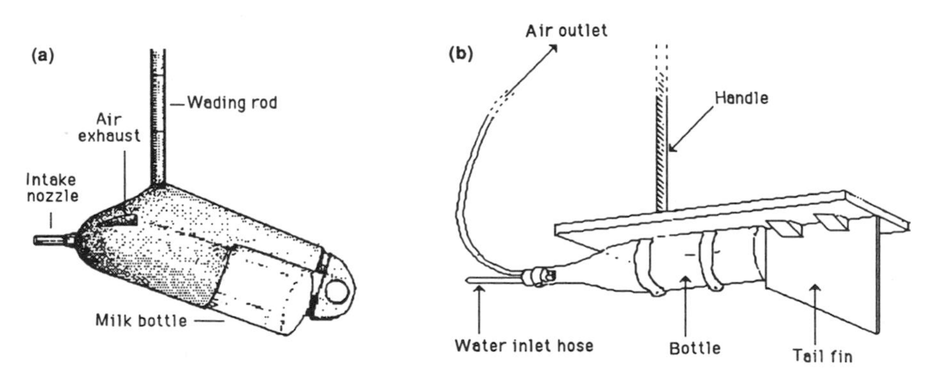

Figure 5.23. (a) Hand-held depth-integrating suspended sediment sampler, US DH-48 model and (b) home-made depth-integrating sampler

disturbing the oncoming flow or the bed sediments. Any sediments disturbed by wading should be allowed to settle before the sample is taken. The sampler is washed with stream water and the nozzle checked for clogging before starting.

The rod is held vertically and the sampler lowered at a constant speed until it just bumps the streambed. It is then raised, also at a constant rate. This transit rate should not exceed 0.4 times the stream velocity. Ideally, the bottle should be between two-thirds and three-quarters full. If it is not full enough, the sampler can be lowered and raised again. If it is too full, the entire sample should be dumped out and the vertical re-sampled. With practice, one can become fairly proficient in judging the speed of the sampler needed to fill the bottle two-thirds full in one "pass".

Brakensiek et al. (1979) and WMO (1981) describe a method for sampling a cross-section called the **equal transit rate (ETR)** method. A measuring tape is first stretched across the stream, usually in conjunction with measurements of current and depth. The ETR method proceeds as follows:

(1) The stream width is divided into six to ten equal-width sections. Fewer verticals will generally be needed in more uniform cross-sections and at lower stages.
(2) The same transit rate is used at all verticals, and the person proceeds from one vertical to the next until a sample bottle is sufficiently full (the last bottle used may be less than two-thirds full). The total number of bottles collected will vary from one cross-section to another.
(3) All the samples from the cross-section are composited to form a single representative discharge-weighted sample.

Methods of computing suspended sediment discharge from these measurements are given in Section 7.5.3.

Point-integrating samplers are similar in design to depth-integrating samplers but are equipped with a valve, end flaps or a door, which can be opened and closed electronically to "trap" a sample at a specific depth in the stream. A more basic approach is to simply uncap a bottle or jar at a specific depth and allow it to fill, although the escaping air bubbles may affect the size of sediment entering the bottle. Point-integrating samplers might be used to develop horizontal or vertical profiles of sediment concentration within a cross-section or to investigate the conditions experienced by a fish just downstream of a boulder or inside a log.

Rising-stage sampler

A rising-stage sampler provides a means of automatically collecting suspended sediment samples from flashy, ephemeral streams. This type of sampler may also be useful for sampling peak concentrations at remote sites since it is usually difficult to be present when runoff peaks occur.

The rising-stage sampler works on the siphon principle and consists of a number of bottles arranged on top of each other in a frame. Each bottle is fitted with a two-hole stopper, one hole for sample inlet and one for air exhaust. In the standard US U-59 (Brakensiek et al., 1979), copper tubing of 4.8–6.4 mm i.d. (inside diameter) is recommended for the inlet and exhaust. Finlayson (1981) also gives a design using flexible tubing for the inlet and air outlet. The inlet tube is connected to a piece of rigid tubing which is mounted on a board and pointed into the flow, as shown in Figure 5.24.

The "kink" formed in the inlet tubing causes the siphon effect. As the

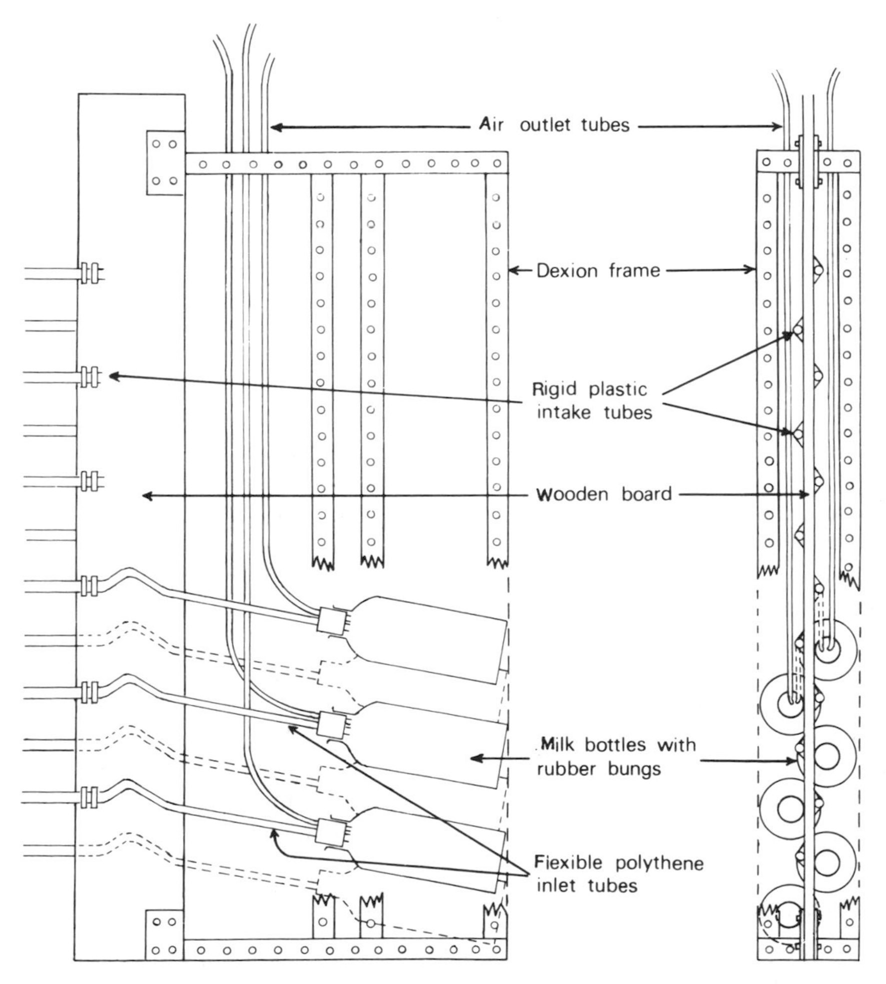

Figure 5.24. Rising-stage sampler, home-made design. Reprinted from Finlayson (1981), by permission of the *Journal of Geography in Higher Education*

water rises, sample bottles fill when the water level reaches the height of the "kink". Thus, samples will be taken just below the water surface. By using more than one sampler, samples can be obtained at several water surface elevations as the water rises.

The rising-stage sampler is most suitable for suspended sediment finer than 0.0625 mm (silts and clays). Larger particles are not sampled adequately with this system since intake velocities are not the same as streamflow velocities. Also, since the sample is taken near the water surface, the coarser particles which are less evenly distributed over the water depth will not be sampled adequately (Section 7.1.4). These limitations should be considered when using concentration data obtained from these samplers.

Automatic pumping samplers

Many different types of automatic pumping samplers are available. Each has its own advantages and disadvantages in terms of mechanical "vigour", ease and expense of repair, run time on one battery charge and ability to extract a representative sample.

These samplers pump water from the stream, a portion of which is retained in a sample bottle. On most models the intake line is given an initial "back flush" to minimize cross-contamination between samples. The samplers can be set to take samples at a given time interval or attached to a flow rate or flow level meter to sample at a proportionately higher rate during higher flows. A float-controlled switch can also be used to activate the sampler when the water stage reaches some predetermined elevation.

For long-term records automatic pumping samplers are usually located near a streamflow gauging station. At these stations an "event marker" can be attached to the water stage recorder to stamp a mark on the chart when a sample is collected.

Bottles should be retrieved after floods or at intervals determined by the time it takes to fill all the sample bottles. Periodic servicing is also required to perform maintenance on the instrument and replace batteries. In areas where temperatures fall below freezing an insulated and/or heated shelter may be required.

Since the sampler intake nozzle is often located near a streambank the samples can give misleading impressions of average stream sediment concentrations. They should thus be calibrated against concentrations taken across the section and a correction factor applied.

Turbidity meters

Attempts have been made to develop methods for continuously monitoring suspended sediment concentrations, which usually rely on the assumption that water with a higher sediment concentration is less transparent. **Turbidity** is a measure of this optical property which inhibits the transmission of light through a sample due to scattering and absorption. To measure turbidity a sample of the streamflow is usually pumped past a photoelectric cell. Attenuance-type meters measure the amount of light transmitted through the sample while nephelometric-type meters measure the amount of light scattered by the sample. These readings are calibrated against measured concentrations.

Turbidity measurements can be affected by factors other than particle concentration, such as the size distribution, shape and absorptivity of the sediment, and, in attenuance-type meters, the colour of the water. Gippel (1989) investigated both types of turbidimeters for the measurement of suspended sediment concentrations. He concluded that turbidimeters gave satisfactory estimates of storm-event sediment loads if sediment particle size was fairly constant and fell within the range 0.0005–0.01 mm mean diameter (fine silts and clays). An infra-red light source was also found to be superior to a visible light source since it was not affected by water colour or algal growth.

5.9.5 SAMPLING BEDLOAD SEDIMENTS

Bedload is the material which generally remains in contact with the streambed when it moves by rolling, sliding or hopping. The distinction between bedload and suspended load is discussed in Section 7.1.3. Bedload movement is important because of its relationship with changes in the bars and bends of a stream's morphology. It is also important to benthic stream biota which can be crushed or disturbed by the moving material.

Accurate sampling of bedload is difficult due to the movement of sediments in bars, ripples and dunes (Section 7.3.4), the "stop and start" nature of sediment movement and the problem of efficiently sampling sediments over a range of sizes. It is important, then, to choose measurement sites where the sampled sediment load is representative of the amount and size of material being carried through the stream. For example, a site downstream from a pond where materials settle out would not be a good location. As with suspended sediments, the highest bedload movement (and in many streams, all of it) will occur during high flows, requiring monitoring during these periods. An adequate bedload sampling programme (e.g. to develop rating curves, Section 7.5.4) can be very time consuming and expensive, and estimates made using simple monitoring methods and/or models are more common.

There are several ways of obtaining estimates of the amount of bedload passing though a particular reach of the stream, including: (1) bedload samplers, (2) tracer particles, (3) measurements of sediment accumulations behind structures, or (4) rule-of-thumb estimations based on the suspended load and type of channel material.

Bedload samplers

Ideally, a bedload sampler should trap large and small particles with the same efficiency yet not alter the natural pattern of flow and sediment movement. An efficiency (ratio of sampled to actual transport rate) of 60–70% can be considered satisfactory (WMO, 1981). Samplers are generally of four types: pit, basket, pan and pressure-difference samplers. Hubbell (1964) reviews and compares several bed material sampler designs.

Pit-type samplers (Figure 5.25(a)) are excavated depressions, lined or unlined, for trapping bedload sediments. A classic example is the permanent installation on the East Fork River in Wyoming, described by Leopold and Emmett (1976). At this site, sediments fall into a trench-like pit across the streambed and are transported by conveyor belt to a storage area for weighing. On smaller streams, a box can be placed in the streambed so that its open top is level with the streambed surface. After a period of time, the material is removed from the box and measured for size and weight.

Basket-type samplers (Figure 5.25(b)) are made of a mesh which is sized such that suspended material passes through but bed material is retained. An opening on the upstream end allows material to enter the basket. These samplers are relatively inexpensive and simple to use. The mesh size will need to be adjusted depending on the chosen dividing line between "suspended" and "bedload" for a given stream and flood event. Since basket samplers will not function well when nearly full, timing of their removal, as well as care in placing and lifting the samplers, is important. They must also be heavy enough or sufficiently well anchored to keep them aligned into the flow and to prevent the baskets themselves from becoming part of the load moved downstream (Goudie, 1981).

Pan-type samplers are usually wedge-shaped, with a pan which contains baffles and slots to trap moving sediments. They are set into the stream so that the wedge points into the oncoming current (WMO, 1981).

Pressure-difference samplers are designed to produce a pressure drop at the exit of the sampler to ensure that the entrance velocity equals that of the stream. The Helley–Smith bedload sampler, shown in Figure 5.25(c), has become something of a standard for bedload measurement. The expanded region causes the streamflow to divert and accelerate around the sampler, leading to separation and a drop in pressure at the exit (see Chapter 6 for theory). A bag is attached to the sampler for collection of sediments. The original design is given by Helley and Smith (1971). Smaller models with a 76-mm orifice can be operated by hand; 152-mm designs can be attached to a backhoe for operation from a bridge. Johnson et al. (1977), in measuring bedload transport in a gravel-boulder streambed in Idaho, USA, reported that the standard-size fine mesh (0.2-mm openings) bags of 0.2 m^2 were easily clogged by organic debris and fine sediments, and recommended the use of larger 0.6 m^2 bags.

Sampling with pressure-difference samplers should be carried out at a uniform cross-section as for suspended sediment sampling. A bedrock surface or concrete "sill", either installed specifically for this purpose or as a road crossing, can act as a platform for sampling sediments travelling over its surface.

Measurements should be taken at three to ten points per cross-section (WMO, 1981). The placement of sampling points should reflect the nature of bedload transport, which, except during floods, will generally take place only over a small fraction of the streambed. Samples are collected by lowering the sampler to the streambed, letting it collect sediment over some time period, then removing the sample and bagging it for later analysis (organic material can first be floated off). Care must be taken to keep the sampler correctly aligned and to avoid scooping up excess material from the bed when placing or removing the sampler. The period of time over which material is collected will vary with the flow conditions and bag size. For example, if no sediment is trapped in 5 min, the time period can be extended to 10 or 20 min rather than accepting a zero measurement as indicative of that section. As pointed out by WMO (1981), statistical analyses of field data from up to 100 repetitions have shown that an impracticably large number of samples must be taken at each point to ensure accuracy. A reasonable compromise would be to take two or three samples at each point and combine them. Methods of computing bedload discharge using this technique are given in Section 7.5.3.

Tracer particles

Several factors will affect the movement of particles as bedload, including the water velocity and turbulence, and the density, shape, and size and compaction of surrounding materials. **Tracer particles** can be used to investigate the types of materials which will move under different conditions. These can be pebbles or boulders which are placed on the streambed and

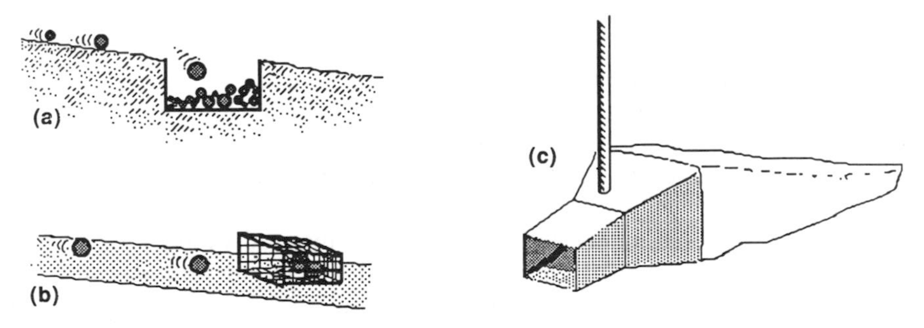

Figure 5.25. Bedload samplers: (a) pit-type, (b) basket-type and (c) Helley–Smith pressure-difference sampler

then located again after a period of high flow. Alternatively, pebbles or sand grains of different weights and shapes can be dropped into a stream at various locations to establish which materials can be transported under the existing streamflow conditions.

Tracer particles to be retrieved after a period of high flow should be weighed, measured, numbered and marked with brightly coloured paint. A range of sizes are normally used. The particles are lined up on the streambed and after the spate, an observer notes which have moved and how far—assuming they have not been "transported" by inquisitive artifact hunters.

Richards (1982) mentions that particles can also be wrapped in aluminium wire and traced with a metal detector in gravel-bed streams. In a large Alaskan gravel-bed river, scientists have even used radio transmitters embedded in large cobbles which were later located (sometimes buried up to a metre deep) with directional antennas (Emmett, 1989).

Statzner and Muller (1989) propose a quick and simple method for flow characterization using standard hemispheres of 39 mm diameter and varying densities (from 1000 kg/m^3 to 10000 kg/m^3). The hemispheres are placed on a level plexiglass plate on the streambed and their movement at differing flows is noted. The authors found that reasonable estimates of shear stress, Froude number and roughness Reynolds number (see Chapter 6 for definitions) could be made by noting the densest hemisphere moved by the flow. Hemispheres could be secured by fishing line for monitoring over a high-flow event.

Measuring accumulated sediments

The pondages behind structures act like "pit-type samplers" on a larger scale. Thus, repeated surveys of accumulated sediment in reservoirs, ponds, excavated depressions or behind debris dams can give an estimate of the amount of sediment transported by the stream over time. In fact, the rate of sediment deposition needs to be known before a dam is built because it will affect the useful life of the reservoir.

Surveying techniques, usually employing a grid (Section 5.3.5), are used to obtain an estimate of the sediment volume. Volume is then converted to weight by multiplying by an estimate of the density (see Section 5.10.7).

The deposited sediments will actually be composed of bedload and the portion of suspended load which has settled out in the stilled waters. By combining surveys of accumulated sediments with suspended sediment sampling both upstream and downstream, acceptable estimates of bedload and total sediment discharge can be obtained (WMO, 1981).

Estimation of bedload

A "rule-of-thumb" approach to the estimation of bedload can be taken by using the guidelines given in Table 5.5. If field work is conducted regularly and observations are compared with measured concentrations, a person can become somewhat skilled at estimating suspended load from the colour of the water. From Table 5.5 a spate of fairly muddy-looking water in what is known to be a gravel-bed stream might be expected to be carrying around 10% bedload. In fact, 10% is a conservative estimate for the middle reaches of many streams (Csermak and Rakoczi, 1987).

5.9.6 EROSION AND SCOUR

Changes in channel form can occur as a result of frost action, floods, trampling, bulldozers or other factors. These changes usually fall under the general category of **erosion**, whereas **scour** refers to the movement of bed materials during a flood (see Section 7.2.4).

Surveying

Aggradation and degradation processes (see Sections 4.1 and 7.2.2) can be monitored by repeating cross-sectional surveys at fixed transect sites. These transects, set perpendicular to the channel, should be permanently located with posts, pegs or other marks which are set far enough away from the

Table 5.5. Estimation of bedload transport. Adapted from Morisawa (1985), after Lane and Borland (1951), by permission of Longman Group, UK

Concentration of suspended load (ppm)	Type of material forming the stream channel	Texture of the suspended sediments	Bedload as a percentage of suspended load
Less than 1000	Sand	Similar to bed material	25–150
Less than 1000	Gravel, rock, or consolidated clay	Small amount of sand	5–12
1000–7500	Sand	Similar to bed material	10–35
1000–7500	Gravel, rock, or consolidated clay	25% sand or less	5–12
Over 7500	Sand	Similar to bed material	5–15
Over 7500	Gravel, rock or consolidated clay	25% sand or less	2–8

bank to prevent loss during floods. If the channel is easily deformed, cross-sections initially lined up perpendicular to the flow may later cut across the stream at odd angles, requiring some "head-scratching" interpretation.

Bank retreat at a specific location can be monitored simply by measuring the distance between a peg or a tree and the edge of a bank. The direction should be guided using a compass or by setting out one reference peg and one or more intermediate pegs which are lined up in the direction of measurement.

The line intercept method (Section 5.3.4) can also be used to assess the extent of active erosion along a streambank or the length of streambank protected by vegetation, rock cover or car bodies. Care should be exercised in conducting surveys so the work itself does not cause bank erosion.

Bank erosion pins

For measuring smaller rates of erosion, erosion pins (Figure 5.26) can be installed in the streambank. These are usually metal rods 1 mm or larger in diameter and from 100 to 300 mm in length (Goudie, 1981). They are inserted horizontally into the bank, leaving only a small fraction exposed. Several pins may be used along a vertical profile, their height in relation to some datum established by surveying. Bank retreat is estimated from the progressive amount of pin exposure, measured with a ruler or callipers.

Some potential problems are that local scour can occur around the base of the exposed pin, and they may catch debris and thus interfere with the natural rate of bank retreat. Erosion may also be so great that the entire pin is swept away. Despite these drawbacks, erosion pins are a common and inexpensive tool used in the monitoring of bank erosion and gully development.

Scour chains

Scour chains can be installed at various points of interest in the channel bed to monitor the amount of scour and fill during floods. These are heavy

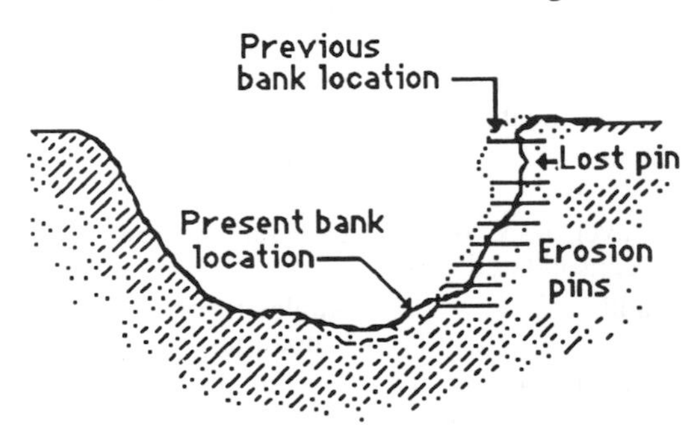

Figure 5.26. Use of bank-erosion pins to monitor the retreat of a streambank

link-type metal chains, cut to a length corresponding to the expected amount of scour. The chains are anchored onto a metal plate, pin or weight of some type, and buried vertically in the gravels or sands of the bed (Figure 5.27).

When a flood scours away the sediments the exposed chain falls flat, forming a "kink". Subsequent filling reburies the chain. The amount of scour can then be found from the original vertical length as compared to the length below the kink, and the amount of fill found from the depth of sediment deposited above the kink. Obviously, it is very important to know the exact location of the scour chain by surveying so that it can be later recovered. Scour chains measure the maximum amount of scour during a flood, which may or may not occur at the same time as the flood peak.

5.10 SUBSTRATES AND SEDIMENTS: ANALYSIS OF PHYSICAL PROPERTIES

5.10.1 GENERAL

The fundamental properties of sediment and soil particles are size, shape, mineralogical composition, surface texture and orientation in space. Additionally, bulk properties include colour, average density, porosity, and permeability. Some measures are of more interest if one is building a house on the soil, whereas others are more important if the intention is to grow plants in it. For the purposes of this text, only the properties related to the erodibility of the streambed and banks and sediment transport will be treated. The reader is referred to texts on soils, sedimentology and engineering soil mechanics for more information such as Brewer (1964) and Craig (1983). Minshall (1984) also provides methods of describing substrate characteristics of biological relevance.

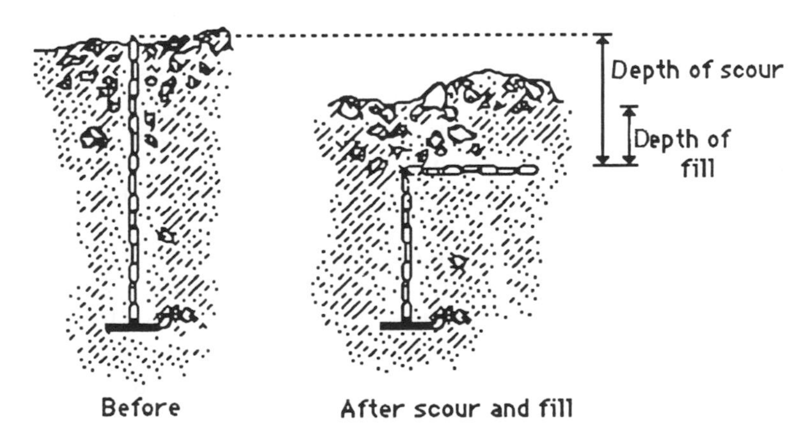

Figure 5.27. Measuring scour and fill with a scour chain

5.10.2 SOIL MOISTURE CONTENT, GRAVIMETRIC METHOD

The gravimetric method is a standard method for soil moisture analysis. A 1–100 g sample is placed into a labelled, pre-weighed container and the "wet weight" of the soil and container is measured using laboratory scales. The sample is then dried at 105°C for 24 h or until the weight stabilizes at a constant value (Brakensiek, 1979). If a soil can is used the lid should be removed before the sample is dried. Other methods of drying a sample using microwave ovens and heat lamps are given by Klute (1986).

Drying time will depend on sample size, the number of samples in the oven, moisture content of the samples, placement within the oven and type of oven. The samples should be cooled before weighing, preferably in a desiccator to prevent re-absorption of moisture from the air. After weighing, the moisture content (as a percentage by mass) is calculated from:

$$\text{Moisture content (\%)} = \frac{\text{Wet mass} - \text{dry mass}}{\text{Wet mass}} \times 100 \tag{5.30}$$

Here, the soil mass is in grams, calculated as:

$$\text{Mass of soil} = (\text{mass of soil} + \text{can}) - (\text{mass of can}) \tag{5.31}$$

5.10.3 SEDIMENT CONCENTRATION

Filtration

In the filtration method a measured volume of well-mixed sediment sample is filtered and the amount trapped on the filter is weighed after drying. The minimum sediment size measured is controlled by the filter size. The choice of filter depends on the type of filter holder available, the particle size, filtration rate and cost.

For vacuum filtration using an apparatus similar to the one shown in Figure 5.28(a) cellulose nitrate membranes with a pore size of 0.45 mm are normally used, although glass fibre filters are also available. An apparatus can also be constructed with a Buchner funnel and a glass vacuum flask (Figure 5.28(b)). Glass fibre or paper filters would be used such as Whatman No. 92 with a pore size of 0.005–0.01 mm or Whatman GF/F which traps particles down to 0.0007 mm.

If the sediments are predominantly sands, coarser filters can be used and vacuum filtration is not necessary. Generally, paper and glass fibre filters have higher flow rates and are cheaper than membrane filters, although membranes are preferable for precise work because their pore size is more precisely controlled. With any filter the filtration rate can be expected to decrease with time as the filter becomes clogged, especially if fine particles are present.

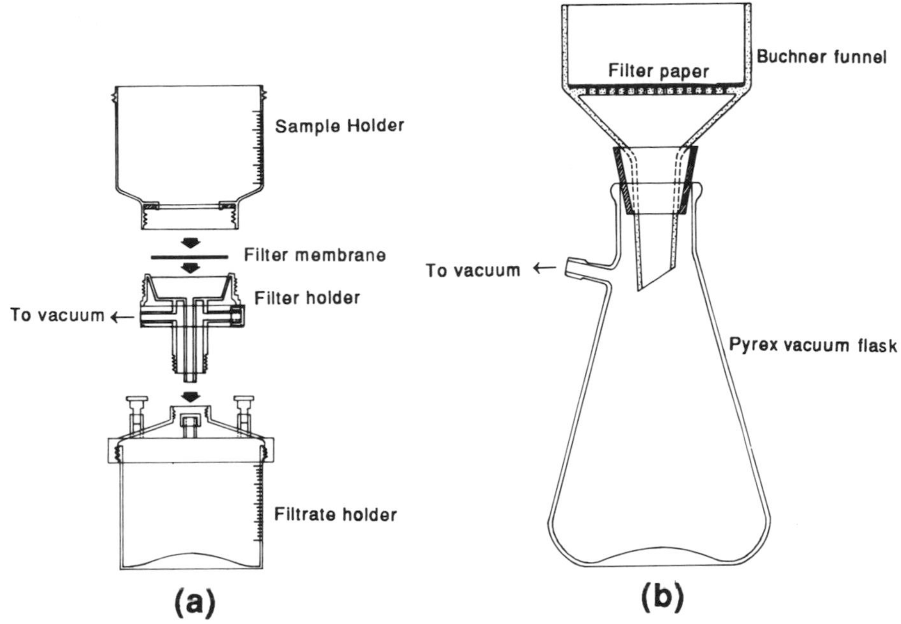

Figure 5.28. Vacuum filtration apparatus for sediment concentration analysis: (a) commercial-type for use with membrane filters and (b) Buchner funnel-type for use with filter paper. Reprinted from Finlayson (1981), by permission of the *Journal of Geography in Higher Education*

A clean, dry filter is weighed prior to use, preferably using a four-figure analytical balance. Filters should only be handled with tweezers. The sediment sample is well agitated and a volume of water–sediment mixture measured with a graduated flask. This is quantitatively transferred to the vacuum funnel using distilled water to wash out any sediment sticking to the flask. The sample is then filtered and the filtrate discarded.

The filter and accumulated sediment should be placed in ceramic crucibles or glass petri dishes and dried overnight at 105°C. After drying, the filter is cooled in a desiccator to room temperature and then weighed again. The sediment concentration (c_s) is expressed in terms of sediment mass per measured volume of water–sediment mix:

$$c_s\,(\mathrm{mg/l}) = \frac{(\text{Mass of filter + sediment}) - (\text{mass of filter})}{\text{Volume of water–sediment mix}} \times 10^3 \quad (5.32)$$

where mass is in grams and the volume is in litres.

A concentration in mg/l is nearly equivalent to ppm (parts per million) by mass for concentrations up to about 16000 ppm. Guy (1969) describes how larger concentrations are converted from mg/l to ppm.

Alternatively, the mass of sample and sample bottle can be determined, and the whole sample filtered. The concentration would then be expressed as a ratio (with all masses in grams):

$$c_s(\text{ppm}) = \frac{(\text{Mass of filter + sediment}) - (\text{mass of filter})}{(\text{Mass of sample + bottle}) - (\text{mass of bottle})} \times 10^6 \quad (5.33)$$

Evaporation

A less involved method is to evaporate away the water by pouring a known volume (or mass) of sediment–water mixture into a container and then drying and weighing it. To save drying time, the measured solution should be allowed to settle for at least 12 h so excess sediment-free liquid can be siphoned or poured off. Sediment concentration is calculated as in Equations 5.32 or 5.33 after replacing "mass of filter" with the mass of the container.

With this method, and the method described by Equation 5.33, the measured "sediment" concentration will also include dissolved solids. Thus, the technique is best for streams where the salt-mineral content is low in comparison with the suspended load, as in headwater streams or during times of heavy sediment runoff. A method of adjusting the concentration is to separately analyse the weight of dissolved solids in the sampled solution. Alternatively, the electrical conductivity (EC) can be measured and a relationship between EC and dissolved solids used to make the correction.

5.10.4 PARTICLE SIZE

Particle size analysis can be applied to any mixture of sediments, whether collected as suspended or bedload sediment samples, bank materials, samples of floodplain deposits or grit from the crop of a cockatoo. Particle size is a somewhat nebulous length parameter which can be defined by various measures. These include the width of the smallest square mesh through which a particle can pass, the diameter of a circle with an area equal to the maximum projected area of the particle, the diameter of a sphere with a volume or settling velocity which equals that of the particle, or simply the longest dimension of the particle (Goudie, 1981).

Size classes have been developed to give a descriptive label to particles grouped within a given size range. Several classifications exist, and standard classes can vary between countries and agencies. Traditionally, the length dimension has been expressed in millimetres, as described by the Wentworth scale in Table 5.6. Sedimentologists typically use the phi (ϕ) scale, also shown in Table 5.6, where phi is equal to the negative logarithm (in base 2) of the particle size in millimetres. For very small particles this method eliminates the inconvenience of unwieldy numbers. Equation 4.7 can be used to convert from mm to ϕ.

The method of particle size analysis should be chosen based on the type and size of material being analysed and the accuracy required. The method will also vary with the amount of sediment available for analysis. For example, many suspended sediment samples will not contain enough sediment for accurate analysis of the larger sizes, and the analysis must be limited to determining the percentage of silts and clays. Methods will be presented in order of their simplicity. For full details of particle size analysis techniques, Allen (1981), BS (1975), Day (1965), Gee and Brauder (1986) or Lewis (1984) can be consulted.

Visual analysis

General classification of surface sediments is often done by eye when assessing the distribution of substrate types in a stream reach. A collapsed version of the grade scale given in Table 5.6 can be used to visually classify

Table 5.6. Grade scales for particle size. Adapted from Brakensiek et al. (1979)

Class (Wentworth)	mm	ϕ
Very large boulder	4096–2048	−12 to −11
Large boulder	2048–1024	−11 to −10
Medium boulder	1024–512	−10 to −9
Small boulder	512–256	−9 to −8
Large cobble	256–128	−8 to −7
Small cobble	128–64	−7 to −6
Very coarse gravel	64–32	−6 to −5
Coarse gravel	32–16	−5 to −4
Medium gravel	16–8	−4 to −3
Fine gravel	8–4	−3 to −2
Very fine gravel	4–2	−2 to −1
Very coarse sand	2–1	−1 to 0
Coarse sand	1–0.5	0–1
Medium sand	0.5–0.25	1–2
Fine sand	0.25–0.125	2–3
Very fine sand	0.125–0.0625	3–4
Coarse silt	0.0625–0.0312	4–5
Medium silt	0.0312–0.0156	5–6
Fine silt	0.0156–0.0078	6–7
Very fine silt	0.0078–0.0039	7–8
Coarse clay	0.0039–0.0020	8–9
Medium clay	0.0020–0.0010	9–10
Fine clay	0.0010–0.0005	10–11
Very fine clay	0.0005–0.00024	11–12

sediments as, for example, boulders, cobbles, gravel, sand, silt or clay. Since silt and clay are not easily distinguished, they are sometimes combined into a single class, "mud". As with texture triangles used for soil classification, the triangle given in Figure 5.29 can assist in the standardization of substrate classes.

Field identification of sand-sized sediments can be assisted by developing a collection of samples of known size in vials, test tubes, or glued onto plastic slides. A hand lens is useful for identifying smaller grains.

The areal survey methods of Section 5.3 can be used to assess the composition of surface sediments. For most streambed materials a one metre square frame is a functional plot size. Thin wire is strung across the frame every 0.1 m. The grid is placed over the sediment with the top of the grid in the upstream direction. The particle sizes in the surface layer can be measured in the field or by taking a photograph and analysing percentage cover using the grid for scale. Photographs should be taken from a consistent height above the grid, as near to vertical as possible (Hamilton and Bergersen, 1984). A card with information on the site and date can be set

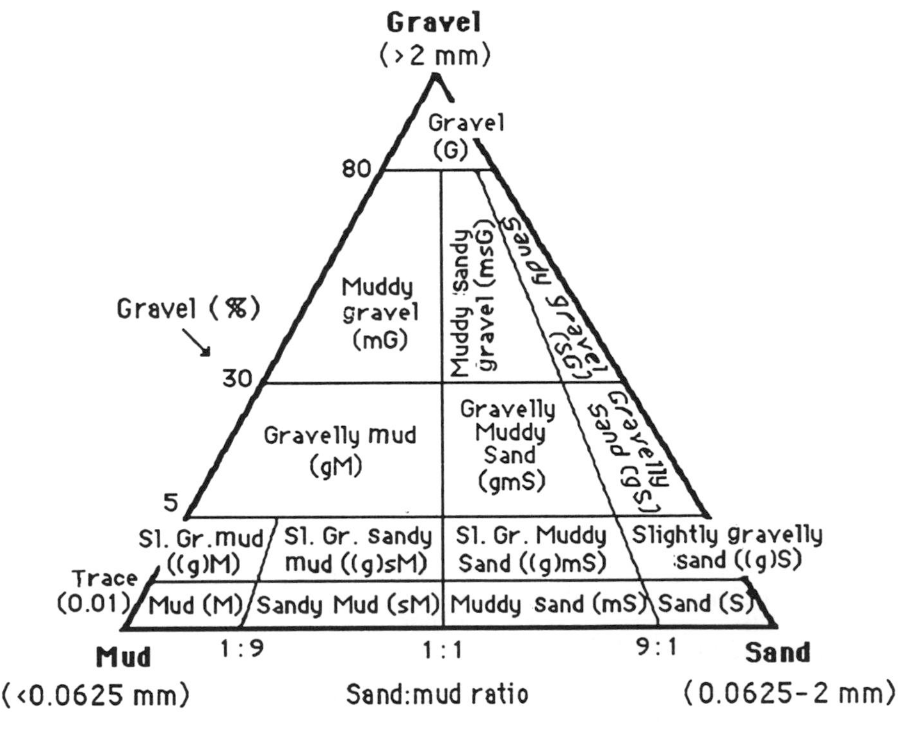

Figure 5.29. Classification of substrates according to gravel, sand and mud composition, *sl* = slightly. Redrawn from Folk (1980), by permission of Hemphill Publishing Co.

on a corner of the frame before the picture is taken. Photographs additionally provide a permanent record of substrate rock type and orientation.

"Hand texturing" of soils

A somewhat subjective but extremely low-effort method of analysing the size composition of soils and finer sediments is by doing a "texture by feel" analysis. Texture is defined as the relative proportions of sand, silt and clay, as shown in Figure 5.30.

The soil composition is estimated from the feel and malleability of a wetted sample (a "bolus"). By working the bolus between the thumb and forefinger a thin "ribbon" can be created with more coherent soils, its length determined by the soil type. Running the sample under a stream of water can help in detecting the sand content by feel. Table 5.7 is a general key for the textural analysis of soils by observation and feel.

To "calibrate" a person to the feel of different soil types, samples can be compared to "type" samples which have been analysed for particle composition by more accurate methods. Such type samples are available in most universities or testing laboratories.

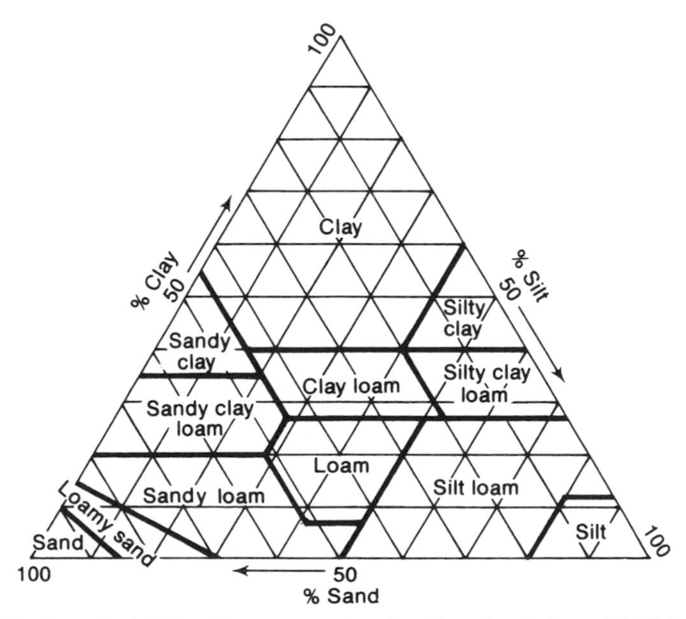

Figure 5.30. Standard US soil texture triangle. Reprinted from McKnight (1990), © 1990, p. 329, by permission of Prentice-Hall, Inc.

Direct measurement

Individual boulders, cobbles and large gravels can be measured directly in the field. Callipers or a rule or survey staff can be used, depending on the size. For smaller particles, graduated eyepieces in microscopes can be employed to make direct measurements.

This method should include multiple measurements along three axes: the *A* or longest axis, the *B* or intermediate axis, and the *C* or shortest axis. Each axis is perpendicular to the other two. If only one axis is measured, the *B* axis length will give an adequate estimate of mean diameter for most particles (Briggs, 1977a).

The short axis (*C*) is relatively uncomplicated to identify since it is simply the shortest axis. The *B* axis is defined as the shortest axis of the "maximum

Table 5.7. Soil texture classification by feel. Adapted in part from Northcote (1979)

Texture class	Behaviour of moist bolus of soil
Sand	Crumbles readily; cannot be moulded; single sand grains adhere to fingers
Loamy sand	Slight coherence; can be sheared between thumb and forefinger to give minimal ribbon of about 6 mm; discolours fingers with dark organic stain
Sandy loam	Bolus just coherent but very sandy to touch; will form a short ribbon; dominant sand grains can be seen, felt or heard
Sandy clay loam	Strongly coherent bolus, sandy to touch; medium size sand grains visible in finer matrix; will form a longer ribbon than sandy loam
Loam	Coherent and rather spongy bolus; smooth feel when manipulated but with no obvious sandiness or "silkiness"; may be somewhat greasy to the touch if much organic matter present; will form a short ribbon
Silt loam	Coherent bolus, very smooth to silky when manipulated; may form short ribbon
Silt	Pure silt will have a smooth, floury or silky feel; bolus can be manipulated without breaking
Silty clay loam	Coherent smooth bolus; plastic and silky to the touch; will form longer ribbon than loam
Clay loam	Coherent plastic bolus; smooth to manipulate; will form ribbon similar to silty clay loam
Sandy clay	Plastic bolus; fine to medium sands can be seen, felt or heard in clayey matrix; will form a thin, long ribbon which breaks easily
Silty clay	Plastic bolus; smooth and silky to manipulate; will form long ribbon
Clay flexible	Handles like plasticine, plastic and sticky; will form a long, ribbon of 5 cm or more

projection plane" (the plane of largest area, perpendicular to the *C* axis). Rather than being the longest axis, the *A* axis is measured perpendicular to the *B* axis. As shown in Figure 5.31 for a "tabular" pebble, the *A* axis is not the corner-to-corner length.

The value in defining *A* and *B* in this way is that it most closely approximates the results which would be obtained if a sieve were used, because it is the *B* axis which determines whether or not an individual pebble falls through a mesh of given size. This is also the reason why, if only one measurement is taken, it is made along the *B* axis.

Dry sieving

Dry sieving is the most commonly used method for the analysis of sand-sized particles. Particles larger than sand are often too bulky and heavy to sieve, and those smaller than sand tend to form aggregates which are not easily sieved in their dry state. If considered significant, carbonates which cement particles together can be removed using dilute hydrochloric acid, and organic matter can be removed with hydrogen peroxide (Goudie, 1981). Fines should be dispersed prior to drying and sieving with a dispersing agent such as sodium hexametaphosphate (Calgon) or with an ultrasonic probe. These fines are washed through a 0.0625 mm sieve and collected for separate analysis by hydrometer or by drying and weighing the total fraction.

Large samples of deposited materials may require splitting before sieving, either by using commercial sample splitters or by successive subsampling. Reducing samples to a manageable size must be done with care, as error can be introduced if a non-representative subsample is obtained. One may also wish to separately pick out and weigh all of the larger-sized rocks. Further subdividing may be desirable to prevent overloading the sieves when

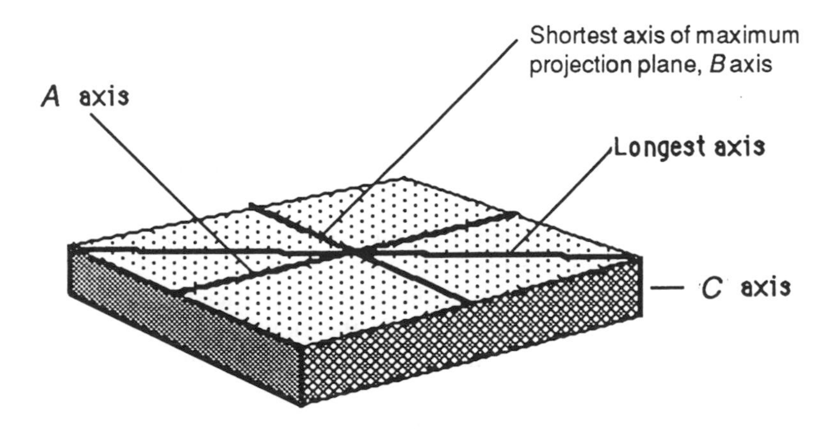

Figure 5.31. Axes on a tabular "pebble"

working down to sieve sizes of 2 mm and finer. The relative size of the subsamples must be used to reconstruct the proportion in each fraction.

Dry sieving requires a completely dry sample. Even if moisture is only 1%, adhesion forces can exceed the weight of grains smaller than 1 mm, preventing them from passing through the smaller sieves. The sample is allowed to air-dry at room temperature by spreading it thinly on trays or newsprint and leaving it for several days. The process can be sped up using a heat lamp or an oven. After drying, aggregates in the sample can be broken up by hand or rolling pin or by grinding, although care should be taken to avoid fragmenting the particles.

A set of sieves of required sizes is stacked together, decreasing in aperture size downwards. Commercially available sieves are sized in a geometric series, normally corresponding to particle size classes (Table 5.6). A regular interval between the sieves should be used. A common convention is to use a 1ϕ interval with a 4ϕ (0.0625 mm) sieve as the smallest size and a -6ϕ (64 mm) sieve as the largest. This would give a "nest" of 11 sieves.

A 100–200 g sample of material is weighed out and placed in the top of the coarsest sieve. The nest of sieves is placed on a shaker with a lid on the top and a tray on the bottom, and the sample is shaken for 10–15 min. If, after shaking, more than a few per cent of the sample is held within the top sieve or in the bottom tray, the procedure should be repeated to further separate the larger and smaller fractions.

The material trapped on each sieve is transferred to a weighing tray, and the sieves are gently brushed to release sediments stuck in the mesh. The mass of the sediment for each size fraction is recorded. The total of all fractions should be checked with the original weight, although some loss is to be expected. The material passing the 0.0625 mm (4ϕ) sieve is normally retained for analysis by hydrometer or pipette.

Wet sieving

Wet sieving is a good method for sizing coarse particles and sand-sized particles when aggregation problems are encountered. If only the sizes of coarser gravels and larger particles are of interest the method is very simple. The coarsest sieve is held over a bucket and a large sample of sediment is poured onto it. The sieve is shaken, with washing as needed, until all particles smaller than the sieve aperture have passed through. The trapped particles are weighed wet because the mass of the water is considered insignificant compared to the mass of these larger particles. The sediments and wash water which has passed through the sieve are poured onto the next smallest sieve and the procedure is repeated until the smallest sieve size is reached.

For particles less than about 8 mm the mass of the water becomes significant and the procedure becomes slightly more involved and time consuming. After each sieving the trapped particles are transferred to a pre-weighed container. Each fraction, including that passing the 0.0625 mm sieve, is separately dried and weighed. The total sample mass is then assumed to be equal to the sum of the masses of the separate fractions. The fraction smaller than 0.0625 mm can be saved for analysis by hydrometer or pipette.

With either wet sieving or dry sieving, the particle shape, sieve opening shape and type and time of shaking will affect the probability of a particle passing through. Standardization of the sieving method is important to ensure reproducibility.

Hydrometer method

Hydrometer analysis is based on Stokes' law (Section 6.5). A hydrometer measures the density of a water–sediment suspension, which changes as the sediment settles out. Hydrometers can be purchased which are calibrated directly in units of concentration, as measured at a specific temperature. Since, from Stokes' law, larger particles fall more rapidly than smaller ones, an estimate of particle diameter can be obtained by measuring the time taken for them to settle. The method given here is a "short-cut" version. Gee and Bauder (1986) provide details of a more standardized method from which a particle size distribution curve (Section 5.10.5) can be developed as well as the percentages of sand, silt and clay.

For either the hydrometer or the pipette method (next section) the sample should be sieved to remove all particles greater than 2 mm. The sample can be pre-treated to remove substances which may affect results such as carbonates, soluble salts, organic matter and iron oxides. A 40 g sample of oven-dry sediment is weighed into a beaker, into which 250 ml of distilled water and 100 ml of sodium hexametaphosphate (Calgon) solution (50 g/l) are added. The solution is allowed to soak overnight. The Calgon solution reduces flocculation of clays, which can have a significant effect on settling times. For fine-textured sediment the sample weight can be reduced, whereas for coarse sands up to 100 g may be needed to ensure reproducible results.

The treated sample is transferred to the cup of an electric stirrer (malted milk-mixer type) and mixed for 5 min. The mixture is then transferred to a sedimentation cylinder and distilled water is added to bring the volume up to a 1-litre mark. A plunger is inserted into the cylinder and the water–sediment solution is vigorously mixed. The stirring is finished with a few slow, smooth strokes before the plunger is removed. Alternatively, the cylinder can be sealed with a rubber stopper and shaken end-over-end for about 1 min.

Timing begins immediately after mixing is stopped. The hydrometer is lowered quickly and smoothly into the suspension and readings are taken at prescribed time intervals. Readings are taken at the upper edge of the meniscus. A drop or two of ethyl alcohol can be added to reduce foaming. The hydrometer is removed, rinsed off (not over the cylinder) and wiped dry. Before each subsequent reading, the hydrometer is carefully re-inserted into the cylinder.

The method given by Foth (1978) involves taking only two readings. One is taken at 40 s to define the amount of silt and clay remaining in suspension and the other at 2 h to measure the amount of clay remaining. It is preferable to keep the cylinders in a room or chamber which maintains a relatively constant temperature.

If the hydrometer is calibrated at 20°C in grams per litre the calculations are straightforward:

$$\% \text{ sand} = \frac{M - R_1}{M} \times 100 \tag{5.34}$$

$$\% \text{ clay} = \frac{R_2}{M} \times 100 \tag{5.35}$$

and

$$\% \text{ silt} = 100 - (\% \text{ sand} + \% \text{ clay}) \tag{5.36}$$

where M is the mass of the sample in grams, R_1 is the first reading (40 s) and R_2 is the second reading (2 h). Since the hydrometer gives a reading in grams per litre and there is 1 litre of solution, R_1 and R_2 are in grams. To correct for temperature effects, readings of both the sediment solution and the blank should be adjusted by adding 0.3 g for every 1°C above 20°C, or subtracting 0.3 g for every 1°C below 20°C (Briggs, 1977b).

Pipette analysis

Pipette analysis, again based on Stokes' law, is perhaps the most accurate method of determining the size distribution of fine sediments. It is often considered the standard method against which other methods are compared. In pipette analysis, changes in concentration of a settling suspension can be found by analysing small subsamples with a pipette at specific settling times.

The standard pipette procedure is very sophisticated, requiring extensive washing and other pre-treatment of the sample and a vacuum apparatus for extracting the subsample with a pipette from a specific depth in the cylinder. Brakensiek et al. (1979) and Gee and Bauder (1986) give details of the standard method and the apparatus required. The method presented here, given by Briggs (1977b), is a much simplified version. A few samples sent to a professional soil testing laboratory for verification will give an indication of the accuracy of this method.

A 20–50 g sample of dry sediment is weighed into a 500 ml measuring cylinder with 0.5 g Calgon. The cylinder is topped up to the 500 ml mark with water. As in the hydrometer method, the sample is well mixed with a plunger or by shaking. As soon as stirring is finished, a pipette is immediately lowered 100 mm into the sample and 10 ml is withdrawn. The aliquot is emptied into a pre-weighed, labelled petri dish, beaker or other weighing container, and the pipette rinsed out into the container.

The procedure is repeated at 3 min 50 s, and then again at 8 h 10 min after the starting time. The 100 mm sampling depth should always be measured from the existing surface and not from the original 500 ml mark. The dishes are dried overnight at 105°C and weighed to obtain the amount of sediment in each sample. Percentages of coarse silt, fine and medium silt, clay and sand are calculated as follows:

$$\%\text{ coarse silt} = \frac{50\,(M_1 - M_2)}{M_T} \times 100 \tag{5.37}$$

$$\%\text{ fine and medium silt} = \frac{50\,(M_2 - M_3)}{M_T} \times 100 \tag{5.38}$$

$$\%\text{ clay} = \frac{50\,(M_3)}{M_T} \times 100 \tag{5.39}$$

and

$$\%\text{ sand} = 100 - (\%\text{ coarse silt} + \%\text{ fine and medium silt} + \%\text{ clay}) \tag{5.40}$$

where M_1, M_2 and M_3 are the masses of the dried sediment in the immediate, 4 min, and 8 h samples, respectively, and M_T is the original dry sample mass, with all masses in grams.

This procedure was developed assuming a constant temperature in the suspension of 20°C. Lower temperatures will reduce viscosity, causing slightly lower settling velocities and biasing the results towards more large particles; higher temperatures will have the reverse effect. Variations in particle density or the effect of Calgon concentration on viscosity may also influence settling times.

Other methods of particle size analysis have used light-scattering principles, such as Coulter counters, laser particle sizers and turbidimeters. Because of their high cost and uncertainties in correction factors, however, they are not routinely used except by commercial laboratories.

5.10.5 PRESENTATION OF PARTICLE SIZE DATA

Sediments can be described simply by the dominant size, by the proportion in each size or by statistical measures which describe the distribution of sizes. The results of a particle size analysis can also be plotted as a histogram to show the percentages in each size grade. This effectively illustrates the size distribution and its skewness and/or bimodality (see Appendix 1). Sediment particle sizes often tend to follow a log-normal size distribution, with a high proportion of particles in the low-to-middle size class and progressively fewer toward the extremes.

If sediment samples are to be compared it is more useful to plot results as a cumulative frequency curve. Commonly, a logarithmic *X*-axis is used for the particle size in millimetres, with a linear *Y*-axis indicating the "percentage finer by mass". This typically yields an S-shaped curve, as shown in Figure 5.32. The curves give an indication of the spread of sizes present: the larger the range, the flatter the distribution curve, and the more uniform the sample, the more vertical the curve. In Figure 5.32 the sample from Site 3 is the most uniform.

If particle sizes are individually obtained by direct measurement it is more appropriate to consider the *number* of particles of a certain size rather than the *mass*. Probability paper is normally used for displaying these *frequency* data, as shown in Figure 5.33. Particles are ranked from smallest to largest and a plotting position is calculated as $100(m/N)$, where m is the rank (0 = smallest) and N is the number of samples (see Section 8.2.4). This gives the "% finer" value for the *X*-axis. For example, if 30 rocks were measured the largest particle would be plotted at $100(29/30) = 96.7\%$. Because the scale does not include 0% the smallest particle is left off.

It becomes difficult to describe mixtures of very different size classes

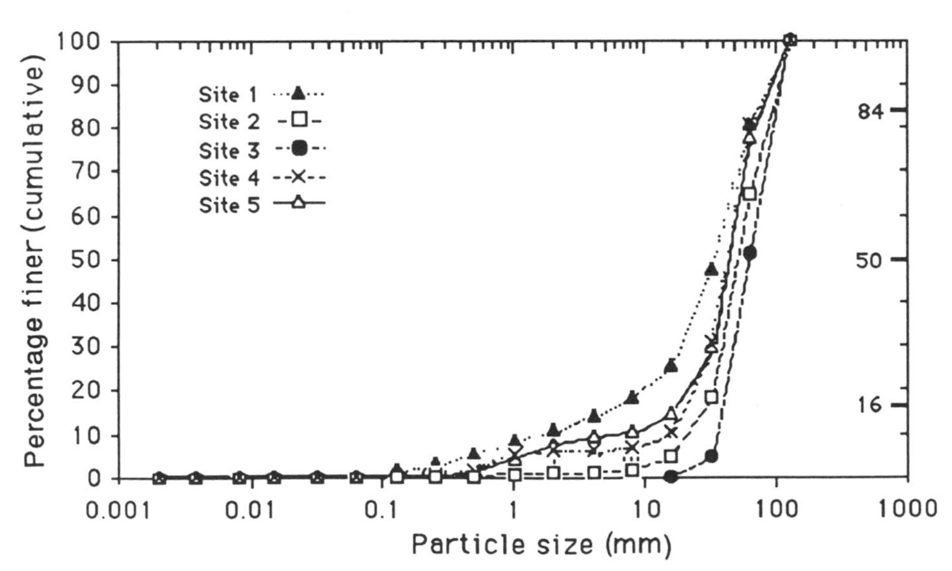

Figure 5.32. Bed surface material particle size distributions, by mass, for five sites on the Acheron River. Sites correspond to numbered locations in Figure 4.7. Composited samples of surface materials <128 mm collected from several points along a stream reach were analysed

because the method of measurement changes over the range of sizes analysed (e.g. hydrometer, sieving, direct measurement). Techniques for combining these data are given by Church et al. (1987) and Griffiths (1967).

Indices for describing the particle size distribution include the mean, median, standard deviation, skewness and other conventional statistical measures (see Appendix 1). These can be calculated from the data or taken directly from a cumulative frequency plot. Sediment particle sizes are often simply described by one percentile value such as their d_{50} or d_{84} or d_{90} values which can be easily read from a cumulative frequency curve. The d_{50}, for example, is the median value, meaning that half the sample (by mass or frequency) is larger, half smaller. The d_{84} and the d_{16}, the diameters for which 84% and 16% of the particles are smaller, respectively, have particular significance since they represent one standard deviation from the mean in a normal distribution (Appendix 1).

The median tends to be a more robust measure than the mean for describing sediment size. The problem with using the mean value is that a few particularly large rocks will bias the mean heavily toward the coarse end. This can be avoided by removing unrepresentative particles from the sample or by using a formula which ignores the "tail ends" of the distribution. However, the preferable approach is to use the mean of the phi values (called the phi mean) on the assumption that the particle sizes are log-

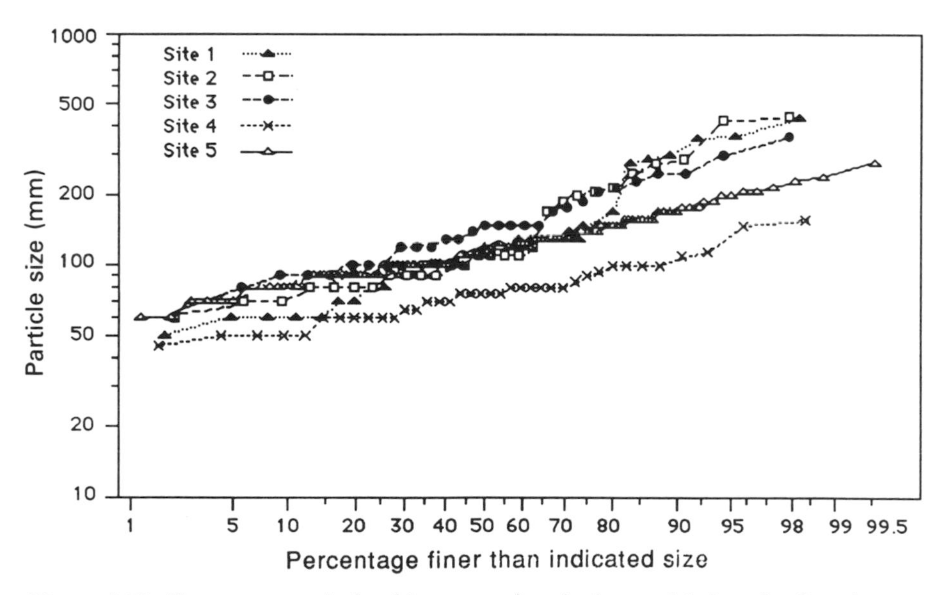

Figure 5.33. Frequency analysis of larger surface bed material sizes for five sites on the Acheron River. Sites correspond to numbered locations in Figure 4.7. Samples were taken by walking up the stream reach and measuring a rock at each step. The particle size is represented by the *B* axis (Section 5.10.4). Lines connecting the points have been drawn in

Table 5.8. Statistical parameters for describing particle size distribution based on ϕ values. Equation 4.7 can be used to convert from millimetres to ϕ and back. After Briggs (1977a)

Parameter	Method of calculation
Median	$\phi 50$
Mean	$\frac{(\phi 84 + \phi 16)}{2}$ or
	$\frac{(\phi 16 + \phi 50 + \phi 84)}{3}$ or
	$\frac{(\phi 10 + \phi 20 + \cdots + \phi 90)}{9}$
Standard deviation	$\frac{(\phi 84 - \phi 16)}{2}$
Skewness	$\frac{(\phi 84 - \phi 50)}{(\phi 84 - \phi 16)} - \frac{(\phi 50 - \phi 10)}{(\phi 90 - \phi 10)}$
Kurtosis	$\frac{\phi 90 - \phi 10}{1.9(\phi 75 - \phi 25)}$

normally distributed. Under this assumption the distribution of particle sizes can be completely described by the phi mean and standard deviation.

Formulae for the mean and other statistical measures reflecting the central tendency and shape of the distribution are given in Table 5.8. In natural sediments the parameters tend to be related; i.e. sediments of a larger mean particle size tend to have a larger range of particle sizes and thus a higher standard deviation, whereas finer sediments are more uniform.

5.10.6 PARTICLE SHAPE: ROUNDNESS, SPHERICITY

The form of a particle (i.e. its three-dimensional configuration) can be described in terms of its shape, sphericity, roundness and surface texture. The description of form should be relevant to the field of interest; i.e. whether sedimentological or biological.

As with many measures, an index of form will be dependent on scale. For example, the overall form of a lava rock pockmarked with air cavities may be fairly round, but its surface texture, its ability to be transported by the stream and the number of nooks for benthic organisms would be very different from a smoother, more solid rock of the same roundness. Surface texture of particles is commonly analysed by scanning electron microscope, and methods will not be covered here. Goudie (1981) provides an interesting discussion on particle form, and is a good place to start pursuing the large and varied literature on this topic.

Two common concepts used in the description of particle shape are sphericity and roundness. The two terms are geometrically distinct; i.e. a pebble with a high sphericity may not necessarily possess a high roundness value. **Sphericity** is a measure of the ratio of a particle's volume to the volume of the sphere which circumscribes it. A sphere is used as a reference form because of the common assumption in many formulae such as Stokes' law (Section 6.5) that sediment particles are spherical. Sphericity is closely related to the surface area to volume ratio, a fundamental measure of the shape of a particle which reaches a minimum when the particle is spherical. The sphericity of a particle is thus important in controlling lift, settling velocity and sediment transport.

Roundness is a measure of "roughness" or "angularity" of the particle, and describes the sharpness of the "corners" on a particle. It is thus strongly affected by abrasion. The roundness of a particle typically changes rapidly during the initial stages of transport, then at a slower rate as the particles continue to be polished. Often, the original shape is reflected in the particle even after a considerable amount of wear.

The techniques given for estimating sphericity and roundness were developed for particles of sizes which are easily handled. The observation of smaller particles can be assisted using a microscope or an overhead projector to increase the viewed size. Scanning equipment and digital image analysis also have potential for describing particle form. For particles of 1 metre size and larger, surveying equipment may be required.

Sphericity and particle shape

Krumbein (1941) developed a definition of sphericity by using a ratio of the volume of an ellipsoid defined by the A, B and C axes to the volume of a sphere circumscribed around the particle defined by the A axis. Sphericity (ψ) can thus be expressed as:

$$\psi = \sqrt[3]{\frac{BC}{A^2}} \tag{5.41}$$

The A, B and C axes are measured as described in Section 5.10.4. Program SPHER in AQUAPAK will calculate sphericity. Krumbein (1941) states that sphericity values less than 0.3 rarely occur in nature.

For individual particles the assumption of a triaxial ellipsoid shape may not be valid, but it is approximated when the average of a group of pebbles is taken. Krumbein therefore recommends that 25 or more pebbles should be used in order to obtain agreement with other methods such as Wadell's (1932, 1933), where the actual particle volume is measured.

Zingg (1935) classified pebbles into four basic shapes: disc, spherical, bladed and rod-like. Brewer (1964) later added three additional classes. Ratios of axis lengths, *B*/*A* and *C*/*B* are used to distinguish the classes, as shown in Figure 5.34.

Roundness

Roundness, a rather vague measure, was defined by Wadell (1932) as the ratio of the curvature of the corners and edges to the average curvature of the particle as a whole. However, Wadell's method of determining particle roundness is very complex. It involves constructing a "planform" image of the particle's maximum projection plane (see Section 5.10.4) and measuring the radii of all of the corners. Roundness is expressed as the ratio of the average radius of curvature of the corners to the radius of the smallest circle which can be inscribed around the maximum projection plane. Semi-circles of known radius can be drawn on a card which can be plastic coated for field estimation of these quantities.

A more practical field method for estimating roundness is to simply visually estimate it using a chart such as the one shown in Figure 5.35. These images were drawn by Krumbein from pebbles classified by Wadell's method. The original drawings (reduced here) were for pebbles of 16–32 mm diameter. Enlargement or reduction can be used to create charts for any size range.

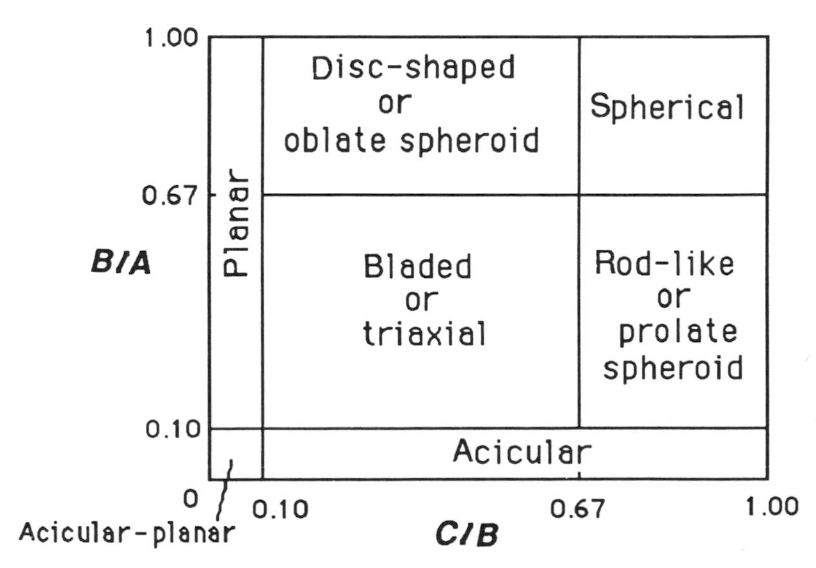

Figure 5.34. Chart for the classification of particle shape. Redrawn from Brewer (1964), adapted from Krumbein (1941), by permission of John Wiley & Sons, Inc.

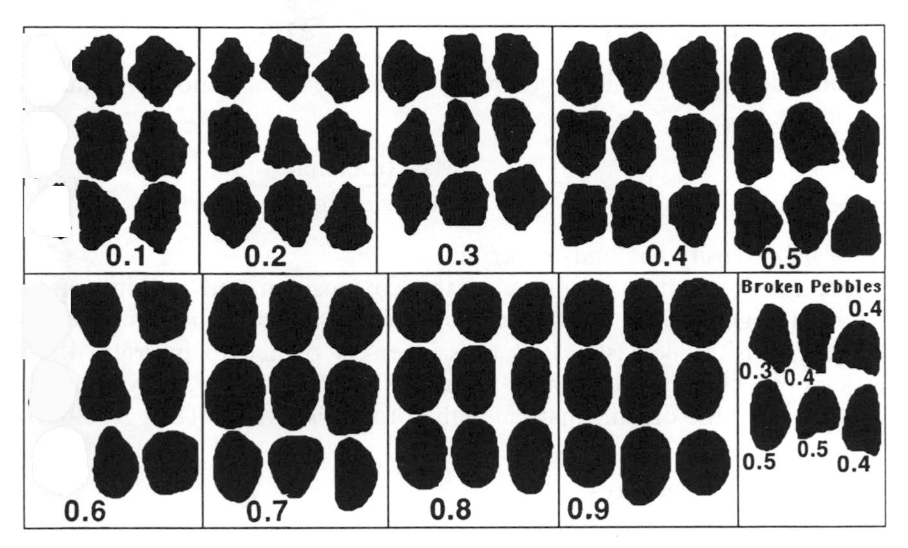

Figure 5.35. Visual comparison chart for the estimation of pebble roundness. Redrawn from Krumbein (1941)

When using the chart the particle should be viewed such that the largest projection is observed. When estimating a roundness value all the smaller corners and edges should be considered as well as the main ones. Broken pebbles are given a roundness rating for the unbroken half and the value is subsequently halved (in the case of an odd value, e.g. 0.7, the value should be rounded down; i.e. 0.3 rather than 0.4).

A set of at least 25 pebbles should be sampled to obtain an average roundness value. It may be advisable to divide pebbles into size classes or even rock type, as there is often a marked change in the roundness of pebbles of different sizes or rock type such as shales and granites. Separation in this manner also makes it easier to compare sediments from different sites, as this can be done on a size or type basis.

5.10.7 PARTICLE ARRANGEMENT AND OTHER MISCELLANEOUS BULK PROPERTIES

The arrangement of sediment particles affects the degree of "packing" of grains, which in turn has an effect on the erodibility of substrates and their permeability to air, water and micro-organisms. Packing is a complex factor involving size, grading, orientation and form (Goudie, 1981). Erodibility of soils will also be affected by moisture content (Section 5.10.2) and particle size (Section 5.10.4). In this section, measures are presented for bulk density, porosity, aggregate stability, embeddedness and sorting.

Bulk density and porosity

The **bulk density** of a soil or bed material sample is the ratio of its mass to its volume:

$$\text{Bulk density} = \frac{\text{Mass of sample}}{\text{Volume of sample}} \qquad (5.42)$$

where bulk density has units of kg/m^3.

Bulk density is thus an index of the composition and packing of the material (Briggs, 1977b). As the volume of voids between particles increases, the bulk density decreases. It is thus inversely related to **porosity**. If the particles can be assumed to have a density of 2650 kg/m^3 (sandy, siliceous particles with little organic matter) then porosity is given as (Briggs, 1977b):

$$\text{Porosity (\%)} = \left(1 - \frac{\text{Bulk density}}{2650}\right) \times 100 \qquad (5.43)$$

Bulk density can simply be measured by taking a sample core of known volume, drying the sample and dividing the dry weight by the core volume. The appropriate volume for a cylindrical core sample is $h(\pi d^2/4)$, where d is the diameter and h is the height of the cylinder. An alternative is the displacement method, which is more applicable for individual rocks or soil clods or other samples of unknown volume, such as those taken by freeze coring. In this method the particle mass is coated with wax or plastic resin (if needed), and the mass is completely submerged in a container filled with water. The volume of water displaced is equal to the volume of the particle mass. The coating is then removed and the sample oven-dried and weighed. Section 5.10.2 provides information on how to oven-dry samples.

For simply obtaining a "feel" for the relative compaction of substrates Gore (1978) suggests observing the penetration of a measuring stick pushed into the sediment under constant pressure over some time. Commercial soil penetrometers can be used in a more standardized version of this technique.

Aggregate stability

Aggregate stability is a measure of the erodibility of soil in terms of its tendency to break down upon wetting. A soil with low aggregate stability tends to form a surface crust which can impede infiltration and hinder plant growth.

The method for measuring aggregate stability is described by Briggs (1977b). Loose particles are first gently sifted from a large soil clod through a garden sieve. The aggregates retained on the sieve are poured into a 1000 ml graduated beaker and gently tapped so that the soil settles. The volume of the soil (V_1) in the beaker is recorded. The beaker is gently filled with water, without damaging the aggregates, and allowed to stand for 30 min. The water is carefully poured off and the remaining volume of soil noted ($\mathcal{V}_2$). Aggregate stability is calculated as:

$$\text{Aggregate stability (\%)} = \left(1 - \frac{(\mathcal{V}_1) - (\mathcal{V}_2)}{(\mathcal{V}_1)}\right) \times 100 \qquad (5.44)$$

Embeddedness

Embeddedness is an index of the degree to which larger particles (boulders, large cobbles) are surrounded or covered by finer sediments (Hamilton and Bergersen, 1984). As embeddedness increases, the biotic productivity of the substrate is considered to decrease.

Platts et al. (1983) use a rating code to describe the percentage of surface area of the largest size particles covered by finer sediments. They give a 5-4-3-2-1 rating to describe channel embeddedness of <5%, 5–25%, 25–50%, 50–75% and >75%, respectively.

The Brusven index is a means of describing both sediment size and percentage embeddedness (Brusven, 1977). The index is composed of a three-digit number (e.g. 51.5) where the digit in the ten's place represents the largest materials in the sample (called the dominant particle size), the figure in the one's place represents the material surrounding the dominant particles and the decimal place is used to describe the percentage embeddedness. A larger particle completely imbedded in fines is assigned a decimal value of 9.

Brusven's original index was modified slightly by Bovee (1982). Rather than embeddedness, the decimal place describes the percentage of sand and smaller size material in the substrate matrix. The index code for this modified method is given in Table 5.9, and is the method of substrate classification used in the model PHABSIM (see Section 9.3.4). Vegetation such as rooted macrophytes or algae can also be treated as a form of substrate and included as a decimal in the hundreds place. The numeral might either represent percentage cover or a rating.

As an example, a substrate mixture of medium cobbles (6) surrounded by small gravels (2) and 40% fines (.4) would have a modified Brusven index of 62.4. One large boulder completely embedded in sand would have an index of 91.9. It should be noted that the index values do not form a continuum from 00.0 to 99.9, because the first number should always be larger than the second; e.g. values of 35.2 or 78.5 would be nonsensical.

Sorting

Sorting is a measure of the spread of particle sizes in the substrate. The degree of sorting can be calculated using the equation for standard deviation

Table 5.9. Substrate code for describing size classes in conjunction with the Brusven index method. After Bovee (1982)

Code	Substrate description
1	Fines (sand and smaller)
2	Small gravel (4–25 mm)
3	Medium gravel (25–50 mm)
4	Large gravel (50–75 mm)
5	Small cobble (75–150 mm)
6	Medium cobble (150–225 mm)
7	Large cobble (225–300 mm)
8	Small boulder (300–600 mm)
9	Large boulder (>600 mm)

in Table 5.8. For descriptive purposes the range of values have been divided into five sorting classes as shown in Figure 5.36, which provides a means of visual identification. The degree of sorting can also be estimated by comparison with photographs of known conditions. For smaller sediments, "standard" vials of sorted sediments can be used.

Andrews (1983) gives an alternative sorting index for bed materials:

$$\text{Sorting index} = \frac{1}{2}\left(\frac{d_{84}}{d_{50}} + \frac{d_{50}}{d_{16}}\right) \tag{5.45}$$

where d is the particle diameter in mm, as described in Section 5.10.5.

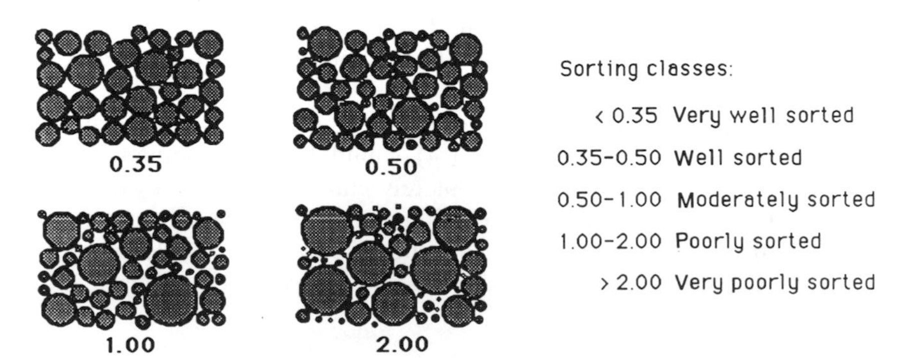

Figure 5.36. Chart for visual estimation of sorting. Redrawn from Dackombe and Gardiner (1983), adapted from Folk (1980), by permission of HarperCollins Publishers and Hemphill Publishing Co., respectively

5.11 WATER QUALITY

Because of its importance to the biota which drink or transpire or live in the water flowing through streams, a large number of attributes grouped under the heading "water quality" are of interest to water researchers. These might include pH, salinity, temperature, sediment concentration, odour, colour, light penetration, dissolved oxygen, levels of nutrients such as nitrogen or phosphorous, concentrations of pesticides or radionuclides or heavy metals, or the presence of pathogenic microbes.

Other than sediment concentration, covered in Section 5.10.3, it is beyond the scope of this book to cover all of the aspects of water quality analysis. The interested reader is referred to APHA (1985), Fresenius et al. (1987), Golterman et al. (1978), Mark and Mattson (1981) and USEPA (1987) and other standard water quality analysis manuals for methods of collecting, preserving and analysing samples for water quality. Some basic recommendations on sampling are given in this section.

Samples can be collected for "spot" readings or taken over a period of time to monitor diurnal, seasonal or yearly changes. If streams are well mixed, a single "grab" sample taken from a bridge or streambank may be adequate for determining an instantaneous measure of the stream's average water quality. Fresenius et al. (1987) recommend that samples be taken by holding a bottle 30 cm under the water surface. Automatic pumping samplers can be used at well-mixed sites to sample on a regular basis.

In some streams, especially those with slow-moving stagnant pools, layers of different water quality may exist. There are specialized sampler designs for collecting samples at specific depths, similar to the point-integrating samplers described in Section 5.9.4. Other designs are described in the above-mentioned references on water quality analysis and in limnology texts.

Sample bottles can be plastic or glass, depending on the parameter to be tested. They should be rinsed once or twice with stream water before a sample is collected. All bottles should be labelled with the date, time and site (including depth) of sampling.

Field analysis is essential for some water quality parameters such as pH, temperature and dissolved oxygen. Various meters have been developed for field measurement of electrical conductivity (EC), pH, dissolved oxygen (DO) and specific ions. Some combined units are also available with dataloggers for automatic data collection.

Field "test kits" for water chemistry analysis are also commercially available. Colourimetric procedures have been developed for many of the water quality parameters, where colour intensity or shade is measured with a spectrophotometer or visually compared to a chart or set of standards. "Test kit" techniques are typically of lower resolution than laboratory methods. However, they may be sufficient for some studies, for example when looking for sudden changes in water quality.

Table 5.10. Recommended sample volumes, bottle types and procedures for preservation of samples for various determinations. Adapted from APHA (1975) and Golterman (1987). Filtration should be done with a 45 μm filter (see Section 5.10.3) immediately after sampling

Determination	Volume (ml)	Bottle type	Preservation
EC, pH, carbonate system (HCO_3^-, CO_3^{-2}).	500	Glass or thick plastic, dark. Fill to top	Refrigerate
Dissolved P, Cl^-, SO_4^{-2}, hardness	100	Glass	Filter; refrigerate
N system, organic C, total P	500–1000	Plastic or glass	5 ml conc. H_2SO_4 (to bring $pH < 2$); refrigerate
Trace elements	1000	Glass	Filter; add 5 ml conc. HNO_3 (to bring $pH < 2$)

If higher accuracy is needed then there is a stronger argument for using more precise laboratory techniques. For most water quality parameters, samples must be preserved in some manner for transport to the laboratory. The method of preservation depends on which parameters are to be measured. Multiple sample bottles may thus be required for each sample, each preserved differently. Table 5.10 gives recommended procedures for the type of sample bottle, sample volumes and methods of preservation. A more detailed list appears in APHA (1985).

Samples should be analysed as soon as possible after collection. This will sometimes require shipping the samples to the laboratory by bus, car, train or aircraft. The bottles should be enclosed in foam containers to prevent breakage and packed in a cooler with ice or dry ice. Keeping the sample in the dark and at low temperature (preferably close to 4°C) retards bacterial growth. Information on the analyses desired, the date and site of collection, and the method of preservation should be written on each bottle.

6 Water at Rest and in Motion

6.1 GENERAL

In contrast to other water bodies, the most noticeable feature of streams is their one-way flow, guided by the influence of gravity. Yet within this unidirectional motion, water molecules follow unpredictable paths seemingly of their own choosing. Like the movement of traffic on a freeway, the general motion can be described in terms of average forward progress. However, the detours, halts and starts of an individual vehicle—or an individual water molecule—defy analysis. Thus one should begin a study of fluid mechanics under the premise that the tools for describing the behaviour of water are only approximations. Nevertheless, by making appropriate simplifying assumptions and generalizations some fascinating traits of the movement of water "en masse" can be examined and described. A very large literature on fluid mechanics exists which has been developed to a level appropriate for engineering applications, and this provides a framework upon which biologists can build to describe the complex interactions of organisms with their flowing environment.

This chapter has been divided into two main streams of study: hydrostatics and hydrodynamics. **Hydrostatics** is the study of water (hydro) at rest (static) and includes the principles of pressure and buoyancy. These principles hold true whether the water is at rest or in motion. However, as soon as a fluid begins to move, viscosity enters the picture, and the study of **hydrodynamics** is therefore more complex. Hydrostatics is covered in Section 6.2, with the remainder of the chapter devoted to the dynamic nature of flowing waters.

Hydrodynamics can be further divided into the study of fluid motion in the microenvironment and in the macroenvironment. In the **microenvironment** near solid surfaces, viscosity has an important effect on fluid behaviour. The focus is on patterns of viscous action as water passes around a surface, creating lift and drag on sediment particles and affecting the lives of small organisms which live on surfaces within the stream. At this scale, turbulence is seen as eddies behind objects and fluctuations in the immediate vicinity of a solid surface. In the **macroenvironment**, gross measures are of interest:

channel discharge, average velocity and the energy or work required to move the fluid. Here, viscosity is not as important, and turbulence is present over the entire depth of flow as large-scale eddies.

For additional information on biologically relevant topics in fluid dynamics, Vogel (1981, 1988) offers a readable, commonsense introduction to fluid properties and their significance, particularly at the microenvironment level. Newbury (1984) provides a concise, straightforward description of channel-scale flow properties. Other references by Davis and Barmuta (1989), Denny (1988), Denny et al. (1985), Hynes (1970), Nowell and Jumars (1984), Silvester and Sleigh (1985), Smith (1975), and Statzner et al. (1988) provide additional information from a biological viewpoint. For general information on fluid mechanics, engineering texts include Daily and Harleman (1966), Douglas et al. (1983), Gerhart and Gross (1985), Roberson and Crowe (1990), Streeter and Wylie (1979), Vennard and Street (1982), and White (1986).

6.2 HYDROSTATICS: THE RESTFUL NATURE OF WATER

6.2.1 PRESSURE

As mentioned in Section 1.2, pressure is defined as force per unit area. Pressure can be specified as **absolute**, with zero pressure (complete vacuum) as a reference, or as a **relative** or **gauge** pressure, with respect to the local atmospheric pressure, where:

$$\text{Absolute pressure} = \text{gauge pressure} + \text{local atmospheric pressure} \tag{6.1}$$

In this text, gauge pressure will be used. Absolute pressure may also be of biological interest (for example, when comparing internal air pressures of organisms at different elevations).

When water pressure acts on a solid surface such as a fish, the side of a levee or the stalk of a water lily, the force acts at right angles to the object's surface, pressing inwards on it from all sides. Pressure increases with water depth. At a depth of about 10.4 m, for example, the pressure increases by about one atmosphere. The relationship between pressure and water depth is linear, and is described by

$$p = \rho g h \tag{6.2}$$

where p = pressure, Pa (N/m^2), ρ = density (kg/m^3), g = acceleration due to gravity (m/s^2) and h = vertical distance below the water surface (m). This relationship is illustrated in Figure 6.1(a). Constant density is usually assumed for bodies of water. This is not a valid assumption, of course, if density changes over depth due to variations in temperature or dissolved solids.

It should be noted that for a liquid of constant density, pressure is dependent only on the depth of water above it. Thus, in Figure 6.1(b), where the two containers both have the same base area and are filled to the same height, the pressure on the base is the same. If force is calculated as (pressure × area) the values are the same for both containers, but if it is calculated as the "weight" of the water (volume × density × acceleration due to gravity), the values are not the same. This phenomenon is called the **hydrostatic paradox** (Douglas et al., 1983). What is important to remember is that it is the **depth**, not the weight of the water, that is important. At some point on a dam at a certain depth, for example, the pressure is the same whether the reservoir stretches 10 m or 1000 m upstream.

The measurement of pressure by **manometry** uses this relationship between pressure and head. In Figure 6.1(c) a simple manometer attached to a fluid-filled chamber indicates the downward pressure imposed by the weight of the animal. The change in height of the manometer can be measured to obtain the change in pressure from Equation 6.2. The density, ρ, would be that of the manometer fluid (e.g. water, oil, antifreeze).

By reverse application of the same principle, a measurement of pressure can be used to determine water depth. For example, pressure bulbs are often used to measure water depth at stream-gauging stations, and "snow pillows" with manometers are employed for measuring the water content of a snowpack. Pressure transducers can be used to translate the pressure into a voltage, which can then be recorded with a datalogger or telemetered to a receiving station.

Example 6.1

Calculate the pressure on the eye of a newt at 1 m depth in 20°C fresh water. From Table 1.2, $\rho = 998.3 \text{ kg/m}^3$, so:

$$\rho g h = \left(998.3 \frac{\text{kg}}{\text{m}^3}\right)\left(9.807 \frac{\text{m}}{\text{s}^2}\right)(1 \text{ m}) = 9790 \text{ N/m}^2$$

6.2.2 BUOYANCY

The principles for finding the pressures and forces on submerged objects can also be used to determine whether objects sink or float—in a word, their **buoyancy**. Objects can have **positive**, **negative** or **neutral** buoyancy depending on whether they tend to float, sink or remain where they are, respectively.

Submarines can alternate between the three states of buoyancy by pumping water in or out of "ballast" tanks. People, too, are close to neutral buoyancy and will sink or float depending on whether their lungs are empty or full. Hippopotami must exhale a large volume of air in order to walk along the

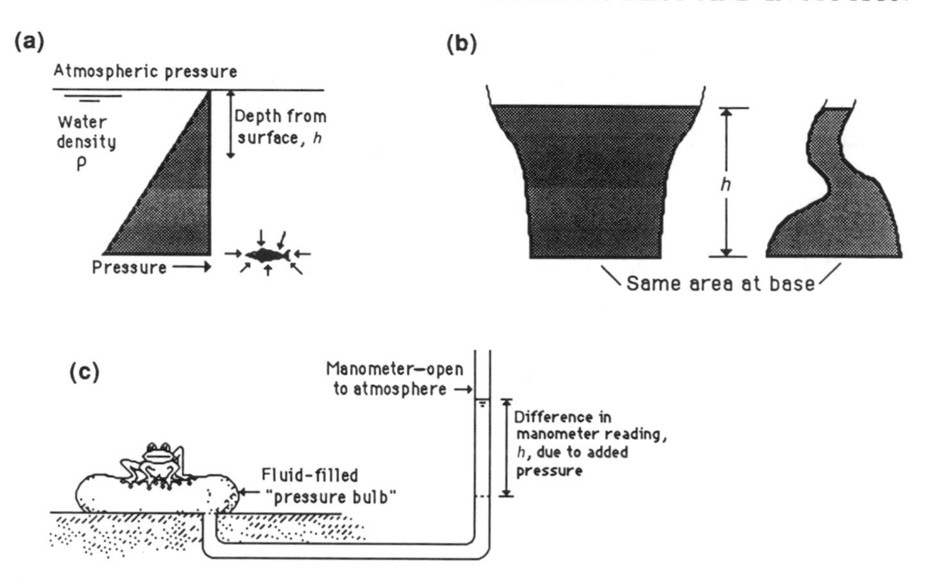

Figure 6.1. Pressure in a body of water: (a) linear variation of pressure with depth, (b) demonstration of the fact that pressure is dependent only on the water depth—the pressure on the base is the same for both containers, and (c) the use of manometry for measuring pressure

bottom of rivers and feed on underwater vegetation (Bolemon, 1989). Most fish have swim bladders containing air to regulate their buoyancy, which comprise about 7% of the volume of freshwater fish (Bone and Marshall, 1982). The common diving beetle which carries a bubble of air under its wings to breathe from also uses it to change its buoyancy. To sink, the beetle squeezes the trapped air (compressing it and thus increasing its density), and to float back up to the surface, it releases the tension to let the air expand and increase its buoyancy. Buoyancy is not only important to submerged or floating organisms or boats, but many flow phenomena such as the mixing of warm and cold regions are dependent on small buoyant forces (White, 1986).

Archimedes proposed the following two laws of buoyancy in the third century BC:

(1) An object totally immersed in a liquid experiences a vertical buoyant force equal to the weight of the liquid displaced; and
(2) A floating object displaces its own weight of liquid.

These laws yield the following general equation for buoyant force:

$$F = \rho g \, \forall \tag{6.3}$$

where F is the buoyant force (Newtons) and $\forall$ is the volume of water displaced (m^3). The buoyant force is caused by the differences in water pressure acting on the upper and lower surfaces of an object due to depth (see Figure 6.2(a)).

For example, if the shark in Figure 6.2(a) has a volume of 0.15 m^3 and the water is 0°C, it would experience a buoyant force of:

$$\left(1028\,\frac{\text{kg}}{\text{m}^3}\right)\left(9.807\,\frac{\text{m}}{\text{s}}\right)(0.15\ \text{m}^3) = 1512\ \text{N}$$

For floating bodies such as sitting ducks or icebergs only part of the object is submerged. The volume in Equation 6.3 is then the volume of water displaced by the underwater part of the object. Knowing this, the relative proportions of an iceberg above and below the water (Figure 6.2(b)) can be estimated. Using the density of sea water at 0°C, and the density for ice from Table 1.2, the calculations are performed as follows:

$$\text{Floating body (iceberg) weight} = \left(917\,\frac{\text{kg}}{\text{m}^3}\right)(\text{g})(\text{volume of iceberg}),\ \text{and}$$

$$\text{Weight of sea water displaced} = \left(1028\,\frac{\text{kg}}{\text{m}^3}\right)(\text{g})(\text{displaced volume})$$

From Archimedes' second principle, the two quantities are equal. By setting the two expressions equal to each other, the ratio (displaced volume/total volume) can be obtained. This is the proportion of ice submerged, 917/1028, or about 89%. The "tip of the iceberg" would then be only 11% of the total volume. In reality, the ice in icebergs may be less dense than the value used because of air spaces between the crystals of snow and ice.

The two examples given illustrate each of the two laws of buoyancy. Both can be applied in concert to solve the riddle: "If a person sitting on a boat on a pond tosses a brick overboard, does it raise the level of the pond?" The answer is no, it lowers the level, but the reasoning will be left to the reader.

6.3 STUDYING THE FLOW OF FLUIDS

6.3.1 STEADY AND UNSTEADY FLOW

The classification of flow as steady or unsteady (Figure 6.3) describes the way it behaves over time. Flow is considered **steady** at a point if its depth and velocity do not change over a given time interval. When waves or eddies travel past the point, the water level and/or velocity change from one moment to the next and the flow is said to be **unsteady**.

Turbulence causes the velocity to continuously fluctuate throughout most

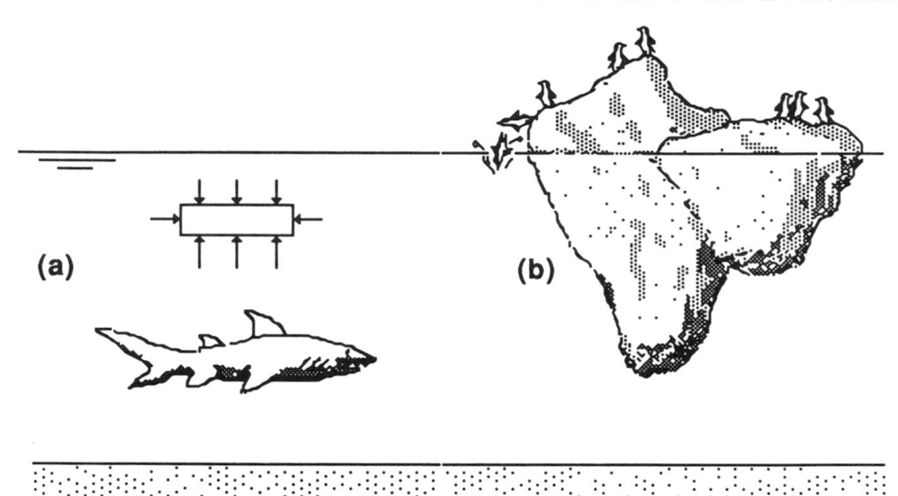

Figure 6.2. Principles of buoyancy. (a) The pressure is greater on the bottom of a submerged object than on the top, creating a net upward buoyant force. The buoyant force on the shark is equal to the weight of the water it displaces. (b) For floating objects the buoyant force is equal to the weight of the water displaced by the submerged portion

of the flow. However, in practice, the flow can be considered steady if values fluctuate equally around some constant value (Smith, 1975). An assumption of steady flow is necessary in the solution of many problems concerning water in motion.

6.3.2 STREAMLINES

A **streamline** is a line indicating the direction of fluid movement at a given instant. Examples are shown in Figure 6.6 (p. 224). Figure 6.4 shows isovels (lines of equal velocity). This is a common way of presenting data from laser doppler anemometers (Section 6.5.6). Isovels are interpreted differently from streamlines and should not be confused (see also Section 6.6.3).

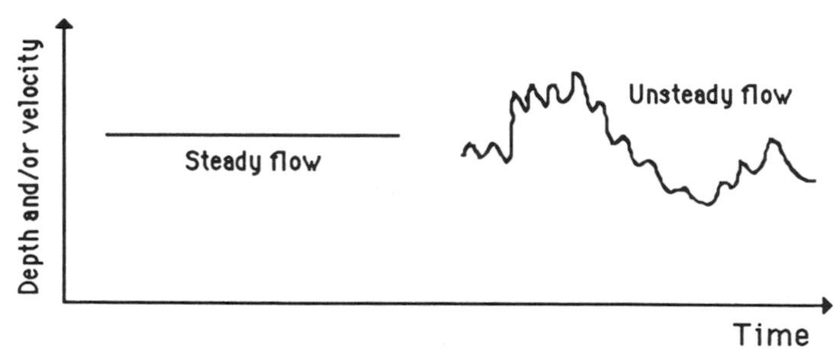

Figure 6.3. Classification of flow as steady or unsteady

Streamlines represent the paths that would be taken by individual fluid "particles" if the flow were steady. Convergence of the lines means the flow is accelerating, divergence that the flow is slowing down. A **streamlined** object is shaped such that the streamlines remain almost parallel as they pass around the object. For example, the snail in Figure 6.4(a) is less streamlined than the mayfly nymph in Figure 6.4(b).

In three dimensions, a **streamtube** is analogous to a "bundle" of streamlines. Flow within a streamtube or between two streamlines follows the principle of continuity (Section 6.3.3). Thus fluid does not pass across the boundaries. Moving water is nearly always turbulent, meaning the direction of movement changes from instant to instant and particles do not travel along regular paths. Therefore, streamlines typically represent average patterns of movement, and are visualized with **streaklines** formed by the trails of dye or bubbles or particles released into the flow.

6.3.3 CONSERVING MASS: THE PRINCIPLE OF CONTINUITY

By the law of conservation of mass, the mass in any system must remain constant with time. For fluids, this becomes the **principle of continuity**. Strictly applied, the principle of continuity only applies to incompressible fluids. For most practical applications, however, water can be considered incompressible.

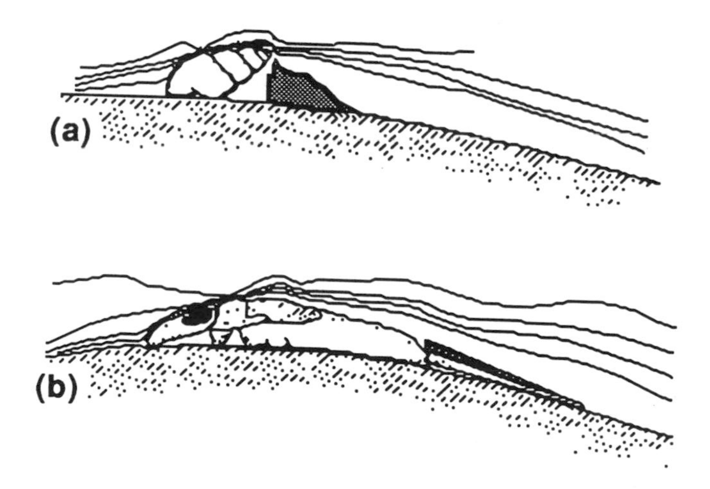

Figure 6.4. Isovels (lines of equal velocity) around: (a) water snail (*Potamopyrgus jenkinsi*) and (b) mayfly nymph (*Ecdyonurus cf. venosus*), shown as a schematic without legs. Flow is from left to right. Free stream velocity is about 0.18 m/s in both figures, and shaded areas represent regions of zero velocity. Redrawn from Statzner and Holm (1989, 1982), by permission of Springer-Verlag and B. Statzner

In the example of steady flow through an iron pipe, where there is no opportunity for fluids to "hide" (no storage), the principle basically states that:

$$\text{Outflow} = \text{inflow} \tag{6.4}$$

However, in a garden hose with a weak spot that "balloons" on filling, the equation would be:

$$\text{Outflow} = \text{inflow} \pm \text{change in storage} \tag{6.5}$$

where the change in storage would be the changing volume of the ballooning region. Water flowing through a reservoir would follow the more general second relationship (Equation 6.5) since the water level is constantly changing.

Equation 6.4 can be written:

$$Q = A_1V_1 = A_2V_2 \tag{6.6}$$

where Q = discharge (m^3/s), A = cross-sectional area (m^2), V = average velocity (m/s) and the subscripts refer to sections 1 and 2, where 1 usually represents the inflow point and 2 the outflow point, as illustrated in Figure 6.5(a).

Equation 6.6 is called the **continuity equation**. It was first derived by Leonardo da Vinci in the year 1500 (White, 1986). From the relationship, it can be seen that if discharge (Q) remains constant, but area (A) is decreased, the velocity (V) will go up. For example, Figure 6.5(a) could represent the tapered nozzle on a firefighter's hose which speeds water towards its destination. In a reverse manner, an increase in velocity can mean a decrease in cross-sectional area. Thus, the flow out of a pitcher contracts as it accelerates from rest in falling towards the earth (Figure 6.5(b)). A counter-intuitive example is that of a stream passing through a constriction, for example where it is partially blocked by sand bars, rock outcrops or bridge piers. At the constriction, the velocity increases and the water depth goes down rather than up as might be expected.

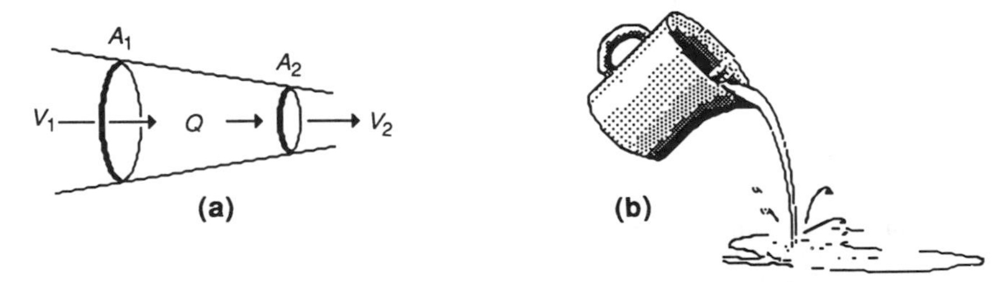

Figure 6.5. The continuity principle applied to: (a) a contracting section; and (b) a free overfall

6.3.4 ENERGY RELATIONSHIPS AND THE BERNOULLI EQUATION

It could be said that water arrives on a catchment with the potential to do great work: to carve smooth pathways across dense bedrock, to carry sediment or simply to rush downstream against the resistance of internal friction. Water contains a certain amount of energy, called **potential energy**, simply due to its vertical location. Water in a prospector's canteen in Death Valley, California, has less potential energy than an equivalent mass of water in a snowfield on Mount Everest. A mass M (kg) of water at a height h (m) above some datum (sea level, a tributary junction or some other reference level), has potential energy (PE) of:

$$PE = Mgh \tag{6.7}$$

where g is the acceleration due to gravity (m/s^2) and PE has units of J (joules).

Potential energy drops as water runs downhill because h decreases. Most of the energy is converted to kinetic energy (KE), the energy of motion. For a mass M (kg) of water, KE is given as:

$$KE = \frac{1}{2}MV^2 \tag{6.8}$$

where V is velocity (m/s) and energy and mass have units as before. In a stream, almost 95% of the kinetic energy is consumed as heat loss through turbulent mixing and friction along the bed and banks (Morisawa, 1968). Because of the high thermal capacity of water (Section 1.3.4), and heat transfer from the water to its surroundings, the water temperature only rises a small amount as a result of this internal heat generation. It is also estimated that a small fraction of the energy, about 0.0001%, is converted to the characteristic sounds—the gurgles and roars—of moving water (Hawkins, 1975). The remainder of the kinetic energy is free to run turbines or move mountains a sand grain at a time. Energy relationships for the bulk flow of streams are covered in Section 6.6.4.

The **Bernoulli equation**, named after Daniel Bernoulli (1700–82), translates the idea of energy conservation into terms applicable to moving fluids. The equation states that energy is conserved along a streamline:

$$KE + PE + \text{pressure energy} = \text{a constant (along a streamline)} \tag{6.9}$$

The terms KE and PE were previously defined. Pressure energy is what people use when they use a hand pump to inflate a bicycle tyre or what archer fish use to squirt water towards aerial prey. In streams it is part of the impetus which motivates a mass of water to continue its work.

Bernoulli expressed the various energy components in terms of **head**, the energy per unit weight (Mg) of fluid:

$$\frac{p}{\rho g} + \frac{V^2}{2g} + z = \text{constant} \tag{6.10}$$

or

$$\frac{\text{Pressure}}{\text{head}} + \frac{\text{velocity}}{\text{head}} + \frac{\text{elevation}}{\text{head}} = \text{constant}$$

where the head terms have units of length (m).

Bernoulli's equation can be applied across streamlines if density is constant and the flow is considered "ideal", meaning that viscous effects are insignificant. Thus the equation can be applied to small objects in the bulk flow region, away from the surfaces of solids. In general, as the effect of viscosity becomes more important, the principle becomes less applicable. Although it applies to an "ideal" situation, the Bernoulli equation provides useful approximations in a variety of situations.

Basically, the equation implies that if one term goes up, another must go down. The equation for manometry (Equation 6.2) and the equations for Pitot tubes (Equations 6.45 and 6.46) are consistent with the Bernoulli equation. It can also be used to explain why lift occurs on airfoils such as the wings of aircraft or birds. From Figure 6.6(a) it can be seen that streamlines are compressed as they travel over the top of an airfoil, meaning that the velocity is higher in that region. From Equation 6.10 the increase in velocity must be balanced by a decrease in pressure (since the elevation is essentially constant). Thus, the pressure on the upper surface is lower than that on the underside, resulting in a net upwards force on the airfoil. The same principle applies when the fluid is water and the "hydrofoil" is a penguin's flipper, a flat fish like a flounder, or a turtle (Figure 6.6(b)). Lift is also generated when water flows over objects resting on the streambed, such as the organisms in Figure 6.4.

When the bulk flow within a stream reach is considered (e.g. for investigating energy losses over some reach length), an apparently similar but fundamentally different equation is used, the one-dimensional energy equation. This equation is discussed in Section 6.6.4.

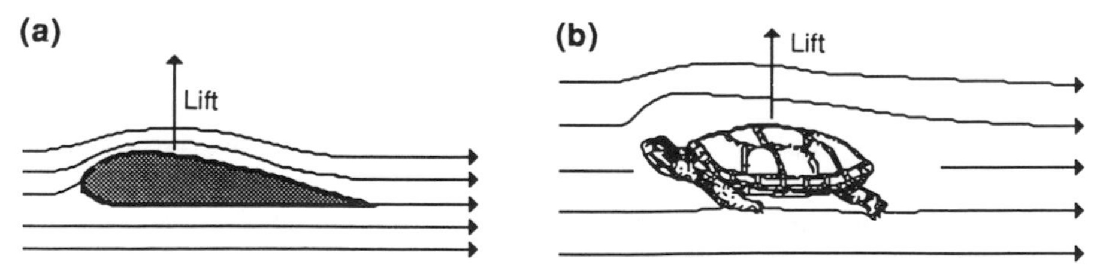

Figure 6.6. Streamlines and lift on: (a) an airfoil and (b) a theoretical turtle

6.4 NARROWING THE FOCUS: FLOW OF A VISCOUS FLUID

6.4.1 LAMINAR AND TURBULENT FLOW

In smoke rising from a lit cigarette the structure of the smoke is at first filamentous, rising straight up. At some point, though, the smooth structure breaks down and the smoke curls and swirls and tumbles in erratic motion (Smith, 1975). The smooth pattern, in which all of the smoke rises upward at a uniform rate, is called **laminar flow**. In **turbulent flow**, elements of fluid follow irregular, chaotic paths and violent mixing occurs, with eddies continuously forming and breaking down. Practical formulae for dealing with the two types of flow are very different.

Laminar flow

In streams, laminar flow may exist as a thin coating over solid surfaces, or where flow moves through the small openings between rocks in a streambed and through dense stands of aquatic weeds. Here, the fluid moves in parallel "layers" which slide past each other at differing speeds but in the same direction. The fluid layer closest to a solid surface is retarded by the no-slip condition (Section 1.3) and decelerates, causing nearby fluid layers to slow empathically.

The process by which one layer encourages its neighbour to slow down is **molecular diffusion**. Binder (1958) likens the process to trains overtaking each other on parallel tracks, with coal tossed from one train to another (Figure 6.7(a)). Higher-speed coal lumps thrown into the slower-moving trains tend to speed them up; conversely, slower-speed lumps thrown into the faster trains slows them down. The resulting velocity profile is illustrated in Figure 6.7(b). As shown, the velocity increases from zero at the solid surface to a maximum value some distance away where the solid boundary no longer influences the movement of the water. Velocity profiles will be discussed further in Section 6.4.2.

If the flow is laminar the amount of force applied to a mass of fluid is

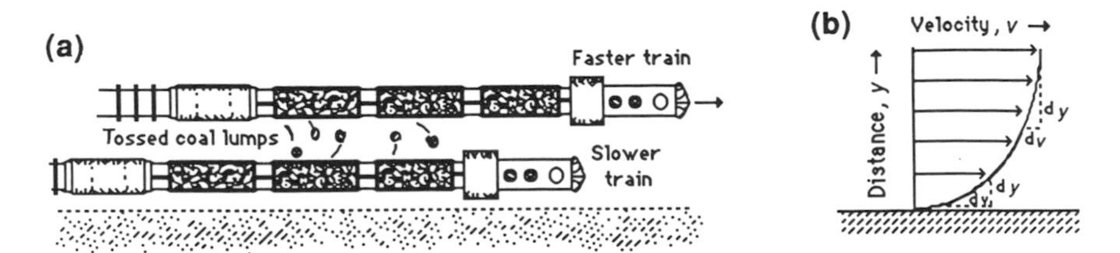

Figure 6.7. Laminar flow: (a) molecular diffusion between "layers" of fluid, represented by coal lumps tossed between parallel trains, and (b) the velocity profile for laminar flow (see text)

directly related to how fast it travels: the more force applied, the greater the speed. This relationship is described by Newton's law of viscosity, stated as:

$$\tau = \mu \frac{dv}{dy} \qquad (6.11)$$

where τ is shear stress (N/m^2), μ is dynamic viscosity (Section 1.3.2) and dv/dy is the velocity gradient, representing the rate of change of velocity, v, with distance, y. Shear stress, as defined in Section 1.2, is a "sideways" force per unit area. Viscosity is the internal friction which causes the fluid to resist being "pushed". The velocity gradient arises as a result of their interaction. Near the solid surface in Figure 6.7(b) it can be seen that the change in velocity, dv, with distance dy is relatively large. Shear stress is highest here. Away from the surface, dv gets smaller in relation to dy, and thus the shear stress decreases.

Fluids which follow the relationship given by Equation 6.11 are called Newtonian fluids, after Sir Isaac Newton, who first proposed this law in 1687 (White, 1986). It is perhaps fortuitous that the two most common fluids in nature—air and water—are virtually perfect Newtonian fluids. Non-Newtonian fluids such as mud, blood and paint are treated in a specialized branch of fluid mechanics called rheology.

Turbulent flow

Whereas laminar flow can be neatly described by a linear equation, turbulent flow can only be defined *statistically*. The motion of an individual water molecule cannot be predicted mathematically, but the harmonized movement of millions of water molecules in turbulent flow can be described by averages. Thus, the velocity measured with a current meter (Section 5.7.3) is only a mean velocity, reflecting the average water motion as it fluctuates across the meter. The accurate mathematical modelling of turbulence remains a frontier research topic, and the description of turbulent flow relies heavily on experimentation.

Turbulence exists at all scales, from the swirling motion created when a salmon scoops out a redd to large whirlpools in a river or cyclones in the atmosphere. Larger-scale eddies tend to generate smaller ones, as in Lewis Richardson's poem (as quoted by Gleick, 1987):

Big whorls have little whorls
which feed on their velocity;
And little whorls have lesser whorls
and so on to viscosity.

The poem itself is a "spin-off" from an earlier work, *The Fleas*, by Augustus deMorgan (1806–71).

In turbulent flow, layers of fluid break up into "globs" which mix with other globs in a chaotic collection of eddies and swirls. It is this turbulent mixing rather than molecular diffusion which is the primary mode of speeding up or slowing down the surrounding fluid. Higher-velocity globs are swept into lower-velocity zones near a solid surface and lower velocity globs are carried into the higher-velocity zones further away. The behaviour is more like what would occur if the "laminar" train cars of Figure 6.7(a) were uncoupled at various points and began acting more like clusters of carnival bumper cars. Diffusion still continues, but the influence of tossed coal lumps or individual water molecules on the overall motion is much less significant.

In turbulent flow, eddying mixes the higher-velocity fluid into the area closer to the solid surface. This causes the velocity profile (Figure 6.8(a)) to be flatter near the solid boundary than for laminar flow (Figure 6.7(b)). In the turbulent region the velocity is erratic, and is depicted by a "fuzzy" line in Figure 6.8(a). A small layer of laminar flow may remain in the region near the solid, as will be described in Section 6.4.2.

In turbulent flow, Newton's law (Equation 6.11) becomes less relevant. The mixing of fluid "globs" into zones of higher or lower velocity tends to augment or retard the local velocity. These fluctuations affect the mean velocity in such a way that it seems as if the viscosity has increased. To account for this effect, a term called eddy viscosity, ϵ, is added to Newton's equation:

$$\tau = (\mu + \epsilon)\frac{dv}{dy} \tag{6.12}$$

The formula encompasses situations of both laminar and turbulent flow. If the flow is entirely laminar, ϵ is zero, and Equation 6.12 reduces to Equation 6.11. Alternatively, for fully turbulent flow, effects due to fluid viscosity are negligible ($\epsilon >> \mu$), and the equation reduces to:

$$\tau = \epsilon\frac{dv}{dy} \tag{6.13}$$

There are no tables giving values for eddy viscosity, ϵ. It is dependent on how vigorous the turbulence is, and must be found by experimentation. It is not a fluid property like μ. Equations 6.12 and 6.13 are used in theoretical developments, and are shown here only to demonstrate the difference in the ways laminar and turbulent flow are analysed.

Turbulence can also be described by a measure called **turbulence intensity**. It is useful to think of a local velocity in the turbulent region as composed of two parts: an average value plus a component which represents the fluctuation about the mean (Figure 6.8(b)). Turbulence intensity is a measure of the strength of the turbulent fluctuations. If N instantaneous velocity measurements are made at a point, the turbulence intensity can be expressed as the root mean square of these measured values:

$$\text{Turbulence intensity} = \sqrt{\left(\frac{\sum_{i=1}^{N} (v-\bar{v})^2}{N}\right)} \qquad (6.14)$$

where $\bar{v}$ is the average of the velocity measurements and all variables have units of m/s. Turbulence intensity can also be expressed as a percentage by dividing the result from Equation 6.14 by $\bar{v}$ and multiplying by 100.

Figure 6.8(b) and Equation 6.14 apply to one component direction. Usually the component in the direction of flow (normally the horizontal component) is of primary interest. In reality, turbulent fluctuations occur in all directions, but reduce to zero at the solid surface because fluid particles are cramped for space in which to fluctuate. A common assumption is that turbulence intensities are the same in both horizontal and vertical directions since they arise from the same sets of eddies. In open channels, turbulence intensity is about 0.10 times the local mean velocity, and decreases gradually towards the water surface (Morisawa, 1985).

The Reynolds number

By comparing Equations 6.11 and 6.13 it can be seen that viscosity is an important factor in laminar flow, but becomes relatively insignificant in turbulent flow. Viscosity tends to dampen turbulence and promote laminar conditions. Acceleration has the opposite effect, promoting instability and turbulence. The resistance of an object or fluid particle to acceleration or deceleration is described by a measure called **inertia**. This is the tendency of an object to maintain its speed along a straight line. It is what keeps a particle of fluid going until it is "aggressed upon by external authority" (Vogel, 1981, p. 67). Whereas high inertial forces promote turbulence, high viscous forces promote laminar flow. The ratio of inertial forces to viscous forces thus gives an indication of whether the flow is laminar or turbulent.

Late in the nineteenth century a famous professor of engineering, Osborne Reynolds, developed such a ratio by investigating the behaviour of flow in

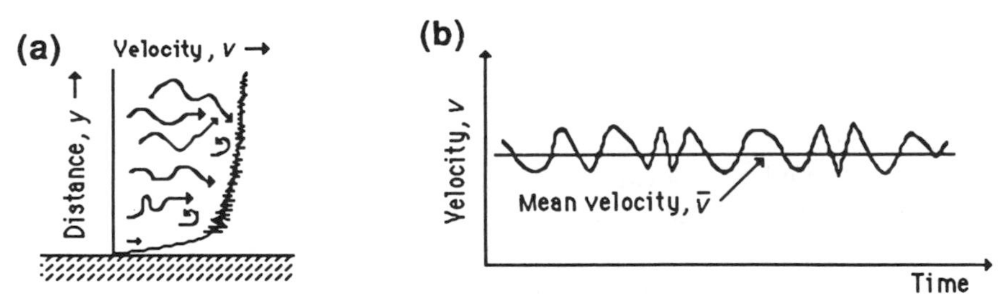

Figure 6.8. Turbulent flow: (a) velocity profile near a solid surface and (b) fluctuations in velocity with time

a glass pipe (Reynolds, 1883). A fine stream of dye was introduced into the pipe so that the flow could be visualized. For slower flows, the dye moved as a straight streak, but as the flow rate was increased, the dye stream began wavering. Laminar flow, which produced the straight streak, was termed "direct" flow by Reynolds, and the turbulent flow which dispersed the dye was termed "sinuous". By varying the speed, the diameter of the pipe and density of the liquid, Reynolds tested the significance of a dimensionless number now known as the **Reynolds number**, *Re*:

$$Re = \frac{VL\rho}{\mu} \quad \text{or} \quad Re = \frac{VL}{\nu} \tag{6.15}$$

with:

V = velocity (m/s),
L = some characteristic length (m),
ρ = fluid density (kg/m^3),
μ = dynamic viscosity ($N{\cdot}s/m^2$), and
ν = kinematic viscosity (m^2/s), where $\nu = \mu/\rho$.

The terms ρ, μ and ν are defined in Section 1.3.2. In Equation 6.15, the terms in the numerator are related to inertial forces and those in the denominator are related to viscous forces. Thus, a large value of *Re* indicates turbulence and a small value, laminar flow. Program REYNOLDS in AQUAPAK is provided for calculating Reynolds numbers.

Reynolds also investigated the transition between the two types of flow in his pipe experiment. Starting with turbulent conditions, he found that the flow always became laminar when the velocity was reduced so that *Re* dropped below 2000. This point of transition is called the *critical Reynolds number*. In pipe flow, the transition will not necessarily occur at this value; in fact, laminar flow has been maintained up to $Re \sim 50\,000$ although it is highly unstable and becomes turbulent at the slightest hint of disturbance.

In pipe flow, the "characteristic length", L, used in calculating the Reynolds number (Equation 6.15) is the pipe diameter. Reynolds numbers can be calculated for other situations by substituting an appropriate characteristic length such as the diameter of a sand grain, the length of a fish or the width of a bird's wing. For solids immersed in a flowing fluid the convention is to use the maximum length of the object in the direction of flow. Vogel (1981) and Purcell (1977) give estimates of Reynolds numbers experienced by aquatic organisms, based on their "typical" values of length and swimming speed, which, in order-of-magnitude terms, are:

10 million — Tuna
10 000 — Olympic swimmer
100 — Goldfish or guppy
0.1 — Invertebrate larvae.

For an aquatic organism, both the movement of the fluid and the movement of the organism within it will govern the Reynolds number. In nature, size and speed tend to work together, where "small" nearly always means "slow" and "large" usually means "fast" (Vogel, 1981). Conditions of both high and low Reynolds number are of interest biologically. Larger creatures such as trout or barramundi live in conditions where viscous forces are less significant and an occasional flip of a tail is sufficient to keep the fish moving through still water. At the low Reynolds numbers experienced by microscopic organisms, however, viscous forces become overwhelming.

Low-*Re* conditions are of little interest to engineers but are a way of life for bacteria or protozoans or other microscopic organisms which, because of their small "characteristic lengths", operate at Reynolds numbers in the range of 10^{-4} to 10^{-5} (Purcell, 1977). Here, inertia is irrelevant in comparison to viscosity, and movement stops immediately when propulsion ceases. Rather than sliding easily through the fluid, the organism essentially carries the fluid along with it, gradually shedding it off to the sides. For a person experiencing the same conditions it would be like swimming in a pool filled with molasses at a speed of a few metres per week (Purcell, 1977).

The advantage of "life at low Reynolds numbers" is that the organism is protected from the action of turbulence by a thick "coating" of highly viscous fluid (as the organism perceives it). However, since mixing is impeded, the transport of energy, nutrients and gases to an organism, and the transport of wastes or current-dispersed gametes away from it, occurs by the slower mode of diffusion. For a sessile species, still water is a hostile environment, and many of these creatures have mechanisms for creating turbulence for improved ventilation and waste dispersal. Mobile micro-organisms need to "move to greener pastures" to feed rather than waiting for food to come to them. Purcell (1977) has written a delightful essay on life at low Reynolds numbers, and the reader is referred to this paper and Vogel's (1981, 1988) works for more information on the biological aspects of low-*Re* conditions.

Aquatic invertebrates may experience "the best of both worlds", both laminar and turbulent. Statzner (1988) points out that these species start life at Reynolds numbers of about 1 to 10, but when they reach their adult form, they may live in conditions of $Re = 1000$ or higher. For these organisms, he concludes that evolution compromises between life at low and high Reynolds numbers.

For the case of flow in stream channels, a measure called the hydraulic radius (see Section 6.6.1) forms an appropriate length parameter for Reynolds numbers. For pipe flow, hydraulic radius is given by:

$$R = \frac{\text{Area}}{\text{Wetted perimeter}} = \frac{\pi d^2/4}{\pi d} = \frac{d}{4}$$

or $d = 4R$

This suggests that the transitional value of $Re = 2000$ might be roughly translated to streams if d is replaced by $4R$. In wide or rectangular streams the average depth of the water is a good approximation for the hydraulic radius. Hence, in this situation, the second form of Equation 6.15 becomes:

$$Re = \frac{VD}{\nu} \tag{6.16}$$

where D is the average depth. If Re is defined in this way for a stream, the transition from laminar to turbulent would be expected at $Re = 500$ (2000/4). From experimental data, the transitional range of Re for open channels is usually considered to be from 500 to 2000, within which the flow can be either laminar or partly turbulent (Chow, 1959). The large variety of shapes and roughnesses of channels make these figures only approximate. Even if the upper limit were taken as 12 500 (50 000/4), a few trial calculations of Re should convince one that the average condition of flow passing through a stream reach would almost never be classified as laminar, except perhaps where the stream is reduced to a thin film of water trickling over a smooth bedrock surface. For example, if $L = 2$ mm, $V = 0.1$ m/s and $\nu = 10^{-6}$ m^2/s, then $Re = 200$, and the flow would be considered laminar. In the field, laminar conditions are best identified with dyes (e.g. food colouring) injected gently into the flow with a syringe.

6.4.2 FLOW PAST SOLID SURFACES: THE BOUNDARY LAYER

The development of velocity profiles from where a fluid contacts a solid surface to where the flow is no longer effectively influenced by the presence of the surface occurs in what is known as a **boundary layer**. Its outer limit is where the speed of the fluid matches the "free stream velocity"—the velocity which would exist if the solid were not there. In a stream the boundary layer caused by the presence of the streambed extends to the water surface. Within its depths smaller boundary layers exist on the surfaces of rocks or snags; fish or aquatic insects; in fact, many organisms live within the boundary layer of other organisms.

As Vogel (1981, pp. 129) says, "most biologists seem to have heard of the boundary layer, but they have the fuzzy notion that it is a discrete region rather than the discrete notion that it's a fuzzy region". Although the boundary layer is somewhat arbitrarily defined, it is a useful concept for explaining interesting phenomena such as why a thin layer of dust sticks to a fan or why a sponge works better than pressurized water for washing the last thin layer of grime off a car.

The term "boundary layer" was originally coined in 1904 by Ludwig Prandtl, a German engineer. His work on theoretical descriptions of fluid behaviour near solid surfaces formed the basis for many of the engineering formulae still used today. It should be remembered that engineering fluid mechanics methods have been developed to suit the needs of engineers; modifications may be needed for application to problems in stream ecology.

"Life in the boundary layer" usually refers to the organisms which live in the relatively slower-velocity region of flow near solid surfaces such as the surface of rocks or the leaves and stems of aquatic plants. Even the swiftest streams have stones covered with algae and vegetation, and mayfly nymphs (Ephemeroptera), caddisfly larvae (Trichoptera), black fly and midge larvae and pupae (Diptera) and others conduct their lives in these swift waters.

The classic approach in most engineering treatments of boundary layer theory is to first discuss the development of boundary layers in the simplest case of flow around a smooth, sharp-nosed, flat plate oriented into the flow. The distributions of velocity and shear stress around the plate are influenced both by the nature of the flow: whether laminar or turbulent, and the nature of the solid: whether rough or smooth. Although flat plates may not have a great deal of ecological significance, the relationships developed are useful in describing the patterns of velocity near surfaces within streams and for calculating skin-friction drag on boats, airfoils and aquatic organisms.

Flow along a sharp, flat plate

On a sharp, flat plate oriented into the flow, the boundary layer begins at its leading edge. The **stagnation point**, which occurs at this leading edge, is a point where the velocity of the oncoming flow is zero (stagnant) because it has collided with the object. Downstream for some distance, the flow across the plate is **laminar**. As the fluid moves further along the plate, layers of fluid at a greater distance away are slowed down and the laminar layer grows. Boundary layer formation across the top surface of a sharp, flat plate is shown in Figure 6.9 for $Re_L \approx 10^7$. The subscript "L" refers to the use of the length of the plate, L, as the characteristic length for computing Re (Equation 6.15). The velocity used in the equation is the approach velocity, or the "free stream" velocity which would exist if the plate were not there.

This thickening of the laminar boundary layer continues with distance from the upstream point until the thickness is so great that the flow becomes unstable and deteriorates into turbulence. The transition point occurs at some critical value of the Reynolds number, given by most authors as:

$$Re_x \approx 500\,000$$

where the subscript x means that x, the distance from the leading edge, is the "characteristic length" used for computing a "local" Reynolds number:

$$Re_x = \frac{Vx}{\nu} \tag{6.17}$$

In the **transition region** the flow is both laminar and turbulent. At the

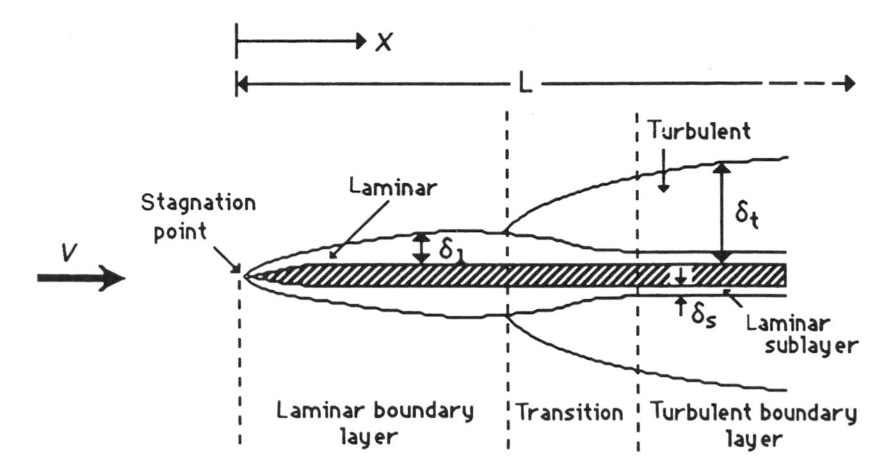

Figure 6.9. Boundary layer formation across the top of a sharp, smooth flat plate (for $Re_L \sim 10^7$, where L is the "characteristic length", in this case the length of the plate). V is the approach velocity, x the distance from the leading edge, and δ the thickness of the boundary layer (shown for laminar (l), turbulent (t) and viscous sublayer (s) regions)

transition a large increase in velocity occurs close to the plate as the velocity distribution shifts from the laminar velocity profile to the flatter-based turbulent profile (see Figures 6.7(b) and 6.8(a)). This principle can be used to locate the transition region by measuring changes in velocity close to the plate with a device such as a pitot tube (Schlichting, 1961).

In the **turbulent** region further downstream the boundary layer continues to grow outward as faster and slower "globs" of fluid mix together. Since mixing is much more effective than molecular diffusion in encouraging neighbouring globs of fluid to behave in the same, way, the turbulent boundary layer grows more rapidly than the laminar layer (see Figure 6.9). In the turbulent region a very thin layer of laminar flow still exists near the solid surface, protecting it from violations of the no-slip condition. This layer is called the **laminar sublayer** or **viscous sublayer**.

This model of boundary layer phenomena is only valid under specific conditions. At higher Reynolds numbers the whole boundary layer shrinks and becomes turbulent closer to the "nose" of the plate. The laminar sublayer also becomes thinner. At lower Reynolds numbers the boundary layer is thicker and the laminar region extends further back on the plate. Below $Re_L \approx 1000$, viscosity plays a more important role, turbulence disappears, and the whole profile becomes laminar. At $Re_L = 10$, White (1986) shows the laminar region extending out in front of the plate, implying that the influence of an object extends forwards at these very low-*Re* conditions, like snow pushed in front of a snow-plough.

The described model of boundary layer development is also valid only when the approaching flow is laminar or the plate itself is moving through still water and the plate is smooth. If the oncoming flow is turbulent or the leading edge of the plate is rough, turbulence will set in much sooner.

Hydraulically rough and hydraulically smooth surfaces

However, what if the flat plate is not smooth? In engineering fluid mechanics "rough" and "smooth" have very exact meanings, linked to the definition of the laminar sublayer. In fact, the very existence of the laminar sublayer is dependent upon how rough the surface is. A surface is said to be **hydraulically smooth** (Figure 6.10(a)) if all surface irregularities are so small that they are totally submerged in the laminar sublayer (smooth plastic pipes are hydraulically smooth). If the roughness height extends above the sublayer it will have an effect on the outside flow, and the surface is said to be **hydraulically rough** (Figure 6.10(b)). Since the thickness of the laminar sublayer varies with flow conditions, both flow velocity and roughness height will determine whether a given surface is "smooth" or "rough".

Hydraulically rough conditions will be most prevalent in streams. However, where the surface irregularities become very small in comparison to the water depth, such as in bedrock streams or in deep lowland rivers with streambeds of fine sand, silt and mud, hydraulically smooth flow can occur. On a finer scale, it can also take place along the surfaces of smooth objects submerged in the flow, such as smooth boulders or the leaves of macrophytes.

Much of the original work on roughness was done by Nikuradse, one of Prandtl's students. He studied frictional head losses in pipes which had been coated with sand grains of uniform sizes. Head loss was found to be a

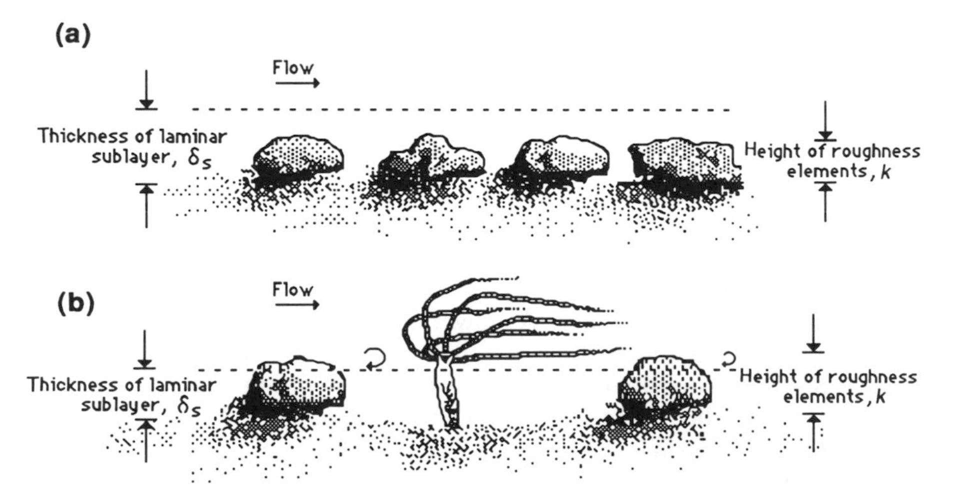

Figure 6.10. Illustration of (a) hydraulically smooth and (b) hydraulically rough surfaces

function of Reynolds number and **relative roughness**, where the latter was described as the ratio of the sand grain diameter to the pipe diameter. The idea has been extended to open channels, where relative roughness, R_{rel}, becomes:

$$R_{rel} = \frac{k}{D} \tag{6.18}$$

where k = some measure of roughness (e.g. particle size) (m) and D = depth of the water (m). In Nikuradse's work, k was defined by the particle diameter, with all grains of the same size. His work provided an important benchmark for determining the "effective" roughness of surfaces such as concrete, where the surface is more irregular. A few approximate values are given in Table 6.1.

In streams, the roughness height, k, is defined by its effect on the flow, and varies not only with the grain size distribution of streambed materials but also with how the roughness elements project into the flow and how they are arranged—including the arrangement of larger "roughness elements" (e.g. sand dunes and piles of woody debris) and smaller roughness elements (e.g. surface irregularities on individual rocks). Typically, some characteristic diameter of the streambed materials such as the d_{50} or d_{85} (see Section 5.10.5) is used as the roughness height.

A "roughness" Reynolds number, Re_*, can be developed using **shear velocity**, V_* and the roughness height, k:

$$Re_* = \frac{V_* k}{\nu} \tag{6.19}$$

(compare with Equation 6.15). Here, V_* is a measure of shear stress expressed in velocity units (m/s). It is computed as:

$$V_* = \sqrt{\frac{\tau}{\rho}} \tag{6.20}$$

Table 6.1. Experimentally derived values of effective roughness height, k. From Chow (1959); reproduced by permission of McGraw-Hill

Material	k (mm)
Brass, copper, lead, glass	0.030–0.90
Galvanized iron	0.15–4.6
Wood stave	0.18–0.90
Cement	0.40–1.2
Concrete	0.46–3.0
Riveted steel	0.90–9.0
Natural river bed	30–900

where τ is the shear stress acting at the surface of a solid (N/m^2) and ρ is the density (kg/m^3) of the fluid (see Section 1.3.1). Other expressions for shear velocity are given in Section 6.5.3.

A surface is considered hydraulically smooth if $Re_* < 5$, hydraulically rough if $Re_* > 70$, and transitional at $5 < Re_* < 70$ (Schlichting, 1961). Thus, the flow near a solid surface will be disturbed if either (1) the roughness elements increase in height or (2) the velocity increases, causing the laminar sublayer to become smaller than the height of the projections. Davis and Barmuta (1989) state that the roughness Reynolds number appears to be an excellent habitat descriptor since it combines the effects of velocity and substrate type.

Arranging the surface roughness

The way in which objects or organisms are spaced can affect the patterns of flow around them. Morris (1955) explored the concept of roughness spacing in pipes. He proposed that the eddies created between roughness elements would have an affect on flow resistance. Both the longitudinal spacing and height of the roughness elements were considered important. Morris classified flow over rough surfaces into three categories, described by the following conditions (parameters are defined in Figure 6.11):

(1) *Isolated-roughness flow* (Figure 6.11(a)): eddies which form behind each element dissipate before the next element is reached. Isolated-roughness flow will occur when k/λ is small.
(2) *Wake-interference flow* (Figure 6.11(b)): roughness elements are closer together, and the eddies from the elements interact, causing intense turbulence. Here, roughness height is relatively unimportant compared to the spacing. The depth of flow above the crests of the elements becomes important since it will limit the vertical extent of increased turbulence. Wake interference flow can occur over surfaces of considerable roughness such as corrugated metal (Morris, 1961).
(3) *Skimming flow* (Figure 6.11(c)): elements are so close together that the flow "skims" over the tops of the elements, with low-velocity eddies occurring in the grooves between the elements. The surface acts almost as if it is hydraulically smooth. Skimming flow occurs when k/λ is high.

The above three categories are applicable when the water depth is much greater than the height of the roughness elements. Davis and Barmuta (1989) introduced a fourth category for the situation where the roughness element height exceeds one-third of the water depth. An additional category could be included for the situation often found in streams where the roughness elements break through the water surface:

(4) *Exposed roughness flow* (Figure 6.11(d)): elements protrude through

the water surface. Flow conditions become very complex as water flows over and around these large obstacles, often forming "whitewater" conditions.

In his study of flow resistance, Morris (1955, 1961) found it convenient to categorize boundary roughness patterns into three broad categories. As suggested by Davis and Barmuta (1989), the classification of boundary roughness patterns may prove to be useful in the study of near-bed environments. It also has relevance in fish-ladder design and the creation of specific flow patterns in stream habitat rehabilitation (Chapter 9).

Although Morris used roughness elements of uniform height, his classifications can be extended to surfaces with variable roughness heights and spacing by using average values of the dimensions (Chow, 1959). Gore (1978) and Wetmore et al. (1990) describe surface profilers for measuring the local roughness of streambed materials. These instruments are similar to the point-frequency frame (Section 5.3.3), where pins are dropped to the surface and the height of each pin recorded. The standard deviation of the pin heights is also used as an indication of surface relief.

A note about reality

In these somewhat theoretical developments the three important concepts to bear in mind are (1) the no-slip condition, which causes flow to stick to solid surfaces and velocity profiles to develop; (2) the fact that shear stress is highest near the bed; and (3) turbulence is highest near the bed. In most parts of a stream, flow can be considered turbulent, with the velocity lowest near the streambed. The velocity profile is actually a continuum—like

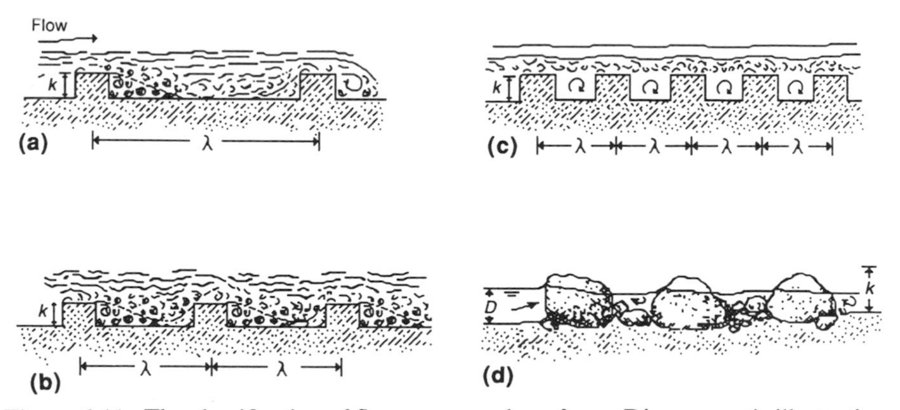

Figure 6.11. The classification of flow near rough surfaces. Diagrammatic illustrations of the flow patterns in (a) isolated roughness flow, (b) wake-interference flow, (c) skimming flow, and (d) exposed roughness flow. In (a)-(c), based on the classification of Morris (1955), water depth is fairly large in relation to roughness height; in (d), roughness elements break through the water surface

ecological zones, the division into laminar and turbulent is a convenient way of viewing regions with unique properties which blend together in a transitional zone. The separation is mathematically convenient.

Even under hydraulically smooth conditions, the viscous sublayer, rather than being stable, continuously fluctuates in thickness. For example, when the flow velocity is increased over a smooth bed of sand grains the laminar sublayer shrinks, and the sand grains are eventually exposed to the turbulence which lifts them from the bed. However, this does not occur uniformly over the whole bed surface. Instead, "patches" of sand are plucked from the bed as the viscous sublayer breaks down in places and the bed is exposed to energetic eddies from the turbulent zone (I. O'Neill, pers. comm., 1990). This "buffeting" effect is no doubt highly significant to organisms living in the high-shear stress, high-turbulence area near the streambed.

Boundary layers are imposed on other boundary layers, and scale determines which one is of interest. For a person picking a spot to sunbathe on a windy beach it pays to find the lee side of a sand dune; a fly landing on the lee side of the person's nose would be similarly sheltered. Within the turbulent boundary layer of a stream, boundary layers form at the surfaces of individual rocks and at those of snails, mayfly larvae or other organisms sitting on the rocks.

Statzner and Holm (1982) used laser-Doppler anemometry to observe flow conditions around *Ecdyonurus* (Mayfly) nymphs as the classic case of "life in the boundary layer". From their studies they surmised that it was unlikely that the boundary layer concept could adequately explain the nature of flow around animals. Rather, the flow patterns may have more to do with morphological structures and/or behaviour mechanisms which allow the animals to direct shearing forces to a point where they can be counteracted (like overlegs or other "anchoring" appendages). Vogel (1981) also made the point that the idea of organisms having flattened shapes to reduce drag and help them stay attached to surfaces was a reasonable theory but an oversimplification.

Although the tools which are presented in the following sections are based on simplifying assumptions, they form a starting point firmly grounded in the field of engineering from which modifications can be developed for biological applications. Questions which might arise at this point are: "How thick is the boundary layer?" "How thick is the laminar sublayer?" "How fast is the flow at a given point?" and "What is the drag force on an object within the layer?" Formulae for these quantities are presented in the next section. Since most of the original work was done with pipes, cylinders and flat plates rather than meandering stream channels, irregular rocks, and leaves of aquatic plants, the equations should not be applied as hard-and-fast rules but as approximations. Because flow patterns over and around objects are difficult to describe, simplifying assumptions are made so that problems can be dealt with empirically and theoretically. If budgets, time and equipment permit, the ideal solution is to directly measure velocity profiles, and a few methods are discussed in Section 6.5.6.

6.5 THE MICROENVIRONMENT: FLOW NEAR SOLID SURFACES

6.5.1 GENERAL

In the microenvironment the focus is on patterns of viscous action as fluid passes by a solid surface. The boundary layers of interest are those which are produced around objects within the flowing fluid. Finer-scale turbulence such as the small eddies behind objects and velocity fluctuations near the surfaces of solids becomes important.

The patterns of flow within the microenvironment form an important component of the physical habitat for aquatic organisms. As water flows by a solid surface it can generate lift and drag on a sediment particle or a benthic invertebrate. Dead-water zones are created on the downstream sides of boulders which affect the speed of water encountered by migrating fish. Flow patterns can influence the behaviour of an organism clinging to a stone or the stalk of a reed. The distribution of flow patterns in streams has also no doubt played a part in the evolution of organisms which are best suited to particular flow environments.

Presented in the next section are formulae for describing the distribution of velocity and the thickness of boundary layers for the somewhat idealized case of flow around a sharp, flat plate. From this point, the description of flow around "bluff bodies" which present a less anorexic profile are developed in Section 6.5.4. Implications in terms of lift and drag are covered in Section 6.5.4 and 6.5.5, and Section 6.5.6 will contain methods of measuring velocity in the microenvironment.

6.5.2 DESCRIBING THE VELOCITY PROFILE AND BOUNDARY LAYER THICKNESS NEAR A SOLID SURFACE

The laminar region

As illustrated in Figure 6.7(b), the velocity profile in a laminar boundary layer has a smooth, parabolic shape. Blasius, a student of Prandtl, developed a relationship for describing the way velocity changes with depth in laminar flow over a flat plate. Tables of values for the Blasius velocity profile can be found in White (1986). Although the profile is not described by an exact formula, it can be approximated by a parabolic equation:

$$v \approx V\left(\frac{2y}{\delta_1} - \frac{y^2}{\delta_1^2}\right) \tag{6.21}$$

where y = distance away from the plate (m), V = free stream velocity (m/s), v = velocity at some distance x along the plate and some distance y above it and δ_l is the laminar boundary layer thickness at some distance x from the leading edge of the plate (see Figure 6.9). Because there is really no "outer limit" to the velocity profile (i.e. the retarding effect on the flow continues for large distances), an arbitrary decision had to be made on where its effect was considered negligible. Blasius assumed that the edge of the boundary layer occurs at $v = 0.99V$, and developed the following equation for boundary layer thickness:

$$\delta_l = \frac{5.0x}{\sqrt{Re_x}} \tag{6.22}$$

where x is, again, the distance from the front of the plate and Re_x is the "local Reynolds number" computed from Equation 6.17. The Blasius solution has been found to correspond well with experimental values (Roberson and Crowe, 1990).

The turbulent region

If we "zoom in" on the turbulent velocity profile of Figure 6.8(a) the average pattern (without the fuzzy lines) will look like either Figure 6.12(a) or 6.12(b), depending on whether the solid surface is hydraulically smooth or rough, respectively. These profiles describe fully developed boundary layers, meaning they are no longer in transition from the laminar conditions at the front of a plate or the entrance to a pipe, flume or channel. In most situations the laminar region is so short in comparison with the total length of the surface that it can be ignored.

The velocity profile in the turbulent zone for either hydraulically smooth or rough surfaces is normally described by a logarithmic equation. Prandtl developed the **universal velocity-distribution law** for describing turbulent velocity profiles by applying a few mathematical sleights of hand such as assuming that "globs" of fluid exchange momentum (to speed others up or slow them down) over some "mixing length". The equation which resulted is of the form:

$$\frac{v}{V_*} = \frac{1}{\kappa}\ln\left(\frac{yV_*}{\nu}\right) + B \tag{6.23}$$

where v is the velocity (m/s) as it varies with distance y (m) away from a solid surface. V_* is the shear velocity (m/s), described in Equation 6.20 as the square root of shear stress divided by fluid density. Thus, the expressions

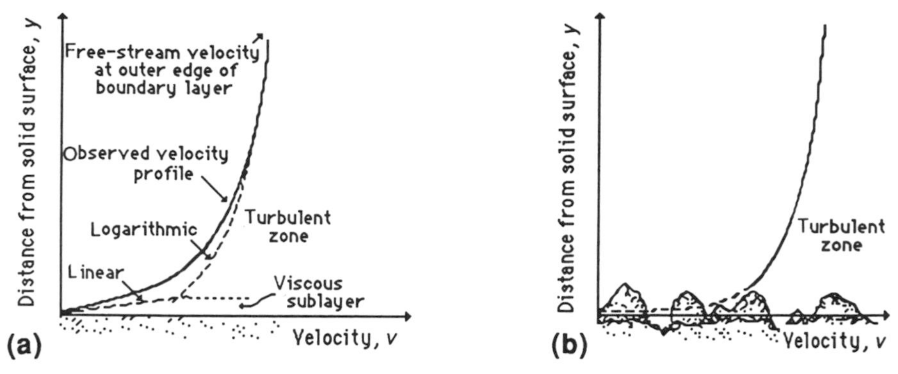

Figure 6.12. Turbulent velocity profiles and their descriptors for: (a) a hydraulically smooth surface and (b) a hydraulically rough one. The curves represent averages; in reality, velocity fluctuates about the average profile in the turbulent zone. Adapted from Daily and Harleman (1966), by permission of D. Harleman

on both sides of Equation 6.23 are dimensionless. The factors κ and B are empirically derived constants, varying with the type, concentration and size variation of roughness and the shape of the pipe or channel conveying the flow. Equation 6.23 is usually attributed to both Prandtl and von Karman (another student of Prandtl's), and κ is sometimes called Karman's universal constant. From experimentation, the nominal value of κ is 0.40, but it can vary widely.

For **hydraulically smooth** boundaries (and base 10 logarithms), Equation 6.23 becomes:

$$\frac{v}{V_*} = 5.75 \log\left(\frac{yV_*}{\nu}\right) + 5.5 \tag{6.24}$$

In practice, the coefficients $1/\kappa$ and B may differ slightly from 5.75 and 5.5, respectively.

For **hydraulically rough** conditions, the equation for the velocity distribution is usually given as:

$$\frac{v}{V_*} = 5.75 \log\left(\frac{y}{k}\right) + 8.5 \quad \text{or} \quad \frac{v}{V_*} = 5.75 \log\left(\frac{30y}{k}\right) \tag{6.25}$$

where k is a roughness measure (see Equation 6.18) and the other terms are as defined for Equation 6.23. Equation 6.25 is given in two forms; in the second version, the constant 8.5 has been incorporated into the log term. The velocity profile is normally assumed to apply down to $y = 0.5k$, a depth of one-half the roughness height. If actual velocity measurements are taken, the equation can be fitted to the measured velocity profile to calculate V_*, the shear velocity (see Section 6.5.3).

For a smooth, flat plate (or similar objects within the flow), the turbulent boundary layer can be described by the **one-seventh power law**, a simple relationship which provides a good fit when the Reynolds number is between about 10^5 and 10^7 (White, 1986):

$$v \approx V\left(\frac{y}{\delta_t}\right)^{1/7} \tag{6.26}$$

where V is the free-stream velocity (m/s) and δ_t the thickness of the turbulent boundary layer, commonly given as:

$$\delta_t = \frac{0.37x}{Re_x^{1/5}} \tag{6.27}$$

Here x is, again, the distance from the leading edge of the flat plate (Figure 6.9). By comparing Equations 6.22 and 6.27 it can be shown that the turbulent boundary layer will grow more rapidly than the laminar layer with distance along the plate. Example 6.3 below illustrates the use of turbulent zone velocity equations.

The viscous sublayer

Perhaps of more interest to organisms living on submerged surfaces is the thickness of the laminar or viscous sublayer, which exists when the surface is hydraulically smooth. In modelling the turbulent velocity profile as a logarithmic function (e.g. Equation 6.23) a problem develops as the curve approaches the solid surface because the logarithm of zero makes no sense. To account for this anomaly, Prandtl used a linear velocity profile near the solid surface to "connect" the logarithmic profile to it. This region was assumed to be the region of laminar flow, in which viscous effects are concentrated.

The velocity profile in this region is described by

$$\frac{v}{V_*} = \frac{V_* y}{\nu} \tag{6.28}$$

where the terms are defined as before. The thickness of the laminar sublayer, δ_s, is defined by the point where the logarithmic and linear profiles intersect, most often given by

$$\delta_s = \frac{11.8\nu}{V_*} \tag{6.29}$$

It is a commonly held belief that stream invertebrates which live in high-velocity environments on substrate surfaces have flattened body shapes to allow them to lead a "sheltered life" in the laminar sublayer. Smith (1975) used Equation 6.29 and an approximation that the mean stream velocity is twenty times the shear velocity to compute several values of the laminar sublayer thickness (for water at 15°C):

Mean velocity, $\bar{v}$ (m/s):	0.01	0.05	0.10	0.50	1.00
δ_s (mm):	27	5.4	2.7	0.54	0.27

He concluded from these calculations that with the mean velocities normally found in streams, "it seems unlikely that the larger invertebrates can be considered as being sheltered in the laminar sub-layer" (Smith, 1975, p. 36). In fact, their presence may actually cause the surface to be considered hydraulically rough, in which case the laminar sublayer does not exist (Davis and Barmuta, 1989).

The separation of the velocity profile into zones allows the separate mathematical description of a region in which viscous effects are important and one where they are no longer significant. From a physical and biological perspective, however, it may be preferable to maintain the view of the profile as a continuum, with a lower-velocity region near a solid surface and eddies of varying scale throughout.

6.5.3 SHEAR STRESS AND DRAG FORCES

The shape of the velocity profile, particularly that part close to a solid surface, will reflect the amount of shear stress the flow exerts on a surface submerged in the flow. We know from Equations 6.11–6.13 that shear stress is related to the steepness of the profile. Considering Figure 6.9 again, it can be seen that the boundary layer thickens with distance along the flat plate. At the plate, the velocity is zero, and at the outer edge of the boundary layer the velocity is nearly equal to the free stream velocity. As the boundary layer thickens, this difference in velocity is "stretched" over increasingly longer distances, and the velocity gradient (the ratio dv to dy) decreases. Thus, shear stress generally reduces with distance back from the leading edge, as shown in Figure 6.13, with a jump at the transition from laminar to turbulent.

By integrating shear stress over the surface area, a value can be obtained for the total shearing force. This is the drag force or "skin friction" force on the plate, symbolized by F_s. For one side of the plate it is:

$$F_s = C_f WL\rho \frac{V^2}{2} \tag{6.30}$$

where W and L are width and length of the plate, respectively, and C_f is a skin friction coefficient which depends on the flow type (e.g. whether laminar or turbulent). To calculate the total drag on both sides of the plate (for example, on both sides of a fish), the 2 in the denominator of Equation 6.30 would be eliminated.

For a fish swimming within the flow, F_s might represent the amount of force that the fish must exert to maintain the same speed. For an object on the streambed such as a boulder, a waterlogged log or a benthic invertebrate,

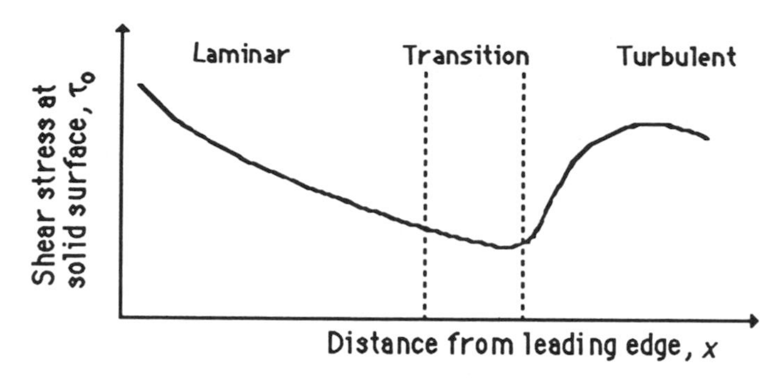

Figure 6.13. Change in shear stress with distance along a flat plate (compare with Figure 6.9). Adapated from Douglas et al. (1983), by permission of Longman Group, UK

it denotes the resistance to flow caused by the object, and is thus related to the likelihood with which it will be swept away. The organisms living on surfaces within the stream must balance the benefits of thinner boundary layers (greater mixing, supply of nutrients and gases, and waste removal) with increased shear stresses and problems of remaining attached. Microbes or suspension-feeding organisms located in the thinner boundary layer at the leading edge of a flat plate (Figure 6.9), for example, would have better supplies of nutrients and gases than those within thicker layers further back, yet the shear stress would be higher (Figure 6.13).

The laminar region

In laminar flow the local shear stress, τ_0, is given by

$$\tau_0 = 0.332\mu V \frac{Re_x^{1/2}}{x} \tag{6.31}$$

where τ_0 is in N/m^2, μ is fluid viscosity and the other terms are as previously defined. This equation can be used to obtain the local shear stress at any point along the plate within the laminar region.

The skin friction force or drag force, F_s, is calculated from Equation 6.30, where:

$$C_f = \frac{1.33}{Re_L^{1/2}} \tag{6.32}$$

Here, Re_L is the Reynolds number in which the "characteristic length" is the plate length, with length measured in the direction of flow.

The turbulent region

Because of turbulent fluctuations, shear stress is increased above what would be predicted from average velocities (see Section 6.4.1). The shear stresses caused by turbulence are referred to as **apparent shear stresses** or **Reynolds stresses**. Assuming that the turbulent boundary layer starts at the leading edge, equations can be developed for shear stress and skin-friction drag. Shear stress at the solid surface, τ_0, is given as:

$$\tau_0 = \rho c_f \frac{V^2}{2} \tag{6.33}$$

where c_f is the "local skin friction coefficient" at a distance x from the leading edge. For **hydraulically smooth** conditions,

$$c_f = \frac{0.058}{Re_x^{1/5}} \tag{6.34}$$

The drag force or overall shear resistance force is equal to the shear stress integrated over the length of the plate. For hydraulically smooth conditions (and for one side of the plate) this is given by Equation 6.30, with:

$$C_f = \frac{0.074}{Re_L^{1/5}} \tag{6.35}$$

As for Equation 6.32, Re_L is the Reynolds number based on the plate length. The relationship between C_f and Re for flat plates is included in Figure 6.20 (in the figure, C_f is given as C_D, the total drag coefficient). For higher Reynolds numbers, more refined analyses are required. A careful comparison of Equations 6.32 and 6.35 will reveal that maintaining a laminar boundary layer as far along the plate as possible is desirable for reducing drag.

If the smooth plate grows barnacles or is otherwise considered **hydraulically rough**, the resistance coefficients are related to the relative roughness (Equation 6.18). White (1986) gives the following equations for resistance coefficients in the fully rough regime:

$$c_f \approx \left(2.87 + 1.58 \log \frac{x}{k}\right)^{-2.5} \tag{6.36}$$

$$C_f \approx \left(1.89 + 1.62 \log \frac{L}{k}\right)^{-2.5} \tag{6.37}$$

where k is a measure of roughness (m) (see Equation 6.18), and the other terms are as previously defined.

The transition region

For the transition from laminar to turbulent flow ($500\,000 < Re_L < 10^7$) the boundary layer will be laminar along part of the upstream section of the plate. Prandtl developed the following "hybridized" equation for the transition region (Streeter and Wylie, 1979):

$$C_f = \frac{0.074}{Re_L^{1/5}} - \frac{1700}{Re_L} \qquad (6.38)$$

Example 6.2

(a) Determine the total drag on a smooth, flat plate of length 0.5 m and width 0.23 m, towed through still water (10°C) at a speed of 2 m/s; and (b) determine the drag if the plate is rough, with an effective roughness height of 1 mm.

(a) The Reynolds number, Re_L is (Equation 5.15):

$$Re_L = \frac{(2)(0.5)}{(1.308 \times 10^{-6})} = 765\,000;$$

thus the flow is transitional. From Equation 6.38

$$C_f = \frac{0.074}{(765\,000)^{1/5}} - \frac{1700}{765\,000} = 0.00270$$

and drag on both sides of the plate (Equation 6.30) is:

$$F_s = (0.00270)(0.23)(0.5)(999.7)(2)^2 = \underline{1.24\text{ N}}$$

(b) From Equation 6.37 (we are assuming that the flow is hydraulically rough and fully turbulent),

$$C_f \approx \left[1.89 + 1.62 \log\left(\frac{0.5}{0.001}\right)\right]^{-2.5} = 0.0102$$

$F_s = (0.0102)(0.23)(0.5)(999.7)(2)^2 = \underline{4.69\text{ N}}$—almost four times the drag force on the smooth plate!

Example 6.3

If a bacterium were released into the flow near a stationary, thin, flat piece of wood oriented into the flow, at what velocity would it travel when it reached a point 140 mm back from the leading edge of the solid and 5 mm away from it? Assume that the boundary layer is turbulent, the surface is rough, the free stream velocity is 0.6 m/s and the water is 15°C. Use (a) Equation 6.25 and (b) the one-seventh power law (Equation 6.26).

(a) Not knowing anything about how rough the wood is, we will start by selecting a value of k within the range given in Table 6.1, say 0.50 mm (0.0005 m). From Equation 6.36

$$c_f = \left[2.87 + 1.58 \log\left(\frac{0.140}{0.0005}\right)\right]^{-2.5} = 0.00849$$

By combining Equations 6.20 and 6.33,

$$V_* = \sqrt{\frac{\tau_0}{\rho}} = \sqrt{c_f \frac{V^2}{2}} = \sqrt{\left(0.00849 \frac{(0.6)^2}{2}\right)} = 0.0391 \text{ m/s}$$

Then, from Equation 6.25,

$$v = (0.0391)\left[5.75 \log\left(\frac{30(0.005)}{0.0005}\right)\right] = \underline{0.557 \text{ m/s}}$$

(b) From Equation 6.17,

$$Re_x = \frac{(0.6)(0.140)}{(1.141 \times 10^{-6})} = 73\,600$$

Then, from Equations 6.27 and 6.26,

$$\delta_t = \frac{0.37(0.140)}{(73\,600)^{1/5}} = 0.00551 \text{ m, and}$$

$$v \approx 0.6\left(\frac{0.005}{0.00551}\right)^{1/7} = \underline{0.592 \text{ m/s}}$$

Although the answers are not exactly identical, they are of the same order of accuracy, and approach (b) is much simpler.

Calculating shear stress from the velocity profile

If the velocity profile is measured near a surface and it plots approximately as a straight line on a linear-log plot such as in Figure 6.14(b) then shear stress can be calculated from the slope of the profile. From Equation 6.20, $\tau_0 = \rho(V_*)^2$, where the equation for V_* along a hydraulically rough streambed is obtained from Equation 6.25 and can be expressed as:

$$V_* = \frac{b}{5.75} \tag{6.39}$$

where b is the slope of the logarithmic velocity profile (velocity versus log(depth)). An example calculation of V_* is shown in Figure 6.14.

6.5.4 FLOW AROUND BLUFF BODIES

The idealized case of flow across a sharp, flat plate is useful in developing a foundation for boundary layer theory. In this section, flow patterns around

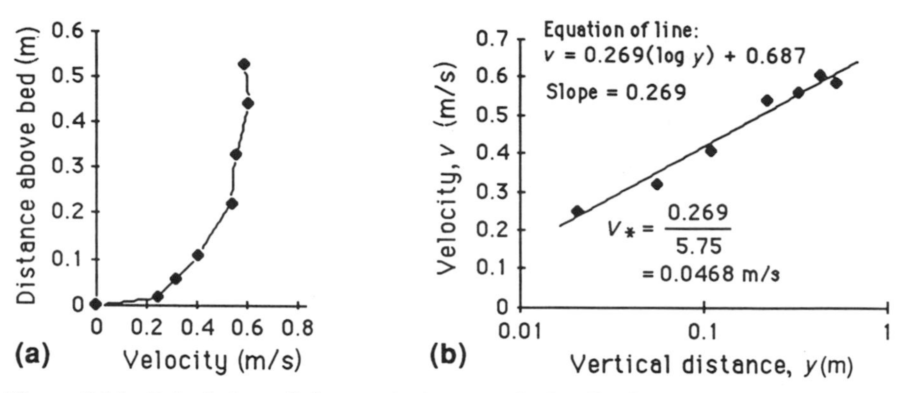

Figure 6.14. Calculation of shear velocity on a hydraulically rough streambed from velocity measurements taken at a vertical at Site 4 on the Acheron River (Figure 4.7): (a) measured velocity profile and (b) regression line (see Appendix 1) fitted to the relationship between velocity and log(depth). Note that the axes are reversed in (b) because v is the dependent variable in velocity distribution equations (see Equations 6.23–6.26)

immersed objects such as sediment particles and crocodiles are examined. In engineering terminology these objects are referred to as **bluff bodies** or **blunt bodies**.

Normally, when flow occurs around a bluff body the velocities and pressures on opposite sides of the object will be different, producing a force perpendicular to the oncoming flow (lift) and/or a force in line with the direction of the flow (drag), as shown in Figure 6.15 (see also Figure 7.17). If viscosity did not exist, drag would not occur—a piece of wood dropped into a flowing stream would remain where it had been dropped rather than floating downstream, and a fish (or other neutrally buoyant examples) set into motion would have difficulty stopping.

Separation and wake

Viscous fluids, however, adhere to solids and lose energy to friction. Several notable features can be observed in the patterns of streamlines around the

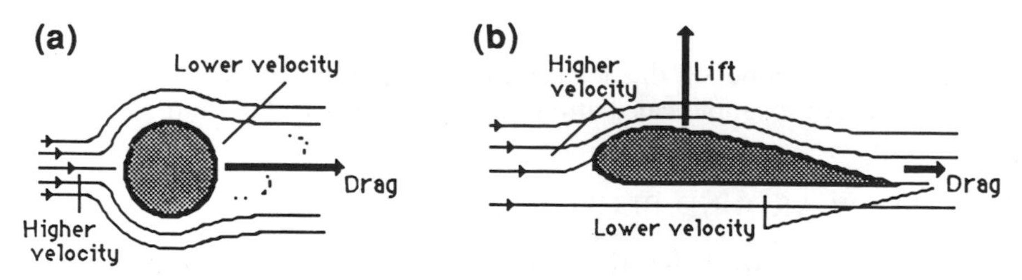

Figure 6.15. Lift and drag on bluff bodies in a viscous fluid

objects in Figure 6.16. On the upstream side of each object a streamline "collides" with the object, creating a **stagnation point** (a point of zero velocity). Boundary layer development begins here. As the fluid progresses around the solid, it loses energy to friction, and at some point the boundary layer and the solid part company. This is called the **separation point** and the phenomenon **boundary layer separation**. Analogous to congested traffic, when one particle stops, others stack up behind it or attempt to move around the "stalled" particle, and the boundary layer thickens. Particles are carried out into the free-stream flow, and the boundary layer as a whole is stretched and drawn away from the solid. A "gap" is left, and downstream particles are pulled up into the region, producing a flow reversal or **vortex**. Intense turbulence and energy dissipation occur in this region, called the **wake**.

Within the wake, a slow, lazy, retarded flow region occurs just behind the solid, called "**dead water**" or the "cavity region". It is here that many aquatic species take refuge from high-speed currents. For objects resembling cylinders (Figure 6.16(b)) a rough rule of thumb is that the length of the dead water region is about one-half of the length of the bluff body. A **stagnation region** of near-zero velocity can also occur in front of a bluff body, such as the flat plate of Figure 6.16(a). In a study of diatom colonization on disc-shaped glass "stubs" placed on a streambed, Korte and Blinn (1983) found increased colonization at the upper leading and trailing edges, and attributed this to interception of drifting oganisms by the upstream face and the presence of micro-eddies (separation) downstream.

Although prediction of the separation point has not been resolved mathematically it is easily recognized by observation. The point of separation is dependent on the shape and roughness of the object and boundary layer properties, including whether the flow is turbulent or laminar. Separation tends to take place at sharp breaks in the surface of solids. In streams it occurs not only around bluff bodies such as irregular rock outcrops or a stream ecologist's waders but also at the edges of the stream where it widens suddenly and at the downstream side of bends (see Figure 6.17).

In general, separation does not occur when the fluid is speeding up but is favoured by deceleration. The more abrupt the deceleration, the more

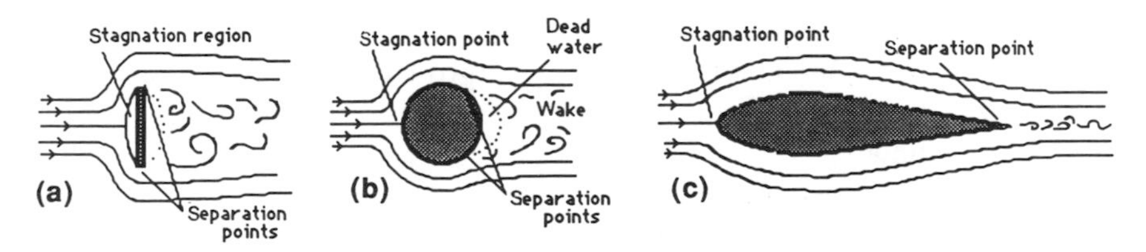

Figure 6.16. Streamlines and flow separation around: (a) a flat plate perpendicular to the flow, (b) a cylinder and (c) a streamlined object

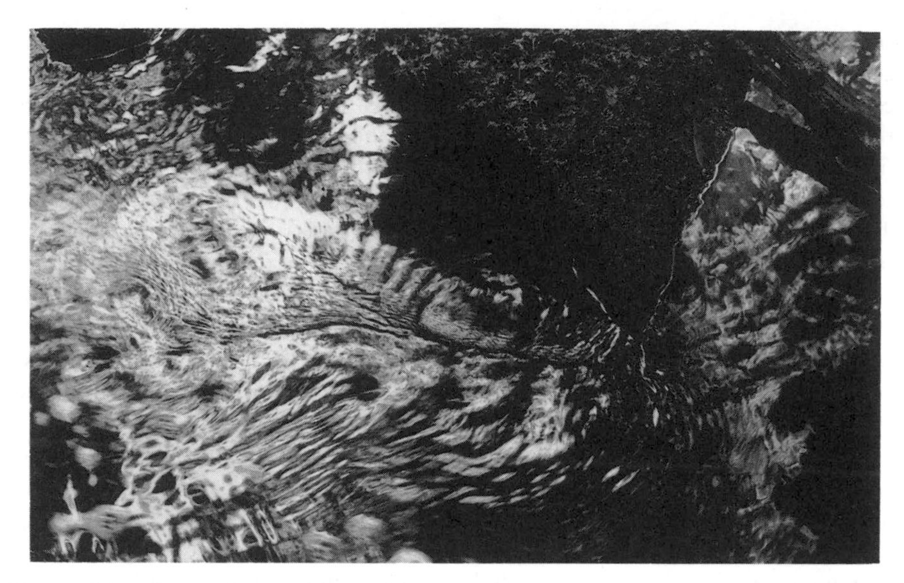

Figure 6.17. Separation and vortex formation at the edge of a rock outcrop. Flow direction is from right to left. Photo: N. Gordon

likely separation is to occur—analogous to the way an abrupt deceleration causes car drivers to be separated from their seats. Streamlining (Figure 6.16(c)) allows flow to decelerate gradually, which delays the onset of separation (Schlichting, 1961). As might be guessed, separation is an important factor in determining drag on an object.

Vortex formation and shedding

The eddying patterns behind an object change in picturesque and dramatic ways with the flow speed. Using the example of a circular cylinder, separation is not considered to occur at Reynolds numbers below about 5 (the value is not known precisely) because high viscous effects inhibit eddying motion (Roberson and Crowe, 1990). Here, nearly all the drag is due to skin friction. Above $Re \approx 5$, separation starts to occur (Figure 6.18(a)). Initially, eddies in the wake region are symmetrical, as shown in Figure 6.18(b)). Downstream of the cylinder the wake begins to oscillate sinusoidally. As the flow speeds up, inevitably one eddy will grow faster than the other until the largest is "shed" into the flow and a turbulent, oscillating wake develops behind the cylinder.

Along with conservation laws pertaining to energy and mass, fluids must also conform to the law of **conservation of vorticity**. This states that circulation created in one place must be balanced by an equal and opposite circulation elsewhere (Vogel, 1981). Thus, vortices are shed from alternating

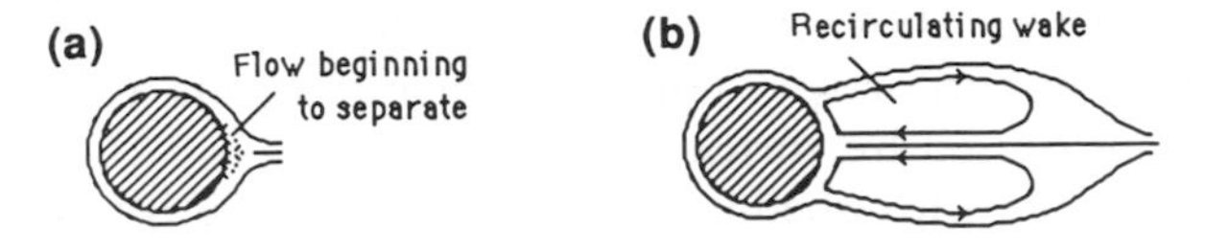

Figure 6.18. Symmetrical eddying around a circular cylinder at: (a) $Re \approx 5$ and (b) $Re \approx 40$

sides of an object and each rotates in an opposite direction to the one just before it. This type of wake region with alternating vortices is called a "von Karman trail" or "Karman vortex street", after von Karman, who gave it a theoretical explanation in 1912. The effect of increasing *Re* can be seen in Figure 6.19.

The elegant flow patterns of vortices around a cylinder are described as *unsteady laminar flow* at Reynolds numbers below about 10^5 (with cylinder diameter the "characteristic length" in Equation 6.15). When *Re* is somewhere between 100 000 and 250 000 the flow becomes turbulent. In the turbulent boundary layer, fluid particles near the solid surface have higher velocities, so a fluid particle will travel further around the cylinder before separation occurs. The wake is thus narrowed and drag is reduced at the critical Reynolds number where the transition from laminar to turbulent flow occurs. This phenomenon will be discussed further in the next subsection.

Shedding of vortices sets up a rhythmic side-to-side force on the cylinder. If this coincides with a natural vibration frequency of the solid, it can produce resonance—which gives rise to the "singing" of wires in wind and which played a part in the failure in 1940 of the Tacoma Narrows suspension

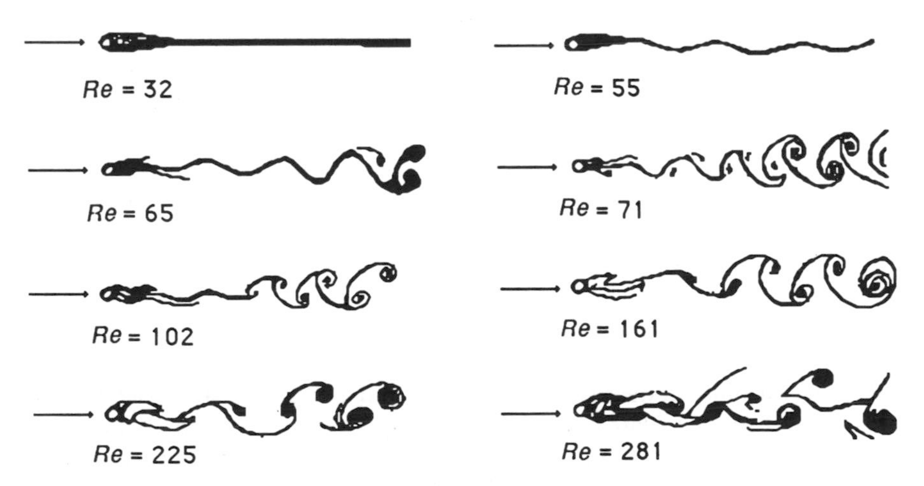

Figure 6.19. The effect of increasing Reynolds number on flow past a circular cylinder. Redrawn from photographs of Schlichting (1961), by permission of V. Streeter

bridge in Washington, USA. A researcher holding a rod in swiftly flowing streams when taking current speed or depth measurements will have a "feel" for the vibrations caused by vortex shedding.

The frequency of vortex shedding is described by the Strouhal number, *St*, named after a physicist who experimented with singing wires in 1878:

$$St = \frac{\eta d}{V} \tag{6.40}$$

where η = frequency of shedding of vortices from **one** side of the cylinder, in Hz (hertz), d = cylinder diameter (m) and V = free-stream velocity (m/s). St is a function of Reynolds number, with d the characteristic length. Vortex shedding occurs in the range $10^2 < Re < 10^7$. Relationships between St and Re are given in most fluid mechanics texts. An average value for St is about 0.21 (White, 1986).

Example 6.4

Calculate the frequency of vortex shedding for the rigid stem of a reed at a point where the stem diameter is 5 mm and the mean flow velocity is 0.3 m/s.

$$St \approx 0.21 = \frac{\eta\,(0.005\text{ m})}{0.3\text{ m/s}}$$

$$\eta = 12.6\text{ vibrations/second}$$

Vortices can also occur in other situations. For example, when birds fly in formation the vertical vortices shed from the wingtips of one bird gives an added lift to the bird flying just next to it. A "kolk" is an upward-spiralling vortex which can occur in streambed depressions, and produces a strong hydrodynamic lift (Morisawa, 1985). A "horseshoe vortex" is a region of circulating fluid motion around an obstacle, wrapped around the upstream side with the arms of the "horseshoe" stretching downstream. Davis (1986) states that a horseshoe vortex system is present around an isolated organism or other object when its height is close to that of the viscous boundary layer thickness. Vortex patterns may affect patterns of colonization on boulders, and may be used by suspension feeders for concentrating food.

Drag

The drag on bluff bodies is composed of two parts: the **skin-friction drag** as described for flat plates plus a drag component due to the shape of the object called **pressure** or **form drag**. Form drag is by far the most important component for bodies which are blunt rather than presenting a slender profile to the flow. It is something which has been experienced by anyone who has opened an umbrella on a windy day or waded in knee-deep water along the ocean's edge. Form drag on bluff or "blunt" bodies has less to do with the "bluntness" of the front side than with separation and pressure differences from front to back (Schlichting, 1961).

If, for example, the flat, sharp plate of the previous section were turned so that the flat side rather than the sharp nose faced the current, as in Figure 6.16(a), one would intuitively expect the drag force to increase considerably. Not only does the drag force increase but the skin-friction drag becomes insignificant since it acts only at the sharp edges. Thus, the drag on the body is totally form drag, and is calculated from

$$F_D = C_D A_p \rho \frac{V^2}{2} \tag{6.41}$$

with A_p the area of the plate (width × length), V and ρ the free-stream velocity and fluid density as before, and C_D a coefficient of drag, which will be discussed shortly.

In fact, Equation 6.41 is applicable for all bluff bodies if F_D is considered the total drag (pressure drag plus skin-friction drag). A_p becomes the projected area of the object as the current "sees" it. Values of A_p for various shapes are given in Table 6.2.

Other "area factors" may be used instead of the projected area, such as the wetted area, the total surface exposed to flow (used for streamlined bodies). C_D is usually determined from experiments with small models in wind tunnels, flumes or towing basins (in which a model is towed through a long body of water). It varies with Reynolds number (based on some characteristic length of the object, e.g. length or diameter), as can be seen in Figure 6.20. The projected area used in Equation 6.41 should be consistent with the area used in the determination of C_D.

C_D is relatively constant for angular bodies at Reynolds numbers over about 10^4. Extensive tables of C_D for bodies at $Re = 10^4$ and above have

Table 6.2. Projected area, A_p, for various shapes (d = diameter)

Type of object	A_p
Plate perpendicular to the flow	Width × length
Plate parallel to the flow	0 (not a blunt body— use Equation 6.30 instead)
Cylinder perpendicular to the flow	Diameter (d) × cylinder length
Cylinder parallel to the flow	$\frac{\pi d^2}{4}$
Sphere	$\frac{\pi d^2}{4}$

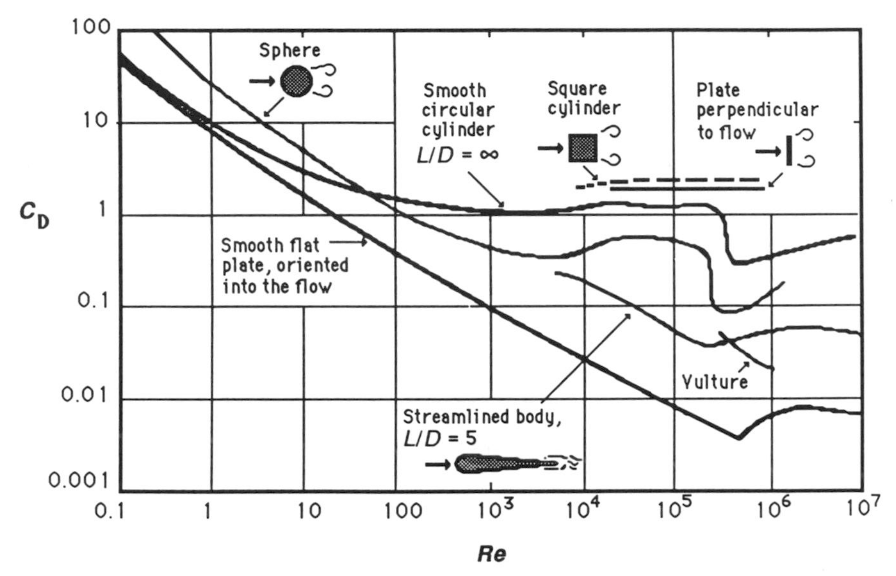

Figure 6.20. The relationship between coefficient of drag (C_D) and Reynolds number (*Re*) for various shapes, L = length; D = diameter. Adapted from White (1986) and Roberson and Crowe (1990), by permission of McGraw-Hill and Houghton Mifflin Company, respectively

been developed; a sampling is given in Table 6.3. The two-dimensional coefficients are used for objects such as cylindrical rods or rectangular posts which are the same shape over a large distance (the dimension extending into and out of the page). When a cylindrical-like caddis fly larvae case is oriented into the flow, it resembles a three-dimensional "flat-faced" cylinder, whereas when it is turned sideways to the flow it is more like a two-dimensional round cylinder. A flounder might be best modelled by a three-dimensional flat ellipsoid.

For circular cylinders and spheres, a characteristic drop in C_D is apparent in Figure 6.20. This drop coincides with the critical Reynolds number where the flow changes from laminar to turbulent. In laminar flow the separation point on a cylinder or sphere occurs close to the halfway mark. In turbulent flow, however, because of the increased mixing, higher-velocity particles are carried further along the solid surface before detaching—at a point about 30° around from the midpoint (White, 1986). The turbulent boundary layer can be "triggered", for example by adding a patch of sand roughness to the front of a ball dropped into water (Figure 6.21(b)). Because separation occurs further back on the sphere, the wake associated with the turbulent boundary layer in Figure 6.21(b) is much smaller than that for the laminar flow conditions of Figure 6.21(a).

As counter-intuitive as it seems, adding roughness at these near-transitional

Table 6.3. Drag coefficients (C_D) for various shapes at $Re > 10^4$. From White (1986) and Roberson and Crowe (1990), by permission of McGraw-Hill and Houghton Mifflin Company, respectively. Drag coefficients are based on the frontal area of the object

Two-dimensional

Shape	C_D Laminar	C_D Turbulent
Plate (facing flow)	1.98	
Rectangular plate (parallel to flow)		
$l/b = 1$	1.18	
$l/b = 5$	1.20	
$l/b = 20$	1.50	
Equilateral triangle	1.60	
Half-tube	2.30	
Round cylinder	1.2	0.3
Elliptical cylinders 2:1	0.6	0.2
Elliptical cylinders 8:1	0.25	0.1

Three-dimensional

Shape	C_D Laminar	C_D Turbulent
Flat-faced circular cylinder (parallel to flow)		
$l/d = 0$ (disc)	1.17	
$l/d = 1$	0.90	
$l/d = 8$	0.99	
60° cone	0.50	
Cup, open towards flow	1.40	
Sphere	0.47	0.20
Ellipsoid		
$l/d = 0.75$	0.50	0.20
$l/d = 2$	0.27	0.13
$l/d = 8$	0.20	0.08

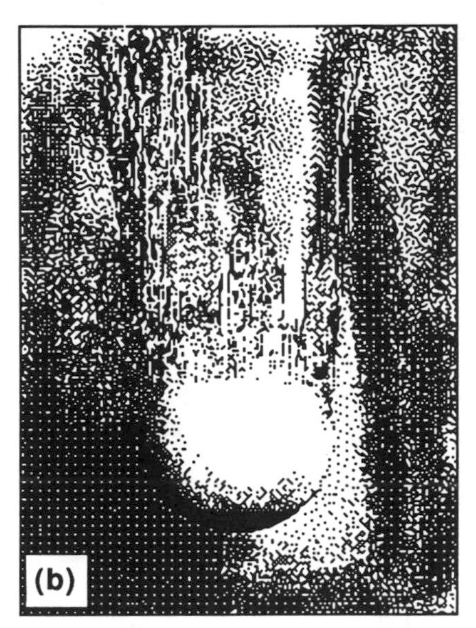

Figure 6.21. Shift in separation point around a sphere dropped into water: (a) with a smooth surface (laminar boundary layer) and (b) with a patch of roughness on its "nose" (turbulent boundary layer). Redrawn from US Navy photographs in White (1986), by permission of the US Navy

Re situations *reduces* rather than increases drag. This is the reason why golf balls, which fly in this range of *Re*, are dimpled. Vogel (1981) points out that cylinders and spheres are rather ideal cases, and more irregular biological objects are likely to yield graphs without this characteristic drop. In fact, for a streamlined shape, roughness would *increase* the drag.

At the lower-*Re* end of the graph (below $Re \approx 1$) it can be seen from Figure 6.20 that the coefficient of drag for a sphere or cylinder varies almost linearly with *Re*. This is the range for droplets of mist floating through the air, the settling of fine sediment particles and the movement of tiny organisms through their highly viscous environment (see Section 6.4.1).

Terminal velocity

Free descent is common to parachuting baby spiders, wind-borne seeds, falling raindrops and settling sediment particles. An object will at first accelerate downwards under the effect of gravity. However, as its speed increases, drag also increases until it reaches a level where all external forces balance. At this point, accleration is zero (by Newton's second law, if the net force is zero and the mass is the same, then in the equation $F = Ma$, a must be 0). Thus, velocity is a constant and is called the **terminal velocity**. The maximum speed reached is dependent on the size, shape and density of the object, as well as fluid properties, which must remain constant over the path of descent if terminal velocity is to occur. A human body has a terminal velocity of about 56 m/s (200 km/h) when falling through the atmosphere.

In 1845 Stokes developed a relationship for the terminal velocity of spheres at very low Reynolds numbers, in the range $Re < 0.1$ (with Re calculated using the particle diameter). In this range the drag coefficient varies linearly with Re, and is given by:

$$C_D = \frac{24}{Re} \tag{6.42}$$

and the terminal velocity for spheres is then computed by Stokes' law:

$$V = \frac{2gr^2}{9\mu}(\rho_s - \rho) \tag{6.43}$$

in which ρ is the density of the fluid, ρ_s the density of the sphere, r the radius of the sphere and μ the viscosity of the fluid (see Table 1.1 for units). Since the equation can be used to determine settling times of particles it is the basis for analysing soil particle size by hydrometer and pipette methods (Section 5.10.4). By assuming that air bubbles in water are spherical, the same equation can be used to approximate the velocity of bubbles rising towards the water surface (although in reality the bubbles will flatten somewhat). In water the law does not apply to particles smaller than 0.0002 mm, since the settling of these minute particles is influenced by Brownian motion.

Example 6.5

Calculate the terminal velocity for a spherical particle of silt with a diameter of 0.05 mm falling within a column of 25° fresh water. Assume that the particle has a density of 2650 kg/m³.

$$V = \frac{2(9.807\ \text{m/s}^2)(0.000025)^2}{9(0.894 \times 10^{-3})}(2650 - 997)$$

$$V = 0.0025\ \text{m/s}$$

Check: $Re = VD/\nu = \dfrac{(0.0025)(0.00005)}{(0.897 \times 10^{-6})} = 0.14$—actually slightly outside the Stokes range.

Thus, in a column of 0.4 m it would take about 159 s or 2.7 min for all this coarse silt to settle out.

For higher-Re conditions where inertial effects are important and C_D is not a linear function of Re, calculating terminal velocity is an iterative process. Roberson and Crowe (1990) give this procedure. An additional complication arises for disc-like objects because they oscillate while falling, making the values of C_D uncertain (Smith, 1975).

Other limitations on Stokes' law include the assumption that the particle is falling in isolation. If the concentration of particles is high, terminal velocities are reduced. Flocculating of small particles also increases the rate of settling above that predicted, which is why a dispersant is used in particle-size analyses. For the opposite effect, alum is added at water-treatment plants to clarify the water. The alum flocculates small particles and makes them settle out more quickly (Smith, 1975).

Reducing drag

In aerodynamic and hydrodynamic engineering designs, "the name of the game is drag reduction" (White, 1986, p. 418). Similarly, for life in a swiftly moving stream it is normally to an organism's advantage to reduce drag and thus decrease the amount of energy needed to either move against the current or maintain position within it. Unlike inanimate objects, organisms usually have some control over how they are oriented with respect to current direction. Vogel (1981) gives the example of ocean limpets which align their longest axis with that of the seaweed leaves to which they are attached. Then, when the seaweed leaf acts like a wind vane blown by the tidal currents, the limpets are also oriented in the proper direction to reduce drag. Other organisms flatten themselves against the substratum; still others have evolved with a sleek, streamlined figure or have appendages which affect drag in complex ways.

Unlike the rigid "bluff bodies" discussed earlier, most organisms are somewhat flexible. The flapping of leaves or the lateral movements of fish further increases drag over what would be predicted from a rigid shape. The complex hydrodynamics of how organisms interact with flow patterns is a fascinating area for research. Webb (1984) provides an interesting article on how fish swim, describing the different body shapes specially adapted for acceleration, manoeuvring and cruising.

Drag is, again, due to both skin friction and form drag, the latter resulting from flow separation. At **low-*Re* conditions** essentially all the drag is skin-friction drag, which acts over the whole surface area of the organism. Here, minimizing drag means minimizing the surface area to volume ratio. In general, smaller objects have higher ratios than larger ones, and flatter objects have higher ratios than more spherical ones. However, the advantages of drag reduction must be balanced with the benefits of having a large surface area-to-volume ratio such as the increased area for absorption of nutrients and gases and the elimination of wastes. Thus, micro-organisms tend to be cylindrical or spherical despite the hydrodynamic disadvantage.

At the higher Reynolds numbers more likely to be encountered by insect nymphs, freshwater mussels and stream ecologists, pressure drag is much more significant than viscous friction. Factors affecting drag at **higher-*Re* conditions** are:

(1) *Shape of the object.* In general, round edges produce less drag than sharp ones, which encourage separation. The best way of reducing drag is through streamlining—tapering the trailing edge to a point and moving the separation point further downstream. The girth measurements of trout, for example, correspond almost exactly with the profiles of low-drag airfoils. Maude and Williams (1983) found that when freshwater crayfish were exposed to increases in current they lowered their bodies closer to the substrate and created a "plough" profile by drawing their front claws together, allowing them to maintain position.

(2) *Orientation*: A streamlined shape is best for reducing drag on an object only if the object can orient itself so that the streamlining does it some good. If the direction of the current is unpredictable and the organism is more attached to its foundation such as a limpet or aquatic plant stem, a cylindrical or hemispherical shape may produce the least overall drag.

    Alternatively, drag can influence the orientation of stream-transported objects, such as rocks which line up in a streambed with their long axes pointed downstream (Section 7.4.2).

(3) *Modifying the wake size*: The principle of *initiating turbulence* to reduce the wake size and thus form drag was mentioned in regard to the dimples on golf balls. Vanes have also been used to *suppress separation* and wake size, reducing drag. For example, vanes on the sides of trucks or slotted wings on aircraft deflect the airflow closer to the surfaces, keeping the flow accelerated and hindering the onset of separation (Schlichting, 1961; White, 1986). Jets which discharge fluid into the boundary layer to increase acceleration or suction mechanisms to remove decelerated flow also help to prevent separation (Schlichting, 1961). The castor oil fish, *Ruvettus*, for example, has structures on its skin which inject fluid into the boundary layer to reduce drag (Bone and Marshall, 1982).

6.5.5 LIFT

Organisms use the principles of lift and drag to move horizontally or vertically and to stay put with the minimum amount of effort. Lift is produced because of differences in pressure above and below an object, as demonstrated in the discussion of Bernoulli's equation (Section 6.3.4). The magnitude of the lift force is computed much the same as drag force, by integrating the pressure difference over the area upon which it acts. In fact, the equation for lift is very similar to Equation 6.30 for drag:

$$F_L = C_L A_p \rho \frac{V^2}{2} \qquad (6.44)$$

where A_p for an airfoil is the "planform area" (as seen from above), ρ and V are as previously defined and C_L, the lift coefficient, is determined from experiment. For airfoils it is common for the planform area to also be used in the calculation of drag because of the relatively large skin-friction drag.

As anyone knows who has "experimented" with the lift on a hand extended out of a car window, the **angle of attack** (the angle of an airfoil to the direction of fluid motion) has a significant effect on lift and drag. From tests on airfoils, C_L increases almost proportionately with the angle of attack until about 15–20° when boundary layer separation occurs. C_L then drops and the aerofoil *stalls* (White, 1986). An inverted airfoil has the opposite effect, producing "negative lift". An example is the vane on the back of a racing car which improves stability and keeps the rear tyres on the ground.

Another means of producing lift is by **rotation** of a cylindrical or spherical object. Note that in Figure 6.22 separation occurs below rather than behind the rotating object. In the illustration the net force is in the direction away from the wake. This is why a cricket ball curves or a tennis ball rises when given an undercut. The phenomenon is called the Magnus effect, named after a German scientist who carried out studies on rotation-produced lift in the nineteenth century (Roberson and Crowe, 1990).

6.5.6 METHODS OF MICROVELOCITY MEASUREMENT

A solution to the problem of estimating velocities in or out of boundary layers is to measure them directly. Muschenheim et al. (1986) provide details on flume design for benthic ecology work. Field current metering techniques are covered in Section 5.7.3. Here, we will be concerned with techniques for measuring point velocities within the "microenvironment". Some of the more common methods are listed in order of complexity.

Flow visualization

Flow patterns around obstacles or organisms and flow phenomena such as separation or the change from laminar to turbulent flow can be observed

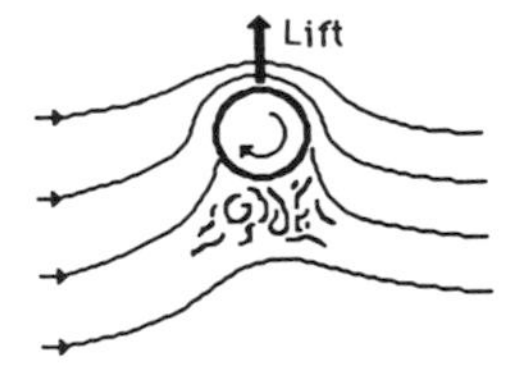

Figure 6.22. Flow past a rotating cylinder. Redrawn from photograph in Schlichting (1961) by permission of V. Streeter

by injecting dye (coloured, fluorescent or luminescent) or India ink into the moving fluid. Some of the first studies of the behaviour of boundary layers were made by sprinkling aluminium particles into the flow to create streaklines around objects. Seeds (e.g. mustard), egg-white and some eggs of invertebrates can also be used for visualization. For some studies the salinity of the water can be adjusted to make the particles neutrally buoyant. Tufts of yarn attached to boundary surfaces will also reveal flow patterns.

Additionally, flow visualization with small, neutrally buoyant particles or hydrogen or air bubbles yields an effective method of estimating flow velocities. By taking photographs with a stroboscope or video camera, point velocities can be determined by dividing the distance travelled by an individual particle by the time interval between the flashes or frames. For this method to be accurate it is important that the fluid motion is truly simulated by the particle motion. Too-large particles may be diverted away from an object by the very boundary layers they are supposed to reveal. For example, when Statzner and Holm (1982) compared their results from laser-Doppler anemometry to previous studies in which velocities were measured using relatively coarse acetyl cellulose particles they found boundary layer thicknesses to be much smaller for equivalent flow conditions.

Since fluid mechanics is such a highly visual subject, much can be gained from qualitative observations. Appendix 2 in Vogel's (1981) book gives several techniques for flow visualization, and films on flow phenomena are available from several sources, i.e.:

(1) Encyclopedia Brittanica Educational Corp; 425 No. Michigan Ave, Chicago, IL 60611, USA;
(2) St Anthony Falls Hydraulic Laboratory, Mississippi River at 3d Ave. SE, Minneapolis, MN 55414, USA; and
(3) University of Iowa, Media Library, Audiovisual Center, Iowa City, IA 52242; Engineering Societies Library, 345 East 47th Street, New York, NY 10017, USA.

Pitot tube

A pitot tube, named after the French engineer who designed it in 1732 (Wilson, 1969), works on the principles given by Bernoulli's equation (Equation 6.10). As shown in Figure 6.23(a), one aperture faces upstream to measure **stagnation pressure** and small holes on the sides of the tube are exposed only to the **static pressure** of the undisturbed fluid (Vogel, 1981). Since, at the stagnation point, all the velocity head has been changed to pressure head it is a simple matter to subtract the static pressure to obtain the velocity term. By connecting a manometer between the two apertures this difference in pressure can be measured directly. The relevant formula is (Vogel, 1981):

$$V = \sqrt{\frac{2gh(\rho_m - \rho_f)}{\rho_f}} \tag{6.45}$$

where g is the acceleration due to gravity, h the difference in manometer readings (m), ρ_m the density of the fluid in the manometer and ρ_f the density of the flowing fluid. If the fluid in the manometer is the same as that being measured (in which case, the tubes are left open to the air—see second diagram in Figure 6.23(a)), the formula becomes:

$$V = \sqrt{2gh} \tag{6.46}$$

Because these equations are based on Bernoulli's relation they perform best when the flow around the probe is considered frictionless, at $Re > 1000$ (the probe diameter, d, is taken as the characteristic length in Equation 6.15).

A disadvantage of Pitot tubes is that they are sensitive to how well they are aligned with the flow direction. The fluid in the manometer acts somewhat like a "stilling well" (Section 5.5.4), so manometer levels do not reflect small-scale turbulent fluctuations. Because the usual velocities measured produce only a small difference in head, pitot tubes tend to be relatively insensitive. They are also physically intrusive, although they can be made very small (even for measuring blood flow in arteries). Advantages of pitot tubes are that "point" measurements can be taken, they require no calibration and they do not need a power source (Vogel, 1981; White, 1986). Their greatest usefulness is for measuring high velocities in chutes or overfalls (Boyer, 1964).

Hot-wire anemometer

A hot-wire anemometer (Figure 6.23(b)) consists of a probe with a fine wire through which electric power is passed to heat it (Boyer, 1964; White, 1986). When the probe is submerged in a flowing fluid the wire is cooled, changing its resistance. This change in resistance is related to the flow velocity. Hot-wire anemometers are available commercially, and Vogel (1981) gives "do-it-yourself" instructions in his book. Although these anemometers work well for air flow measurements they are not well suited to measurement of water flow because of hysteresis effects and frailty of the thin wire. It can also be difficult to get a stable calibration. An improvement is the hot-film anemometer (Figure 6.23(c)), which follows the same principles as a hot-wire anemometer but is less fragile and more stable (White, 1986).

Laser-Doppler anemometer (LDA)

A superior device for measuring velocities in water is the laser-Doppler anemometer (LDA). This has the ability to measure point velocities quickly and continuously without disturbing the flow field. With an LDA (Figure 6.23(d)) velocity is measured by detecting the Doppler shift of light scattered by particles moving through the flow (the change in the sound of a train as it moves away is a Doppler effect). Commonly, two laser beams of different frequencies are used, which are focused on a test section. When particles move through this test section the Doppler shift is detected as the difference frequency between the two beams. The Doppler signal is directly proportional to the velocity of the particle. For an LDA to work there must be a sufficient concentration of particles in the fluid. Normal impurities in water will often suffice, but it is sometimes necessary to add "seed" particles which are usually of the order 0.0005–0.005 mm. The inertia of the scattering particles is considered negligible so that their movement reflects that of the surrounding fluid.

LDA systems are extremely expensive and not well suited to field work. The advantages are that (1) the flow is not disturbed during measurement, (2) detailed measurements can be taken very close to surfaces to investigate boundary-layer phenomena, (3) there is no need for calibration and (4) the LDA is capable of measuring turbulent fluctuations. LDA systems are available which can measure multiple components (vertical and horizontal) of velocity. Hino et al. (1986a,b) have also reported the development of a scan-type LDA which will measure near-instantaneous velocity profiles through a cross-section rather than at a point, although this type of instrument is still in the experimental phase.

6.6 OPEN-CHANNEL HYDRAULICS: THE MACRO-ENVIRONMENT

6.6.1 FIRST, A FEW DEFINITIONS

Depth (D): the vertical distance between the water surface and some point on the streambed.

Stage (y): the vertical distance from some fixed datum to the water surface. The datum might be the elevation of zero flow (e.g. the thalweg point) or mean sea level.

Discharge (Q): the volume of water passing through a stream cross-section per unit time.

Top width (W): the width of the stream at the water surface. Except in channels with vertical walls, it will vary with stream depth.

Cross-sectional area (A): the area of water across a given section of the stream. If a "slice" is made through the flow at right angles to the flow direction, this is the area "exposed".

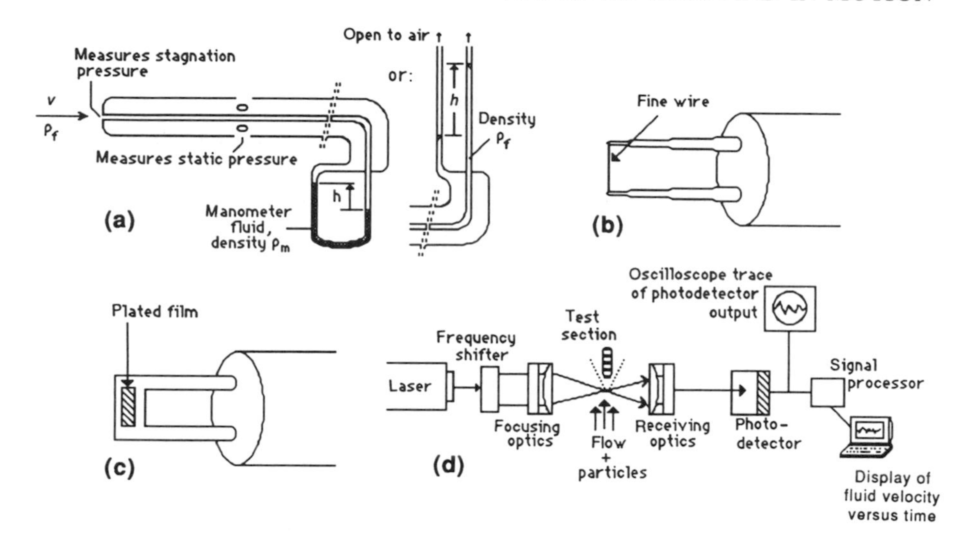

Figure 6.23. Four types of micro-velocity meters: (a) Pitot tube, (b) hot-wire anemometer, (c) hot-film anemometer, and (d) laser-Doppler anemometer. Adapted from White (1986), by permission of McGraw-Hill

Wetted perimeter (P): the distance along the stream bed and banks at a cross-section where they contact the water.

Hydraulic radius (R): the ratio of the cross-sectional area to the wetted perimeter:

$$R = \frac{A}{P} \tag{6.47}$$

Hydraulic depth (D): the ratio of the cross-sectional area to the top width:

$$D = \frac{A}{W} \tag{6.48}$$

In streams which are very wide in relation to their depth (a width-to-depth ratio of about 20:1 or more) the hydraulic radius and hydraulic depth are almost equal and approximate the average depth of the stream.

The symbol D will be used interchangeably for depth at a vertical, mean depth across a section, and hydraulic depth. Its interpretation will be explained in the equations where it appears. Some of the terms which describe cross-sectional dimensions are shown in Figure 6.24.

6.6.2 INTRODUCTION TO HYDRAULICS

The analysis of bulk flow patterns of water surface shape, velocity, shear stress and discharge through a stream reach comes under the heading of

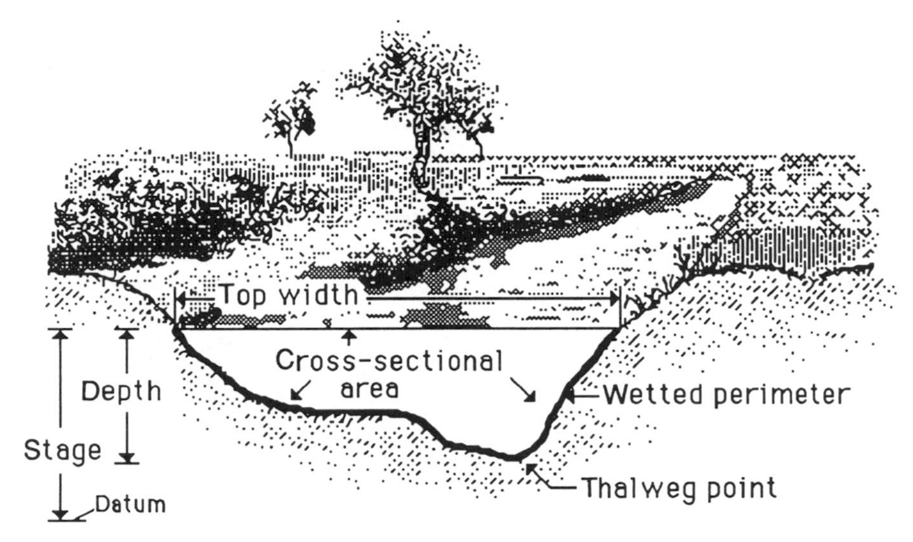

Figure 6.24. Terminology describing the geometry of open channels

open-channel hydraulics. Here, the effects of viscosity are not as important as the forces causing the water to move and stay in motion. The vertical distribution of velocity is dependent on the size of roughness elements in the flow path. Turbulence, as large-scale eddies and spiralling flow, occurs over the entire depth of flow.

In comparison to flow through pipes, in open-channel flow the water surface is exposed to atmospheric pressure. Because the water depth and channel geometry can change in natural stream channels, problems are much more complex. However, it is the mathematical description of pipe flow from which many of the formulae for describing open-channel flow have been developed. The description of pipe flow will not be covered here, although it can occur in natural pipes in the soil and where streams flow through karst areas. For information on pipe flow and further information on open-channel flow, readers can refer to fluid mechanics texts and texts on open-channel flow such as Chow (1959) and Henderson (1966).

Flow in open channels can be classified as:

- Steady/unsteady;
- Uniform/varied;
- Laminar/turbulent; or
- Supercritical/critical/subcritical

The classification of steady/unsteady depends on whether the flow depth and velocity change with time at a point, and is discussed in Section 6.3.1. In contrast, the classification uniform/varied is based on whether depth and velocity vary with respect to distance (Figure 6.25). If depth and velocity remain constant over some length of channel of constant cross-section and slope (a "prismatic" channel), then the water surface is parallel to the streambed and the flow is said to be **uniform** in that reach. If the flow in a stream is uniform it is said to be moving at its **normal depth**, an important parameter in engineering design. The assumption of uniform flow conditions considerably simplifies the analysis of water movement in streams.

In **non-uniform** or **varied** flow the water depth and/or velocity change over distance. Examples are where the flow moves through a bedrock constriction or passes from a pool to a riffle. Varied flow can be subdivided further into the categories **rapidly varied** and **gradually varied.** If the depth changes abruptly over a relatively short distance, as at a waterfall or wave, the flow is rapidly varied; when changes are more widely spread the flow is gradually varied. Gradually varied flow can be analysed relatively easily (Section 6.6.6), whereas the description of rapidly varying flow (Section 6.6.7) requires experimentation.

Conditions favouring uniform flow in natural streams are rare compared with the well-controlled flow conditions in concrete canals or irrigation ditches. However, uniform flow can be approximated by long, straight runs of constant slope and cross-section, and it is under these conditions that the uniform flow equations of Section 6.6.5 provide reasonable estimates.

The third classification, laminar/turbulent (see Section 6.4.1), is somewhat irrelevant at the "macroenvironment" level. Although regions of laminar flow can exist near the surfaces of rocks or organisms within the stream, the bulk flow is nearly always turbulent except perhaps in the rare case where it flows slowly as a thin film of a few millimetres. The last classification, supercritical/critical/subcritical, is related to the combined patterns of velocity and depth, and will be discussed in Section 6.6.4.

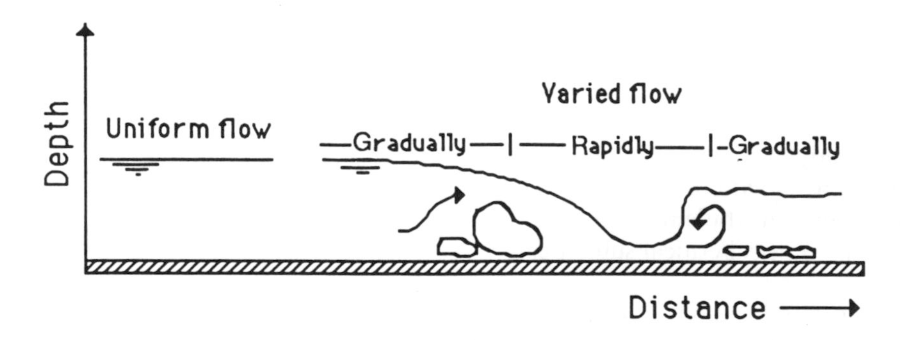

Figure 6.25. Classification of open-channel flow as uniform or varied

6.6.3 THE VARIATIONS OF VELOCITY IN NATURAL CHANNELS

Velocity, as the rate of movement of a fluid particle from one place to another, varies with both space and time in a stream. The effects of velocity on stream organisms are discussed in Section 2.1. From continuity (Equation 6.6), mean velocity across any section is simply:

$$V = Q/A \tag{6.49}$$

the discharge divided by the cross-sectional area. The flow velocity tends to increase as slope increases and as bed roughness decreases. In the extreme case of a waterfall, where slope is maximal and "bed roughness" is non-existent, velocities of up to 8.1 m/s have been reported (Hynes, 1970). Water in natural channels tends to remain below about 3.0 m/s. To put speeds into context, a flow velocity of 0.1 m/s is barely perceptible, whereas at 3.0 m/s it is difficult to stand up in knee-deep water.

Because of frictional resistance (see Section 6.4), the flow is retarded near the streambed, streambanks and even slightly at the water surface. Along with flow turbulence, this resistance causes variations in the distribution of velocity (1) with time; (2) with depth; (3) across a section; and (4) longitudinally. Additionally, it leads to spiralling cross-currents which impose another degree of complexity on the main downstream movement of water.

Variation with time

At any given point in a stream the flow velocity fluctuates rapidly because of surges and turbulent eddies. As mentioned earlier, measurements of point velocities taken with current meters or probes are actually time-averaged values of velocity. Instantaneous velocities can be much larger or smaller than the average because of the variability in turbulent flow. These fluctuations often appear to have a cyclical or "pulsing" behaviour, rather than a random trend (Morisawa, 1985). Although merely a statistical nuisance to hydrologists, the nature of turbulence may have profound implications for the organisms which are exposed to it.

Velocities also change in response to changes in discharge. In general, velocities will increase most where resistance is least, at the centre of the stream. Even during floods, low-velocity zones will still remain near the stream edges. The change in velocity with discharge falls under the topic of hydraulic geometry, which is covered in Section 7.2.6.

Variation with depth

If current meter readings are taken over a vertical (Section 5.7.3) a graph of the measured velocities plotted against depth illustrates the vertical velocity profile. Figure 6.26 shows some of the forms which a measured velocity profile could assume. From zero velocity at the bed the velocity increases with vertical distance, at first rapidly but levelling off as it reaches its maximum value. Channel shape, bed roughness and the intensity of turbulence all influence velocity profiles.

In a "typical" velocity profile (Figure 6.26(a)) the maximum velocity tends to occur just beneath the water surface because of resistance with the air. The closer the measurements are taken to the streambanks, the deeper is this maximum velocity (Chow, 1959). Slower surface velocities will occur where the flow is retarded by the floating leaves of aquatic plants. In the centre of broad, rapid streams the profile will look more like Figure 6.26(b), with the maximum velocity found right at the free surface. When the depth of rocks, boulders, plants, dead logs, sand dunes and other "roughness elements" is high in relation to the depth of the water, water velocities within the roughness elements may differ substantially from velocities over the top of the protrusions. For example, Jarrett (1984) mentions that shallow, steep, cobble and boulder-bed streams in mountainous areas can have S-shaped profiles such as in Figure 6.26(c). Thus, the velocity profile may alter in shape as the water level changes.

A logarithmic function is conventionally used for describing vertical velocity profiles, and in most instances it provides a good approximation. If the velocity varies logarithmically with distance from the streambed it can be demonstrated mathematically that the mean value of velocity, $\bar{v}$, occurs at about 0.4 of the water depth (measured upwards from the bed, as shown in Figure 6.26(a)). This is the point at which velocities are measured if only one reading is taken (see Section 5.7.3). For the average velocity at a vertical, Equation 6.25 for hydraulically rough conditions becomes:

$$\bar{v} = 5.75\, V_* \log\left(\frac{12.3R}{k}\right) \tag{6.50}$$

from which V_* can be calculated for a vertical if the mean velocity and relative roughness (see Equation 6.18) are known. Smith (1975) indicates that the value of relative roughness, k/R, varies from more than 0.2 for a

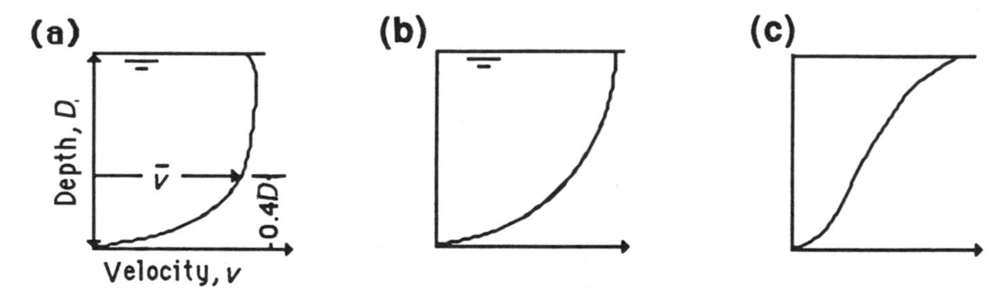

Figure 6.26. Three variations on the vertical velocity profile

shallow stream flowing over a shingle bed to less than 0.0002 for a deep flow over fine clay sediments. Thus, in rocky streams, the shear velocity is approximately 1/10 of the mean velocity but only about 1/30 of the mean velocity when the streambed is of fine sediments.

Using the information in Figure 6.14 we can back-calculate a value of k from Equation 6.50 using the water depth as an approximation for R. This gives $k \approx 0.09$ m and $k/R \approx 0.17$. This profile was taken near the thalweg of the stream, where the median bed material size (see Figure 5.33) was on the order of 100 mm (0.1 m)—quite similar to the computed value of k.

Other velocity profile equations have also been developed for describing velocity distributions over irregular streambeds, based on particle sizes. Recent developments are discussed by Coleman and Alonso (1983) and Nezu and Rodi (1986).

Variation across a section

Velocities tend to increase towards the centre of a stream and decrease towards the perimeter because of frictional resistance at the bed and banks. **Isovels**, lines joining points of equal velocity, can be plotted as a "map" of a stream cross-section, like elevation contours plotted on a topographic map. The distribution of velocity is easily visualized in the plots shown in Figure 6.27. Where isovels are crowded, velocity gradients, and thus shear stresses, are higher. At a bend in the river (Figure 6.27(b)) isovels tend to be closer together on the outer bank where the main current is forced against it by centrifugal force.

Longitudinal variation

Patterns of velocity variation within a channel reach can also be shown as a "bird's eye view", as in Figure 6.28. If velocities (surface velocities or mean vertical velocities) are plotted from a number of sections this can give an indication of how close the reach approximates uniform flow conditions, as well as identifying, for example, areas of high bank erosion potential or

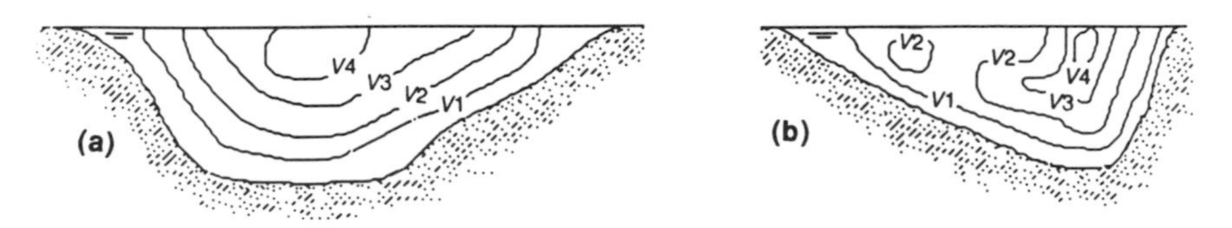

Figure 6.27. Velocity contours or isovels at stream cross-sections: (a) in a relatively straight section and (b) at a bend. In both diagrams, V4 > V3 > V2 > V1. Adapted from Morisawa (1985), by permission of Longman Group, UK

good fish habitat. Areal "maps" showing zones of different velocities can be created from a large number of measurements, much like soils or vegetation maps (Section 3.3; see also Figure 9.4).

Spiral flow

For a complete description of velocity distributions in streams, the concept of **spiral**, **helical** or **secondary flow** should not go without mention. Spiral flow (Figure 6.29) is a consequence of frictional resistance and centrifugal force. When a cup of billy tea is stirred it is the reason the leaves congregate in the middle rather than at the edges. In a stream, water is hurled against the outside banks at bends, causing the water surface to be "superelevated". This increase in elevation creates a gradient, causing flow movement from the outer to the inner bank. A spiralling motion is generated along the general direction of flow (Petts and Foster, 1985). The first observation of spiral flow is credited to Thomson in 1876 (Chow, 1959). Compared to the forward, downstream currents, secondary lateral and vertical currents are relatively small, yet they cause the mainstream current to vary from a predictable course and contribute to energy losses and bank erosion at bends.

On the outside of a bend the rotary motion is from top to bottom, which tends to scour the bank. On the inside the flow is upward and decelerating, and any material carried tends to deposit out, creating point bars (Vennard and Street, 1982). Spiral flow also occurs in straight sections, where the upwelling of waters near the stream's centre can increase the local surface elevation, resulting in a curved water surface (Leopold et al., 1964). Spiral flow is particularly pronounced where channel boundaries are irregular.

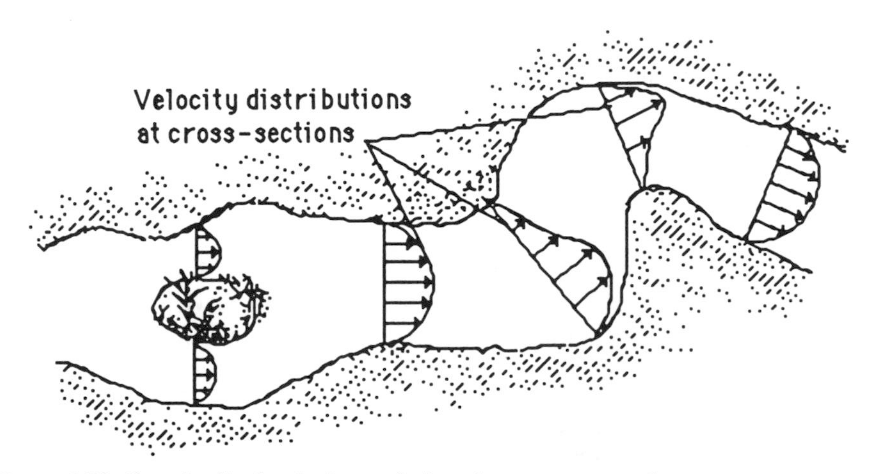

Figure 6.28. Longitudinal velocity variations in a stream reach

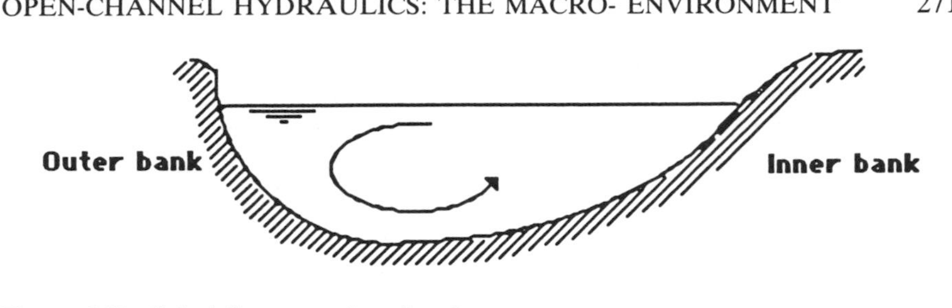

Figure 6.29. Spiral flow at a river bend

Chow (1959) gives a formula for the strength of a spiral flow as a percentage ratio of the mean kinetic energy of the lateral flow to the total kinetic energy of flow at a cross-section (see Section 6.3.4 for definition of kinetic energy). This ratio is relatively high if the Reynolds number of the approach flow is low. It decreases as the channel becomes deeper in relation to its width and as bend angles become less severe.

Circulation patterns in streams are very complex and unpredictable, yet an awareness of their existence can provide a starting point from which to study their effects as well as material for contemplation while watching a fishing line snaking from side to side in a stream. For ecological purposes, velocities near the bottom of a channel or at the water surface will often be of greater significance than averages across a cross-section or a reach. Variations of velocity with time, neglected in gross measures of mean velocities, may be critical to the aquatic insects or sand grains buffeted by pockets of turbulence.

6.6.4 ENERGY RELATIONSHIPS IN STREAMS

The one-dimensional energy equation

The Bernoulli equation (Equation 6.10) describes the conservation of energy along streamlines. It can be assumed to apply to objects within a mass of flowing water when viscosity can effectively be ignored. However, in developing a general equation for describing energy relationships in streams the frictional resistance along the channel bed and banks due to viscosity must be taken into account.

For this purpose, we use cross-sectional values of elevation, pressure (depth) and velocity, and the **one-dimensional energy equation**:

$$z_1 + D_1 + \frac{V_1^2}{2g} = z_2 + D_2 + \frac{V_2^2}{2g} + h_1 \tag{6.51}$$

where z = elevation (m), D = mean water depth (m), V = mean velocity (m/s), g = acceleration due to gravity (m/s^2) and the subscripts 1 and 2 refer to upstream and downstream sections, respectively. A derivation of this equation can be found in fluid mechanics texts. As in Equation 6.10, all terms have dimensions of length and are called **head** terms (e.g. elevation head, pressure head and velocity head, respectively). The **head loss** term, h_l, accounts for energy loss in the form of flow separation, turbulence, heat generated as a result of frictional resistance and other forms of internal energy. Empirical formulae have been developed for calculating head loss which take into account the fluid viscosity and boundary roughness effects. Head losses can occur due to local effects such as hydraulic jumps (Section 6.6.7), sudden expansions or contractions, or rapid changes in flow direction at bends.

Water depth, D, is the average water depth in rectangular channels, but is measured at the thalweg point (Figure 6.24) in irregular channels. The use of water depth as the pressure head term is valid for slopes less than about 10°. Chow (1959, p. 39) should be consulted for a more accurate equation if slopes are greater.

A diagram showing the terms in Equation 6.51 for non-uniform flow conditions is given in Figure 6.30. The **total energy line** represents the total energy head of the flow, and its slope is called the **energy slope** or **gradient**. In uniform flow the slope of the streambed, water surface and energy line are equal, and thus the head loss is simply equal to the difference in elevations of the streambed or water surface from the top of a reach to the lower end.

Specific energy

Specific energy is the energy of a section relative to the streambed; i.e. the distance between the streambed and the total energy line in Figure 6.30. For a channel of small slope the specific energy at a section, E_s, is:

$$E_s = D + \frac{V^2}{2g} \tag{6.52}$$

where the terms are the same as in Equation 6.51. Again, all terms, including E_s, have dimensions of length. For a given channel section and discharge, specific energy is a function of water depth, and the relationship is described by a specific energy curve. Specific energy curves for three different discharges are shown in Figure 6.31(a).

For each value of discharge, there is a minimum depth—the **critical depth**, D_c. If the water depth, D, is higher than D_c, the flow is considered subcritical, and if $D < D_c$, it is considered supercritical, as will be discussed later in this section. Theoretically, the critical depth is the minimum depth which can occur as water flows over the top of a boulder or log (Figure 6.32(a)) or the crest of a waterfall or spillway (Figure 6.32(b)). This minimum depth occurs slightly upstream from the brink because of curvature of the water surface.

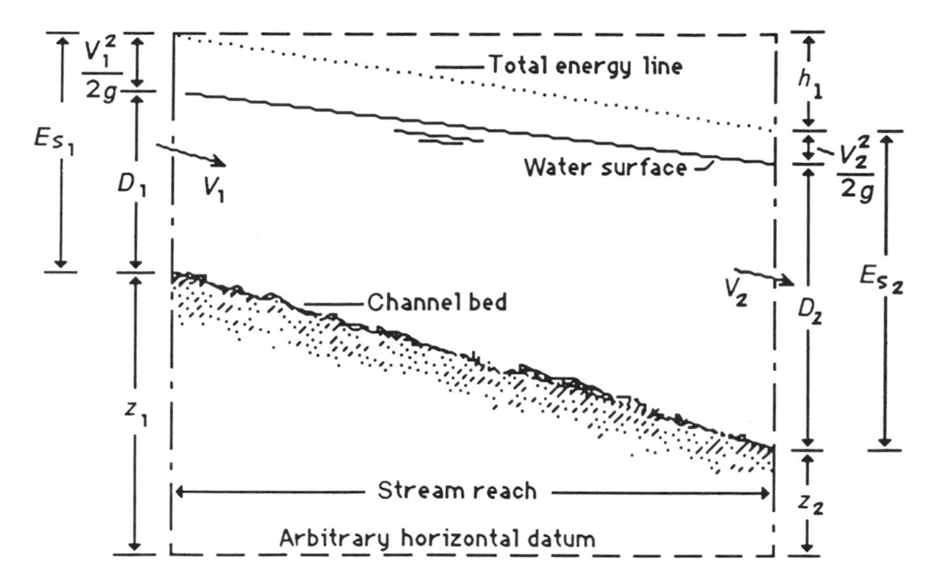

Figure 6.30. Schematic showing the terms in the energy equation. z = the elevation above some datum, V = mean velocity, D = water depth, h_l = head loss over a reach and E_s = specific energy, the sum of velocity and pressure head terms. Subscripts 1 and 2 refer to upstream and downstream cross-sections at a stream reach

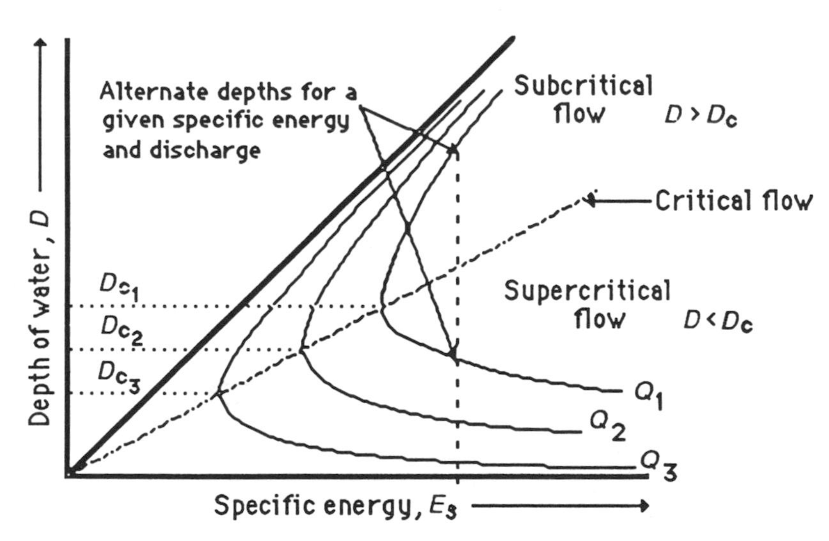

Figure 6.31. Generalized graph of specific energy for different discharges, where $Q_1 > Q_2 > Q_3$. Critical depths D_{c1}, D_{c2} and D_{c3} correspond to discharges Q_1, Q_2 and Q_3, respectively

If critical flow can be located in a stream (or created with a weir) then the flow rate can be determined from the critical depth. For this application the weir must extend across the section (with no flow underneath or around the edges). At critical flow,

$$E_c = \frac{3}{2} D_c \tag{6.53}$$

and thus from Equation 6.52:

$$\frac{V_c^2}{2g} \text{ (the velocity head)} = \frac{D_c}{2}$$

yielding the equations:

$$V_c = \sqrt{gD_c} \tag{6.54}$$

and

$$Q = A\sqrt{gD_c} \tag{6.55}$$

where D_c is the hydraulic depth at critical flow. In wide or rectangular channels, D_c can be approximated by the average water depth, but in irregular channels, the definition of hydraulic depth, A/W, should be used. Texts on hydraulics should be consulted for a more complete derivation of these equations.

Example 6.6

Calculate the discharge flowing over a broad, level concrete sill if the critical depth is 0.31 m and the width of the water surface is 8.2 m. Assume the flow area is rectangular, so $A = (8.2)(0.31) = 2.54\ \text{m}^2$.

Answer: $Q = 2.54\sqrt{[9.807\,(0.31)]} = 4.43\ \text{m}^3/\text{s}$

If the overfalls in Figure 6.32 extend across the whole stream, they function somewhat like **broad-crested weirs.** Because flows near critical depth tend to be unstable (from Figure 6.31, it can be seen that a slight

Figure 6.32. Critical depth at overfalls: (a) at a log and (b) at a spillway crest. D_c is the critical depth, V is the mean velocity, and H is the water depth above the crest measured some distance upstream (see also illustration of weir, Figure 5.20)

change in energy will cause the depth to change markedly), measurements at weirs are usually taken some distance upstream of the crest (see Figure 5.20). If it is assumed that the velocity head term is negligible at this point in relation to depth, and that the energy is the same as at the crest, then this depth above the crest (H) is equal to E_c. From Equations 6.53 and 6.55, and assuming $A = WD_c$, this equation for discharge can be derived:

$$Q = 1.70WH^{3/2} \tag{6.56}$$

For broad-crested weirs used in stream gauging the coefficient 1.70 is usually somewhat less due to energy losses. Equation 6.56 is often verified in flumes as a laboratory exercise in engineering courses.

Flow velocity and depth "balance" each other for a given specific energy and discharge. Figure 6.33 illustrates this principle for an idealized "longitudinal slice" along a stream. Upstream of the boulder at section 1, a **stagnation point** is created. The velocity increases where the water flows around the boulder, with a corresponding drop in depth. Velocity also increases if the channel contracts or the bed is elevated by cobbles, as at section 2. As shown, the depth drops rather than increasing. Sections 3 and 4 show a region of rapidly varied flow, accompanied by a large drop in the energy line.

Equations 6.52–6.55 are based on the assumption of uniform, steady flow. They are approximately correct for the gradually varied flow shown in Figure 6.33 upstream of section 3. In the rapidly varied flow of sections 3 and 4 they are not applicable. It should also be realized that in a stream, the energy line would be determined by cross-sectional values of velocity and depth rather than those at individual verticals.

If a stagnation point occurs in a reach, such as at a boulder, a bridge pier or a tree trunk, it can be used to give a rough indication of the flow velocity. Since the velocity there is zero, the depth of flow is equal to the specific energy. The depth of the "piled up" water, D_s, is thus approximately equal to the velocity head of the approaching flow (see insert in Figure 6.33):

$$D_s = \frac{V^2}{2g} \quad \text{or} \quad V = \sqrt{2gD_s} \tag{6.57}$$

The water depth at the stagnation point fluctuates greatly, and a visual average should be made if this depth is measured with a rule or staff gauge.

Froude numbers and flow classification

The line connecting the critical depths in Figure 6.31 separates the portions of the specific energy curves representing **supercritical** and **subcritical flow.** Thus, except at critical flow, flow of a given specific energy can have one of two depths, called **alternate depths**. As mentioned previously, this provides a means of classifying the flow. The classes are separated on the basis of a

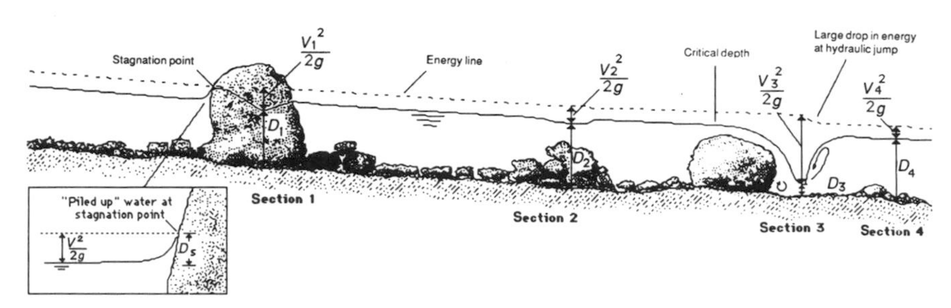

Figure 6.33. Schematic showing the balance of energy in a longitudinal "slice" through a stream reach. Scale is exaggerated for the velocity head term. The flow is subcritical in sections 1, 2 and 4, and supercritical in section 3. Rapidly varied flow is discussed further in Section 6.6.7 and by Chow (1959). Drawn with the assistance of R. Newbury

dimensionless quantity, the **Froude number.** This number represents the ratio of inertial to gravitational forces, where the gravitational forces encourage water to move downhill and inertial forces reflect the water's compulsion to go along or not. The Reynolds number (Equation 6.15) is a better measure of "internal" conditions whereas the Froude number is a better descriptor of bulk flow characteristics such as surface waves, sand bedforms, and the interaction between flow depth and velocity at a given cross-section or between boulders.

Froude number, *Fr*, is defined as:

$$Fr = \frac{V}{\sqrt{gD}} \tag{6.58}$$

where V is the mean velocity (m/s), g the acceleration due to gravity (m/s^2) and D the hydraulic depth (m). Again, for rectangular or very wide channels, the hydraulic depth can be replaced with the average water depth. "Local" Froude numbers can also be calculated, for example, where water flows over or between boulders, where the local depth and velocity at a vertical are used.

From Equation 6.58 it can be seen that for critical flow, $V_c/\sqrt{gD_c} = 1$. Three flow classes can now be designated:

- $Fr < 1$ subcritical (or slow or tranquil) flow
- $Fr = 1$ critical flow
- $Fr > 1$ supercritical (or fast or rapid) flow

The analogy can be drawn between Froude number and Mach number, where a value of 1 separates supersonic from subsonic, based on whether the speed of sound is exceeded or not. As with Mach number, the Froude number also deals with speed. The denominator, $\sqrt{gD}$, represents the speed of a small wave on the water surface relative to the speed of the water, called **wave celerity**. At critical flow the wave celerity is equal to the flow velocity. Any disturbance to the surface will remain stationary. In subcritical flow the flow is controlled from a downstream point and any disturbances are transmitted upstream. By comparison, supercritical flow is controlled from an upstream point and any disturbances are transmitted downstream (see Section 5.6).

The direction of wave propagation can be used to locate regions of subcritical, critical and supercritical flow in a stream. A pencil, stick or finger (we will call it a "Froude indicator") contacting the water surface will generate a "V" pattern of waves downstream. If the flow is subcritical, waves will appear upstream of this "indicator" as in Figure 6.34(a), whereas they do not appear when the flow is supercritical (Figure 6.34(b)). By passing the "indicator" through a region of transition from subcritical to supercritical the location of critical flow can be identified as the point where all upstream waves disappear or the downstream angle of the "V" is 45°.

In streams most of the flow will be subcritical. Supercritical flow can be found where water passes over and around boulders, and in the spillway chutes of hydraulic structures. Usually, it is accompanied by a quick transition back to subcritical flow (a hydraulic jump—see Section 6.6.7), which appears as a wave on the water surface.

The Froude number is gaining acceptance as an index for the characterization of local-scale habitats. In a study of the distribution of the filter-feeding caddisfly, *Brachycentrus occidentalis* Banks, for example, Wetmore et al. (1990) found that the Froude number was a better predictor of larval habitat than depth or velocity taken individually. The larvae preferred regions where streamlines converged over and around boulders and cobbles, exposing them to higher concentrations of food materials. The authors describe a laboratory profiling device modified for the field measurement

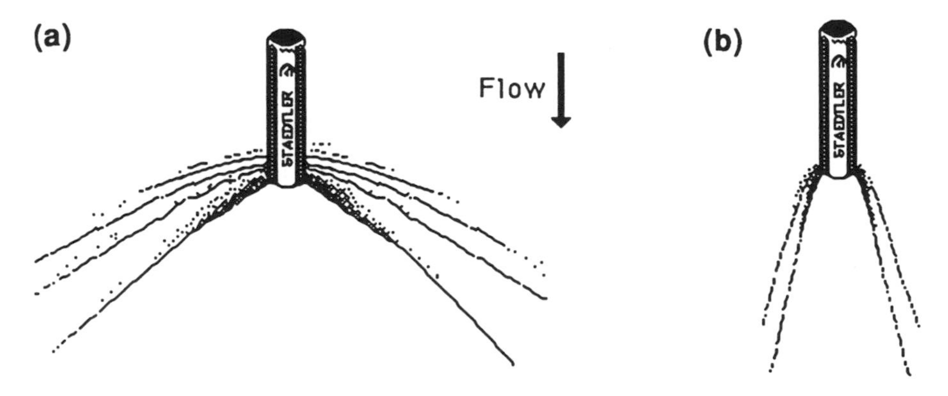

Figure 6.34. Detection of (a) subcritical and (b) supercritical flow at the water surface

of local water depths and velocities in shallow riffles. A small current meter with a 10 mm diameter propellor was used for measuring velocities at 0.4 depth (subcritical regions) and 0.5 depth (supercritical regions), as measured upwards from the streambed. In this application in which point measurements rather than cross-sectional averages are used, critical flow is not necessarily defined by $Fr = 1$.

6.6.5 SHEAR STRESS AND THE UNIFORM FLOW EQUATIONS OF CHEZY AND MANNING

Much theoretical and empirical research has been done in an attempt to develop a relationship between velocity and stream channel characteristics. If the flow is uniform, then depth and velocity are constant along the channel. Under these conditions, gravity (as a force causing motion) and friction (as a force opposing it) are in balance.

The frictional force causing flow resistance along the channel boundary can be expressed per unit area as a **shear stress** (see definition, Section 1.2). For channels, shear stress, τ, is given as:

$$\tau = \rho g R S \tag{6.59}$$

where τ is in N/m^2, R is hydraulic radius (m) and S is the slope of the energy line, which in the special case of uniform flow is equal to the bed slope and the water surface slope. If the channel is wide or rectangular, R can be replaced by the average flow depth (see Section 6.6.1). Shear stress generally declines in the downstream direction since bed slope decreases, and thus organisms are more likely to have adaptations against dislodgement in headwater streams (Townsend, 1980).

By setting the gravitational force in the direction of flow equal to the frictional resistance (assumed approximately proportional to the square of velocity), the following equation can be derived:

$$V = \sqrt{\left(\frac{\rho g}{\alpha}\right) R S} \tag{6.60}$$

where V = mean velocity (m/s), R = hydraulic radius (m), S = energy slope, α = a coefficient, mainly dependent on boundary roughness, and ρ and g are water density and gravitational acceleration, as previously defined. A French engineer, Antoine Chezy, introduced this equation in 1768 when given the task of designing a canal for water supply in Paris (Henderson, 1966). The term $\sqrt{\rho g/\alpha}$ is normally made into a single coefficient, Chezy's "C", and the equation written in the form:

$$V = C\sqrt{RS} \tag{6.61}$$

Chezy's C has units of m$^{1/2}$/s, and varies from about 30 in small, rough channels to 90 in large, smooth ones (White, 1986). It is considered to be a function of relative roughness and Reynolds number (Section 6.4). A variety of methods for obtaining C have been developed, and details are provided by Chow (1959). The Chezy equation is still popular in many European countries.

Robert Manning, an Irish engineer, first introduced an alternative to Chezy's equation in 1889. It was later simplified to its commonly used form, given in SI units as:

$$V = \frac{1}{n} R^{2/3} S^{1/2} \tag{6.62}$$

or

$$Q = \frac{1}{n} A R^{2/3} S^{1/2} \tag{6.63}$$

where V is the mean channel velocity (m/s), Q is the discharge (m^3/s), S is the slope of the energy line (in uniform reaches, equal to the bed and water surface slopes), R is the hydraulic radius (m) and n is a coefficient referred to as "Manning's n". Manning's equation was derived from data collected on artificial and natural channels with a range of shapes and boundary roughness. The exponent on R was actually found to vary from 0.6499 to 0.8395, and Manning adopted the value 2/3 as an approximation. Because of its practicality, Manning's equation is probably the most widely used uniform flow equation for open-channel flow calculations in English-speaking countries. An example calculation using Manning's equation is given in Section 5.8.2. Program MANNING in AQUAPAK is provided for applying the equation.

To balance the dimensions of the equation, the $1/n$ term must have units of $m^{1/3}$/s. The normal preference is to leave n dimensionless and attach all the remaining units to a coefficient. This has a value of 1 in SI units, as in Equations 6.62 and 6.63, but becomes 1.486 in the Imperial system. Thus, a book of n values is applicable in America or Australia or Antarctica. In general, n increases as turbulence and flow retardance effects increase. Sediment load can have a variable effect on flow resistance, as it dampens turbulence which reduces resistance in some circumstances, but the consumption of energy in transporting the load acts like an increase in resistance (Trieste and Jarrett, 1987). Manning's n can thus be thought of as a "tuning" or "calibration" factor which integrates the effects of flow resistance caused by bed roughness, the presence (and flexibility) of vegetation, the amount of sediment or debris carried by the flow and other factors (Chow, 1959; Trieste and Jarrett, 1987).

Information on the field estimation of Manning's n is given in Section

5.8.2. Attempts have also been made to give n more of a physical basis. For example, Strickler (1923) developed the following equation for gravel-bed streams with median grain size d:

$$n = sd^{1/6} \tag{6.64}$$

where n = Manning's n and s is a coefficient. When d is the median diameter (d_{50}, in mm), $s \approx 0.013$ (Henderson, 1966).

Manning's n has also been found to vary with flow depth. Equation 6.64 may be applicable when the roughness elements are very small in comparison to the water depth, but n typically rises as the relative roughness (Equation 6.18) increases. According to Chow (1959), Manning's equation is not considered to be applicable when the roughness height exceeds 1/3 of the water depth ($R_{rel} > 1/3$). Relative roughness is included in an equation for estimating Manning's n developed by Limerinos (1970) from data on lower gradient streams with bed material of small gravel to medium-size boulders. The equation is given in SI units as:

$$n = \frac{0.1129R^{1/6}}{1.16 + 2.0\log\left(\dfrac{R}{d_{84}}\right)} \tag{6.65}$$

where R is hydraulic radius (m) and d_{84} is the diameter (m) for which 84% of the streambed particles are smaller (see Section 5.10.5). For higher-gradient streams (slope 0.002–0.04), Jarrett (1985) gives

$$n = 0.39S^{0.38}R^{-0.16} \tag{6.66}$$

where the terms are as defined in Equation 6.63. This equation is applicable to stable channels with hydraulic radii between 0.15 m and 2.1 m and similar in character to the Colorado streams for which the equation was derived. This approach is similar to that of Riggs (1976) (Equation 5.28), requiring no estimate of roughness. For the data used in deriving the equation, the standard error of estimate was 28%.

Equations 6.64 and 6.65 are provided as options in program MANNING for the estimation of Manning's n. These n values can be checked with estimates made from other methods (e.g. Cowan's, Section 5.8.2) or used for guidance when adjusting n for different flow levels.

When the stream overtops its banks the water velocity in the main channel is typically much higher than that on the floodplains. For these compound sections Manning's equation is applied to subsections (e.g. the main channel and one or more side channels) to obtain subsection mean velocities and discharges. These discharges are then summed to get the total discharge for the section. The mean velocity for the entire section is equal to the total discharge divided by the total cross-section area (Chow, 1959), although this may have little relevance in compound channels. It is assumed that the water surface is horizontal across the whole section, and that the energy slope is the same in all subsections. Example 6.7 below illustrates the procedure. It should also be noted that wetted perimeter is calculated only over the wetted distance; it does not include the dividing lines between the subsections.

In natural streams which are irregular in cross-section, and which wind and bend, drop over falls and into pools, bed roughness is not the only source of energy loss. In fact, these uniform flow equations may be inapplicable in very steep streams with pool-step structures (Section 7.3.2), as the flow often passes through critical depth (Section 6.6.4). Jarrett (1985) suggests 0.04 as an upper limit for slope. An understanding of the assumptions behind the equation and the factors which affect Manning's n is helpful in narrowing the error margins around the estimates of velocity or discharge. Although the equations are based on an assumption of uniform flow, it is common practice to use them when describing gradually varied flow. For conditions which deviate substantially from uniform flow, adjustments must be made to account for head losses (see Equation 6.51). These procedures are discussed by Chow (1959), Henderson (1966), Jarrett (1985) and Jarrett and Petsch (1985).

As semi-empirical equations, the equations of Chezy and Manning are safely used only within the range over which they were developed—trapezoidal channels with uniform flow and clear water. For practical purposes, flow through a straight stream section of fairly constant cross-section, mean velocity and roughness characteristics can be considered approximately uniform, and the equations will perform best if these conditions are met.

Example 6.7

Calculate the discharge through a section where the stream has overflowed onto the floodplain and the dimensions of the water area are as shown. For both subsections, $S = 0.005$. In subsection 1, $n = 0.060$, and in subsection 2, $n = 0.035$.

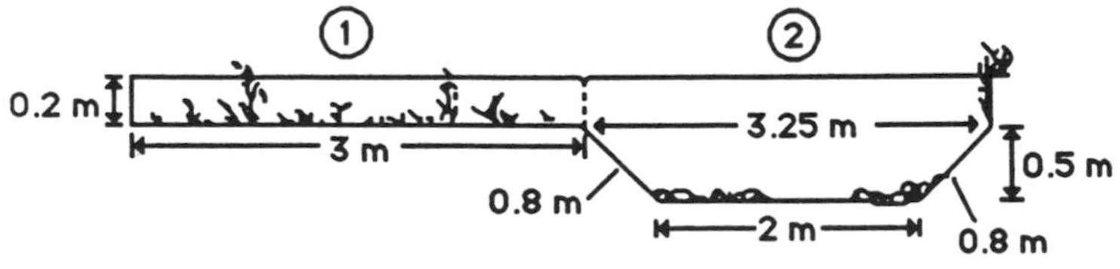

①
②
0.2 m
3 m
3.25 m
0.5 m
0.8 m
2 m
0.8 m

Subsection 1: $A = 3(0.2) = 0.6\ \text{m}^2$

$P = 0.2 + 3 = 3.2\ \text{m}$

$$R = 0.6/3.2 = 0.188\ \text{m} \Rightarrow V_1 = \frac{1}{0.060}(0.188)^{2/3}(0.005)^{1/2} = 0.39\ \text{m/s}$$

Subsection 2:

$$A = \left(\frac{2 + 3.25}{2}\right)(0.5) + 0.2(3.25) = 1.96 \text{ m}^2$$

$$P = 0.2 + 2(0.8) + 2 = 3.8 \text{ m}$$

$$R = 1.96/3.8 = 0.516 \text{ m} \Rightarrow V_2 = \frac{1}{0.035}(0.515)^{2/3}(0.005)^{1/2}$$

$$= 1.30 \text{ m/s}$$

Therefore total discharge for section:

$$Q = A_1V_1 + A_2V_2 = (0.6)(0.39) + (1.96)(1.30) = 2.78 \text{ m}^3\text{/s}.$$

6.6.6 WATER-SURFACE PROFILES IN GRADUALLY VARIED FLOW

Since true uniform flow seldom occurs in streams most problems involve the treatment of gradually or rapidly varied flow. The former is usually considered to occur over a relatively long length of channel. The longitudinal shape of the water surface within a section of gradually varied flow can be predicted from the "energy budget" of the flowing water. These profiles are also known as **backwater curves** when they describe subcritical flow conditions.

Classifying flow profiles

The type of water surface profile exhibited (i.e. how the flow depth changes longitudinally) depends on the relationship between the actual water depth (D), the normal depth (D_n) and the critical depth (D_c). The normal depth is that which would occur if the flow were uniform and steady, and can be predicted from the equations of Chezy or Manning. Critical depth is, again, the depth at which specific energy is minimum for a particular discharge.

A **critical slope** is one which sustains uniform critical flow ($D_n = D_c$). Because a small difference in energy will cause the flow to fluctuate to either supercritical or subcritical (see Figure 6.31), the flow across a critical slope tends to have an unstable, undulating surface. A **mild slope** is one which is less than the critical slope, the normal depth is subcritical ($D_n > D_c$) and the flow has a downstream control. A **steep slope** is steeper than the critical slope, the normal depth is supercritical ($D_n < D_c$) and the control is upstream. Long steep slopes are uncommon in natural channels; instead, they are usually broken up by pools and drops (Section 7.3.2). It is possible for a given channel slope to be classified as mild, steep or critical at different streamflows since both D_n and D_c vary with discharge.

Based on the relationship between D, D_n and D_c, the resulting profiles can be categorized. These have conventionally been used for describing the entire flow within a reach, but may have potential use for describing local water surface patterns around and over boulders and branches. Labels for each category consist of a letter designating the type of slope and a number which indicates the relative position of the flow depth:

(1) Type 1 curve: the actual depth is greater than both D_c and D_n. Flow is subcritical.
(2) Type 2 curve: the actual depth is between D_c and D_n. Flow can be either subcritical or supercritical.
(3) Type 3 curve: the actual depth is smaller than both D_c and D_n. Flow is supercritical.

Only mild and steep profiles will be described here. There are other categories for critical, horizontal and adverse (uphill) slopes, which can be found in Chow (1959) or in chapters on gradually varied flow in fluid mechanics texts. The diagrams shown below are drawn with exaggerated slopes since most streambed slopes are very low.

M profiles: The M1 profile is one of the more well-known curves, and can be seen where a mild-sloped stream enters a pond. The M2 profile may occur upstream of a sudden enlargement in a channel reach or where the slope becomes steeper. M3 profiles are found where supercritical flow enters a mild-sloped channel such as under a sluice gate (e.g. at a dam) or under a log. They usually end in a hydraulic jump (Section 6.6.7).

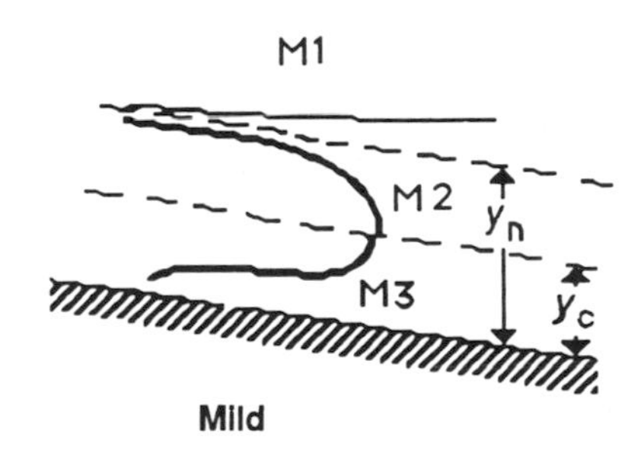

S profiles: The S1 profile begins with a rise at the upstream end, then becomes horizontal. It can occur where a steep channel empties into a pool. S2 is called a drawdown curve, and may be found at the downstream end of a channel enlargement. S3 can occur where water enters a steep channel from under a gate. It is considered "transitional" and is usually very short.

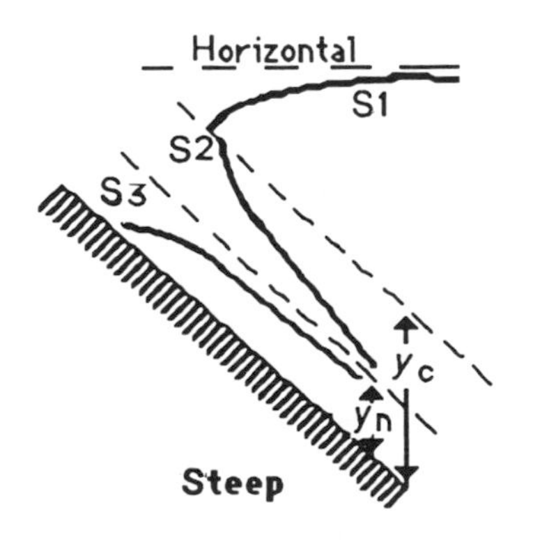

The starting point for water surface profiles is a **control section** (see Section 5.6), which can occur at sluice gates and weirs (or their natural equivalent), bedrock constrictions, at sharp changes in channel slope or any location where critical flow occurs. Larger discharges can "drown out" features which might act as controls at lower stages. Thus, the flow profile is affected by control sections, yet the flow itself determines whether a section is controlling or not (Henderson, 1966). A combination of a few of the flow profile categories is illustrated in Figure 6.35.

Synthesizing the shapes of gradually varied flow profiles

In gradually varied flow, depth, area, roughness, and/or slope change slowly along the channel. A mathematical description of the water surface shape can be derived from principles of energy (Section 6.6.4) and continuity (Section 6.3.3). In practice, the **standard step method** is most commonly used. This is applicable to natural channels with their changing slopes and irregular banks, under gradually varied flow conditions. In this method the channel section is divided into short reaches and computations are carried step by step from one end of a reach to another. If the control point is downstream (subcritical flow), calculations are performed upstream (to derive a backwater curve), whereas calculations are carried downstream if the flow is supercritical.

To apply the standard step method in natural streams a field survey should be conducted to collect data on the cross-sectional geometry, channel slope, roughness and present water surface profile (see Chapter 5 for survey techniques). The computation method requires an iterative solution, which is described by Chow (1959) and Henderson (1966).

A simple program for deriving profiles in rectangular channels, BACK-

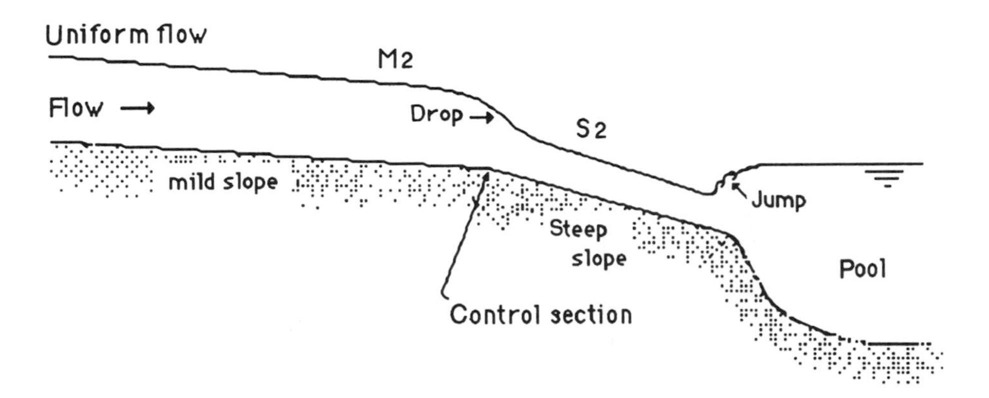

Figure 6.35. Flow profiles in combination along a reach

WATR, is provided in AQUAPAK for illustrative purposes. In natural streams, adjustments must be made to account for energy losses under non-uniform conditions. PHABSIM, a physical habitat simulation model (Chapter 9) has an option for developing backwater curves, and the HEC-2 computer program developed by the US Corps of Engineers (HEC, 1982) is commonly used in practice for deriving water surface profiles in natural streams. Davidian (1984) provides a comprehensive description of the technique.

By assuming different values of discharge at a cross-section, a "family" of flow profiles can be generated for various conditions of stage and discharge. In this manner, stage–discharge relationships (Section 5.8.1) can be developed for ungauged sections. Some other applications include deriving water surface profiles upstream from snags or bridges, or describing the profiles in a tributary stream as the depth in the main stream level fluctuates.

6.6.7 HYDRAULIC JUMPS AND DROPS, ALIAS RAPIDLY VARIED FLOW

Rapidly varied flow occurs over relatively short lengths of channel and it is typically a location of high energy loss. In these situations of intense turbulence a sketch is preferable to an analytical solution for describing the surface profile. Examples are **hydraulic jumps**, where the flow changes from supercritical to subcritical, and **hydraulic drops**, where the reverse occurs. Because of the energy loss, supercritical flow will not "jump" all the way up to its alternate depth (Figure 6.31) in a hydraulic jump, and subcritical flow will not "drop" all the way down to its alternate depth in a hydraulic drop.

Hydraulic drops occur where flow accelerates—for example, as it passes over an obstacle, through a chute, or from a mild slope to a steep slope. Hydraulic jumps take place where upstream supercritical flow meets subcritical flow, such as at the downstream side of large boulders, below narrows created by rock outcrops or where the slope changes from steep to mild. Because of the sudden reduction in velocity, hydraulic jumps are associated with highly turbulent conditions, whitewater and large losses of energy. Since they are such effective energy dissipators they are often encouraged in the design of spillway chutes and structures for dissipating the erosive power of water. They are also effective for mixing fluids in wastewater treatment and for the mixing of nutrients and aeration of water in streams. Fish capitalize on the backflow in the standing waves of hydraulic jumps to give them a "boost" upstream (Hynes, 1970).

The length of flow affected by the hydraulic jump ranges from four to six times the downstream depth. Its appearance is influenced primarily by the upstream Froude number, with the channel geometry having a secondary effect. Some of the patterns exhibited by hydraulic jumps are illustrated in Figure 6.36. The range of upstream Froude numbers given for each type

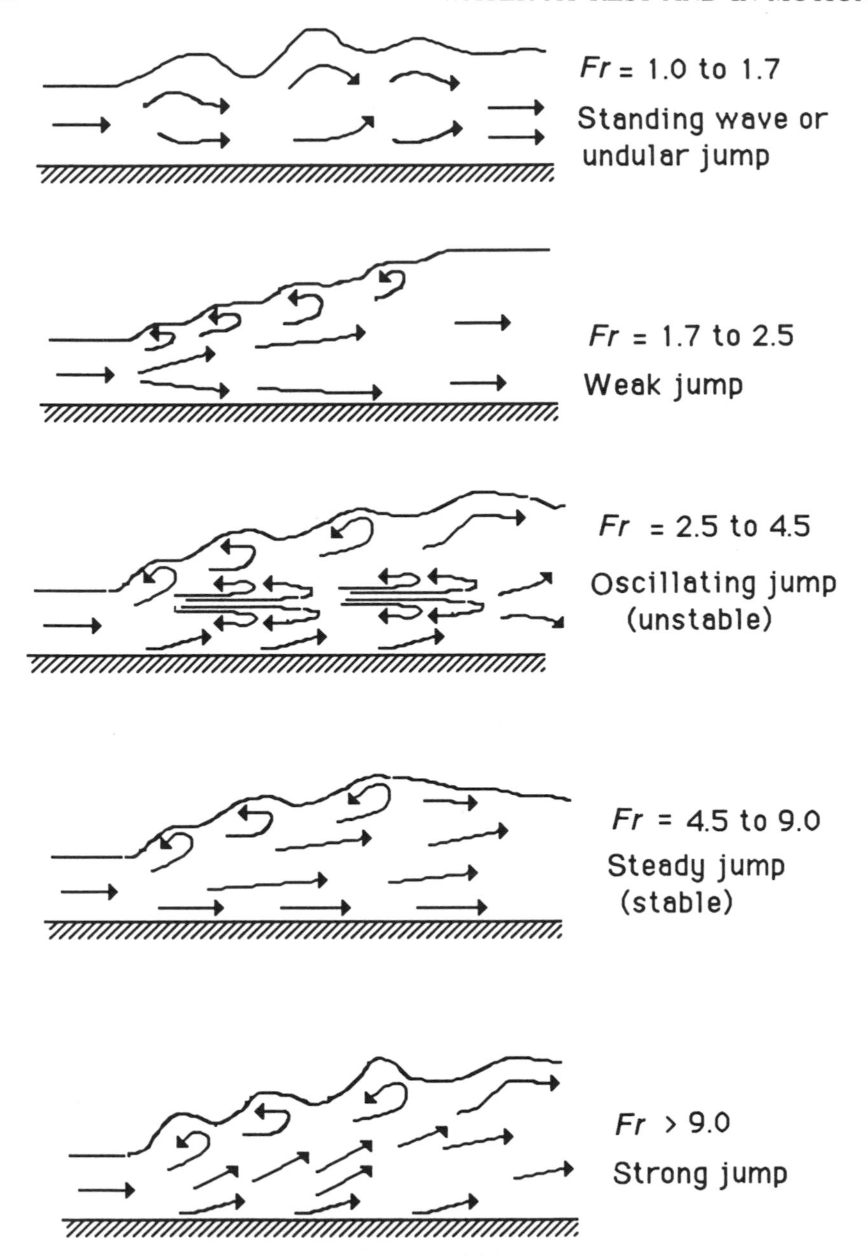

Figure 6.36. Hydraulic jump patterns based on upstream Froude number (*Fr*). Adapted from White (1986), by permission of McGraw-Hill

can serve as a basis for classifying hydraulic jumps. Energy dissipation in hydraulic jumps increases from less than 5% in standing waves to over 70% in strong jumps. Oscillating jumps tend to be unstable, generating large waves which can travel long distances and damage streambanks (White, 1986). It should be noted that hydraulic jumps are not possible if the upstream flow is subcritical ($Fr < 1$).

The upstream Froude number can be simply determined at a jump by measuring the upstream and downstream water depths and using the formula (Henderson, 1966):

$$\frac{D_2}{D_1} = \frac{1}{2}\left[\sqrt{(1 + 8Fr^2)} - 1\right] \tag{6.67}$$

where the subscripts 1 and 2 refer to upstream and downstream depths, respectively, and Fr is the upstream Froude number.

Example 6.8

Calculate the upstream Froude number and give the classification for the hydraulic jump where the upstream depth is 0.15 m and the downstream depth is 0.27 m.
Answer:

$$\frac{0.27}{0.15} = \frac{1}{2}\left[\sqrt{(1 + 8Fr^2)} - 1\right]$$

$$2(1.8) + 1 = \sqrt{(1 + 8Fr^2)}$$

$$(4.6)^2 = 1 + 8Fr^2$$

$$Fr = \sqrt{\frac{20.16}{8}} = 1.59$$

This is an undular jump.

Flow in natural channels is typically varied, unsteady, turbulent and subcritical. However, uniform, steady, and laminar conditions are often assumed in order to simplify the equations which describe flow. The various categories are useful for classifying the flow environment experienced by aquatic organisms, and they give insight into the usefulness and limitations of equations which have been based on theoretical definitions of flow conditions. It should be kept in mind that the theory of open channel flow assumes flow in prismatic channels (constant cross-section and slope). In applying the theory to irregular natural channels we are stretching thin the boundaries of truth, and must interpret results with judgement and caution.

7 Patterns in Shifting Sands

7.1 INTRODUCTION TO STREAM CHANNELS, STREAMBEDS AND TRANSPORTED MATERIALS

7.1.1 GENERAL

As water works its way downstream some of its energy is expended on the transport and rearrangement of materials in the stream's bed and banks. Its ability to create geometric patterns such as braided deltas, rippled sand bars or regular meanders is as fascinating to the casual observer as it is to the scientist attempting to explain or model the processes involved.

Stream channels display more or less regular downstream changes in width, depth, velocity and sediment load, accompanied by changes in the distribution of stream biota. Streams are considered "open systems" because they experience continuous inflows and outflows of energy and matter. More importantly, they are dynamic systems, with changes occurring over a range of time scales from instantaneous to geological. Stream levels shift and sediment loads fluctuate, meanders migrate, floods scour and deposit, banks collapse, sand bars grow and the effects of change at one point are reflected elsewhere in the system.

Adjustment of channel form is of interest both to geomorphologists studying the behaviour of natural rivers over long time periods and to hydraulic engineers concerned with shorter-term changes affecting channel stability near bridges, dams and property boundaries. Ecologists, who study the evolution, distribution and interaction of organisms and their adaptation to the environment, are concerned with both long-term changes which form and re-form habitats and the short-term fluctuations which have a more immediate impact.

In comparison to the study of fluid mechanics (Chapter 6), **fluvial geomorphology**, the study of water-shaped landforms, tends to be more qualitative. The focus is on trends and descriptions rather than precise predictions. General texts for further information on the topics discussed in this chapter include Gregory and Walling (1973), Knapp (1979), Knighton (1984), Leopold et al. (1964), Morisawa (1985), Petts and Foster (1985), Richards (1982, 1987), and Schumm (1977).

7.1.2 MAKING UP A CHANNEL BED

Effects of geology and hydrology

In bedrock streams, channels are eroded as a result of mass failure of large rock slabs and the slow chipping and grinding of the channel bed by stream-transported debris. **Potholes** are a common feature in bedrock channels, which result when stones are ground against the bedrock by spiralling flow.

In alluvial streams the flow moves across beds of material deposited as a result of riverine or glacial processes. The distribution, composition and shapes of rocks, pebbles, sands and clays in an alluvial channel bed reflect the ease with which its source materials are broken up and rounded, and the hydraulics of both high and low flows. Weak rocks such as mudstones or shales or sandstone, for example, are easily broken down, whereas quartzites resist erosion. The size of sediments and the distance they have travelled from their sources thus becomes a geological record of the stream's evolution and hydrology.

In headwater regions, bed materials are usually large, often exceeding the sediment-carrying capability of the flow. These large, often angular rocks disintegrate in the channel near to the original source. Progressing downstream, the mean grain size of substrate materials generally decreases as sediments are fragmented and abraded and smaller sediments are sorted out and carried off. This occurs rapidly in the first few kilometres of a stream and more slowly thereafter.

Local flow conditions can also have a sorting effect on sediments. Coarse materials may line riffles and other regions of high shear stress. Finer sediments are found in depositional regions such as large pools, between boulders in headwater streams, or in the "flow shadow" of bends, confluences, tree roots or other obstructions. The patterns of substrate size may only change at high flows, especially in streams with coarser, more heterogeneous materials. Thus, the spatial distribution of materials will be more closely related to previous flood events than the "typical" flows carried by the stream.

For the Acheron River, the stream system used as an example throughout this text (see map, Figure 4.7), a downstream trend is not apparent in the particle-size data of Figures 5.32 and 5.33. The third-order site has the coarsest materials due to armouring (Section 7.4.2), whereas at the fifth-order site the increase in the d_{50} is due to an influx of large materials from rock outcrops along the Little River.

Flow resistance

Bed materials offer resistance to the flow, which influences the rate of energy loss along a stream and, in turn, has a strong relationship with channel patterns. Flow resistance is caused by (1) grain or surface resistance and (2) form resistance, somewhat analogous to the categories of surface and form drag described in Section 6.5.4.

Grain resistance is the resistance offered by individual grains. In gravel streambeds it tends to be the major component of flow resistance. Grain resistance is considered a function of relative roughness, the ratio of roughness height to water depth (Equation 6.18). Thus, its effect diminishes as the roughness elements become submerged at higher discharges. To reflect the influence of the largest particle sizes, roughness height is commonly described in terms of the d_{84} or d_{90} of the bed materials (see Section 5.10.5). The spacing and arrangement of these larger particles can also affect flow resistance (Section 6.4.2).

Form resistance is associated with the topography of the channel bed. Bedforms result from the interaction of streamflow patterns and bed sediments, particularly in sand-bed streams (Section 7.3.4). However, form roughness can also refer to the troughs, potholes and plunge pools of bedrock channels and the larger-scale forms of pool–riffle or pool–step sequences (Section 7.3.2).

Progressing downstream, channel topography typically changes from a poorly defined pool–step structure in headwater streams to more developed pool–riffle sequences in gravel-bed reaches, and finally to sand bedforms as the grain sizes become smaller and more uniform. In conjunction with channel patterns, streambed forms seem to regulate resistance in a self-adjusting manner so as to effectively dissipate energy over a wide range of flow and bed material conditions (Morisawa, 1985).

7.1.3 WHAT SORT OF DEBRIS IS TRANSPORTED?

Total load refers to the amount of dissolved and particulate organic and inorganic material carried by the stream. Although it should be realized that sharp boundaries do not exist, the total load can be divided into three groupings:

- Flotation load
- Dissolved load
- Sediment load

The last category can be further subdivided on the basis of particle transport rate, size distribution, density, or chemical and mineralogical composition. Sediment is usually considered to be the solid inorganic material, which is separated from particulate organic matter during analysis. Commonly, sediment load is separated into the following categories:

- Washload
- Bed-material load—which can be transported as:
  Suspended load or
  Bedload.

The subdivision of the various loads within a stream is illustrated in Figure 7.1.

Flotation load

The **flotation load** consists of the logs, leaves, branches and other organic debris which are generally lighter than water (until they become waterlogged). The amount of organic debris supplied to a stream depends on the density and type of vegetation along the banks, the amount of bank failure and tree fall, and the floating debris picked up from the floodplain by flood waters.

The organic debris is the foodstuff of decomposers and some aquatic invertebrates. Large trees and saplings provide shelter for fish, and form debris dams which trap sediments and modify the channel shape by redirecting swift currents. The larger woody debris may play an important part in channel stabilization (Section 7.5.1). Removal of riparian vegetation reduces the supply of organic material to the stream, which may have long-term effects on both its ecology and morphology. Flotation load, however, can also cause problems when it impinges against bridges or other structures.

Dissolved load

A stream's **dissolved load** is the material transported in solution. Local geology, land use and weathering processes affect the amount of dissolved

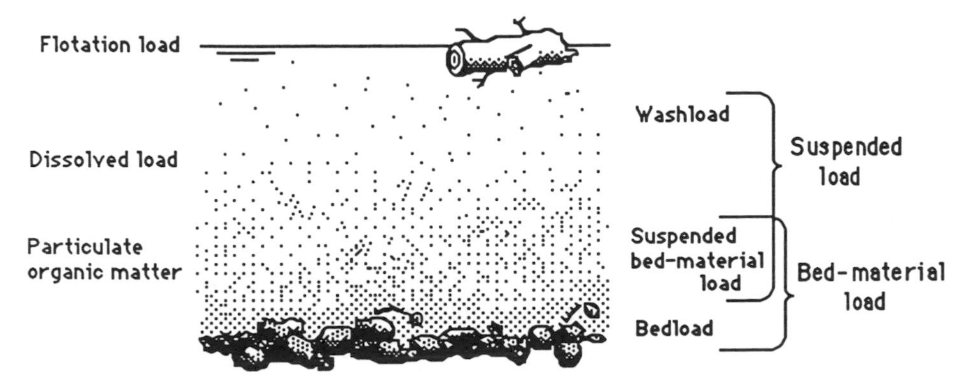

Figure 7.1. Categories of transported materials in a stream

load, which can often exceed the sediment load by total weight. Natural origins of dissolved loads include sea salts dissolved in the rainwater of coastal areas, and the chemical weathering of rocks—sometimes enhanced by organic acids from the decay of vegetation. In general, water originating as groundwater tends to have a higher soluble load than surface-derived runoff. Industrial effluents and agricultural fertilizers and pesticides are becoming more significant sources of solutes. Dissolved load is therefore associated with chemical water quality (Section 5.11). Although it will vary from river to river, an average of 38% of the total load of the world's rivers is dissolved material (Knighton, 1984).

Because of turbulence, dissolved loads will generally be uniformly distributed over a stream cross-section. Exceptions exist in localized areas of saline stratification and in reaches where groundwater of high dissolved load enters the stream. Dissolved loads also change with changes in discharge. During a runoff event, concentrations generally decrease at first as rainwater dilutes the stream water, but later increase as groundwater reaches the stream, bringing in dissolved materials (Hjulstrom, 1939).

Washload

Washload refers to the smaller sediments, primarily clays, silts and fine sands, which are readily carried in suspension by the stream. This load is "washed" into the stream from the banks and upland areas and carried at essentially the same speed as the water. Only low velocities and minor turbulence are required to keep it in suspension, and it may never settle out. Its concentration is considered constant over the depth of a stream. For practical purposes, the smallest washload grain is considered to be about one-half micron (0.0005 mm), to separate it from dissolved materials. The largest size is usually taken as 0.0625 mm. This division is somewhat arbitrary, but objective.

It is the *rate of supply* from uplands or streambanks which determines the amount of washload transported, rather than the ability of the stream to carry it. According to theory, streams have an almost unlimited capacity for transporting washload. Streams cannot become "saturated" with sediment as they can with dissolved solids (Hjulstrom, 1939). The washload can constitute a large percentage of the total volume and mass carried by a stream or river.

High washloads may typify streams with banks of high silt-clay content. However, washloads can also be contributed from fire-denuded slopes, the ash from volcanic eruptions or other disturbances in the catchment from road or dam building and agricultural practices.

Bed-material load

Bed-material load is the material in motion which has approximately the same size range as streambed particles. In alluvial streams the amount of bed-material load transported is controlled by flow conditions. This load may be further subdivided according to whether it is transported in suspension or remains in touch with the bed.

Suspended bed-material load is the portion which is carried with the washload, remaining in suspension for an appreciable length of time. It is supported by the fluid and kept aloft by turbulent eddies, but will settle out quickly when velocities drop. Over a vertical profile its concentration is highest near the streambed. **Bedload** is that portion which moves by rolling, sliding or "hopping" (saltation), and is partly supported by the streambed. It is thus only found in a narrow region near the bottom of the stream (see Figure 7.2).

For field measurements, the distinction between sediment load types is commonly based on the method of data collection (see Section 5.9). Suspended sediment samplers will capture both washload and suspended bed-material load, which are often grouped into the single category, **suspended load**. Most depth-integrating suspended sediment samplers (Section 5.9.4) can measure to within about 90 mm of the streambed. The amount of material travelling along in contact with the bed can be measured with bedload samplers or by surveying the amount of sediment accumulated over time in depressions or behind logs (Section 5.9.5).

Other criteria for separating suspended load and bedload include using the d_{10} (CSU, 1977) or a grain diameter of 0.0625 mm (as for washload) as the point of division. The division into suspended load and bedload is convenient for the development of sediment transport equations based on the different modes of movement. In reality, a particle of a given size can move in suspension, hop along the streambed or stay put, depending on the flow rate. Density and shape as well as size can also affect the way in which a particle is transported. On average, less bedload than suspended load is transported over a year. The ratio of bedload to suspended load is typically in the range 1:5 to 1:50 (Csermak and Rakozki, 1987). This ratio will be higher in headwater streams and during floods, when bedload can often exceed the suspended load.

7.1.4 SEDIMENT DISTRIBUTION AND DISCHARGE

Figure 7.2 gives a general illustration of the vertical distribution of sediment. Larger sediments which move only as bedload are the most highly concentrated at the bed, whereas silts and clays are distributed more or less uniformly from bed to water surface. The actual distribution over a particular vertical in a stream is dependent on the particle sizes and the velocity and turbulence intensity (Section 6.4.1) of the water.

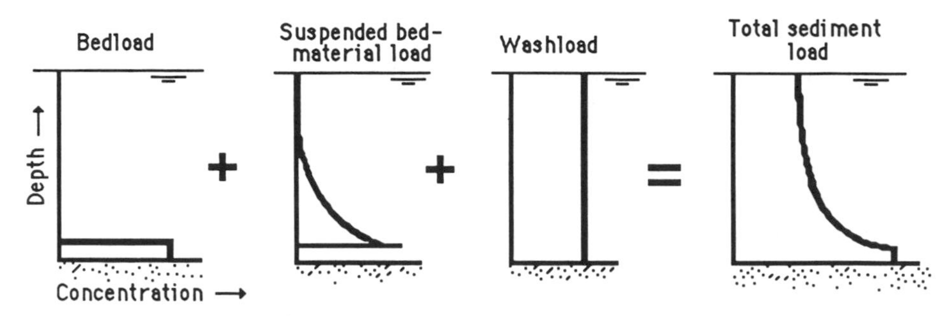

Figure 7.2. Schematic diagram of the vertical distribution of sediment load concentrations

Considering all particle sizes together, both concentration and mean grain size increase towards the bed. Variations can occur across stream cross-sections and with increases in discharge. The largest quantities of sediment and the coarsest fractions tend to be transported in the path of maximum velocity, which does not always coincide with the thalweg (the path of deepest flow). If velocity and turbulence increase, the larger particles are distributed more evenly.

Sediment discharge is the amount of sediment moving past a cross-section over some period of time. It is usually reported in units of mass per unit time such as kg/s or tonnes per day or year. Vertical and lateral patterns of both sediment concentration and velocity determine the distribution of sediment discharge at a cross-section, as shown in Figure 7.3. This distribution must be taken into account when sampling sediments (Section 5.9). A depth-integrating sediment sampler, for example, takes a velocity-weighted sample, which integrates the distributions of both velocity and sediment concentration. Additionally, if an automatic sampler intake nozzle is located near the streambank, its samples may be sufficiently representative in a stream which carries mostly washload, but will give misleading information in streams which transport coarser materials.

The pattern of sediment discharge does not necessarily coincide with the

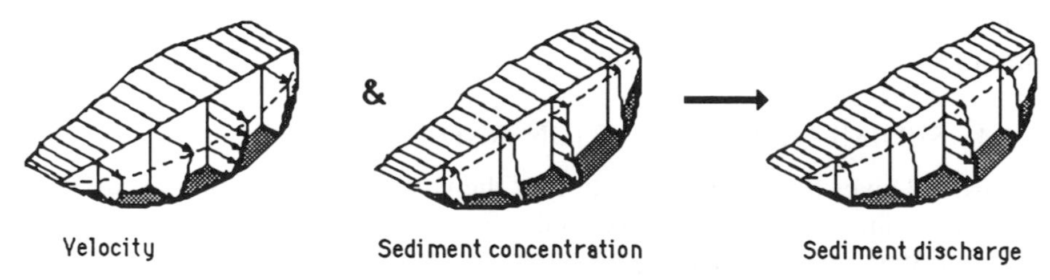

Figure 7.3. Graphical representation of sediment discharge at a cross-section as a result of velocity and sediment concentration patterns. Adapted from Nordin and Richardson (1971)

pattern of runoff. At sediment source areas, erosion is usually highest at the beginning of a rainstorm when sediments are more readily available. These are washed into the stream with the first rivulets of overland flow, and as the sediment supply is exhausted, the concentration drops quickly. In these streams, sediment concentrations will ordinarily be greater when the stage is rising, with coarse particles depositing out quickly as the hydrograph starts to fall. Thus, the concentration will be different at the same discharge on the rising and falling limb of a hydrograph (see Figure 7.20). This effect is called **hysteresis**, and complicates the prediction of sediment concentration from streamflow (Section 7.5.4). In headwater areas the sediment concentration will typically peak before discharge does. Downstream, in large basins, the peak sediment load may match or even lag behind the discharge peak when the site has to "wait" for the sediment to be delivered from upstream.

Patterns of *total* sediment concentration do not give the whole story, however, as the *sizes* of sediments transported will also change with time. Washload fines typically occur in higher concentration at the beginning of an event. Bedload movement and entrainment of bed materials increases as the discharge reaches its peak, and then a reduction in particle size occurs as the hydrograph tapers off. This pattern is dependent on the types of sediment and dissolved materials available from uplands and channel. As was shown by Johnson et al. (1985), sediment loads also tend to be different for snowmelt and rainfall runoff events.

The amount of sediment carried through the outlet of a catchment depends on two factors:

(1) The amount of sediment eroded and transported to the stream from upland sources; and
(2) The ability of a stream to carry the washed-in sediments and to re-work and transport bed and bank materials.

Streams can therefore be considered either **supply limited** or **capacity limited,** depending on whether their ability to carry sediment exceeds the amount available or vice versa. In regard to the type of sediment carried, washload is considered supply limited and bed-material load capacity limited.

The first of the above two factors is dependent on climate, land use and the geology and topography of the catchment. Hillslope erosion can result from raindrop splash, the hydraulic action of overload flow, rill and gully formation, mass movement such as landslides, the scuff of a boot, or the turn of a tractor tyre. A large amount of effort has gone into the modelling of soil loss and the effects of land management on hillslope erosion. Upland processes are, however, beyond the scope of this book, and the interested reader is referred to Branson et al. (1981), Finlayson and Statham (1980) and geomorphology texts.

The second of the above factors is dependent on the hydraulic and hydrological properties of the channel and the erodibility of its bed and banks. Relevant field measurements are covered in Chapter 5 and the processes which cause individual grains to be lifted into the flow are described in Section 7.4. The larger-scale processes which contribute to the amount of sediment delivered from a catchment, its sediment yield, are discussed in Section 7.5.

7.1.5 ECOLOGICAL IMPLICATIONS

For many aquatic organisms the channel bed is a "substrate" to be used as a foothold, as a site to deposit or incubate eggs, as "grit" for grinding food or as a refuge from floods (Minshall, 1984; Statzner et al., 1988; see also Section 2.1). In engineering practice the channel bed is normally considered the boundary between solid and fluid—a section of the "wetted perimeter". For many organisms the boundary is not as distinct. The streambed surface is rich with organic matter trapped in the pits between grains, which provides nutrients for organisms near the base of the food chain. Below the surface, the hyporheic zone (Section 2.1) forms an interface between stream and groundwater systems. This region can be extremely active biologically. Many organisms reside temporarily in the interstices of the surface materials, whereas some may carry out their whole life cycle deep within stony streambeds (Hynes, 1970). The streambed acts as a refuge for benthic organisms, providing shelter from floods, drought and extremes of temperature. Ward and Stanford (1983) call it a "faunal reservoir", capable of recolonizing the stream if stream populations are depleted by adverse conditions.

Species differ in their substrate preferences and requirements. The suitability of a substrate for colonization by aquatic flora and fauna depends on its average particle size, its mix of sizes, the size of pore spaces, degree of packing and embeddedness, and its surface topography. Freshwater crayfish and some aquatic insect species such as dragonfly and stonefly larvae live in the crevices between and beneath rocks. Others, such as the purse-case caddisfly, require unstable fine-grained sands where moss cannot grow. Still others such as midge larvae need mud into which they can burrow. Salmonids require a mix of gravels with small amounts of fine sediments and rubble as an optimum spawning substrate mix (Beschta and Platts, 1986). Algae, mosses and other aquatic plants also have specific substrate requirements. The plants, in turn, provide substrates for other organisms to cling to or shelter behind.

Thus, the distribution of sediment sizes along a stream will be one of the physical habitat factors influencing the distribution of organisms. In general, the highest productivity and diversity of aquatic invertebrates seems to occur in riffle habitats with medium cobble and gravel substrate (Gore, 1985). Areas of shifting sands commonly have reduced species abundance and richness (Minshall, 1984).

Biological activity in coarser substrates is dependent upon the maintenance of inter-gravel flow rates for the replenishment of nutrients and oxygen and the removal of metabolic wastes. If excessive fines are washed into a stream, as, for example, from road or dam construction, they can form a "mat" on top of the coarser bed materials. Fines can also work down between the coarser grains to form a type of "hardpan" layer. The infilling of gravels with finer sediments can reduce inter-gravel flow rates, suffocate eggs, limit burrowing activity and trap emerging young. Gravel-bed streams which become filled with silt may show a shift in the insect species compositions from mayflies (Ephemeroptera) and caddisflies (Trichoptera) towards midgefly larvae (Diptera), which, in turn, can affect fish species compositions (Milhous, 1982). Because of these ecological effects, there has been some interest in estimating the flow required to remove fines from the streambed. Methods for estimating these "flushing flows" are discussed in Sections 7.4.3 and 9.3.5.

The transport of particulate matter is both bane and benefit to aquatic organisms. Organic particulate matter is a source of food for downstream organisms. However, when the flow quickens, the larger grains can become deadly projectiles. Fine silts and clays clog gills like a particulate "smog", reduce light needed for photosynthesis and periphyton production, and interfere with the foraging success of sight feeders and filterers. The shifting of whole segments of the streambed uproots and scours away benthic organisms. Heavy metals and other toxic substances can also be adsorbed onto sediment and are thus transported and despositied along with it.

Vogel (1981) implies that abrasion and alterations of form of the stream bottom during floods have more critical impacts on biota than velocity *per se*. Jowett and Richardson (1989) cite a study on rivers in New Zealand where the abundance of trout decreased significantly after a major flood, although the coarsening of substrate, removal of excessive algal growth and deepening of pools improved habitat for future use. Aquatic vegetation such as algal mats and macrophytes will affect bed stability during high flows. After floods, recolonization by bacteria, fungi and algae binds and stabilizes the substrate, improving conditions so that other organisms can come back more quickly.

The movement of sediments and the composition of streambeds will thus have different effects on different species. The ecological advantages and disadvantages of sediment movement and streambed composition should therefore be weighed carefully in studies of instream flow needs, channel changes, or the effects of land-use practices (Chapter 9).

7.2 STREAM-SHAPING PROCESSES

7.2.1 A NOTE ABOUT STREAM POWER

As described in Section 1.2, power is the amount of work done per unit time, where work and energy have the same units. **Stream power** has a number of definitions, related to the time rate at which either work is done or energy is expended. It is a useful index for describing the erosive capacity of streams, and has been related to the shape of the longitudinal profile, channel pattern, the development of bed forms, and sediment transport.

In studies of sediment transport, Bagnold (1966) originally defined **stream power per unit of streambed area** (ω_a) as:

$$\omega_a = \tau_o V \tag{7.1}$$

where τ_o is the shear stress at the bed (N/m^2) and V is the mean velocity (m/s) in the stream cross-section. Thus, ω_a has units of N/m·s (watts/m^2).

Perhaps a more useful definition of stream power is as the rate of potential energy expenditure over a reach or **stream power per unit of stream length** (this equation can easily be developed from Equation 1.5 as an exercise):

$$\omega_\ell = \rho g Q S \tag{7.2}$$

where S is the energy slope of the reach (see Section 6.6.4) and ω_ℓ has units of kg·m/s^3 (watts/m). Another form is **stream power per unit mass** of water (a mass of 1 kg):

$$\omega_m = gVS \tag{7.3}$$

where ω_m has units of m^2/s^3 (watts/kg). Alternatively, it can also be expressed as a **stream power per unit weight** (a weight of 1 N):

$$\omega_w = VS \tag{7.4}$$

where ω_w has units of velocity (m/s) (watts/N). This measure can also be considered the time rate of head loss over a reach, where head is energy per unit weight (Section 6.3.4). Any of the terms ω_a, ω_ℓ, ω_m or ω_w can be referred to as a "unit stream power", because the stream power is expressed per unit area, length, mass or weight, respectively. In the literature, however, ω_ℓ is commonly given the symbol Ω and called **total stream power**.

As slopes become steeper and/or velocities increase, stream power goes up and more energy is available for re-working channel materials. In a bedrock stream with no sediment transport all the stream power is spent in the frictional dissipation of energy. In alluvial channels with mobile boundaries part of this stream power is used for transporting sediment. Figure 7.4 compares high and low stream power situations. It can be seen that straightening and clearing a channel would increase its slope and velocity, and thus its stream power. This increases the amount available for erosion and sediment transport. Alterations to the stream power at a point can therefore initiate changes in sediment transport and channel shape.

Stream power also increases as discharge increases at a site. However, even though discharge increases in the downstream direction, stream power per unit area (ω_a) typically decreases because the slopes decrease. Flash floods in steep ephemeral channels, for example, can generate very large values of stream power. Costa (1987) ascertained that a 1973 flood on a tributary to the Humboldt River, Nevada, had a unit stream power (ω_a) of 8160 N/m·s, as compared to 12 N/m·s for floods on the Mississippi and Amazon Rivers. Jarrett and Malde (1987) estimated a unit stream power (ω_a) of 75 000 N/m·s for the prehistoric Bonneville Flood, which catastrophically discharged a very large volume of water down the Snake River in southern Idaho. The estimated peak discharge of approximately 935 000 m^3/s inundated the Snake River Canyon to depths greater than 130 m in places, and deposited large gravel bars and boulders in its path.

The relationship between a channel and its floodplain is important in determining stream power. In a study of alluvial rivers in Victoria, Australia, Brizga and Finlayson (1990) found that rivers which remained within their banks at high flows tended to have high stream power and relatively coarse bed materials. In comparison, rivers which flooded over their banks at high flows had lower stream power, transported finer sediments and had more stable channels.

Langbein and Leopold (1964) proposed that the shape of streams is a compromise between two opposing tendencies: (1) for energy to be expended uniformly over the length of a stream (implying constant stream power), and (2) for the total expenditure of energy to be minimized over the length of a stream. The "typical" concave shape of the longitudinal profile (Section 4.2.6) may be partly explained by the first theory. In headwater streams,

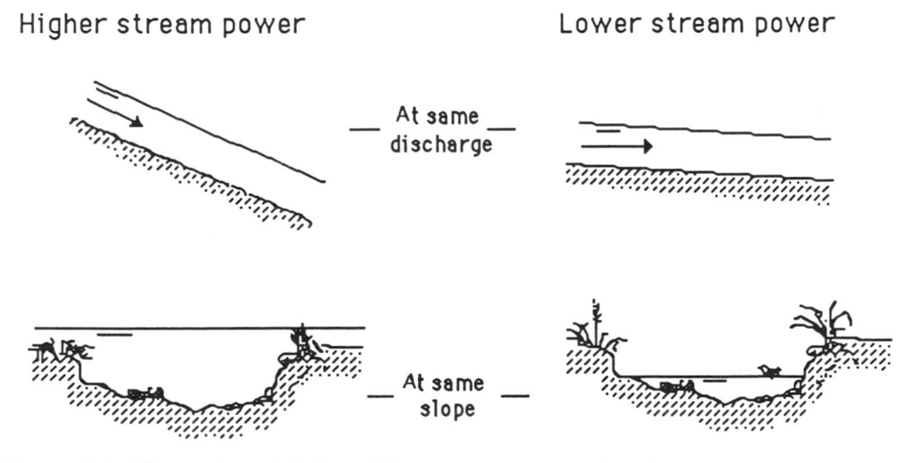

Figure 7.4. Illustration of high and low stream power situations

discharge is low and slope is high, whereas in valley streams, discharge is high and slope is low; thus the product QS remains relatively constant over the length of a stream. The reader is referred to the fluvial geomorphology texts mentioned in Section 7.1 for further discussion on the use of stream power in the description of stream channel form and formation.

7.2.2 ADJUSTMENTS AND EQUILIBRIUM

One of the most interesting characteristics of open systems such as streams is their capacity for self-regulation. In terms of channel adjustment a stream has several variables to "play with". It can change local slopes and velocities, rearrange bed materials, transport more or less sediment, and change its channel pattern—within certain constraints. "Feedback" mechanisms act to stabilize the system so that some degree of equilibrium can be established. This discussion applies primarily to alluvial streams; in bedrock streams, channel adjustment is not quickly achieved.

If a stream is in equilibrium condition it is considered both stable and graded. A stream is described as **stable** if its cross-sectional geometry remains relatively constant over some time scale. It is considered **graded** if its slope is just sufficient to transport all the material delivered to the stream. If a stream's ability to transport sediments (its sediment transport capacity) is less than that required to move sediments arriving from upstream, then some of the sediment is deposited, leading to **aggradation**. In contrast, if not enough sediment enters to "consume" the transport capacity, the flow will erode the river bed and/or banks to pick up sediment, leading to **degradation**.

Aggradation is commonly seen where a stream enters a reservoir or pond. The "delta" slowly grows downstream, eventually filling the impoundment (see Figure 7.6). In stream channels, aggradation can elevate the channel bed within natural levees, causing adjacent areas to become swamp-like wetlands. Degrading streams, on the other hand, can cause a lowering of the surrounding groundwater level, leaving both native and agricultural plants high and dry. Degradation can occur where the local transport capacity has increased, e.g. from the straightening of a meander. The erosion will often propagate back upstream—a process called **headcutting**. A steep knickpoint (Chapter 4) usually exists at the upper end of the cut. Erosion is more pronounced in this region, and the swirling action of water in the downstream pool can create an undercut which eventually leads to failure of the overlying material and upstream migration of the knickpoint.

Sites of aggradation and degradation can be revealed by comparing the longitudinal profiles from two different time periods. Over time, aggrading sections tend to become steeper and degrading sections flatter. Because changes in slope affect velocity and sediment transport, "feedback" can occur to slow or reverse the trend.

A channel may undergo gradual change in form, steepness and sediment transport, which can occur as an average trend or cycle. Oscillations about the average state can also take place. Furthermore, systems may be influenced by **thresholds**; i.e. abrupt changes may occur when some critical value is exceeded. As an example of a threshold response, a rock can remain perched on a small pedestal until progressive weathering eats away at the base and the rock suddenly tumbles. In fluvial systems, thresholds may exist for sediment movement, bank collapse or the sudden cutoff of a meander bend. Thus, changes in erosion and deposition may not necessarily be due to external influences such as land-use change but may have appeared suddenly when long-term processes reached a threshold.

Schumm (1977) suggests that drainage basin evolution can be considered at four time scales, as illustrated in Figure 7.5. Over a major period of geological time an uplifted terrain is gradually worn down and slopes progressively become less severe. Within this time frame the channel is continuously adjusting to changes in discharge and sediment load, causing fluctuations about some average trend. **Dynamic equilibrium** describes this type of behaviour. Abrupt periods of adjustment can also occur within this time frame, and the term **dynamic metastable equilibrium** was introduced to allow for the influence of thresholds. **Steady-state equilibrium** refers to a generally stable form, about which seasonal and other short-term fluctuations occur, such as scour and fill during floods. At the time scale of one day the channel form is essentially in a **static equilibrium**, unless a flood or a bulldozer is passing through.

Fluctuations and thresholds can result from the complex interactions of many variables. A major event such as a flood, a bushfire, channelization or tectonic activity can cause a chain reaction of responses which continue for many years. For example, Schumm (1977) describes the complex response of a tributary to the lowering of the elevation at its outlet. This can occur when the stream into which it flows degrades to a lower base level. Predictably, headcutting will progress upstream from the mouth. However, the pattern of erosion, deposition, changes in sediment load and renewed incision within the tributary as it adjusts to the new base level can be extremely complex.

At the time scale appropriate to most stream research a stream system will most likely exist in a state of "meta-stability" rather than equilibrium. Like wars interrupting relatively peaceful phases to change the direction of history, channel developments may also be characterized by relatively sudden changes between quiescent periods. Vannote et al. (1980) propose that the structure and function of stream communities adjust to changes in physical

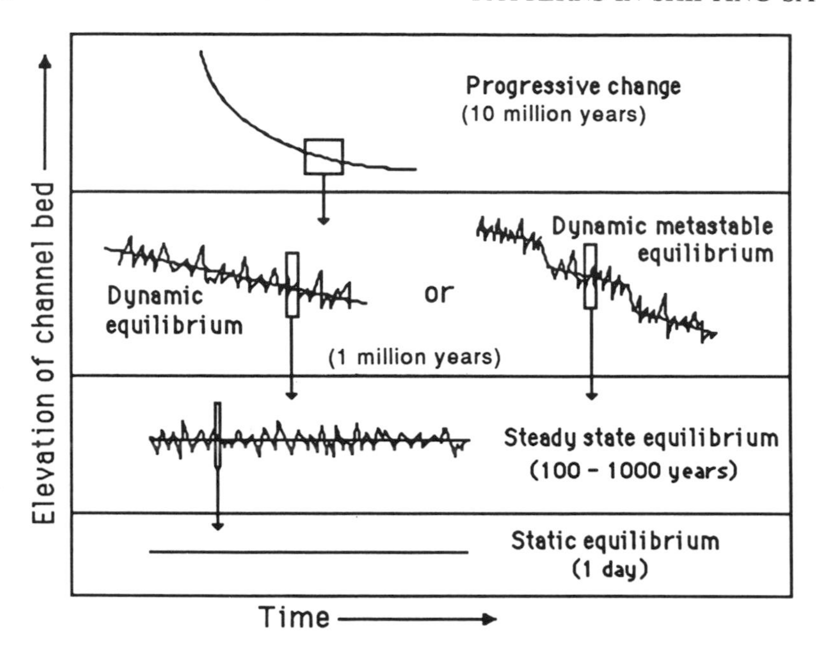

Figure 7.5. Drainage basin evolution at varying time scales. Adapted from Schumm (1977), by permission of John Wiley & Sons, Inc.

habitat. It is thus important to consider the type and direction of change when evaluating the condition of a stream channel. As Bovee (1982) mentions, a habitat and instream flow evaluation (Section 9.3) based on the assumption that the stream will remain in its present form will be invalid if the stream is not in an equilibrium condition.

7.2.3 BALANCING SLOPE, STREAMFLOW, AND SEDIMENT SIZE AND LOAD

The shapes of a stream—its snake-like meanders, its tumultuous drops, its stretches of calm water—are controlled by climate and landscape geology. Over time, channel form is adjusted to accommodate the discharge and sediment loads of a catchment. Complex interrelationships exist between channel dimensions, channel patterns, sediment supply, streambed roughness and steepness of the valley floor and stream channel. Alterations to any of these components, either natural or human-induced, will have an impact on others. Effects are difficult to predict because of the difficulty in isolating the role of a single variable.

A general, qualitative expression for the balance between sediment discharge (Q_s), stream discharge (Q), particle size (d_{50}) and stream slope (S) was presented by Lane (1955), and states that:

$$Q_s d_{50} \sim QS \tag{7.5}$$

where d_{50} is the median sediment particle size (see Section 5.10.5). Because this equation is qualitative, no units are given. However, it can be used to obtain a general sense of the way a stream will respond to changes. For example, by shifting the relative "weights" of variables on either side of the equation, it can be seen that:

- A channel will remain in equilibrium (neither aggrading nor degrading) if changes in sediment load and particle size are balanced by changes in water discharge and slope.
- A reduction in the sediment load (Q_s) can result in a decrease in slope if other factors remain constant. When a dam is installed, accumulation of sediment in the reservoir often results in the release of clearer water downstream (lower Q_s and possibly lower d_{50}). This can lead to degradation (lower S) downstream, as shown in Figure 7.6.
- An increase in sediment load (Q_s) can result in aggradation (higher S) if other factors are constant. At the point where a small but highly sediment-laden tributary enters a clearer main stream, for example, aggradation occurs at the confluence and upstream.
- Larger sediment particles can be transported by steeper slopes and/or higher discharges.

Changes in streamflow (Q) can result from climatic variations or diversions either into or out of a stream. Interbasin transfers made for hydropower or mining operations or municipal supplies can affect both the source stream and the stream into which the water is diverted. The large increase in flow in the receiving stream can result in rapid degradation downstream from the transfer point. Sediment load (Q_s) can be increased by natural failures of mountain sides or streambanks or by human-caused disturbances. In most situations, however, streamflow and sediment load tend to increase and decrease together rather than in isolation.

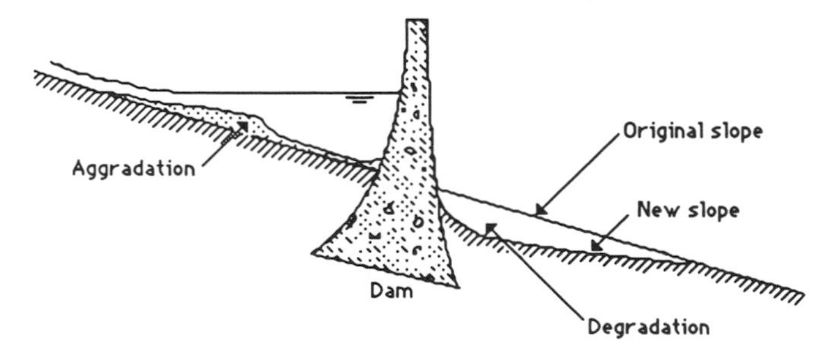

Figure 7.6. Aggradation and degradation at a dam

7.2.4 FLOODS AND FLOODPLAIN FORMATION

To a geomorphologist, a **floodplain** is the relatively flat valley floor formed by floods which extends to the valley walls. To engineers, it may have a more precise definition as the region covered by the 100- or 200-year flood (see Chapter 8). In this discussion it is used more loosely as the valley floor adjacent to the stream channel which becomes inundated at high flows.

Floods are commonly defined as those flows which overtop the banks of a stream. They are renowned for their awesome power and disastrous effects on the cities and farmlands which happen to lie in their paths. Their benefits include the ability to replenish topsoil and nutrient supplies on floodplains, to provide water to seedlings and trees requiring periodic inundation, to flush out anoxic or saline waters or deposits of fine sediments, and to permit aquatic animals to migrate to calmer, nutrient-rich shallows in the floodplains to feed and breed. The ecological impacts of floods are further discussed in Chapter 8.

At low flow, the stream may follow a winding path between rocks and around bends and bars. With increasing discharge, the path of travel is shortened as the flow "shortcuts" across the tops of bars and envelops meanders. This increases the slope of the water, delivering it downstream more efficiently. As floodwaters rise, they begin to erode the more susceptible parts of the bed and banks. In ephemeral streams and large rivers in semi-arid areas, for example, it is characteristic for the whole width of the bed to be downcut as the stage rises. When the water spills over onto the floodplain its slope approaches that of the surrounding valley. The hydraulics of overbank flow become very complex as the faster-moving water of the main stream interacts with the slower-moving water on the floodplain. "Rollers"—water spinning around a vertical axis—may develop alongside the channel banks, accelerating erosion in these areas. Bank vegetation may be uprooted and added to the heavy debris load carried by the rushing water. Extremely large floods can leave a lasting imprint on streams; others may have little effect, especially if they have followed on the heels of an even larger flood which has already rearranged the channel.

When floods subside, sediment deposits on the channel floor, filling in the scoured areas. This **scour and fill** process is a well-known phenomenon at gauged cross-sections, where adjustments must be made to the stage–discharge relationship (Section 5.8.1) to account for its effects. This "classic" picture will not apply to all sites, however. Even during floods, sediment deposition still takes place behind obstructions and in other areas of slack water at the channel perimeter. Leopold et al. (1964) state that when scour occurs at a pool there seems to be simultaneous filling on downstream bars or riffles.

Because the water spilling over onto the floodplain moves at a lower velocity it cannot carry as much sediment. Coarser materials tend to deposit out close to the channel rim, and large levees can be naturally constructed by this mechanism. Finer sediments are widely deposited over the floodplain, concentrated in the wake region behind obstacles such as fences or vegetation. Valley floors are thus gradually built up of layers of coarse material from old streambeds and glacial deposits, and finer silts and clays which have dropped out of suspension onto the floodplain.

The floodplains of braided and meandering streams (Section 7.3.1) form differently. Braided streams migrate widely across the valley floor, leaving isolated bars behind. In meandering streams, bars grow from deposits on the inside of river bends and the meanders migrate outwards and down the valley. Although floodplain formation is considered a long-term process, channel shifts may occur frequently enough to cause changes in habitat for aquatic and wetland species. In a floodplain of the Little Missouri River, for example, a study of the distribution of trees in different age groups led Everitt (1968) to conclude that half of the floodplain had been re-worked over a period of only 69 years.

7.2.5 CHANNEL-FORMING DISCHARGES

The concept of a "channel-forming" or "dominant" discharge is a convenient one for analytical and conceptual purposes. The **dominant discharge** is considered to be a single discharge equivalent in its effect to the range of discharges which govern the shape and size of the channel. Whereas the gross form of the channel may be shaped by larger, rarer discharges, the maintenance of that form and its smaller-scale features such as gravel bars and bedforms may be more closely related to more frequent discharges (Harvey, 1975). This will also be affected by the relative stability of the stream's bed and banks.

The discharge which just fills the stream to its banks, sometimes termed the **bankfull discharge**, is often assumed to control the form of alluvial channels. Flow resistance reaches a minimum at bankfull stage, and thus the channel operates most efficiently for the transport of water at this level (Petts and Foster, 1985).

In the USA, Leopold et al. (1964) report that bankfull discharge occurs approximately every one to two years, although a wide range of values have been found for this average recurrence interval. For 72 rivers in New Zealand, for example, Mosley (1981) reported average recurrence intervals of 1 to 10 years, with a median value of about 1.5 years. In streams with fairly constant flows the bankfull dimensions may be controlled by a discharge close to the mean annual flood, but in streams characterized by sharp flood peaks amid long periods of low flows the channel capacity will be related to the higher, less frequent events (Gregory and Walling, 1973). High values of indices such as the coefficient of variation (C_v) of annual flows or the index of variation (I_v) of annual peak flows (see Appendix 1) may indicate that a river is less likely to develop a form in equilibrium with "average" discharge and sediment load conditions. In a study of 15 sand-bed streams in the midwestern USA, for example, Pizzuto (1986) found that bankfull depth increased with greater flow variability. Additionally, if the channel has become incised below the floodplain it will not fill to its banks as frequently (Gregory, 1976). In recently incised channels in Australia, Woodyer (1968) used bench levels corresponding to the present floodplain level to derive average recurrence intervals between 1.2 and 2.7 years, fairly close to the interval given by Leopold.

Bankfull stage has been defined in a number of ways. The popular concept is the definition mentioned previously, the height of the floodplain surface. However, the definition of "bankfull" becomes much more difficult if the section is not well defined, for example: (1) where the bank tops are not the same elevation, (2) in braided streams, (3) where the break between the channel banks and floodplain is not obvious, and (4) at complex cross-sections where benches or terraces are present. Slight differences in interpreting bankfull elevation can mean large differences in discharge and thus the associated average recurrence interval.

Various authors have developed criteria for defining bankfull stage. Ridley (1972) utilized a "bench index" to find the maximum break in slope on the banks, and Wolman (1955) suggests using the minimum width-to-depth ratio. Another definition is the average banktop elevation. These morphological definitions of bankfull stage at a cross-section are shown in Figure 7.7.

Vegetal clues such as the lower limit of grasses, mosses, liverworts and forbs may also give an indication of bankfull stage. Gregory (1975) suggests the use of lichen limits, which were found to be fairly consistent and associated with less than a 2-year average recurrence interval. Limits may be associated with the sensitivity of plants to inundation or levels of bedload abrasion which clear the rock surface. This latter approach may be of more ecological relevance, and was suggested by Newbury (1989) for the evaluation

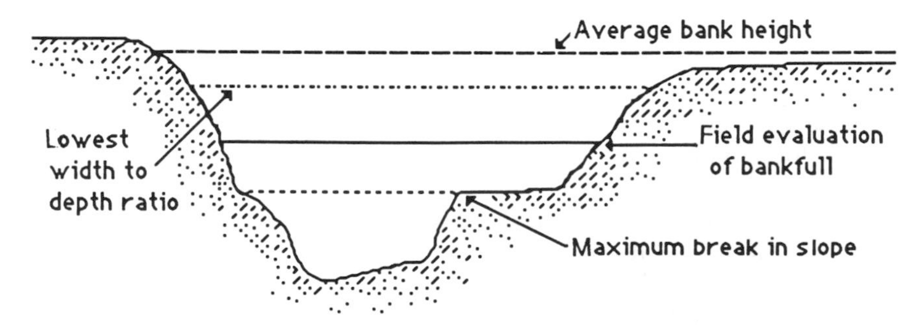

Figure 7.7. Bankfull stage as defined by various indices for a cross-section at Site 4 on the Acheron River (see Figure 4.7). Vertical scale is exaggerated

of hydraulic habitats in streams. It may also prove more useful for incised or human-modified channels. Rather than being defined by an interpretation of channel geometry, it is defined by the channel which is regularly maintained by scouring every year or so. Field identification of this "channel maintenance" stage using clues derived from scour lines and vegetation limits was used in the development of hydraulic geometry relationships for the Acheron River (see Figure 7.9). Figure 7.7 shows the interpretation of bankfull using field identification at one cross-section. From field interpretation at the gauging station, the channel maintenance discharge for the Acheron River at Taggerty is about 25.2 m^3/s, with a recurrence interval of approximately 0.3 years (partial duration series, see Figure 8.6).

Unless the cross-section has a gauging station it will usually be necessary to estimate bankfull discharge by indirect methods. Chapter 5 gives procedures for measuring bankfull width and depth and for calculating discharge by the slope–area method.

7.2.6 FLUVIAL GEOMETRY

A stream can adjust its channel dimensions to accommodate the amount of water and sediment carried. The description of cross-sections and general trends in channel geometry over the length of a stream are valuable for defining patterns of aquatic habitats. The width-to-depth ratio and hydraulic geometry relationships provide methods for quantitatively describing channel shape. Hydraulic geometry additionally describes the way in which factors vary with discharge.

Width-to-depth ratio

The **width-to-depth ratio** (*W/D*) is often used as an index of cross-sectional shape, where both width and depth are usually measured at the bankfull level (Section 7.2.5). *D* is the mean depth across the section. Examples from the Acheron River are shown in Figure 7.8.

The width-to-depth ratio generally increases in the downstream direction

Figure 7.8. Examples of bankfull cross-section from sites 1 to 5 on the Acheron River (see map of Figure 4.7). Cross-sections are plotted on the same scale, with the vertical scale exaggerated by a factor of 5

in most streams. However, it is strongly dependent on the composition of the stream banks. Channels in easily worked sand beds are wider and shallower with sloping sides. In comparison, channels carved into bedrock or silt-clay deposits are narrower and deeper with more vertical banks (Morisawa, 1985). Schumm (1977) used a silt-clay index as a surrogate for the complex influences on bank stability to develop the relationship:

$$(W/D) = 255I^{-1.08} \tag{7.6}$$

where I is the percentage of particles in the channel bed and banks less than 0.074 mm. Both (W/D) and I are dimensionless. This relationship was derived from data on channels in the Great Plains of the USA and the Riverine Plains of New South Wales, Australia. Schumm suggests that stable channels should plot close to the line described by Equation 7.6. The scatter in the original data should be taken into account rather than regarding this as a precise boundary. Aggrading streams would be wider and shallower, plotting above the line, and degrading streams would be deeper and narrower, plotting below the line. For example, at Site 4 on the Acheron River, the silt-clay index from Figure 5.32 is about 5%. From Equation 7.6 $W/D = 255(5)^{-1.08} = 44.8$. Since the actual W/D ratio is 26 (Figure 7.8), this would imply that the site may be slightly degrading.

Both depth and width can respond rapidly to changes in sediment load and/or discharge. Whether a stream erodes downwards or outwards is influenced by both local shear stresses and whether the bed or banks are the most easily eroded. Bank vegetation also increases the resistance to erosion through its binding effect on banks, with erosion decreasing as the percentage (by weight) of roots in the soil increases (Richards, 1982), and this leads to narrower channels than would otherwise be expected. The effect of vegetation on channel shape is more pronounced in smaller streams.

Hydraulic geometry

Hydraulic geometry describes the way in which channel properties change with streamflow. A stream's cross-sectional area, for example, is generally determined by the amount of water it must carry; i.e. headwater streams are smaller than the rivers into which they flow.

In a study on a large sample of rivers in the Great Plains and the Southwest area of the United States, Leopold and Maddock (1953) proposed that mean depth (D), width (W), mean velocity (V) and sediment load (Q_s) varied with discharge (Q). These interrelationships, which they termed the hydraulic geometry of streams, are described by the following equations:

$$W = aQ^b \tag{7.7}$$

$$D = cQ^f \tag{7.8}$$

$$V = kQ^m \tag{7.9}$$

$$Q_s = pQ^j \tag{7.10}$$

The coefficients a, c, k, and p, and the exponents b, f, m and j are empirically derived. The hydraulic geometry equations should be recognized as power functions. When plotted on log-log paper the slopes of the curves, described by the exponents b, f, m and j, indicate the average change of width, depth, velocity and sediment load, respectively, with changes in discharge, and do not vary with the units used. The coefficients a, c, k and p do have dimensions.

Since, by the continuity equation (Equation 6.6),

$$Q = VA = WDV \text{ (with } D \text{ the hydraulic depth, } A/W)$$

it follows from Equations 7.7 to 7.9 that:

$$WDV = aQ^b \cdot cQ^f \cdot kQ^m = Q$$

and thus:

$$b + f + m = 1, \text{ and}$$

$$ack = 1$$

Hydraulic geometry relationships can be applied to the description of how variables change with discharge (1) at a particular location ("at-a-station") or (2) over a drainage basin ("downstream").

At-a-station variations are due to the local configuration of the channel and the way in which water flows through the section. For example, velocity will increase more rapidly with discharge in narrow channels constrained by vertical cliff walls than in broad, shallow channels. In general, at higher discharges meandering or braided streams (Section 7.3.1) spread out (higher b) and straight streams speed up (higher m). Both cross-sectional shape and flow velocity tend to change abruptly when the stream begins to flow over its banks. At-a-station relationships are of interest particularly if streamflows at that section are critical to some biological behaviour such as fish migration.

Leopold and Maddock (1953) cite average at-a-station coefficients of $b = 0.26$, $f = 0.40$ and $m = 0.34$. Thus, as discharge increases at a cross-section, velocity goes up and depth increases faster than width (the width-to-depth ratio drops). These coefficients can be expected to differ considerably from reach to reach.

At a station, Equation 7.8 is the inverse of the familiar stage–discharge equation (Section 5.8.1) if stage is used rather than mean depth. Equation 7.10 is a rating curve for sediment yield, which will be discussed further in Section 7.5.4.

Downstream changes in channel geometry can be investigated by linking information from a number of sites within a stream system. This method is only valid if the discharges used for comparison are of the same average recurrence interval (see Section 8.2.2). Values of mean annual flow or bankfull discharge are commonly used, under the assumption that these flows occur at approximately the same frequency on a large number of rivers.

Using mean annual flows, Leopold and Maddock (1953) found that downstream increases in depth, width and velocity relative to discharge were similar for rivers of varying drainage basin size and setting. Average values of the hydraulic geometry exponents for the rivers studied were $b = 0.5$, $f = 0.4$ and $m = 0.1$. From these exponents it can be seen that large rivers tend to be wider and shallower than smaller streams, and that velocity increases slightly in the downstream direction. The latter conclusion may be somewhat surprising, because whitewater mountain streams give the impression of flowing faster than meandering valley streams. Although the headwater streams are steeper, the lower roughness in the valley streams due to reduced particle sizes can lead to increases in velocity (see Equation 6.6.3).

Discharge typically increases with distance downstream because of the increasing area of drainage. Thus, catchment area, as a more readily measured factor, is sometimes used in place of discharge in hydraulic geometry relationships. The downstream changes in width, depth and channel maintenance discharge with catchment area for the Acheron River system are shown in Figure 7.9, along with data from rivers in the USA.

The patterns of discharge as well as the total amount will have an influence on channel size and shape. "Flashy" rivers of semi-arid areas with quick, large peak flows may develop wider channels than those in areas where streamflow is more constant (Gregory and Walling, 1973). Park (1977) summarized and examined worldwide hydraulic geometry data from several studies. Although he found that the at-a-station exponents showed considerable scatter, the downstream exponents were more consistent and clustered near the original values of Leopold and Maddock. In some streams of humid temperate areas, negative values of the velocity exponent, m, were found, indicating that mean velocity decreased in the downstream direction. Park also pointed out that differences in the relationships could be due to the flow level used (e.g. bankfull, mean annual flow, etc.), the differences between gauging station sites and field sites, and the methods used for fitting a line through the data.

Hydraulic geometry relationships have been used for the quantitative description of riverine habitat by Hogan and Church (1989) and Kellerhalls and Church (1989). They can also be useful in studies of how land-use changes affect channel shape and size. For example, increased discharges and stream channel enlargement may accompany the urbanization of a catchment (Morisawa, 1985). Departures from general downstream trends may reveal points of impact, e.g. as a result of dams, channelization or diversions. In stream-rehabilitation work (Section 9.4) the hydraulic geometry

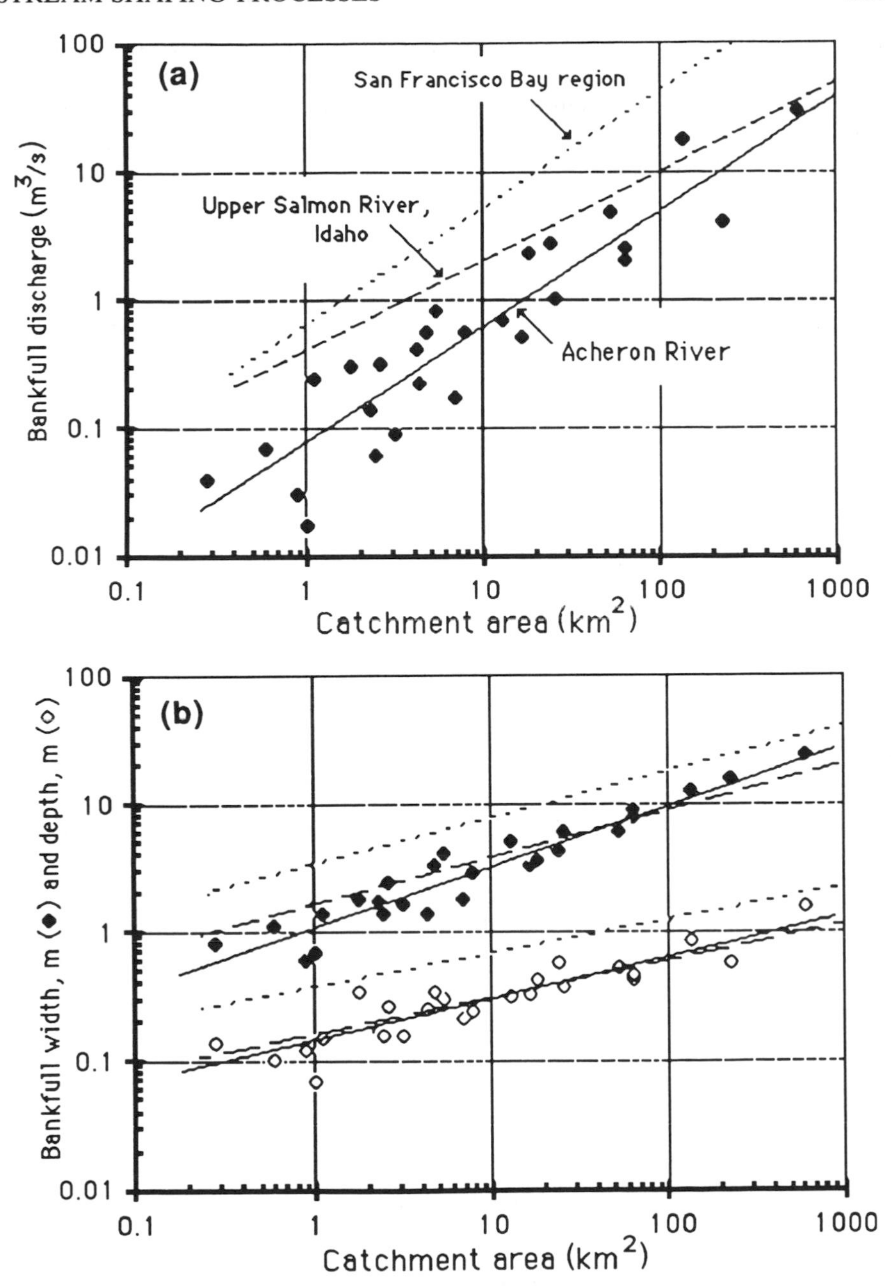

Figure 7.9. Downstream hydraulic geometry relationships for (a) bankfull discharge and (b) bankfull width and depth. Data from the 26 sites on the Acheron River (see map of Figure 4.7) are shown as individual points with a fitted (solid) regression line. Trend lines for the Upper Salmon River are from Emmett (1975), and for the San Francisco Bay region, from Dunne and Leopold (1978). Line symbols are the same in (a) and (b)

relationships from undisturbed areas can be extrapolated downstream or to other basins as a guide for reconstructing the natural geometry of degraded streams (Newbury and Gaboury, 1988). Another implication is that the curves can be used in reverse to estimate discharge from channel geometry. For example, Wolman (pers. comm., 1989) states that the mean annual flow can be reasonably estimated for ephemeral stream channels from the height of naturally formed levees.

7.3 THE INS AND OUTS OF CHANNEL TOPOGRAPHY

7.3.1 CHANNEL PATTERNS

The term "channel pattern" describes the planimetric form of streams. Channel patterns can be classified as straight, meandering, braided or anastomosing, as illustrated in Figure 7.10. These categories are an arbitrary means of classifying a continuum of forms. Variations from straight to braided can occur from one stream to another or even over the length of one stream, due to differences in geological history, stream slope, discharge patterns and sediment load. Channel patterns can also change with time; for example, streams which are braided at one flow level may meander at higher or lower stages.

In upland areas the channel pattern closely follows that of the incised valley which threads its way between hillslopes. Downstream, however, where accumulated deposits of alluvial material create wide valleys, the stream's slope is not as strongly influenced by that of the landscape. In general, a braided pattern tends to coincide with high slopes and high stream power and a meandering one with lower slopes and lower stream power. Braided streams are also associated with coarse bed and bank materials and the movement of sediment as bedload. In comparison, meandering streams tend to have more cohesive bed and bank materials and suspended sediment loads.

The channel patterns are distinguished on the basis of channel multiplicity and sinuosity. **Sinuosity** is a measure of the "wiggliness" of a watercourse, and has a number of definitions. The most commonly used measure is the sinuosity index (SI), given as:

$$\mathrm{SI} = \frac{\text{Channel (thalweg) distance}}{\text{Downvalley distance}} \tag{7.11}$$

The SI is normally computed from measurements of stream and valley lengths taken from maps or aerial photographs. The reach length should be at least twenty times the average width of the channel (Bell and Vorst, 1981). Stream length is measured by methods described in Section 4.2.3. "Valley length" presents some difficulty, since, by strict definition, streams

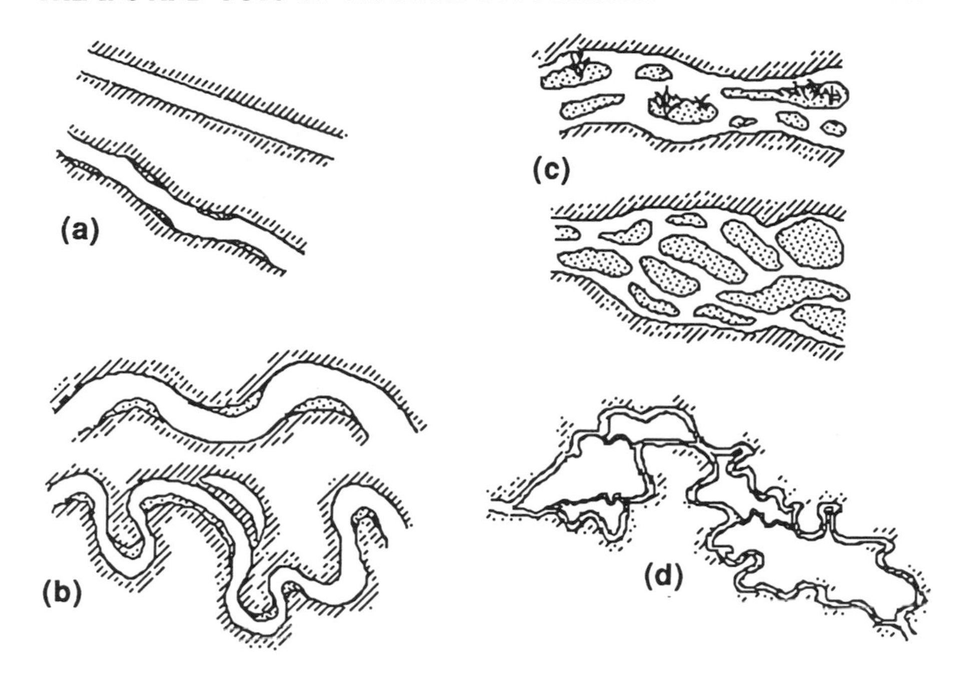

Figure 7.10. Channel patterns: (a) straight, (b) meandering, (c) braided, and (d) anastomosing

tightly confined within V-shaped valleys have the same length as the valley, giving an SI of 1 whether or not the streams appear "straight" to the eye. In practice, straight-line segments which follow the broad-scale changes in channel direction can be used as a measure of valley length. For the section of the Acheron River in Figure 3.4 upstream of the junction with the Little Steavenson to the edge of the map, SI ≈ 154 mm/115 mm = 1.34. The fractal dimension (Section 4.2.3) may be a more consistent measure of a river's "crinkliness" as it is independent of map scale; however, it is more difficult to compute than SI.

In straight streams, SI = 1, whereas a value of 4.0 is considered to be highly intricate meandering. Meandering streams are somewhat arbitrarily defined as those with an SI value of 1.5 or more. The term **sinuous** is sometimes given to stream patterns which are intermediate between straight and meandering. A description of channel patterns based on their sinuosity, bank characteristics, sediment loads, relative dimensions, bankfull velocity and stream power is given in Table 7.1 (pp. 316–17) and the relationship between channel form, slope and sediment load shown in Figure 7.11.

The channel pattern represents an adjustment of shape in the horizontal plane, and is one of the variables which can be modified by a stream to improve the efficiency with which it conveys water and sediment. Changes

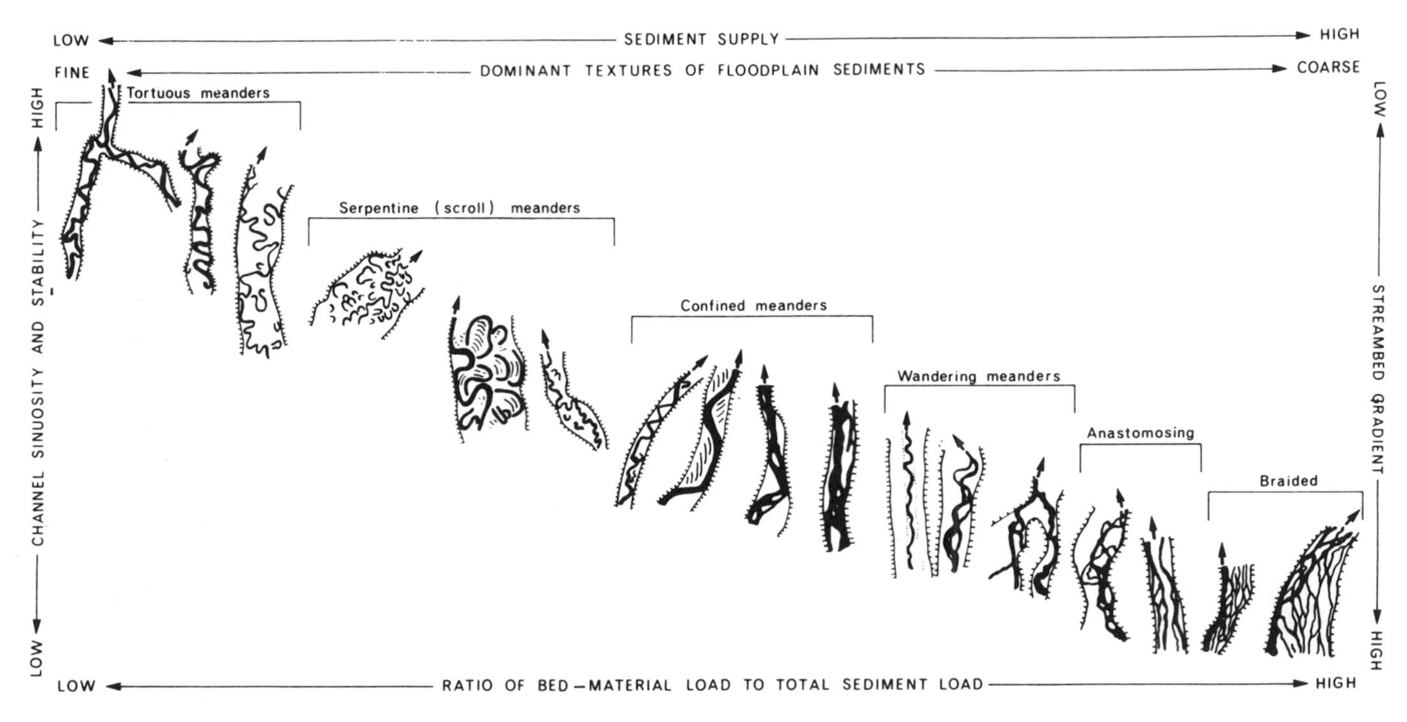

Figure 7.11. The relationship between the form and gradient of alluvial channels and the type, supply and dominant textures (particle sizes) of sediments. Reprinted from Selby (1985), © M. J. Selby, 1985, by permission of Oxford University Press

in slope, and type and amount of sediment load can lead to variations in channel patterns and thus stream habitats. Thresholds may also exist, meaning that in one reach a large change in slope and/or sediment load may have little effect on the channel pattern, but if the reach is close to some threshold level, a slight change in slope or sediment load can have "striking repercussions" (Gregory and Walling, 1973). Channels may also change from non-braided to braided in association with alterations from dense riparian vegetation to sparsely vegetated banks (Leopold et al., 1964).

7.3.2 POOLS AND RIFFLES

The words "pool" and "riffle" immediately bring to mind an image of the features they represent, especially to trout anglers, from whom the terms originated. Loosely defined, a **pool** is a region of deeper, slower-moving water with fine bed materials, whereas a **riffle** has coarser bed materials and shallower, faster-moving water, often associated with whitewater. At riffles, the cross-sectional profile is more rectangular, whereas pools have more asymmetric profiles. The term **run** is sometimes given to an intermediate category in which the flow is less turbulent than in riffles but moves faster than in pools. Pools and riffles alternate in a "pseudo-cyclic" manner (Knighton, 1984), with the depth of the pool controlled by the elevation of the riffle just downstream (see Figure 7.12). They can be thought of as "vertical meanders". This **pool–riffle** periodicity may be important in the cycling of nutrients along a stream (Goldman and Horne, 1983).

Riffles tend to support higher densities of benthic invertebrates, and are thus important food-producing areas for fish. Due to competition and predation as well as size limitations, young fish and small fish tend to inhabit riffles. Since riffle areas are first affected by reduced discharges, Bovee (1974) suggests that riffle-inhabiting species should be used as indicators in determining low-flow requirements of streams (Section 9.3). Deeper pools with overhanging banks and vegetation support larger fish. During low flows these pools can become isolated pockets of water which allow the survival of aquatic organisms. The most productive streams have a combination of pool sizes (Hamilton and Bergersen, 1984). In terms of physical habitat, the pool–riffle structure provides a great diversity of bedforms, substrate materials and local velocities. Brussock et al. (1985) propose that the reason biotic diversity is greatest in mid-reaches is because they typically possess a pool–riffle morphology.

Pools and riffles are fairly easily distinguished by eye at low flow. However, where one begins and the other leaves off becomes a matter of interpretation. A close-up survey of the longitudinal profile (see Figures 4.14 and 5.11) can be divided at sharp breaks in slope to distinguish the forms. A plan view may be helpful in distinguishing changes not only in bed material between riffles and pools but also in width, since riffle areas

Table 7.1. Classification of channel patterns. Based in part on Leopold et al. (1964), Morisawa (1985) and Selby (1985)

Stream type	Description	Width-to-depth ratio	Sinuosity	Bankfull velocity (m/s)	Stream power
Straight	Single channel with meandering thalweg. Well-defined banks, often containing bedrock. Channel is typically stable, with minor widening or incision. Sediment load is suspended and/or bedload; load is usually small in comparison to transport capacity. Cross-sections tend to have marked central "hump". Found in short reaches (also channelized streams).	Low <40	Low 1.0–1.5	High >3	High
Meandering	Single winding channel, usually with well-defined banks. Channel shifts mainly due to erosion by undercutting on outside of bends, causing the outward growth and downvalley migration of meanders. Pools form in this region of high velocity and turbulence. "Cutoffs" can occur across the base of a meander loop, leaving crescent-shaped "oxbow lakes" (billabongs),	Low <40	Moderate to high 1.5–4.0	Low to moderate 1–3	Low to moderate

	which support an ecology different from that of the river. Point bars of sand and gravel form on the inside of bends. Sediment load is mainly suspended load, approximately balanced with transport capacity. Meander wavelength, λ, averages ten to fourteen times channel width.				
Braided	Multiple channels with bars and islands, often with poorly defined banks of non-cohesive materials. Channel and bank erosion follows a fairly random pattern. Flow is concentrated into flanking chutes, increasing velocity and sediment transport capacity. Sediment load is primarily bedload, and load is large relative to transport capacity. Found in glacial streams, alluvial fans and deltas.	High >40	Low to moderate 1.0–2.0	Varied, depending on slope and straightness of individual channels.	All ranges, from high in straight streams to low in sinuous streams with islands, channel bars, or deltas.
Anastomosing	Multiple channels with relatively permanent, stable vegetated islands, in comparison to braided streams where channelways are constantly shifting. Banks are cohesive, and sediment load is primarily suspended load.	Low <10	Varied	Varied	Moderate

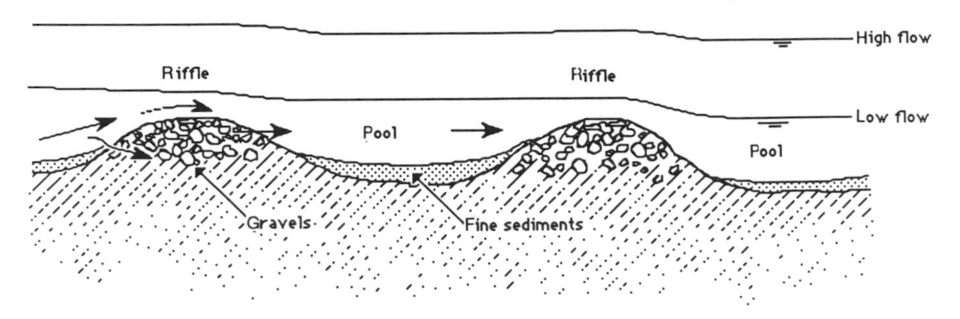

Figure 7.12. Pool–riffle sequences at low and high flow

tend to be about 12% wider than pools on average (Richards, 1982). Richards (1976) defines riffles and pools more formally as positive and negative residuals, respectively, from a regression line fitted through the bed profile. Lisle (1987) proposed the use of "residual depths" to define the extent of pools from a horizontal line through the crest of the downstream riffle. Alternatively, Bren (pers. comm., 1990) suggests using Froude numbers to distinguish between pools ($Fr < 0.1$) and non-pools ($Fr > 0.1$), a definition which will change with flow levels.

The alternating pool–riffle bedform is most common in streams with mixed bed materials ranging from pea to watermelon size (2–256 mm) (Knighton, 1984). Jowett and Duncan (1990) demonstrate that it is more pronounced in streams with high flow variability. Pool–riffle sequences are often found in meandering streams, where pools occur at meander bends and riffles at crossover stretches. As would be expected from the average meander wavelength (Table 7.1), the deeps and shallows both follow a more or less regular spacing of five to seven channel widths (Leopold et al., 1964). "Noise" in the rhythmic patterns can result from local controls such as bedrock constrictions or tree roots, or an increased supply of coarse sediments. This semi-regular pattern is also found in straight channels, bedrock streams and the dried remains of semi-arid ephemeral streambeds.

In steep, boulder-bed mountain streams the pool–riffle sequence is replaced by a **pool–step sequence**, where water tumbles over accumulations of boulders as short waterfalls plunging into small scour pools. "Organic" steps or riffles can also be created by large fallen trees or debris dams. These debris-created steps also tend to be regularly spaced (Morisawa, 1985).

The pool–riffle bedform is considered a means of self-adjustment in gravel-bed streams which acts to regulate energy expenditure. In meandering reaches, energy loss is high at channel bends because of the curvature. This may be balanced by energy losses to turbulence in the straight riffled sections where roughness is greater.

Falls, riffles and pools become more important at low flows when they become a more dominant component of channel geometry. At low flows, the sites immediately downstream of riffles are locations of intense energy loss, and energy loss becomes concentrated over a very small percentage of the stream's length. For example, the riffles in the Grand Canyon, Arizona, which provide thrills and spills for river rafters, concentrate most of the fall over only 10% of the downstream distance (Leopold et al., 1964).

At high flows, energy loss becomes more uniform, where such uniformity may be needed to move the sediment load (Morisawa, 1985). As kayakers are well aware, the water surface slope, depth of flow and speed of the current become more uniform over the stream reach at high flows (see Figure 7.12). At these times, it becomes questionable whether the terms "pool" and "riffle" are even applicable. As discharge rises, velocity and depth increase more rapidly in pools than in riffles. The shear stress in pools can eventually exceed that in riffles, which may be part of a sorting mechanism for concentrating coarser materials in riffles (Knighton, 1984). Sediment movement from riffles is postponed until very high flows occur, at which time the coarse particles move from one riffle to another. Very coarse fragments, however, will still tend to collect in the deepest part of pools.

Of significance to the resident flora and fauna, the pool–riffle bedforms remain relatively fixed in location, unlike sand bars and dunes (Section 7.3.4) which tend to migrate. Pool–riffle structures are usually formed by rare, large historic events, and pool–step systems may relate to even rarer, higher-intensity discharges (Petts and Foster, 1985). Thus, the flows which form and maintain these structures are different, and they remain relatively stable under all but extreme flow conditions (Knighton, 1984). Pools may deepen as a result of localized scour during low to moderate flows, especially at bends or at the downstream side of logs (Beschta and Platts, 1986). They may also fill with sediment if the sediment supply increases.

7.3.3 BARS

Bars are fairly large bedform features created by the deposition of sediments. Whereas submerged sand bars can pose a threat to the unskilled river navigator, exposed bars of sand or gravel can become stabilized by vegetation to create island refuges for migrating waterfowl. Like pool–riffle structures, bars also tend to be formed at higher discharges and then remain in place to define the path of low flows. They have a variety of shapes, and can be composed of a wide range of grain sizes. Bars can be classified by their location in the stream as shown in Figure 7.13 and described by Knighton (1984) as follows:

Point bars primarily form on the inner bank of meanders and often create sandy beaches which slope gradually into the water.

Alternate bars occur periodically first along one bank and then along the opposite one, with a winding thalweg running between the bars. These can form in relatively straight sections of sand-bed streams, creating a meandering pattern at low flow.

Channel junction bars develop where tributaries enter a main channel.

Transverse bars cross the width of the stream (for example, at riffles) often at an angle diagonal to the flow. In sand-bed streams these tend to be flat-topped and covered with smaller bedforms such as ripples (Section 7.3.4).

Mid-channel bars are characteristic of braided reaches, often existing as diamond- or lozenge-shaped gravel mounds. These are aligned with the flow, separating it into smaller rivulets. Mid-channel bars tend to grow from the downstream end. They typically have coarser materials on the upstream side and finer materials on the downstream one.

Bars are stable features in many locations, with erosion equalling deposition during floods. Even when bars change constantly in form or location, as in braided streams, the amount of area covered can remain fairly constant. The removal of bars may actually lead to instability and an attempt by the river to "heal" itself. Hooke (1986), for example, observed that bars and shoals removed from the River Dane in England redeveloped in two to three years.

In a gravel-bed stream in Northern California, Lisle (1986) found that bars tended to form three to four bed-widths downstream and one bed-width upstream of large obstructions and bends. The downstream bars typically formed on the same side of the stream as the obstructions. He proposed that "non-alluvial boundaries" of woody debris, bedrock, fabricated material or root-defended bank promontories actually stabilized the locations of gravel bars and pools by affecting downstream secondary currents and through backwater effects. Thus, the introduction of large obstructions into a stream can be expected to modify the channel shape both upstream and downstream.

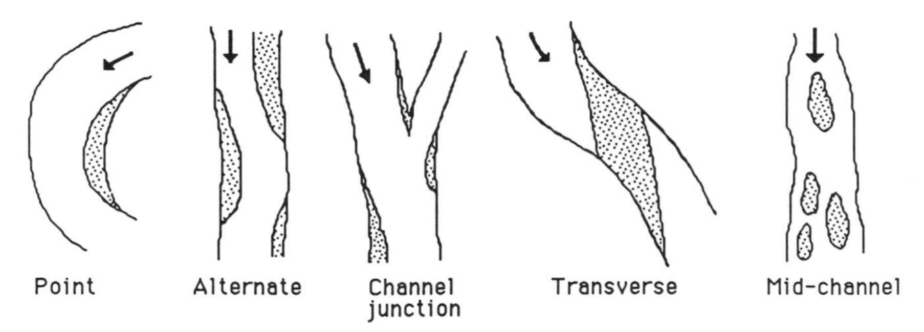

Figure 7.13. Classification of bars (see text)

7.3.4 DUNES, RIPPLES AND FLAT BEDS OF SAND

Sand-bed streams present a "Sahara-like" landscape to the benthic invertebrates and aquatic plants seeking a foothold. Because of the smaller, more uniform grain sizes, sand beds are highly mobile and readily moulded into different bedform shapes under the sculpting effects of flow patterns. In turn, the bedforms provide feedback through form resistance to affect local velocities, shear stresses and sediment transport. Local scouring and deposition within troughs and across crests causes the bedforms to be somewhat self-perpetuating.

From extensive studies in both flumes and natural channels a fairly predictable sequence of bedforms has been found to occur as stream power increases. The forms are divided into two categories, based on a Froude number less than or greater than one: the lower flow and the upper flow regimes. In the **lower flow regime** sediment transport is relatively low and flow resistance high due to the large separation zone behind the crests of dunes or ripples. In the **upper flow regime** resistance is reduced and sediment transport is increased. The progression of bedforms typically follows the sequence given as follows and shown in Figure 7.14. Forms illustrated in Figure 7.14 but not listed are transition or intermediate phases in which smaller forms can be superimposed on larger ones.

0 *Initial flat bed:* Flat sand beds are an oddity outside laboratory environments, and finding one in a natural stream is unlikely. For flumes and theoretical discussions it serves as a starting point. When the water velocity is increased, sand grains begin moving singly at first, then in patches. As the grains move together over the streambed, particles tend to accumulate in clusters. Then, suddenly, the clusters of particles orient in a series of regular waves and hollows longitudinal to the flow (Hjulstrom, 1939).

1 *Ripples*: Ripples are small corrugations in the bed with relatively sharp crests. They form under **hydraulically smooth** conditions. The more fine-grained the sediment, the more well-developed the ripple (Hjulstrom, 1939). The distance between ripples is fairly uniform. This ripple "wavelength" is dependent on particle size and is independent of flow depth (Smith, 1975). It typically ranges from 150 to 450 mm (Simons and Richardson, 1961).

2–3 *Dunes*: If the flow is **hydraulically rough**, dunes will form rather than ripples. Dunes have rounded crests and are larger than ripples. Their dimensions are related to flow depth and are only slightly dependent on particle size (Smith, 1975). Dunes enlarge to a point where further growth is impossible because of high velocities and transport rates at the crest (Leopold et al., 1964). In the Mississippi River, this type of dune is represented by long sand bars up to hundreds of metres long.

Both ripples and dunes migrate *downstream* as sand grains move up the more gradually sloped upstream side and fall down the steeper downstream face. Sand deposited on the upstream face is closely packed whereas the material on the downstream side is unstable. Researchers who wade in sand-bed streams will have an appreciation of the denseness of the upstream sides of dunes which will support their weight, as compared to the quicksand-like material on the lee side.

4 *Transition zone*: As stream power increases and the Froude number approaches 1, ripples and dunes are washed out. This constitutes the transition from lower to upper regime.

5 *Plane bed*: In beds of finer sediments (< 0.4 mm) a plane bed develops. In comparison to the initial flat bed, at this step the bed and fluid have less distinct boundaries and the high sediment transport creates a dust storm-like environment (Leopold et al., 1964). Suspension of sediments further decreases flow resistance by dampening turbulence and thus reducing energy loss (Smith, 1975).

6 *Standing waves*: With larger sediment sizes, standing waves develop rather than a plane bed (Simons and Richardson, 1961). Standing waves are those in which the water surface and bed surface profiles are synchronized and both sand and water waves are stationary. These can begin forming at a Froude number (Fr) of about 0.84.

7–8 *Antidunes*: At this step, velocities and sediment transport are both high. Waves form on the water surface, accompanied by the formation of rounded bed waves called antidunes. These are extremely unstable and constantly form, disintegrate and re-form. They progressively move *upstream* as sand erodes from the downstream side of a dune and deposits out on the upstream face of the next dune downstream. Antidune wavelengths range from 230 mm to 6 m (Hjulstrom, 1939), and are approximately twice the depth of the water. They have been observed in natural streams with beds of fine sand to coarse gravel (Simons and Richardson, 1961).

8 *Chutes and pools*: At even higher stream powers, the bed rearranges itself to create a series of hydraulic jumps for energy dissipation. These high-Fr situations may occur when flash floods sweep down steep sandy gullies of semi-arid regions. Large quantities of sediment are suspended in the breaking wave. WMO (1980) gives a method for computing discharge from the distance and time between these "translatory waves".

Bedform shape acts as another type of self-regulating mechanism in streams, ensuring efficient transport of both water and sediment. In less-uniform sands, armouring (Section 7.4.2) can stabilize bedform size and shape until the flow is great enough to remove the coarse surface layer.

As indicated in Figure 7.15, flow resistance increases through the lower

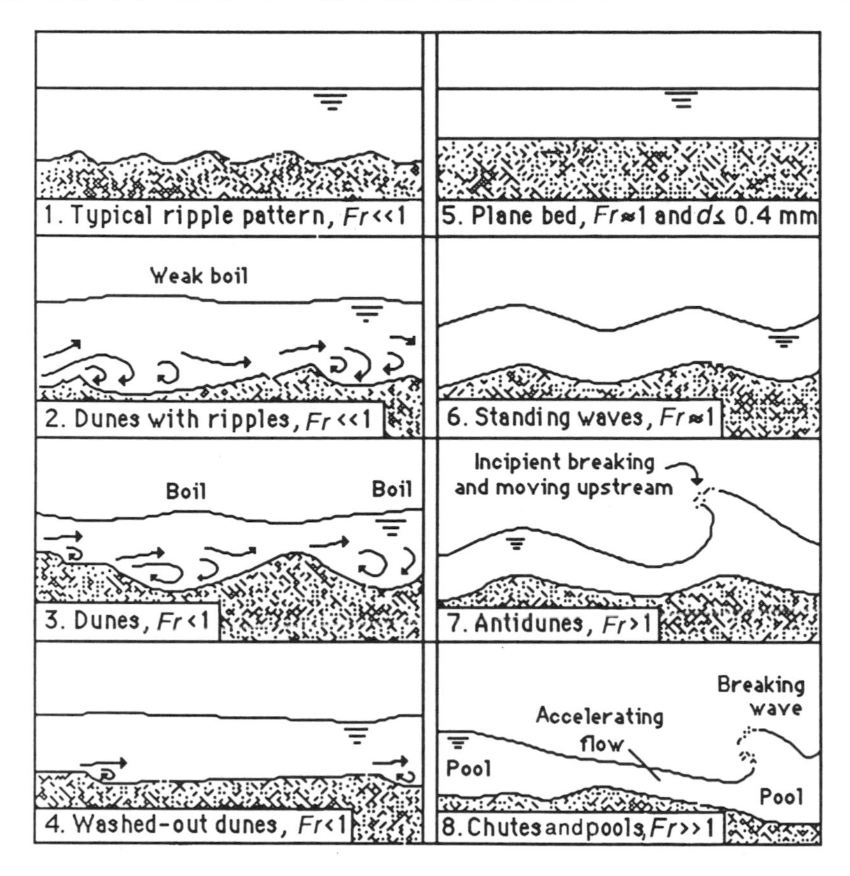

Figure 7.14. Progressive changes in bedform shape. Adapted from Simons and Richardson (1961), by permission of the American Society of Civil Engineers. *Fr* = Froude number

regime until dune formation, then decreases through the transition to a plane bed, increasing again in the upper regime. Thus, in a sand-bed river, Manning's *n* (Sections 5.8.2 and 6.6.5) can vary fourfold simply as a result of changes in bedforms (Schumm, 1977). Ripples and dunes slow the increase in velocity until, at higher stages, the dunes wash out, producing a more efficient channel and possibly reducing the height of flood peaks (Schumm, 1977). Bedforms, therefore, can wreak havoc with at-a-station hydraulic geometry relations (Section 7.2.6) such as the relationship between depth and discharge.

Although somewhat unpredictable, the self-adusting mechanisms of streams have fascinating implications for those studying habitat dynamics and the movement of water, sediments, nutrients and biological inhabitants of streams.

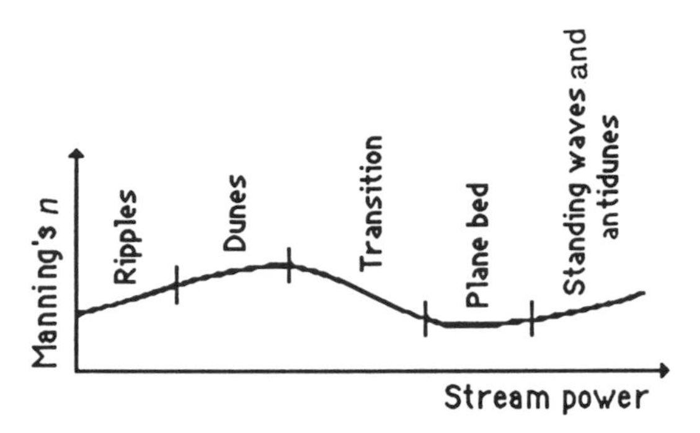

Figure 7.15. The variation of flow resistance with bedform. After CSU (1977)

7.4 SEDIMENT MOTION

7.4.1 EROSION, TRANSPORT AND DEPOSITION

Sediment motion consists of three stages: initiation of motion, downstream transport, and deposition. At the extremes of particles size, some of the finer sediments may never settle out, and large boulders may never get off the ground. For the large majority of particles, however, motion consists of periods of rest alternating with periods of activity, and any given particle will be eroded, transported and deposited many times *en route* to some accumulation point. The rests, stops and starts of an individual particle have a random nature, but as with fluid motion, there are techniques for addressing the overall movement by relating it to easily measured, average flow properties.

Since most of the stream's energy goes into overcoming frictional resistance along its bed and banks, a critical energy level must be reached before a stream can erode or transport channel materials. Thus, the concept of some "threshold" or "critical" value is fundamental to studies of particle motion. Several thresholds may exist for a particle of a given size and shape: a bed-erosion threshold at which the particle just begins to move; a "lift-off" threshold which must be exceeded to suspend the particle; and another "settling" threshold, below which the particle will drop out of suspension. Threshold values can be specified by a number of hydrodynamic factors such as velocity, shear stress and stream power.

At low discharges, only washload will be transported. When disharge is increased to a certain level, particles on the bed and/or banks will begin to **erode**. Observations of particle movement in both flume and field situations indicate that particles begin to vibrate as the flow intensity is increased. Lighter organic materials such as leaves and bark will be transported first. Initially, a few grains are entrained, then whole patches of surface material are lifted off into the flow. The forces which "pluck" a particle from the streambed and keep it aloft are mainly due to upward surges of water—"gusts" from small, turbulent eddies.

Sediment transport is considered to occur once the threshold for movement has been crossed. Larger particles tend to roll in the direction of the bed slope, whereas fines usually follow a longer path dictated by the spiralling movement of the fluid. Eventually, an equilibrium is reached between the amount of sediment and the energy available to carry it. **Flow competence** refers to the maximum particle size which can be carried at a given flow state. *Apparent* competence, or the maximum particle size actually carried by a stream, may be less than the *true* competence if the largest transportable particle sizes are not available.

Numerous approaches to the problem of estimating sediment transport rates have been taken. Most are based on some relationship between the amount of sediment carried and the difference between "actual" and "critical" values of some factor such as stream discharge, shear stress, mean flow velocity or stream power. Some equations are physically based; some probabilistic in an attempt to account for the intermittency of particle movement; but all are empirical to some degree. Because of the extreme complexity of the problem due to channel configurations, velocity variations, and mixtures of particle sizes and packing, the performance of most sediment discharge equations "does not inspire confidence in results based on them" (Dawdy and Vanoni, 1986, p. 73S). If sediment transport equations are used, they should ideally be calibrated with field data and not used outside the range of calibration. Some of the fluvial geomorphology texts listed in Section 7.1.1, and texts on sediment transport such as Graf (1971), Raudkivi (1967), Thorne et al. (1987) and Yalin (1972) will provide further information on the subject.

A stream basically carries its load until it lacks the energy to do so, at which time **deposition** takes place. For bedload materials, deposition occurs when the material stops rolling, sliding or hopping; for suspended material, it happens when the material settles out of suspension. In still water, a particle will settle out at a rate dependent on its terminal velocity (Section 6.5.4).

Briggs (1977a) describes three major processes which lead to deposition:

(1) The orientation of the particle in the flow changes, increasing resistance. The lift and drag on a non-spherical particle will change with its orientation. Like an aircraft wing, a particle can "stall" at a certain angle. The forces acting on the particle will also affect its orientation on the bed (see **imbrication**, this section).
(2) The competence of the stream decreases, meaning that the energy available for transporting sediments decreases. Suspended sands and gravels settle out when the streamflow drops, and when local velocities

decrease, such as at pools, the insides of meander bends, tributary confluences, canyon mouths or other widenings, or the downstream side of gravel bars or islands, sand dunes, rocks or tree trunks. At ponds and reservoirs, the coarser particles deposit out at the upstream end whereas finer particles move farther downstream. Larger bedload materials may cluster together at riffles or steps.

(3) The quantity or size of the sediment load suddenly increases. This type of deposition may occur if a stream suddenly receives an influx of sediment from a landslide or bank collapse. Particles will remain oriented as deposited until they are re-worked by the stream.

The "size selectivity" of deposition helps to create and shape large and small bedforms. These, in turn, affect local flow resistance, velocities, and turbulence which encourage further erosion or deposition. Their interaction is part of the self-regulating feedback which enables a stream to adjust to changing discharges and sediment loads.

7.4.2 DEVIATIONS FROM "IDEAL"

The picture presented in the previous section is muddied by the great variability and unpredictability of sediment motion. In natural streams, bed materials are made up of a conglomeration of particle sizes, especially in cobble or gravel-bed streams and rocky headwater reaches. The mixture of particle sizes, their arrangement and the amount in suspension can have significant effects on local velocities and thus sediment transport.

Concentration of suspended materials

The velocity required to lift and transport sediment can be affected by the amount of sediment already in suspension. Suspended fines tend to reduce the turbulence and thus lessen the eroding power of the flowing water (Hjulstrom, 1939). For example, Bayly and Williams (1973) provide data which show that the flow velocity required to entrain clays in muddy water is about 1.7 times that for clean water. However, because suspended fines also increase the density and viscosity of the water, coarser particles will not settle out as quickly in muddy water as they would in clean water.

In contrast, larger materials such as suspended sands or rock fragments may increase erosion if they are hurled against banks composed of fine materials such as clay, loam or volcanic ash. An additional consideration is the effect of suspended materials on flow resistance. Since turbulence is decreased, flow resistance is reduced and thus higher velocities and sediment movement are possible. However, Parker (1982) found that high concentrations of suspended sand in gravel-bed streams seemed to suppress erosion. The effect, therefore, will depend both on the type and concentration of materials in suspension and the composition of the streambed (see also Table 5.5).

Mixtures of particle sizes

The mixture of particle sizes found in natural streams creates interactions which would not occur if materials were of a uniform size. One example is the scouring of sand hollows around the base of boulders by horseshoe vortices (Section 6.5.4). Interactions also affect the ease with which an individual particle will be eroded and transported. The filling of interstices between larger particles can act to "cement" them in place. Particles which are partly buried or in the wake of larger particles or bedforms will move less readily than isolated particles sitting on top of a flat streambed. Smaller particles can also "hide" in the lower-velocity region of the boundary layer, whereas larger ones will protrude farther into the velocity profile (Section 6.4.2).

Recent work by Parker (1982) and Andrews (1983) in self-formed rivers with naturally sorted gravel and cobble bed material has indicated that mixed-size particles move over a much narrower range of discharge than expected. This has led to the controversial theory of "equal mobility", which says that nearly all of the grain sizes begin moving at nearly the same discharge. The theory does not imply that the entire bed surface begins moving at the same time. Instead, at any instant, the sediment load may consist of particles over a range of sizes, and the bed "selectively unravels" from different locations as discharge increases (Prestegaard, 1989).

Thus, the movement of coarse particles probably occurs more frequently than is commonly assumed. Andrews (1983), for example, found that in nine Colorado rivers, particles as large as the d_{90} of the bed material were entrained by bankfull discharges.

Armouring

Armouring is the development of a surface layer which is coarser than the bed material beneath it (Figure 7.16a). It "protects" the finer materials underneath, which are not mobilized until the armour layer is removed. Armouring may or may not occur in streams. If present, it may take place over the whole bed or only in patches where scour is greatest. In some areas, thick, erosion-resistant streambeds can develop which are rearranged only during extreme floods.

Explanations for the mechanics of armouring are subject to controversy. One is that the layer results from the "winnowing away" of finer materials from the surface layer. On arid lands, wind and surface wash on upland slopes may act in a similar manner to leave a protective gravel layer called "desert pavement". Some armour layers may result from the accumulation of large materials rather than the removal of fines. However, there seems to be an inherent tendency for large particles to find their way to the surface. Leopold et al. (1964) give the example of a truck dumping dry gravel, where the largest particles roll across the smaller materials and down the outside of the pile.

Once a channel is armoured, the subsurface materials are protected from erosion until the armour layer is broken up. Dawdy and Vanoni (1986) cite recent studies which propose that armouring causes a restructuring of the sediment transported, acting as a type of "regulator" to enable a stream to transport more of its coarse sediments. The few sites stabilized by armouring may be critical to the stability of the entire channel.

An index of the strength of an armour layer (A) can be defined as the ratio of the armour layer grain size to the subsurface grain size:

$$A = \frac{d_{\mathrm{a}}}{d_{\mathrm{sub}}} \tag{7.12}$$

where, commonly, d_{a} is the d_{50} of the surface, armoured layer, and d_{sub} the d_{50} of the subsurface bed materials. The index, A, is typically between 1.5 and 3 in gravel-bed streams (Parker, 1982). The influence of armouring should be considered not only in predictions of sediment movement but also in the collection of bed-material samples.

Imbrication

The orientation and angle of deposited particles depends on the particle shape and the forces affecting the particles at the time of deposition. If deposition is caused by a sudden fall in the stream's competence, particles tend to be deposited in their position of transport. In fairly steady flow, deposited particles tend to fall into a more stable position. For example, rod-shaped particles will line up with their long axes parallel to the direction of flow. Many gravelly stream deposits, particularly those with disc-shaped pebbles, will show particle **imbrication**, where particles are stacked against each other, nose-down into the oncoming current (Figure 7.16(b)). Briggs (1977) states that imbrication may be a position of maximum resistance to movement for large bedload sediments.

7.4.3 PREDICTING A PARTICLE'S "GET UP AND GO"

In studies of aquatic systems it may be of interest to predict the flow level which will shift sediments, upon or under which benthic organisms reside. For streambeds which have become filled with fines the same principles can be used to calculate the "flushing flow" required to remove sediments from

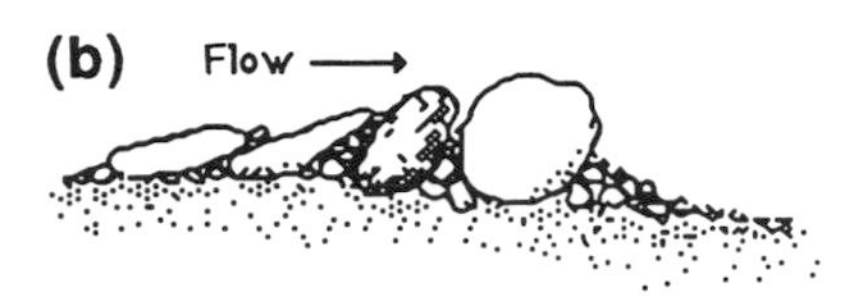

Figure 7.16. Arrangement of surface bed materials: (a) armour layer over finer subsurface materials and (b) imbrication of disc-shaped particles

the surface of the bed or from beneath the armour layer (see Section 9.3.5). The magnitudes of past floods are also sometimes reconstructed using the maximum size of deposited rocks. Gregory (1983) is a source of information on these "paleohydrological" techniques.

As mentioned in Section 7.4.1, prediction of sediment entrainment is usually based on some "critical" state, above which particles begin to move. The initiation of motion of a sediment particle can be described by a variety of factors: (1) lift and drag forces, (2) a critical velocity, or (3) a critical shear stress. Critical stream power has also been used in sediment transport equations by Bagnold (1980) and Yang (1973), but will not be discussed because it is not normally used for the application presented here. These are all slightly different but related ways of looking at the same phenomenon.

Because of the intermittency with which particles are lifted into the flow, the visual detection and interpretation of the "critical" state is highly subjective. In some approaches this threshold is treated as a statistical property, and methods of extrapolating back to "zero transport" have been employed. Most of the background work has been done in flumes with grains of uniform size. Adjustments must be made to apply the techniques to the mixed particle sizes and varying flow patterns of natural streams.

Lift and drag forces

Lift and drag forces (Chapter 6) act on a sediment particle when differences in pressure and velocity exist from top to bottom or front to back of the grain (Figure 7.17). These forces tend to jostle the particle in place, or, if the forces are strong enough, they can start it rolling or lift it into the flow. The Bernoulli-type concept of lift (Section 6.3.4), however, probably has less bearing on particle movement than the instantaneous upward velocity components of turbulent flow. Because of the difficulty in modelling the effect of turbulence, other approaches using averages are favoured for predicting particle movement.

Critical velocity

Hjulstrom (1939) developed the graph shown in Figure 7.18, which relates average velocity to particle size. The curves show the limiting velocities for

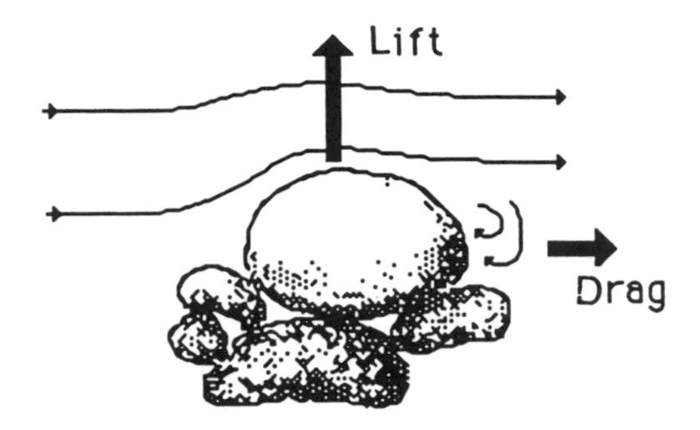

Figure 7.17. Lift and drag forces on a sediment particle

the three states of erosion, transportation and deposition. As illustrated, fine to medium sand between 0.3 and 0.6 mm is the easiest to erode. For larger particles the erosion velocity increases as it becomes more difficult to lift the heavier grains. Finer materials present a smoother profile to the flow and are thus less affected by turbulence. However, in silts and clays, a cohesive electrochemical force binds the finer particles together, making erosion more difficult. Clays tend to erode as aggregates rather than as individual particles. Unconsolidated silts and clays, however, may erode at lower velocities than indicated by the curve.

The Hjulstrom curves demonstrate that the velocity needed to entrain a

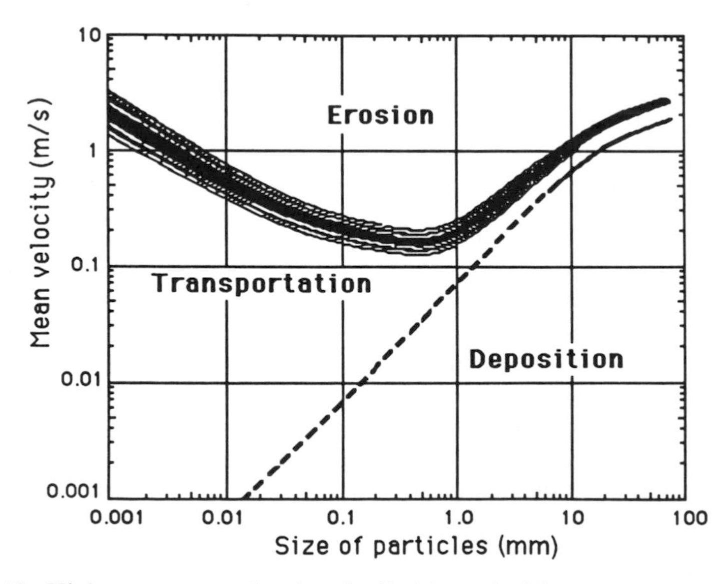

Figure 7.18. Hjulstrom curves showing the limiting velocities required for erosion, transportation and deposition of uniform material. Adapted from Hjulstrom (1939), by permission of the American Association of Petroleum Geologists

particle is greater than that required to keep it moving. Once set into motion, particles continue to be transported until the velocity drops below some speed, indicated by the dotted line in Figure 7.18. Fine materials will remain in suspension at very small velocities as long as there is sufficient turbulence. The difference between erosion and deposition velocities is much less for larger particles; thus the velocities setting them into motion must be maintained or the particles will drop out again. This is consistent with the concept that larger materials tend to travel as bedload (Section 7.14).

A limitation of the Hjulstrom curves is that they were developed from data collected on sediments of uniform grain size and streams of depths greater than 1 m (Hjulstrom, 1939). In reality, a large range of particle sizes will move at any given average stream velocity depending on local currents and particle characteristics. For example, Smith (1975) states that the critical velocity of organic matter with a density of 1050 kg/m^3 is about 1/6 that of an equivalent-sized mineral particle. The Hjulstrom curves are, however, useful for general estimates of sediment entrainment.

In the development of a model for predicting the amount of usable habitat in a stream, Jowett (1989) states that the suitability of a streambed for instream flora and fauna depends on its stability. He defines relative bed stability (RBS) as the ratio of the critical velocity required to just move a particle (V_c) to the actual or predicted water velocity near the bed (V_b):

$$\text{Relative bed stability (RBS)} = \frac{V_c}{V_b} \tag{7.13}$$

with both V_c and V_b in m/s. V_c can be obtained from the Hjulstrom curves. Additionally, the US Bureau of Reclamation (USBR, 1977) gives simple equations for both critical velocity and bed velocity, presented in SI units as:

$$V_c = 0.155\sqrt{d} \tag{7.14}$$

where d is the average particle diameter (mm) and

$$V_b = 0.7V \tag{7.15}$$

with V the mean stream velocity (m/s). Equation 7.14 is considered applicable for uniform particles of sizes greater than 1 mm. In place of Equation 7.15, Jowett also uses a variation on the logarithmic velocity profile (Section 6.5.2) with the assumption that V_b occurs at a distance of 0.01 m above the bed.

Example 7.1

Calculate the relative bed stability (RBS) for Site 3 on the Acheron River (Figure 4.7) at bankfull discharge, if average bankfull velocity is estimated as 0.37 m/s and the relevant particle sizes are those in Figure 5.33.

Answer: From Figure 5.33, $d_{50} \approx 150$ mm. Thus, from Equation 7.14,

$$V_c = 0.155\sqrt{150} = 1.90 \text{ m/s}$$

and from Equation 7.15 for bankfull conditions,

$$V_b = 0.7(0.37) = 0.26 \text{ m/s},$$

and therefore:

$$\text{RBS} = \frac{1.90}{0.26} = 7.3$$

This is much higher than 1.0, the value at which particles would be expected to move. Thus, the bed would be considered highly stable at bankfull discharge.

Critical shear stress

An alternative to critical velocity is the concept that a critical shear stress is required to set a particle into motion. In terms of sediment erosion and movement, the term **tractive force** is commonly used as a synonym for shear stress (Dingman, 1984). Shear stress typically increases with discharge, but as with velocity, it is unevenly distributed within a channel.

In deriving an equation for critical shear stress it is assumed that when a particle is just about ready to hop out of bed the shear force acting to overturn it is balanced with the submerged weight of the particle, which holds it in place. By equating the two forces at the threshold of movement, an equation for critical shear stress (τ_c) can be obtained:

$$\tau_c = \theta_c g d(\rho_s - \rho) \tag{7.16}$$

Here, d is a "representative" particle size in metres, τ_c is in N/m^2, g is the acceleration due to gravity, and ρ_s and ρ are particle and water densities, respectively, in kg/m^3. The dimensionless constant, θ_c, is a function of particle shape, fluid properties, and arrangement of the surface particles. It is commonly termed the **dimensionless critical shear stress**.

Equation 7.16 was first proposed by an American engineer, Shields, in 1936. He related θ_c to another dimensionless factor, the "grain" or "shear" Reynolds number, Re_* (Equation 6.19). The Shields curve, showing the relationship between θ_c and Re_* is illustrated in Figure 7.19.

The value of Re_* gives an indication of whether the flow is considered hydraulically rough or smooth. In reference to Figure 7.19, Yalin and Karahan (1979) give the transitional region as approximately $1.5 \leqslant Re_* \leqslant 40$. Under their assumption that the roughness height is approximately $2d_{50}$, this gives boundaries of 3 and 80 for the roughness Reynolds number, which corresponds fairly closely with the range given in Section 6.4.2. Different values for this range will be obtained depending on the representative grain size used (e.g. d_{65}, d_{85}, etc.) and the interpretation of "incipient motion".

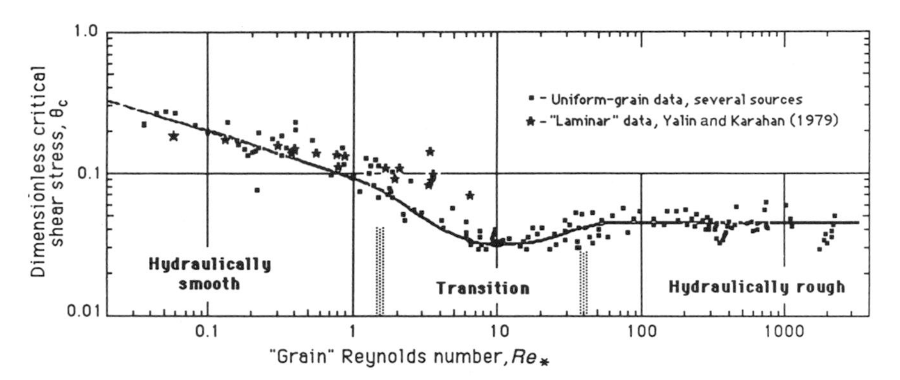

Figure 7.19. "Shields curve", showing the relationship between dimensionless critical shear stress, θ_c, and the grain Reynolds number, Re_*. Shaded bars indicate approximate boundary between hydraulically rough, smooth and transition zones. Adapted from Yalin and Karahan (1979), by permission of the American Society of Civil Engineers

Under hydraulically smooth conditions, the particles are enveloped by the laminar sublayer. Grains in the uppermost layer of the bed are dragged together as a "grain carpet". Under hydraulically rough conditions, grains are exposed to the turbulent flow, and grains detach individually and randomly as a result of instantaneous shear stresses. Thus, it takes more "mobility" in laminar flow for a grain to lift off, as indicated in Figure 7.19 by the left end of the curve and the laminar data of Yalin and Karahan derived using sand particles moving through glycerole.

For larger values of Re_* the Shields curve levels off. Experimenters have established different values for this constant, typically within the range 0.040 to 0.060. Shields (1936) originally found 0.056, and the graph in Figure 7.19 gives an average value of about 0.044. Because of the considerable amount of scatter in the experimental data, it should be stressed that these values are not exact.

For $\theta_c \approx 0.06$, and assuming a sediment density of 2650 kg/m^3, Equation 7.16 becomes:

$$\tau_c = 0.06(9.807)(d)(2650 - 1000), \text{ or}$$

$$\tau_c = 970d \tag{7.17}$$

with d in metres, or:

$$\tau_c = 0.97d \tag{7.18}$$

with d in millimetres. Thus, as a "rule of thumb", the critical shear stress required to move a particle (N/m^2) is approximately the same as the particle's diameter in millimetres.

Although the Shields curve may work well for the uniform and fine sediments for which it was developed, its application to mixed gravel-bed materials is more difficult because of the interactions discussed in Section 7.4.2. An implication of the recent work on equal mobility (see "Mixtures of particle sizes", Section 7.4.2) is that since all particles move at essentially the same shear stress, τ_c can be calculated based on one characteristic grain diameter rather than computing individual values of τ_c for each size fraction. For streambeds, the median size of the bed materials (d_{50}) is often taken as the representative diameter, d.

Andrews (1983), for example, recommended a value of 0.020 for gravel-bed streams, based on field studies, and suggested that a completely exposed particle may be entrained at values of θ_c as low as 0.010. In a study of seven steep mountain streams in California, however, Kondolf et al. (1987) used a value of 0.060, but found that gravels did not move at the predicted shear stresses because of the non-uniform flow patterns in the irregular, boulder-cascade channels and the shielding of gravel deposits behind boulders. Church (1978) cites a value of 0.11 for a streambed with imbricated and closely packed materials. Thus, order of magnitude variations in the Shield's parameter can be expected in natural channels. Suggested values are given in Table 7.2.

The effect of armouring on the Shields parameter was addressed by Parker (1982) and Milhous (1986, 1989). For armoured streams, both recommend using the d_{50} of the armour layer as the appropriate grain size in Equation 7.16. From research on Oak Creek, an armoured stream in Oregon, USA, Milhous found that if enough shear stress was applied to just move a small portion of the larger particles in the surface layer, the fines deposited among the armour particles were flushed out. He recommends two values of θ_c: (1) 0.021 for flushing fines from surface sediments and (2) 0.035 for movement of 30% of the armour layer and "deep" flushing of trapped fines.

Table 7.2. Suggested values for the Shields parameter, θ_c, for mixed bed sediments. Based on Andrews (1983), Carson and Griffiths (1987) and Church (1978)

Condition of streambed	θ_c
Loosely packed: "quicksands" and gravels with large voids filled with water	0.01–0.035
Normal: uniform materials or a "settled" bed with fairly random grain arrangements	0.035–0.065
Closely packed: smaller materials fill the voids between larger particles	0.065–0.10
Highly imbricated	>0.10

He points out that the parameters are somewhat subjective and applicable only if it is assumed that the Oak Creek results can be extrapolated to other streams.

For a particle to move, the actual shear stress must exceed the critical value. If velocity profiles are available, the local shear stress at a vertical can be calculated from the procedure given in Figure 6.14. Alternatively, the shear stress over the whole channel perimeter can be estimated from Equation 6.59. Andrews (1983) suggests that the actual depth of the zone of maximum bedload transport should be used in Equation 6.59 rather than the hydraulic radius. However, without bedload measurements, this may be difficult to determine.

Example 7.2

Determine the particle size which will be entrained by bankfull flow at Site 3 on the Acheron River (Figure 4.7) using critical shear stress methods (see also Example 7.1). Relevant measures are: channel slope = 1.2% and bankfull depth = 0.40 m. At this site, the cross-sections are fairly wide and rectangular. The reach has a slight pool–riffle topography and is heavily armoured. Assume that particles are spherical with $\rho_s = 2650$ kg/m^3.

Answer: From the sieve analysis of the smaller bed materials (Figure 5.32) it can be seen that there are few fines in the surface layer, meaning that it is not closely packed. We will choose a value of 0.04 for θ_c, to be consistent with both Milhous's recommendation and the data of Table 7.2.

Letting $\rho = 1000$, $S = 0.012$, and, since the channel is wide and rectangular, $R \approx$ mean depth = 0.40 m, then, from Equation 6.59, the shear stress at bankfull flow is:

$$\tau_{bkf} = 1000(9.807)(0.40)(0.012)$$
$$= 47\ \text{N/m}^2$$

This value is set equal to τ_c, the critical shear stress for movement. Then, from Equation 7.16,

$$\tau_c = 47 = 0.04(9.807)(d)(2650 - 1000)$$

giving

$$d = 0.073\ \text{m} = 73\ \text{mm}$$

This is approximately equal to the d_{50} of the smaller surface materials (Figure 5.32), but from Figure 5.33 it represents less than 5% of the larger materials sampled from the bed surface. Assuming that the latter is indicative of the armour layer sizes, this would mean that only a small fraction of the bed surface could be mobilized at bankfull flow in this armoured reach.

Additionally, if it is assumed that $\theta_c = 0.010$ for isolated particles (e.g.

tracer particles), then the largest isolated particle which will move is four times the above value, or 280 mm. This corresponds to nearly the d_{95} of the larger materials. Thus, large isolated particles may move over the armoured surface at bankfull flows.

7.5 SEDIMENT YIELD FROM A CATCHMENT

7.5.1 SEDIMENT SOURCES AND SINKS

Sediment yield is defined as the total sediment outflow from a catchment over some unit of time, usually one year. It can be calculated from measurements of suspended and bedload sediments (Section 7.5.3) or estimated from relationships between sediment yield and water discharge (Section 7.5.4).

The amount of sediment arriving at a downstream site represents the net balance between the amount of sediment stored in the channel and the amount contributed to it from upland sources, bank erosion and transportation of streambed materials. Schumm (1977) states that the majority of sediment yield is explained by storage and periodic flushing of alluvium. The relative contributions of sediment from channel and non-channel sources varies with basin size and is difficult to assess. In general, the steeper, shorter slopes of upper catchments supply more sediment from hillslopes. Further downstream, the storage of eroded material increases and channel erosion becomes more important (Knighton, 1984).

The relationship between the amount of sediment carried into a stream and the amount measured at some point downstream is described by the **sediment delivery ratio** (SDR), where:

$$\mathrm{SDR} = \frac{\text{Sediment yield at a measurement point along the stream}}{\text{Total amount of eroded material contributed from slopes above the measurement point}} \tag{7.19}$$

Sediment yield and upland erosion are usually calculated on an annual basis. The value of the SDR is normally less than 1, although values range widely. In general, it is larger for small catchments where drainage dissection is higher and sediment delivery more efficient. The ratio may exceed 1.0 during individual events if hillslope input is minimal and stream hydraulics favour the removal of channel materials, (for example, during snowmelt or low-intensity rainfall events when water mainly reaches the channel as interflow). Although the SDR is difficult to evaluate, it forces investigators to consider the amount of sediment stored on the hillslope or in the channel before making generalizations about upland erosion rates from sediment yields measured at some point in the stream.

Bank erosion

A large number of factors control the rate of bank erosion, including the composition of bed materials, streamflow patterns and amounts, soil moisture, frost action, channel geometry, vegetation type and cover, and activity of burrowing animals (Knighton, 1984). In bedrock, the channel banks erode slowly under the grinding action of swirling rocks or from erosion along fracture lines, which can cause large chunks of rock to fall away. In alluvial channels, banks tend to be composed of finer materials than the streambed. Weak banks of gravel or other unconsolidated alluvium collapse easily, forming wide, shallow channels, whereas banks of more cohesive materials form deep, narrow ones (see Section 7.2.6). In composite banks the erodibility will be controlled by the strength of the weakest layer.

Bank erosion in alluvial channels commonly results from the slumping of saturated soils or the "pseudo-cyclical" process of undercutting, failure of the overhanging bank and gradual removal of the fallen material (Knighton, 1984, p. 62). Bank undercutting is caused by the combined actions of large-scale eddies, spiral flow, and waves from wind or passing boats, which gradually loosen and remove bank materials. Undercutting may eventually cause trees to topple into the stream, releasing large quantities of sediment. However, the downed trees provide habitat for aquatic flora and fauna, retard the downstream movement of sediments and may protect the bank from further erosion.

Bank erosion is highly variable and episodic, and is usually associated with flood flows. However, the condition of the bank will affect its susceptibility to erosion. For example, deep saturation, frost action or trampling can cause soils to be more easily washed away. Multi-peaked flows may thus be more effective in bank erosion than single-peaked ones because the bank is saturated when subsequent peaks arrive.

Streambank stability refers to a bank's resistance to change in shape or position, whether attacked by flood flows or ice floes. Streambank condition and quality of aquatic habitat are closely linked. Stable undercut banks provide shade and cover for fish or burrow sites for river-dwelling mammals such as platypus or muskrats. Swallows and water ouzels may nest higher up on the face of steep banks. A one-metre loss of bank (measured horizontally) from the collapse of an overhang will thus alter the physical habitat differently than a similar loss from erosion of a vertical or sloping bank (Bohn, 1986). Therefore, for ecological purposes it may be more important to define bank loss in terms of bank function rather than the distance it shifts.

The health and composition of riparian vegetation will influence streambank stability. Smith (1976), for example, found that a bank with a 50 mm-thick root mat of 16–18% root volume afforded 20 000 times more protection from erosion than a comparable bank without vegetation. Excess trampling of banks removes protective vegetation and reduces bank stability. Platts and Nelson (1985) observed that large floods badly damaged channel banks in heavily grazed sections of a river in Utah, USA, but actually improved bank form in a protected ungrazed enclosure, with more undercut area and lower angles on exposed bank slopes.

Field techniques in Chapter 5 relevant to bank stability include stream surveys, methods of measuring areal extent of vegetation, and methods of analysing soil properties such as moisture content, bulk density, aggregate stability, soil texture and particle size.

Channel storage of sediment

If the amount of sediment leaving a stream system is less than the amount entering, then the difference must be due to storage within the channel. Channel storage can be divided into three categories: (1) temporary storage in the channel bed and bedforms; (2) longer-term storage behind obstructions; and (3) very long-term storage in valley floodplain deposits (Megahan, 1982).

In reservoirs and canals, the trapping of sediments is referred to as **sedimentation**. This normally has a negative connotation since it reduces the life of reservoirs and can clog irrigation ditches. In natural streams, numerous sites for storage exist behind "mini-dams" created by boulders, rock outcrops, logs, roots or accumulations of debris. These storage sites may be so effective that the impacts of landslide deposits, gravel-mining operations, road construction or forest harvest may not be felt downstream until sediments are removed by a large event—possibly many years later.

In a 6-year study on forested catchments in Idaho, USA, Megahan (1982) concluded that logs were the most important type of obstruction because of their longevity and the large volume of sediment trapped behind them. Streamflow variations were also found to affect the amount of sediment stored. Only large, stable obstructions remained in the channel during a high-flow year, whereas lower flows appeared to favour more obstructions with smaller storage volumes. On average, fifteen times more sediment was stored behind obstructions than was delivered to the drainage outlets. The study illustrates the need to consider all components of the sediment budget both under natural conditions and when monitoring the effects of land-use change.

When organic debris no longer enters a stream the banks become unstable, streamside erosion accelerates and the channel topography can become smoothed from the filling of pools and flattening of riffles. Lisle (1983,

p. 46) states that riparian trees and large woody debris should be treated as if they "belong to the aquatic ecosystem". Further, Beschta and Platts (1986) point out that researchers are starting to recognize that many streams are relatively "starved" of large organic material in regard to channel stability.

7.5.2 SEDIMENT YIELD VARIATIONS

Variability in sediment yield can result from intermittent bank or hillslope collapse, channel incision into alluvium of varying composition, the impact of fire or volcanic activity, change in land use, variable patterns of streamflow, and activity in the channel such as dredging, the cavorting of cattle, or the stirring up of bed sediments by large bottom-feeding organisms. Sediment yield also changes with drainage basin size. As the size of the drainage basin increases, there are more sites for permanent or temporary storage of sediments and the sediment yield per unit area decreases (Schumm, 1977).

In general, streams in semi-arid areas have more highly variable sediment loads than those in humid regions (Nordin, 1985). It is characteristic of streams, especially those with high flow variability, for most of the sediment transport to occur during a few days of high flow. Concentrations also tend to be higher at the beginning of the snowmelt or rainfall season when there is more erodible material on catchment surfaces. Subsequent runoff events carry less and less sediment over the runoff season. Sediment loads also change from year to year. Changes in the amount of sediment delivered during low- and high-flow years can have a considerable impact on the nature of the streambed and its inhabitants. For example, in a pool of the Mississippi River, Bhomik and Adams (1986) found that the low turbidity during a low-flow year caused a shift in the benthic community from clams to plants.

Regional variation of average annual sediment yield as a function of climate was studied by Langbein and Schumm (1958). They produced the classic curve shown in Figure 7.20 using data from approximately 100 sediment gauging stations in the USA. The "effective precipitation" is the average annual precipitation, adjusted to a value which produces the same runoff as from regions having a mean annual temperature of 10°C.

Sediment yield per unit area was found to reach a maximum at an effective precipitation of about 300 mm, which corresponds to areas receiving intense, infrequent rainfalls. Langbein and Schumm reasoned that lower precipitation levels would not produce as much runoff for transporting sediments, whereas higher precipitation encouraged vegetation growth, protecting soils from erosion. Subsequent studies with global data have yielded two more "peaks", one representing Mediterranean climates with annual precipitation from

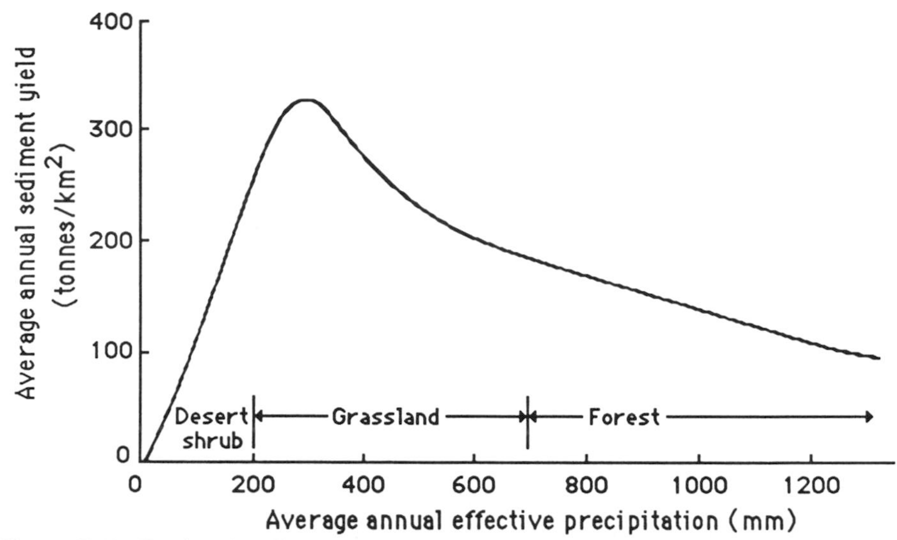

Figure 7.20. Regional sediment yield, as related to effective precipitation. Adapted from Morisawa (1985) after Langbein and Schumm (1968), by permission of Longman Group, UK. "Effective precipitation" is adjusted to a value producing the same amount of runoff as from regions with a mean annual temperature of 10°C

1250 to 1350 mm and another tropical monsoon conditions (>2500 mm) (Walling and Kleo, 1979). Table 7.3 presents average annual sediment yields for some of the world's larger rivers.

7.5.3 COMPUTING SEDIMENT DISCHARGE AND YIELD FROM MEASURED CONCENTRATIONS

Field methods for the collection and analysis of suspended and bedload sediment samples and dissolved solids are covered in Chapter 5. The "daily" sediment concentrations published by water authorities are commonly instantaneous suspended sediment values from samples taken at the same time each day, rather than an average concentration for the day. In large rivers where concentrations usually change little over the day, these values are probably adequate for calculating daily suspended sediment discharge as:

$$Q_s = 0.0864\, Q_D\, c_t \qquad (7.20)$$

where Q_s = suspended sediment discharge (tonnes/day), Q_D = average daily discharge (m^3/s) and c_t = daily suspended sediment concentration (mg/l—or ppm if the water-sediment mixture is assumed to have a density of 1000 kg/m^3). The 0.0864 converts from seconds to days and mg to tonnes. Daily dissolved load can be computed using the same equation.

Table 7.3. Sediment yields of some major rivers, as measured at the mouth (compare with Table 4.2). Adapted from Holeman (1968)

River	Mean sediment yield (10^6 tonnes/yr)
Amazon, Brazil	498
Congo, Congo	65
Orinoco, Venezuela	87
Yangtze, China	500
Brahmaputra, Bangladesh	726
Mississippi, USA	312
Mekong, Thailand	170
Parana, Argentina	82
St Lawrence, Canada	4
Ganges, Bangladesh	1450
Danube, USSR	19
Nile, Egypt	111
Murray-Darling, Australia	32

If a peak event occurs between sampling periods, as will be common in flashy or ephemeral streams, the sample will not be particularly representative of the day's sediment load. Thus, records will generally be biased towards low flows and low estimates of sediment load. In a monitoring programme samples should be taken more often so that a "sediment hydrograph" can be constructed, as illustrated in Figure 7.21. Since the analysis of sediment samples is time consuming, various sampling strategies have been developed to minimize the number of samples collected (Nordin, 1985). The general strategy is to collect more samples when the sediment concentration is high, as indicated in Figure 7.21. Some "smoothing" of the line drawn through the data points may be required because of the large fluctuations in sediment concentration.

Once a hydrograph has been constructed, sediment discharge for individual time steps is calculated from Equation 7.20, where time steps can coincide with sampling times or "breakpoints" where the slope of the sediment or runoff curve changes.

Quantities are calculated for each time interval as:

$$\left(\frac{(Q_s \text{ at time 1}) + (Q_s \text{ at time 2})}{2}\right)(\text{time 2} - \text{time 1}) \tag{7.21}$$

Daily sediment discharge totals, in units of tonnes/day can then be calculated by summing up values over individual time intervals within the day. Porterfield (1972) is a source of more information on computing sediment discharge. Again, the same procedure can be used for daily dissolved load.

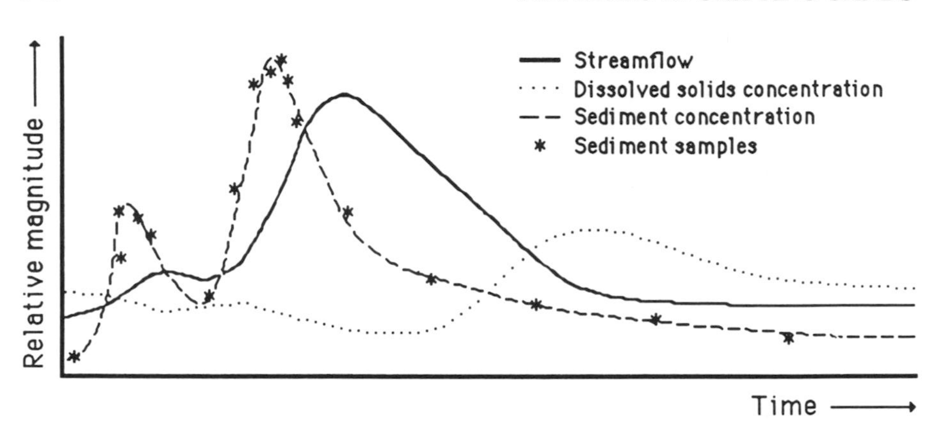

Figure 7.21. Generalized graphs showing the change in streamflow, sediment load and dissolved load during a runoff event

Bedload discharge (Q_{bl}) is calculated from the amount of sediment sampled over some unit of time. Often, a particle size analysis is done to determine the bedload discharge for each size fraction. For pit-type samplers or reservoir surveys, it is simply calculated from the mass of sediment trapped over some time interval (e.g. one year). If volume is measured, it must first be converted to mass using an estimate of the bulk density of the deposited sediment (Section 5.10.7).

Using Helley–Smith-type samplers (Section 5.9.5), bedload discharge per unit width of stream (Q_{bl}, kg/m·s) is calculated for each measurement point as:

$$Q_{bl_i} = \frac{M_i}{Wt} \tag{7.22}$$

where t is the measurement time interval (s), M is the mass of the sample (kg), W is the width of the sampler entrance (m) and i refers to the ith measurement point. As with the velocity–area method of computing discharge (Section 5.7.3), each measurement is assumed to represent a certain segment of the cross-section. Total bedload discharge for a cross-section is obtained by multiplying each measurement by the width interval it represents and summing over the width of the stream.

7.5.4 THE ESTIMATION OF SEDIMENT DISCHARGE FROM STREAMFLOW

Sediment concentrations are often poorly related to discharge. This is partly due to the hysteresis effects discussed in Section 7.1.4 or varying amounts of washload. Therefore, it is always preferable to measure sediment concentrations rather than attempting to estimate them from streamflow data. However, sediment data will not always be available during particular years, seasons, or events of interest. Estimating sediment concentrations from streamflow data is a widely used approach in hydrological practice.

Sediment concentrations obtained over a range of discharges can be used to develop a sediment rating curve, given previously as Equation 7.10:

$$Q_s = pQ^j \tag{7.23}$$

where Q_s and Q are sediment and water discharge, as before, and p and j are empirical coefficients. The value of j, which does not depend on the units used, typically lies between 1.5 and 3.0 (Knighton, 1984). From hydraulic geometry relationships (Section 7.2.6), Leopold and Maddock (1953) found that an increase in sediment discharge is related to an increase in velocity and a reduction in depth.

Depending on their intended use, sediment rating curves can be developed for instantaneous, daily, monthly or yearly data, with the scatter in the relationship decreasing as the time interval increases. In large, stable rivers the relationships will be fairly well defined. However, short-term rating curves for most streams, especially those which are supply-limited, will exhibit a considerable amount of scatter. Estimating instantaneous or event concentrations from discharge in these streams is risky at best. USDA (1979) further discusses the preparation of sediment rating curves and methods of treating hysteresis effects.

Annual or seasonal rating curves may still be fairly useful for streams where instantaneous, daily, or event rating curves are unacceptable. For example, Nordin (1985) found that the relation between annual sediment load and annual flow was "reasonably well defined" for the Rio Puerco, an ephemeral tributary of the Rio Grande in New Mexico, USA.

To develop a rating curve, sediment concentrations should be measured over a large range of discharges. Sediment discharge is calculated for individual samples by Equation 7.20 and a log-log relationship fitted to the sediment discharge and streamflow data. If measurements allow, curves can be developed separately for suspended load and for bed-material load, as shown in Figure 7.22. It may also be appropriate to develop separate curves for each season or for rainfall and snowmelt events.

From the rating curve, a record of daily, monthly or yearly sediment discharges can be generated from streamflow data. These can be used to derive yearly or seasonal sediment yields which may be useful in locating regions of high (or low) sediment production, e.g. for targeting erosion control works. Program TRANSFRM in AQUAPAK can be used to generate a file of sediment discharge data from streamflows by applying a rating equation (Equation 7.23). However, the limitations of the data should be taken into account. For example, a rating curve developed during an "average" year would not be applicable to an extremely wet year in which

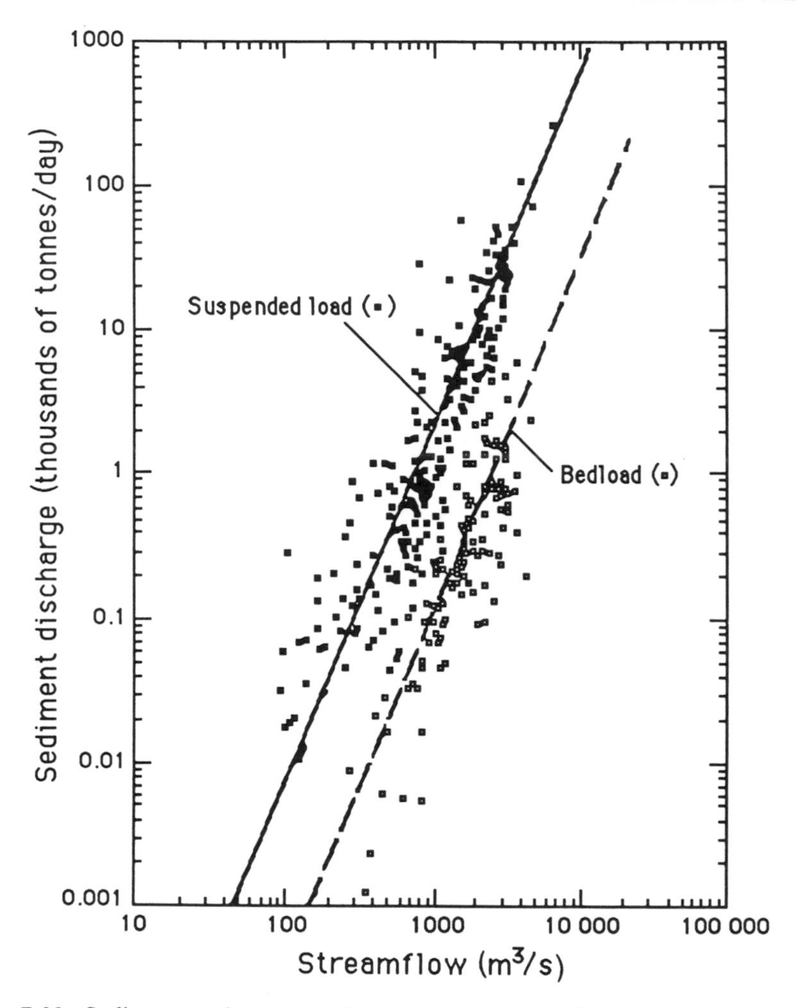

Figure 7.22. Sediment rating curve for the Snake and Clearwater Rivers in the vicinity of Lewiston, Idaho. Based on data of Emmett (1977); redrawn from Csermak and Rakoczi (1987), by permission of Water Resources Publications

mudflows and bank collapses were common. There is also a danger in extrapolating beyond the data; for example, a rating curve developed for low flows should not be applied to a streamflow record in which large floods occur. It should be realized that the rating curve is an example of spurious correlation (Appendix 1), and should be interpreted only in reference to sediment discharge, not sediment concentration. An evaluation of the error involved should always be included when estimating sediment discharge using a rating curve.

With a sufficiently long record of either estimated or measured sediment data, the same techniques used for the analysis of streamflow data can be applied; e.g. basic statistics and histograms (Section 4.4), and sediment duration or frequency curves (Section 8.3). However, care should be taken in interpreting the sediment record to infer the future from the past. Gradual trends are common as a result of changes in grazing or forestry practices, or in the amount of cultivated or urbanized area. Sudden changes due to fire, floods, or the construction of reservoirs may impact sediment yields for many years afterwards—or permanently. This effectively alters the population (see Chapter 1) from which subsequent samples are drawn. An example is the impact on the Toutle River from the 1980 eruption of Mount St Helens in the USA. Concentrations of suspended sediment up to 200 000 mg/l were measured shortly after the eruption, and the long-term effect on the estuary of the Columbia River into which the Toutle drains can only be guessed (Nordin, 1985).

Channel patterns, channel geometry and movement of sediment all have an important role in shaping the habitats of aquatic organisms. These patterns are affected by the geological and hydrological setting of the stream. This chapter has focused on the dynamic nature of the stream channel, to which the stream biota must adapt. Methods for analysing and describing patterns of streamflow are given in Chapter 8.

8 Dissecting Data with a Statistical Scope

The importance of information is directly proportional to its improbability (Raudkivi, 1979, p. 396).

8.1 INTRODUCTION

8.1.1 GENERAL

How often will a river overtop its banks? How long will this drought last? If the stream's flow is such and such today, what is the probability it will be so and so tomorrow? How often will a platypus burrow be inundated? How many days out of the year will the flow be below five cumecs? How large does a reservoir need to be if a town's water supply is to be maintained with a 99% reliability?

These are the types of questions addressed by hydrologists using methods in statistics and probability. Increasingly sophisticated satellites for monitoring the weather and complex global computer models are improving the prediction of near-future streamflows. Yet, forecasts of the streamflows which will occur next year or over the next 20 years still cannot be modelled with much certainty. However, if records are available for a sufficient length of time and we assume that these past records are an indication of what will happen in the future, we can estimate with some confidence the average yearly streamflow, the variability of the stream or the probability that a flood of a certain magnitude will occur within the next year.

A streamflow record is a sample out of time of all the flows which will ever course through a stream, past and future. Statistics calculated from this sample are used to predict what might occur in the future under the same conditions. The actual sequence of high and low flows in the past record is not expected to repeat itself exactly—in fact, it would be very extraordinary if it did. Variations over time occur more or less randomly, although some "persistence" is seen in streamflow data; e.g. low-flow days tend to follow other low-flow days. Scientists continually look for cyclical trends in streamflow data which might correlate with a measurable phenomenon such as sunspot cycles or the El Niño–Southern Oscillation variation in sea temperatures. Success, however, has been limited. Thus, streamflow data will continue to be treated statistically until such time as the weather can be predicted—or controlled—with greater reliability.

In Chapter 2 definitions are given for various statistical terms, and Chapter 4 provides methods of exploratory statistical analysis for describing the pattern, relative magnitude and variability of streamflow. Appendix 1 includes a summary of basic statistical formulae. In this chapter, methods of analysing probabilities of streamflow magnitudes and durations are presented. The methods given in Sections 8.1–8.4 require streamflow records from gauged sites. However, in many cases it is the stream behaviour at ungauged sites which is of most interest. If gauging stations are located nearby, it may be possible to adjust the records slightly for the ungauged site. Alternatively, a regional analysis can be conducted to develop estimated records for stream sites within the region using catchment or channel characteristics. Section 8.5 covers regionalization methods and Section 8.6 describes methods of numerical taxonomy which can bc uscd to classify streams and catchments into hydrologically homogeneous groups.

8.1.2 FLOODS AND DROUGHTS

In engineering hydrology the emphasis is primarily on the extremes—the floods and droughts—because these are of the most importance *economically*. These extremes are also important *ecologically* since they affect the populations and distributions of aquatic organisms. Hydrologists are probably most well known for their work on predicting the magnitude of floods of a certain probability (e.g. the "100-year flood"). Knowledge of flood behaviour is needed in the design of dams, bridges and highway culverts, and in the development of flood-insurance rates and flood-zoning maps.

A flood can be defined simply as a flow which overtops the streambanks. However, this definition is difficult to apply in mountain gorge streams or others which have poorly defined banks (see discussion of bankfull flow, Section 7.2.5). An ecological definition was given by Gray (1981), who considered a flood to be a discharge which scoured substrates and disrupted the biota. Ecologists also use the term "spate" for high flows.

Floods affect the ecology of a stream by rearranging streambed habitats, scouring away aquatic or riparian plants and increasing the drift of aquatic insects. Most biota avoid high velocities and hurtling particles by sheltering behind rocks or snags, burrowing into the streambed and banks, moving to slower water along the stream's edges and in backwaters, or by having life cycles which are terrestrial or aerial during flood-prone seasons. To some fish, floods act as a cue for spawning, and they may migrate upstream or downstream during flood periods. When floods occur at unusual times the fauna may be severely depleted, requiring many years to recover (Goldman and Horne, 1983). In desert streams which are often subject to sudden, unpredictable floods, aquatic insects have developed specialized coping mechanisms such as rapid development to the adult stage (Gray, 1981).

Peak flows are caused by various factors such as heavy rainfall on saturated or frozen soils or rapid snowmelt from warm Chinook winds. Even though the highest flow in a year or the peak of a given hydrograph may or may not constitute a "flood" in the strictest sense, these flows are still important in an analysis of how often floods can be expected to occur. Techniques for flood-frequency analysis are given in Section 8.2.

The length of time (duration) an area is under water is also important in a flood analysis. For a riverside cottage the damage may be the same whether it is underwater for 5 minutes or 5 days; however, for a tree, it can mean the difference between renewed vigour and root suffocation. The period of inundation will affect the length of time during which nutrients and biota pass between the stream and floodplain. Inundation for long periods of time can also raise groundwater levels. To assess the duration of floods, techniques such as flow-duration curves (Section 8.3) or flow-spell analysis (Section 8.4) can be applied.

Flood frequency and flow duration are related to the channel form, as shown in Figure 8.1. In this "typical" river the average flow is very low, and the stream's level is much less than bankfull level for most of the year. About once each year there is enough runoff to fill the channel. Overflow onto the floodplain accommodates the waters of larger, rarer floods. This pattern will vary from stream to stream and from one cross-section to another on the same stream.

At the opposite extreme from floods are droughts and other periods of low flow. Very low and even zero flows are normal in streams during years of low precipitation, particularly if the stream is intermittent or ephemeral. They may also be artificially imposed by diversions or upstream storages. No uniform definition of "drought period" has yet been established, but it generally refers to long-duration periods of low flow. Like "flood", "drought" is usually defined in economic terms, based on water supply needs. Droughts can have immediate and lasting effects on the animal and plant populations of streams. Unless aquatic organisms have developed special adaptations such as long dormant phases, rapid development and prolific reproduction (e.g. the fairy shrimp and toads which live in the temporary lakes of dryland regions), the ecological impacts of drought maybe more long-lasting than the effects of floods.

An analysis of low flows is necessary before a stream can be used as a reliable source of water supply (McMahon and Arenas, 1982). Dry spells put pressure on managers who must allocate limited water supplies between irrigators, cities and instream biota. There has been increasing interest in

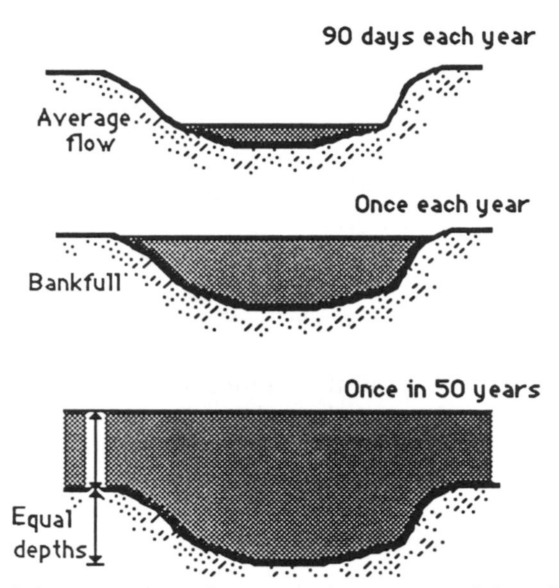

Figure 8.1. Typical river stages at various rates of occurrence. Adapted from Leopold and Langbein (1960)

defining "minimum instream flows" for sustaining aquatic life and/or diluting pollutants. Instream flow analysis is covered in Section 9.3.

Various drought indices have been developed for describing the severity of droughts, including: (1) the lowest flow during the drought period, (2) the mean discharge during the drought, (3) the volume of water "deficiency" below some flow-threshold level; and (4) the duration of the drought (WMO, 1983). Low-flow frequency analysis (Section 8.2), flow-duration curves (Section 8.3), flow-spell analysis (Section 8.4) and recession-curve analysis (Section 4.3.4) are methods applied to the analysis of low flows.

8.1.3 DATA CONSIDERATIONS

As mentioned in Section 3.2, hydrological data must be accurate, representative, homogeneous and of sufficient length if they are to be statistically analysed to provide useful answers. To ensure accuracy, the data should first be examined for gross errors and missing data. Representativeness means that the data represent the conditions of interest. Many studies will require natural ("unregulated" or "virgin") streamflows. If a dam has been installed on the stream, the natural flows must be reconstructed by adjusting the recorded values by the amount of storage, difference in evaporation, and release. In contrast, if regulated flows are of interest, reservoir operating rules can be applied to past "unregulated" records to create a record which simulates the presence of a reservoir. Homogeneity implies that the streamflows have occurred under uniform conditions; i.e. stream channelization, a doubling of the forest density, or a change in the gauging station location may lead to a non-homogeneous record. The double-mass curve, described in Section 3.2.3, is a technique for detecting non-homogeneity. Additionally, to use past records to describe the future, the data must be **stationary**, meaning that statistical properties remain constant over time; e.g. climate changes do not occur.

A streamflow record is only a sample in time. The "population" consists of all streamflows over an indefinitely long time period. A typical record of 20 years in length is usually made up of mostly medium-sized occurrences, with a few large events and a few small ones. From this information inferences can be made about the future and past behaviour of the stream. The value of these inferences depends on how well the sample represents the range of high, low and medium values occurring over time. Since the true population is unknown, it can only be assumed from statistics that more records will bring us closer to the "truth". In fact, a rigorous mathematical treatment of very short records may be very inferior to "guesstimates" made by old-timers who have lived along a river for generations.

To improve the reliability of short records, streams should not be treated in isolation but with knowledge of the behaviour of similar streams. Several short-term records from stations within hydrologically similar basins can be combined using regionalization methods to increase the effective length of record. These methods are presented in Section 8.5.3.

8.1.4 PUTTING STATISTICS INTO A PROPER PERSPECTIVE

In an ideal situation, we would be able to apply uniform methods and obtain consistent results under all situations—short or long records, small or large catchments, stable or flashy streams. However, as Dalrymple (1960, p. 20) cautions, "A method is not better because it leads to uniform answers if those answers are uniformly wrong". Therefore, in this chapter several different methods may be presented for the same type of analysis, some more appropriate to specific situations than others.

The treatment of hydrological data with statistics has led to considerable debate about which techniques are best and the further refinement of the statistical methods used. When reading the extensive literature on the testing of these methods one is often left with the impression that researchers have forgotten that the numbers were generated by a catchment rather than a computer. In fact, many validations of methods have been made using synthetic data. It is easy to lose sight of the fact that streams were running at varying rates of flow long before anyone developed statistics, and will continue to do so in spite of our attempts to pin them down numerically. Klemes (1986) has written a refreshing treatise on "mathematistry" in hydrology which is well worth reading by those needing a healthy dose of reality about the limitations of statistics.

In methods such as flood-frequency analysis and regression analysis where a line is fitted through the data there is always the temptation to define the line in its geometric sense—that it has infinite length. However, extrapolation beyond the data should be done cautiously, with knowledge of the error margins on the extrapolated values. Overmire (1986, p. 35) gives the following quote from Mark Twain's *Life on the Mississippi* to illustrate the perils of reckless extrapolation:

> In the space of one hundred and seventy-six years the Lower Mississippi has shortened itself two hundred and forty-two miles. That is an average of a trifle over one mile and a third per year. Therefore, any calm person, who is not blind or idiotic, can see that in the Old Oolitic Silurian Period, just a million years ago next November, the Lower Mississippi River was upward of one million three hundred thousand miles long, and stuck out over the Gulf of Mexico like a fishing rod. . . . There is something fascinating about science. One gets such wholesale returns of conjecture out of such a trifling investment of fact.

For ecological work the consequences of an inaccurate estimate may not be as significant as in engineering. However, the error margin on an estimate and its consequences should still be evaluated as an essential part of any statistical analysis. As Haan (1977, p. 5) says, "statistics should be regarded as a tool, an aid to understanding, but never as a replacement for useful thought".

8.2 STREAMFLOW FREQUENCY ANALYSIS

8.2.1 GENERAL CONCEPTS

Frequency analysis is a method for assigning probabilities to events of a given size. Appendix 1 gives a description of the concepts of frequency and probability, and methods of displaying their distribution. In streamflow frequency analysis we most often use cumulative distributions and speak of probabilities of exceedance. For high flows, "exceedance" usually means a streamflow value is *greater* than some amount, whereas for low flows, it is usually *less*; e.g. the probability that the flow is greater (or less) than 12.3 m^3/s.

In Section 5.10.5 particle-size data are analysed with a type of frequency analysis. However, there is a subtle difference in that the frequencies are *percentages* rather than *probabilities*. In sizing sediments the idea was to simply *describe* the bucket-full of sediment which was passed through a sieve. In flood or low-flow frequency analysis the object is to use data from a short time period not only to describe the "bucket-full" of past flows but also to evaluate the probability of future events.

In a streamflow frequency analysis the basic procedure begins with a ranking of the most extreme events of the past. Probabilities can be determined either by graphical methods or by fitting a theoretical probability distribution to the data. The fitted curve or distribution can then be used to find probabilities associated with floods or low flows of a given size, within the limits of prudent extrapolation. The techniques will be discussed with a focus on flood-frequency analysis. Low-flow frequency analysis is presented in Section 8.2.6.

With the 20 years of record typically available for most streams, estimation of extreme events rarer than about the 1-in-50-year event should only be considered rough guesses, and extrapolation beyond the data calls for professional judgement by "experienced individuals who are knowledgeable about the subject and the impacts of their decisions" (Hagen, 1989, p. 18). Regionalization (Section 8.5.3) can help to increase the accuracy of extrapolated values.

8.2.2 PROBABILITY AND AVERAGE RECURRENCE INTERVALS

In flood-frequency analysis, probability is often expressed as a "1-in-*N*-year" chance. For example, a 1-in-100-year flood is one which would be expected to occur, on average, once in every 100 years. This *average* length of time between two floods *of a given size or larger* is called the **average recurrence interval** or **return period**. The 1-in-100-year flood would have a **probability** of 0.01 or 1% of being equalled or exceeded in any one year. Probability (P) and average recurrence (T) are thus reciprocals:

$$P = \frac{1}{T} \tag{8.1}$$

For example, if $T = 100$ years, $P = 1/100 = 0.01$.

A 1-in-6-year flood occurs with the same frequency as rolling a 7 with two (non-loaded) dice. A 1-in-50-year flood occurs with almost the same probability as drawing an ace of spades from a pack of playing cards. An extremely rare flood (of Noah magnitude or larger) may only occur with the same chance as that of a monkey randomly typing out the word "Ecology" on a typewriter.

Probability or recurrence interval only tells us how *likely* a flood (or drought or zero flow) is. It says nothing about *when* it will actually come. As someone once said, "one out of 100 Alaskan bears bite . . . the trouble is, they don't come in numerical order". Thus, a 100-year event might occur this year, next year, several times or not at all during our lifetime.

8.2.3 THE DATA SERIES

For flood-frequency analysis it is important to have a record which is representative of the total population of floods which are likely to occur on a stream. Records should be relevant to the problem and based on a consistent set of stream, catchment and climatic conditions. Peak flows resulting from different meteorological causes (e.g. snowmelt, rainfall, typhoons, dam failures) can represent different populations, and it may be preferable to analyse the data separately, as suggested by Jarrett (1987).

For probabilistic analysis, streamflows are also required to be **independent**. For example, if a storm produces a double-peaked hydrograph, only the highest peak should be used in a flood-frequency analysis. A problem can also arise when a large event in December spills over into January, becoming the largest "event" in the two adjacent calendar years. For this reason, water years rather than calendar years are preferred (see Section 4.4.2).

The data series must also be **adequate**. If the sample is too small the probabilities will not be reliable. From Table 8.1 it can be seen that 18 years of data would allow the 10-year flood to be estimated with an error of about 25%, whereas over 100 years of data are needed to estimate either the 50- or 100-year flood within plus or minus 10%. The table was developed from synthetically generated data, and errors will undoubtedly be higher for real data. These limitations are not meant to be discouraging, but to point out the need for extracting as much information as possible from gauged data and field observations, and for using the most appropriate statistical methods.

Since a longer record (larger sample size) is more likely to be representative of the "true" population than a short one, some searching may be warranted to obtain information for years outside the published streamflow record. Old newspapers can be read for historical information on the dates of floods and approximate depths and discharges. Discovering a lack of evidence of historical floods can be just as important. Botanical information from flood-

Table 8.1. Length of record (years) required to estimate floods of various average recurrence intervals with 95% confidence. Adapted from Linsley et al. (1975b), by permission of McGraw-Hill

Average recurrence interval (years)	Error	
	10%	25%
10	90	18
50	110	39
100	115	48

damaged trees such as the number of tree rings in a re-grown branch or a tree scar can also give an indication of both the height of a flood and the year of occurrence. Hupp and Bryan (1989), for example, used tree-ring analysis of flood-damaged vegetation to reduce the standard error of estimation of flash flood frequencies by as much as an order of magnitude. Harrison and Reid (1967) also cited good correspondence between a frequency analysis based on tree scars and a traditional analysis using gauged data. They found that most of the scars were caused by floating ice, and that summer floods which lacked debris did not scar trees. Special procedures must be used to incorporate historical data into a frequency analysis, and the reader is referred to IAC (1982) and IEA (1987).

Flood frequencies can be analysed using data on the stage, volume or discharge of flood events. **Discharge frequency** is analysed most often, and this is the approach taken in this text. **Stage frequency** may be of value in determining how often a stream will overtop its banks or have sufficient depth for migrating fish. However, a stage frequency analysis only applies to the *actual gauge location* since the shape and stage of the stream can change considerably over a short distance. It may also change over time if the channel adjusts its shape. Stage frequency analysis is best accomplished by analysing discharge frequencies and then converting back to stage with the most recent stage–discharge relationship at a particular site of interest (WMO, 1983). **Volume frequency** has potential use in the prediction of water supply, and can, with caution, be extrapolated to reaches away from the gauging site. Dalrymple (1960) should be consulted for techniques of stage and volume frequency analysis.

Instantaneous flows are normally used for analysis, particularly for flood frequencies. Daily flows may be used, but with the knowledge that they represent an average over the day or for a specific hour (e.g. an 8 a.m. reading) rather than the highest streamflow occurring within that day.

Flood-frequency analysis is generally performed on a data series composed of the single highest peak flow in each year, the **annual maximum series**. For low flows we can instead speak of an **annual minimum series**, which is composed of the single lowest flow (instantaneous or daily) from each year. The value of low-flow frequency analysis is often improved by using periods of significant length, e.g. the flow over 7 or 10 days. The annual minimum series would then be made up of the minimum 7- or 10-day flow from each year of record.

In the analysis of flood flows, frequent events (occurring every few years or several times a year) are sometimes of more interest. A **partial-duration series** composed of flows above some "threshold level" is preferable for this situation. A special case of the partial-duration series is the **annual exceedance series**, where the N highest peaks within the N years of data are chosen, irrespective of when they occur (Shaw, 1988). These concepts are illustrated in Figure 8.2.

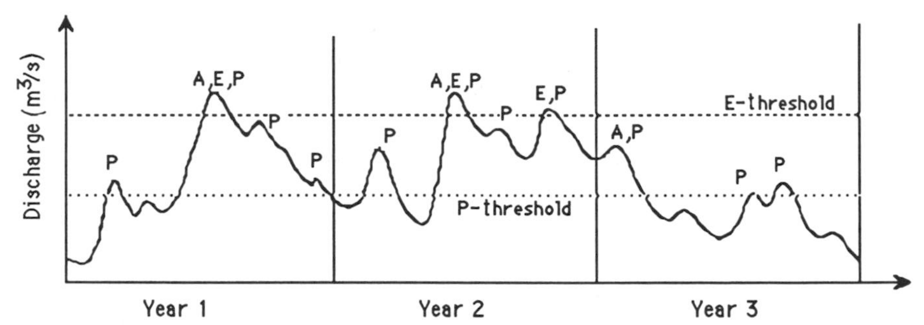

Figure 8.2. Flood peak data series: annual maximum (A), partial duration (P) and annual exceedance (E). The threshold value for the partial duration series ("P-threshold") is set at a chosen value; for the annual exceedance series ("E-threshold"), it is set to give N peaks in N years of data (here, 3 peaks for 3 years of data)

The threshold value for a partial duration series can be set arbitrarily, depending on which flow levels are of the most interest. One method is to set the threshold equal to the smallest annual peak flood. Another possibility is to set it to bankfull flow (Section 7.2.5). For calculating average recurrence intervals of one year or less, the base is usually set such that, on average, three to four floods per year are included in the series (Dalrymple, 1960).

To visualize the difference between the various types of series, one could consider the movements of a cockatoo soaring above a forest. The highest altitude reached each year by the cockatoo would represent the annual maximum series. In the partial-duration series the "threshold level" could be represented by the treetop elevation. Each time the cockatoo flew above the trees the highest point reached would be recorded. This record, collected over a period of time (say, 20 years) would constitute the partial duration series. The 20 highest points reached in that 20 years would represent the annual exceedance series.

From the annual maximum series we could calculate the probability that the cockatoo would fly up to a certain altitude or higher in any given year. From the partial-duration series we could make a statement such as "on average, the cockatoo flies to at least a height of 500 metres above the ground 4.3 times per year". The interpretation of the annual maximum, annual exceedance and partial duration series is different, and is discussed in Section 8.2.9.

8.2.4 GRAPHICAL METHODS: THE PROBABILITY PLOT

In graphical methods of flood-frequency analysis the peak flow data are ranked from highest to lowest and given a **plotting position**. For an annual series this position is a sample estimate of probability. The data and their respective plotting positions are plotted as points on graph paper and a line is drawn to interpret the points. The result is called a **probability plot** and the fitted line a **flood-frequency curve**. Even with the increasing use of computer packages for fitting probability distributions, the probability plot is still a common practice in hydrology as a quick and simple approach.

Plotting the data on arithmetic paper will usually result in a S-shaped curve which is difficult to interpret. Instead, the data are usually plotted on special **probability paper**. The Y-axis represents the peak flow value and the X-axis either the probability of a flow being equalled or exceeded, or the average recurrence interval (e.g. see Figure 8.4).

Probability paper is designed to produce a straight-line plot when the appropriate theoretical probability distribution is plotted as a cumulative distribution function (see Appendix 1). Sample data which plot more or less as a straight line on a specific paper can be judged to be of that distribution type. Commercially available probability paper is most commonly based on the normal distribution, and normally distributed data will plot as a straight line on this paper. The same paper with a logarithmic scale is available for plotting log-normally distributed data.

To plot peak discharge data (either the annual maxima or the peaks above some threshold level), the data values are first ordered by size. The largest value is given a rank of 1, the second largest a rank of 2, and so on until the lowest value has a rank equal to N, the total number of data points. If two values are equal, they should still be assigned different ranks.

A plotting position is assigned to each data value using a plotting position formula, three of which are given in Table 8.2. Theoretically, the largest flood should plot at 0 (there would be no chance of it ever being exceeded) and the smallest one at 1 (every flood would be equal to or greater than this value). However, there is no guarantee that the sample record contains the largest and smallest values; hence, plotting positions should lie between 0 and 1 (or 0% and 100%). The different plotting position formulae tend to give similar values near the middle of the data, but can vary considerably at the tail ends.

The Weibull plotting position is most commonly applied. This has been adopted by the US Geological Survey and is applicable for both annual flood data and the partial-duration series (Dalrymple, 1960). The choice of a plotting position formula is, in fact, the same as choosing an underlying probability distribution. Hosking et al. (1985) recommends the second formula for the general extreme value distribution (Section 8.2.6). Cunnane (1978) reviewed several plotting position formulae and found that some of the traditional plotting formulae yielded plots which were biased, leading, on average, to overestimation of flood peaks at high recurrence intervals. He suggested the third formula and recommended $\alpha = 0.40$ as a single compromise value for use with all distributions. Other values of α are given

Table 8.2. Plotting position formulas (N = number of years of record, m is the rank of the event, the largest event having rank 1, α = constant, P = probability, and T = average recurrence interval, years). After Cunnane (1978) and Hosking (1985)

Name	Probability of exceedance, P	Average recurrence interval, T
Weibull	$\frac{m}{N+1}$	$\frac{N+1}{m}$
GEV	$\frac{m-0.35}{N}$	$\frac{N}{m-0.35}$
Cunnane	$\frac{m-\alpha}{[(N+1)-2\alpha]}$	$\frac{[(N+1)-2\alpha]}{(m-\alpha)}$

in the discussion of individual probability distributions (Section 8.2.6). A probability plot using the Cunnane plotting position with $\alpha = 0.40$ can be constructed with an option in program FLOODFRQ in AQUAPAK.

There can be a great amount of uncertainty in the plotting positions for the largest events. The extreme events may plot as "outliers", far off the line defined by the more frequent events. There are statistical tests available for testing outliers, as given by IAC (1982) and IEA (1987). Data from another stream with a longer record can be used to obtain a better estimate of the average recurrence interval for these larger events, as described by Dalrymple (1960). Regional analysis (Section 8.5.3) is also recommended rather than studying a single site in isolation.

A line is fitted through the plotted data to develop a flood-frequency curve. Linsley et al. (1975b) recommend as a general rule of thumb that a curve can be fitted by eye if one intends to use the frequency analysis for information on floods with recurrence intervals less than $N/5$. For larger recurrence intervals a theoretical probability distribution should be fitted to the data to enable more consistent, objective estimates. Straight lines can be drawn between consecutive data points for interpolating values directly from the data, although this should not be attempted for the larger flood flows. More often, a straight line is fitted through the entire data set. When eye-fitting this line more weight should be given to the middle and upper values since these will normally be the flood magnitudes of interest. Extrapolation beyond the range of data should never be done with an eye-fitted curve. Errors in the estimates of extreme events can be very large using graphical procedures, especially if the eye-fitted line is poorly judged or if the data do not follow a straight line with the chosen probability paper and plotting position.

8.2.5 FITTING PROBABILITY DISTRIBUTIONS

General information on probability distributions and descriptions of statistics such as mean, variance and skewness are given in Appendix 1 and Section 2.3.1. The Normal or Gaussian distribution is probably familiar to most scientists and engineers. The probability distribution of flood peak data, however, is rarely bell-shaped. One reason for this is that flood data have some finite lower limit. This will be zero in many streams; in others, there will be some theoretical lower limit which represents the smallest expected annual flood. Second, flood data are usually skewed to the right (positively), with a few very high values creating a smoothly tapering "tail" on the upper side. The flood data for the Acheron River are shown as a frequency histogram in Figure 8.3.

The "true" underlying mathematical law to which a stream's flood data conform will never be known absolutely, but we can *assume* that the data closely follow a certain distribution and *test* how well they are described by it. By fitting a theoretical distribution to flood data, statistical methods such as confidence limits and tests of hypothesis can be employed to assist in the interpretation of flood-frequency curves. Inferences can thus be made about how "good" a flood estimate is. In general, a probability distribution will fit the data best and provide the most accurate estimates near the middle values; less so near the tail ends.

Although no one distribution will fit all flood data, specifying the distribution used and the method of fitting it will allow other researchers to obtain the same results from the same set of data. The procedure is thus much more objective than graphical methods using eye-fitted curves. An additional benefit of using probability distributions is that the estimated parameter values compactly summarize the characteristics of the distribution. Often, these parameters are related to factors such as catchment area, rainfall, topography and other physiographical and meteorological measures (Haan, 1977). The estimated parameter values can be used in a regional analysis (Section 8.5.3) or as factors for stream classification (Section 9.2).

A large variety of distributions have been investigated for application to

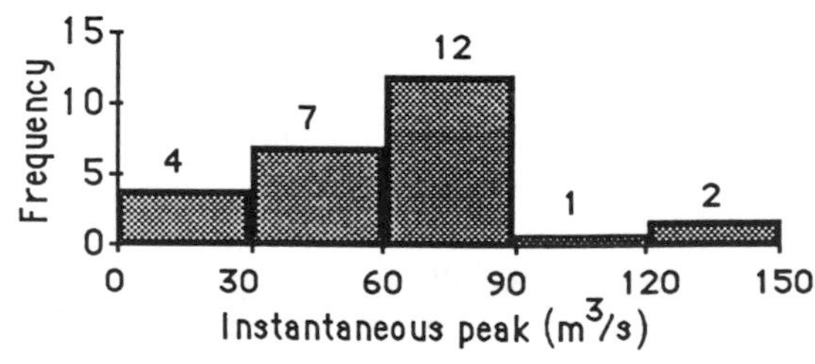

Figure 8.3. Frequency histogram for the annual maximum instantaneous flows on the Acheron River at Taggerty (see Figure 4.7), years 1963, 1965–1989 (26 years of data)

flood data, and, as with plotting positions, hydrologists constantly debate the relative merits of each. Some distributions will be more appropriate for some streams than for others. In the United States and Australia the **log Pearson Type III (LPIII)** distribution has been selected as a standard by federal agencies. Benson (1968) and IEA (1987) describe the criteria considered in making the decision. The LPIII distribution includes the **log-Normal (LN)** distribution as a special case. The **general extreme value (GEV)** distribution is the standard method for flood-frequency analysis in the UK (Chow et al., 1988). The **Weibull** distribution is often used for the description of low-flow frequencies. These distributions are discussed in Section 8.2.6.

Probability distributions are defined by their parameters. To fit a distribution to flood peak data one must estimate the "true" parameter values from information contained in the sample data series. The three approaches most often used for fitting probability distributions are:

- Graphical
- Method of moments (including probability-weighted moments)
- Method of maximum likelihood

In the **graphical method** plotting positions are assigned as described in Section 8.2.2. The plotting position formula and probability paper must be compatible with the distribution to be fitted. If a straight line can be fitted through the data, the parameters can be directly estimated. For example, if normal probability paper is used the 50% mark represents the mean and the 15.9% and 84.1% marks are one standard deviation from the mean.

Probability paper can be developed for particular distributions with specific parameter values by scaling the axes appropriately. For distributions in which skew is a parameter, such as the LPIII or Weibull distributions, a different probability paper must be constructed for each value of skewness. For the interested reader, Haan (1977) gives further information on constructing probability paper.

In the **method of moments**, moments of the "true" probability distribution are estimated from the data series. An explanation of moments, which are measures describing the location, scale and shape of a distribution, is given in Appendix 1. These are derived from sample statistics such as the mean, standard deviation and coefficient of skewness. For hydrological variables, moments greater than third order are not generally computed. Higher-order moments are very unreliable unless extremely large sample sizes are used—much greater than the 20 years or so of streamflow record typically available.

In most frequency analyses using the method of moments, the relation between flood magnitude and probability can be reduced to a simple equation given by Chow (1964b) as:

$$x_T = \bar{x} + K_T s \tag{8.2}$$

where x_T is the magnitude of an event with an average recurrence interval T years, $\bar{x}$ and s are the mean and standard deviation of sample values, respectively, and K_T is a **frequency factor**. To fit a distribution, the mean and standard deviation are calculated from the sample data and entered into Equation 8.2 with the appropriate K_T value. If an appropriately scaled probability paper is used only two values of x_T need to be calculated (three to be safe) to define the distribution by a straight line.

The **method of maximum likelihood** was first introduced in the 1920s by Fisher, a geneticist and statistician (Devore, 1982). For large sample sizes this method is superior to others since the resulting estimators of population parameters are considered to be more efficient and accurate. In this method a likelihood function is derived which indicates how likely the observed sample is, assuming that it is from a certain distribution with a range of possible parameter values. Maximizing this likelihood function yields parameter values which agree most closely with the observed data (Devore, 1982).

Solving the maximum likelihood equations to obtain parameter values normally requires an iterative procedure and can need a substantial amount of computer time. Since an efficient estimate will not necessarily exist, a solution may or may not be found.

Probability weighted moments (PWMs) are similar to conventional moments in that they summarize and describe the characteristics of a probability distribution. Hosking (1989) defines a type of PWMs called L-moments, in which the moments are linear functions of the data values, hence the "L". In comparison to conventional moments in which the data values are squared, cubed, etc., with L-moments it is the probabilities which are manipulated. This gives less weight to the very high or very low data values.

The advantage of L-moments is that they are less sensitive to sampling variability and less subject to bias. They are robust in the presence of outliers, meaning that they give consistent results even if the extreme values contain measurement errors. For small samples they produce parameter estimates which are sometimes more accurate than even maximum likelihood estimates (Hosking, 1989).

8.2.6 A FEW GOOD PROBABILITY DISTRIBUTIONS

In general, the more parameters a distribution has, the better it will fit a set of data and the more flexibility it has for fitting many different sets of data. However, for the amount of data normally available the reliability in estimating more than two parameters may be very low. "Thus a compromise must be made between flexibility of the distribution and reliability of the parameters" (Haan, 1977, p. 146).

The distributions presented are those with the widest use in hydrology, and some of their more common applications will be discussed. NERC (1975) provides a more comprehensive summary for further information. In AQUAPAK, flood-frequency program FLOODFRQ will fit a selection of probability distributions to a data series.

Normal distribution

The normal distribution is a two-parameter distribution, defined by its mean and variance. The K_T value in Equation 8.2 is simply the standardized normal variate Z tabulated in most statistical texts. In the Cunnane plotting position (Table 8.2) $\alpha = 0.375$ for the normal distribution.

The mean of a normal distribution corresponds to a 0.5 probability, indicating that 50% of the values are higher, 50% lower. For an annual series this would correspond to the 2-year average recurrence interval. Haan and Read (1970) found that annual water yield from small watersheds in Kentucky followed a normal distribution. In program FLOODFRQ in AQUAPAK, both the normal and log-normal distributions are fitted by the method of moments, which for these distributions is the same as the method of maximum likelihood. Appendix 1 provides further information on thc normal distribution.

Log-normal distribution (LN)

Like the normal distribution, the two-parameter log-normal distribution is defined by its mean and variance. However, in this case it is the logarithms of the data which follow a normal distribution. The data are usually transformed by taking either natural (base-e) or base-10 logarithms. The mean and standard deviation would be calculated using these transformed values, and Equation 8.2 becomes:

$$\log x_T = \overline{\log x} + K_T s_{\log x} \tag{8.3}$$

where "log" can represent either base-10 logarithms ($\log_{10}$) or natural logarithms (ln). The K_T value is the same as for the normal distribution. When a value of $\log x_T$ is computed, the antilog is taken to convert back to original units:

$$x = 10^{\log_{10} x} \text{ (for base-10 logarithms)} \tag{8.4}$$

or

$$x = e^{\ln x} \text{ (for natural logarithms)} \tag{8.5}$$

As mentioned in Section 5.10.5, the log-normal distribution is often used in the description of sediment size data; it also applies to the distribution of raindrop sizes (Chow et al., 1988). Flood peak data and low flow data also often follow a log-normal distribution.

In the Cunnane plotting position (Table 8.2), $\alpha = 0.375$ for this distribution. FLOODFRQ fits the log-normal distribution by the method of moments. A log-normal distribution has been fitted to the Acheron River flood data in Figure 8.4.

Log-Pearson Type III distribution (LPIII)

The log-Pearson Type III distribution is a three-parameter distribution. It lends itself well to flood data which tend to have a lower limit but no upper limit and also contains the log-normal distribution as a special case when the skew coefficient is zero. These reasons as well as the flexibility of a three-parameter distribution have made the LPIII popular for flood-frequency analysis.

Estimates of the moments are computed from the logarithms of the data series similar to the procedure for the LN distribution. K_T values are provided by Haan (1977), IEA (1987) and Linsley et al. (1975b). Equation 8.4 or 8.5 is used to convert back from the log domain. For the Cunnane plotting position, $\alpha = 0.4$ (Srikanthan and McMahon, 1981).

A negative skew value, which is common for Australian flood data, means that the LPIII distribution has an upper bound. This can theoretically cause difficulties if one wants to estimate, say, the 1000-year flood. However, for most applications it is not a problem (IEA, 1987). Since small sample sizes tend to give unreliable estimates of the coefficient of skew, the US Water Resources Council (IAC, 1982) recommends using a regionalized estimate. This regional skew coefficient would be used to obtain the K_T value for Equation 8.2. Chow et al. (1988) give this procedure as well as a US map showing generalized skew coefficients of annual maximum streamflows. For Australia, the method is given by IEA (1987).

In FLOODFRQ the LPIII distribution is fitted by the method of moments. A sample output is shown in Figure 8.4. For these data, the skew is slightly negative.

The extreme value (EV) distribution

The largest or smallest values—the extremes—are of particular importance in hydrology. Examples would be the maximum annual peak flow, the minimum 10-day flow or the minimum monthly flow. The extreme value distribution is based on the idea that these extreme values are taken from the ends of some "parent" distribution. If the fur thickness on water rats were considered as an analogy, the "parent" population might be the fur lengths over the whole rat and the "extreme values" the lengths on just the tails or noses. The distribution of these extreme values is dependent on the sample size as well as the parent distribution.

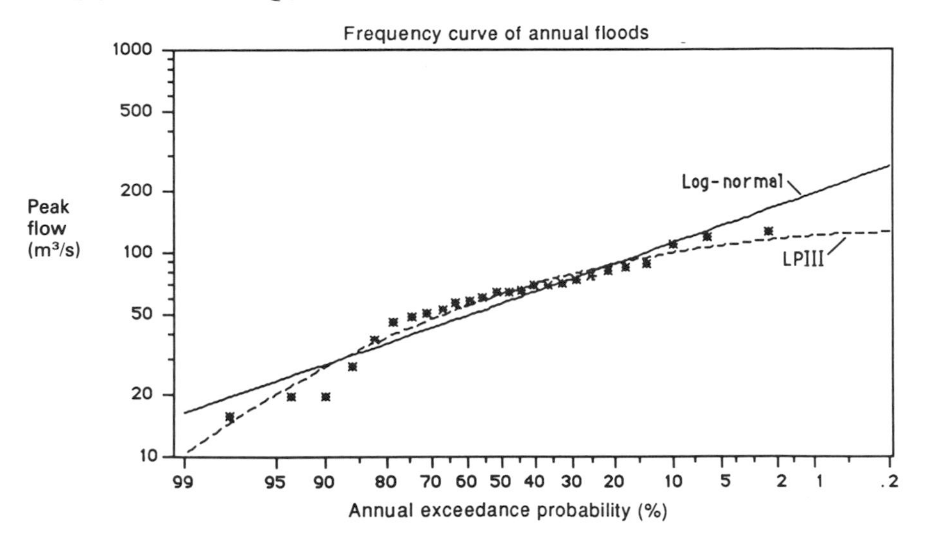

Figure 8.4. Frequency analysis for the annual maximum series, Acheron River at Taggerty, years 1963, 1965–89. Log-normal and LPIII distributions have been fitted to the data and plotted on log-normal probability axes. Peak flows are in m^3/s

Three types of extreme value distributions have been developed (Haan, 1977):

EVI: distribution has no upper or lower limits;
EVII: distribution is bounded on the lower end;
EVIII: distribution is bounded on the upper end.

These three forms are special cases of a three-parameter distribution called the **general extreme value (GEV)** distribution (Jenkinson, 1955). This is included as an option in program FLOODFRQ, and the distribution is fitted using probability-weighted moments. The plotting position for the GEV is given in Table 8.2.

The EVI distribution has a constant skewness of 1.1396; for the EVII, it is higher and for the EVIII, it is lower. There has been little interest in the EVII distribution in hydrology. The EVI distribution is often used in flood-frequency analysis, and a form of the EVIII distribution is commonly used in the analysis of low flows. Only the EVI and EVIII distributions will be discussed here.

EVI distribution

An EVI distribution is also called the Gumbel or double-exponential distribution. It is described by two parameters, a scale parameter and a location parameter, where the latter is the mode of the distribution. As mentioned earlier, the coefficient of skew is a constant, 1.1396. The log-normal distribution is a special case of the EVI distribution when the coefficient of variation is 0.364. Also, the average recurrence interval for the mean value is a constant 2.33 years. It is from this relationship that the **mean annual flood** is often defined as a flood with an average recurrence interval of 2.33 years.

The EVI has been found to be satisfactory for describing the distribution of yearly maximum daily discharges (Haan, 1977). Frequency factors (K_T values) for the EVI are given by Haan (1977) and Chow (1964b), and Gumbel paper is commercially available. For the Cunnane plotting position, $\alpha = 0.44$ for the EVI distribution. FLOODFRQ fits this distribution by the method of probability-weighted moments.

EVIII distribution

In hydrology, the EVIII distribution has most often been applied to the analysis of low flows, particularly the annual minimum flows. This distribution is attractive because of its fixed limit. By "turning over" the curve the limit becomes a lower rather than an upper limit. This is accomplished by fitting the distribution to $(-x)$ rather than (x). For this special case, it is known as the **Weibull** distribution, after Weibull, who first used it in an analysis of the strength of materials.

The distribution has three parameters: (1) a lower limit (zero or some low finite value), (2) a location parameter and (3) a skewness parameter. The lower limit has been used as the *probable minimum flow*; i.e. the stream is never expected to drop below this value if the same hydrological conditions are maintained. The location parameter defines the low-flow values which will be exceeded about 63% of the time (or a recurrence interval of 1.58 years), and is sometimes called the **characteristic drought**.

The data used in a low-flow analysis are the annual low flows of some duration (e.g. the lowest 1-day or lowest monthly flow in the year). If seasonal flows are of concern, the low-flow values can be selected only from a particular season within each year. Generally, the number of consecutive low-flow days is set equal to 1, 7, 15, 30, 60, 120, 183 and 284 days to cover the range of possible durations. Program WEIBULL in AQUAPAK is provided for low-flow analysis. In this, the Weibull distribution is fitted by the method of moments.

As for flood-flow analysis, it is important that the data are independent. For long durations they will inevitably contain serial dependence. Again, a calendar year may not be an appropriate time base because a drought period may cross the December–January boundary, especially in the southern hemisphere. The streamflow data should be divided into climatic years which start at the time of year when the flow is likely to be high and seasonal low-flow periods are not bisected (WMO, 1983). Using a "climatic year" which is 6 months out of phase with the water year is one approach. The starting month for the year can be specified as an option in program WEIBULL.

The first few (most frequent) values in a low-flow frequency curve often plot as a much steeper line than the remaining values (IH, 1980). It can be assumed that the break in the curve indicates the point where flows are no longer considered "drought flows", and should be excluded from the frequency analysis. IH (1980) indicates that this break tends to occur at an exceedance probability of around 65%. However, for Victorian streams, Nathan and McMahon (1990c) adopted a value of 80% as being more appropriate. The distribution is then fitted only to those flows with exceedance probabilities less than this threshold value. In the same manner as for zero flows (Section 8.2.7), a correction is applied to the probabilities to account for the excluded flows. Program WEIBULL fits the Weibull distribution only to the lower 80%. It should also be noted that the probability axis is truncated so that the more common low flows are not shown.

As with other types of analysis, extrapolation beyond the record should be done with caution. However, Gumbel (1963) found that satisfactory forecasts could be made up to an average recurrence interval of $2N$ on several American rivers.

Some countries base water quality standards on low flow conditions such as the 7-day, 2-year low flow (McMahon and Mein, 1986). Bovee (1982) also states that the 7-day, 10-year low flow is often considered a "minimum streamflow" for determining reservoir storage for water supply and in the design of wastewater treatment. However, it is considered much too low for supporting aquatic ecosystems if maintained as an instream flow all year.

A sample of the output from WEIBULL is shown in Figure 8.5. From the graph, the 7-day, 2-year low flow (probability 50%) is about 0.25% of the mean = (0.0025)(3785.7) = 9.46 m^3/s (as a 7-day total) or an average daily flow of 9.46/7 = 1.35 m^3/s. This means that a 7-day period in which the flow averages 1.35 m^3/s occurs as a minimum about every 2 years (or conversely, there is a 50% chance of this taking place in any one year).

8.2.7 ZEROS AND HOW TO TREAT THEM

In hydrology, data often have zero as a lower limit—negative streamflows making little sense except perhaps where water runs upstream in an estuary. However, zeros present a problem because they cause the distribution to be truncated; i.e. P (0) is not zero: 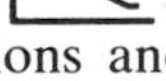 . One approach is to add a small, constant value to all the observations and then make a correction to the final solution. However, this can greatly affect estimates of extreme events, and as a result, the method should not be used.

Another approach uses the theorem of total probability as described by

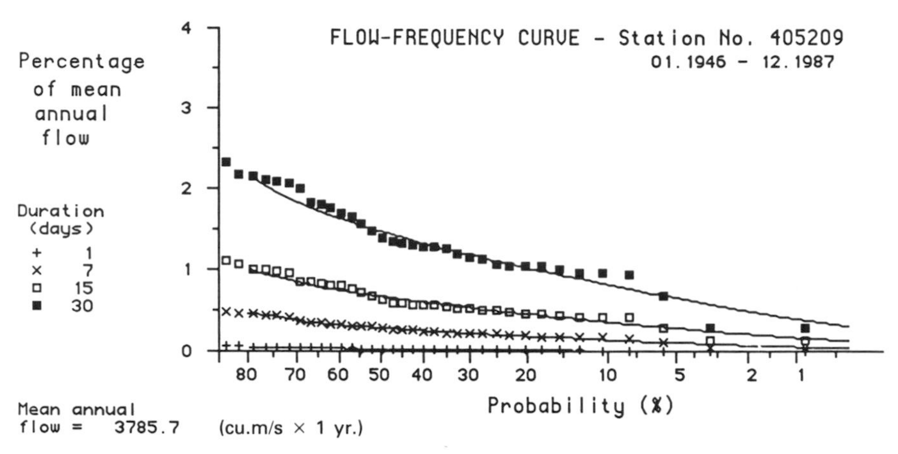

Figure 8.5. Sample output from program WEIBULL for low flow-frequency analysis

Haan (1977). For non-negative values this theorem simply says that the probability of flows equalling zero plus the probability that they are greater than zero accounts for all of the probability: or, as more concisely stated,

$$P(x = 0) + P(x > 0) = 1 \quad (\text{if all } x \geq 0)$$

To account for the proportion of zero values all non-zero observations are analysed, and the resulting probabilities are multiplied by the fraction of non-zero values in the data. For example, if one quarter of the data were zero values, and from the remaining data the probability of a flood being exceeded was calculated as 0.2, the actual probability of exceedance would be (0.75)(0.2) = 0.15. This relationship can be written as:

$$P(x) = cP^*(x) \tag{8.6}$$

where $P(x)$ is the probability of exceedance for all values, c is the probability that x is not zero ($P(x \neq 0)$), and $P^*(x)$ is probability of exceedance for the non-zero values.

For those who prefer their water glasses empty rather than full, the probability that x equals 0 can be used instead:

$$P(x) = (1 - c')P^*(x) \tag{8.7}$$

where c' is now the probability that x is equal to zero, $P[x = 0]$. This technique can be easily adapted for threshold values other than zero, as mentioned in regard to low-flow frequency analysis (Section 8.2.6).

All data above zero or some threshold are analysed by graphical or numerical frequency analysis techniques and the calculated exceedance probabilities are adjusted using Equations 8.6 or 8.7. A new flood-frequency curve is then re-plotted with the adjusted values. The zero values have the effect of shifting the frequency curve to the right, making large events even more rare and discharges lower for the same probability.

Example 8.1

We will consider the effect of zero values on the flood-frequency curve for the Acheron River (Figure 8.4). Let us say that 6 years of very severe drought occurred after the actual period of record, during which the flow was zero. This would give a total of 32 years of record. For the original observations we will assume that they are distributed log-normally with $\overline{\ln x} = 4.041$ and $s_{\ln x} = 0.5335$. Taking the zero flows into account:

(a) What is the probability of an annual peak flow which exceeds 150 m^3/s?
(b) What is the magnitude of the 20-year flood?

Answer:

(a) The probability of non-zero values is estimated as $c = 26/32 = 0.8125$. Then Equation 8.6 becomes:

$$P(150) = 0.8125 P^*(150)$$

$P^*(150)$ can be obtained by solving Equation 8.3 for K_T and then finding the probability from a table of values for a standardized normal distribution.

$$\begin{aligned} \ln x_T &= \overline{\ln x} + K_T s_{\ln x} \\ \ln(150) &= 4.041 + K_T(0.5335) \\ 0.969 &= 0.5335 K_T \\ K_T &= 1.82 \end{aligned}$$

Using this value in place of the Z values in a standard normal table, this gives a probability of 0.9656. This represents the cumulative area  . However, we want the exceedance probability, which is the area  . Because the total area is 1, the probability of exceedance for the non-zero values is:

$$P^* = 1 - 0.9656 = 0.0344 \text{ (the inverse of this is 29, the average recurrence interval in years)}$$

Thus,

$$P(150) = (0.8125)(0.0344) = 0.028$$

or the probability that a flood of 150 m^3/s will be exceeded in any year is 2.8%. The recurrence interval is $1/0.028 = 36$ years, as compared to 29 years for the non-zero data.

(b) The probability of a 20-year flood being equalled or exceeded in any year is $1/20 = 0.05$. This value is entered into Equation 8.6 along with the previous value of c to obtain:

$$P(x) = 0.05 = (0.8125)(P^*(x))$$

giving $P^*(x) = 0.06154$. Again, to enter the standard normal tables, a cumulative value is needed: $(1 - 0.06154) = 0.938$. This yields a K_T value of about 1.54. Then, from Equation 8.3, the peak flow associated with a 20-year average recurrence interval is:

$$\ln(x_{20}) = 4.0414 + 1.54(0.5335) = 4.86$$

and

$x_{20} = e^{4.86} = 129\ m^3/s$ (as compared with $136\ m^3/s$ for the original data).

8.2.8 GOODNESS, THE DISTRIBUTION FITS!

From a statistical point of view we often want to know how well the chosen distribution fits the observed data. With graphical methods this is done by visually assessing the goodness of fit. Somewhat less subjectively, statistical goodness-of-fit tests are available to test the hypothesis that the observed data are actually from the fitted probability distribution. Two common methods are the chi-square test and the Kolmogorov–Smirnov test.

According to Haan (1977), many hydrologists discourage the use of these goodness-of-fit tests when fitting distributions to streamflow data. One main reason is that these statistical tests tend to be insensitive in the tails of the distributions—the areas of most importance for prediction of extreme events. Neither of the tests given are very powerful, especially with small samples; i.e. the probability of accepting the hypothesis that the distribution fits, when in fact it does not, is very high. This may be of little comfort to those looking for an objective basis for deciding whether a distribution fits the data or not. However, the statistics can be of assistance when comparing the relative merit of one distribution or one set of parameters against others.

Chi-square test

The chi-square test is one of the most commonly used goodness-of-fit tests. The actual number of observations falling into "class intervals" (e.g. the bars in the frequency histogram of Figure 8.3) are compared with the number of observations expected from a given distribution. The chi-square test statistic (χ^2) is calculated as (Haan, 1977):

$$\chi^2 = \sum_{i=1}^{n} \frac{(O_i - E_i)^2}{E_i} \tag{8.8}$$

where n is the number of class intervals and O_i and E_i are the observed and expected number of observations, respectively, in the ith class interval. As a generality, the number of class intervals should be 5 or greater, but for hydrological data with sample sizes on the order of 30–60, 6 or 7 class intervals is a good number (Yevjevich, 1972). The calculated χ^2 statistic is compared with tabulated values found in most statistical texts, and the hypothesis that the data are from the probability distribution is rejected if:

$$\chi^2(\text{calculated}) \geq \chi^2_{1-\alpha,\, n-p-1}$$

where p is the number of distribution parameters estimated (e.g. two for log-normal, three for LPIII) and α is the level of significance.

The χ^2 test can also be conducted using unequal-width class intervals. Details can be found in Haan (1977). Haan also suggests combining classes if the expected number of observations in the class is less than three.

Example 8.2

Test the hypothesis H_o: the Acheron River flood frequency data (Figure 8.4) are log-normally distributed. Again, $\overline{\ln x} = 4.0414$, and $s_{\ln x} = 0.5335$, with 26 years of data.

Answer:

Figure 8.3 shows the distribution of the observations. To calculate the expected number of observations in each class interval we use probabilities obtained from a standard normal distribution table, multiplied by 26. The tables are entered using the value of K_T from Equation 8.3 as for Example 8.1. The first two calculations are shown, and the remainder are tabulated below.

Class interval 0–30: $\ln(30) = 4.014 + K_T(0.5335) \Rightarrow K_T = -1.20$
Probability from the normal table: 0.1151

Class interval 30–60: $\ln(60) = 4.014 + K_T(0.5335) \Rightarrow K_T = 0.09$
Probability from the normal table: 0.5359, so for the interval it is $0.5359 - 0.1151 = 0.4208$

Class interval (m^3/s)	K_T	$P(Z \leq z)$ from std normal table	Prob. for class interval	Expected	Observed
0–30	−1.20	0.1151	0.1151	3.0	4
30–60	0.09	0.5359	0.4208	10.9	7
60–90	0.86	0.8051	0.2692	7.0	12
90–120	1.40	0.9192	0.1141	3.0	1
>120	—	1.0000	0.0808	2.1	2

Then, from Equation 8.8,

$$\chi^2 = \frac{(4-3)^2}{3} + \frac{(7-10.9)^2}{10.9} + \cdots + \frac{(2-2.1)^2}{2.1} = 6.64$$

To obtain a tabulated value of χ^2 we use $n-p-1 = 5 - 2 - 1 = 2$ and $\alpha = 0.05$, to obtain $\chi^2_{0.95,2} = 5.99$. Because the calculated value is higher, we conclude that the log-normal distribution does not adequately describe the data at a significance level of 0.05.

Kolmogorov–Smirnov test

This is a quick and simple goodness-of-fit test. Whereas the chi-square test is an example of a parametric method (where the distribution is known or assumed), the Kolmogorov–Smirnov test is a non-parametric method. The only condition required is that the assumed distribution is continuous; i.e. it can be of any form. This means that the method can be used for "eye-fitted" lines through plotted points as well as analytically fitted distributions.

To visualize this test, the assumed distribution and the ranked data are plotted on probability paper. The *maximum deviation* between the data and the fitted line is the test statistic, k_s:

$$k_s = \max |F(x) - O(x)| \tag{8.9}$$

where $F(x)$ corresponds to the cumulative distribution function of the assumed distribution and $O(x)$ to the plotting position for observation x. These values can be tabulated rather than plotted, and the maximum deviation calculated directly. Values of $F(x)$ would be computed to correspond with each value of $O(x)$, as for Example 8.3.

For a chosen level of significance, the calculated value of k_s is compared with tabulated values of the Kolmogorov–Smirnov statistic (Table 8.3). If k_s is equal to or greater than the critical tabulated value the hypothesis that the sample data are from the given distribution is rejected. This is simply a formal method of evaluating the largest "gap" between the sample values and the fitted line. Care should be taken when using probability paper to evaluate the largest deviation and not the largest distance. For example, in Figure 8.4 the maximum deviation between the data and the log-normal curve is about $(0.61-0.45) = 0.16$. For 26 values and a significance level of 0.05, the critical tabulated value ≈ 0.26. Since k_s is less than this value, this indicates that the LN distribution adequately describes the data at a significance level of 0.05.

8.2.9 INTERPRETING FREQUENCY CURVES

Annual maximum frequency curves are analysed and interpreted differently than partial-duration curves. Although more points can be produced from

Table 8.3. Critical values of the Kolmogorov–Smirnov test statistic for a significance level of 0.05. After Birnbaum (1952), by permission of the American Statistical Association

Sample size	Critical value
5	0.563
10	0.409
15	0.338
20	0.294
25	0.264
30	0.242
40	0.21
50	0.19
70	0.16
100	0.14

the partial-duration series for defining the flood-frequency relationship, more points does not necessarily guarantee better estimates (NERC, 1975). One drawback of including more floods in the analysis is the increased possibility that flood peaks will be interdependent, since the conditions affecting one flood (i.e. weather patterns or soil moisture conditions) may also affect others occurring during the same season or year.

However, smaller floods will occur more frequently than indicated by the annual maximum series, because several peaks in one year may be higher than the highest flood in others. Also, some of the smaller peaks selected as annual maxima may not really be floods. Thus, for these smaller, more frequent events the partial-duration series is more appropriate.

For average recurrence intervals (ARI) of about 10 years or more the difference between the annual maximum series and the partial-duration series becomes very small. As can be seen in Figure 8.6 for the Acheron River, the two series converge at an ARI of about 6 years. It can also be seen that the partial-duration series is simply an extension of the annual exceedance series. IEA (1987) recommends that the partial duration series should be used for ARI < 10 years and the annual maximum series for ARI ≥ 10 years. For the partial-duration series, with ARI < 10 years, graphical interpolation between data points is considered sufficiently accurate. However, a probability distribution should be fitted for making inferences about the probability of rarer events.

For the annual maximum series the exceedance probability is the probability that a flood of a given size or larger will occur in any one year. The average recurrence interval is the average interval in which a flood of a given size or larger will occur as an annual maximum. Referring to the LPIII curve in Figure 8.4, the exceedance probability of a flood of 100 m^3/s

is about 20%. This is the probability that the highest flood in a year will exceed 100 m^3/s. Alternatively, the average recurrence interval is 1/0.2 = 5 years, meaning that, on average, the flow will exceed 100 m^3/s once every 5 years.

Probabilities obtained from the annual maximum series can be used to answer questions of the form: "What is the probability of the 10-year flood being exceeded in the next 20 years?" This is done by simple "dice-rolling" probability methods. For example, the probability of one or more events exceeding the 10-year flood in the next 20 years is:

$$1 - (0.90)^{20} = 0.88$$

The general formula is given by Chow et al. (1988) as:

$$P(X \geq x_T \text{ at least once in } N \text{ years}) = 1 - \left(1 - \frac{1}{T}\right)^N \tag{8.10}$$

In the partial-duration series, of which the annual exceedance series is a special case, the "average recurrence interval" is still the average interval between floods exceeding a given size. However, the floods are no longer *annual* maxima, and probability is no longer a useful concept (IEA, 1987). The annual exceedance series (N values in N years of record) can still be plotted on probability paper, but the probability scale should be converted to average recurrence intervals (ARI) by taking the inverse (Equation 8.1). For example, in Figure 8.6 the "2-year flood" from the annual exceedance series is about 70 m^3/s. Thus, in the 25 years of record used in the analysis this discharge was exceeded 25/2 = 12.5 times. Alternatively, we can say that we would expect a flood of this size to be exceeded 50 times (100/2) within a 100-year period.

If the threshold for constructing the partial-duration series is reduced to include more than N values for N years the series can no longer be plotted on probability paper. Plotting positions for the average recurrence interval can still be calculated from the formulae in Table 8.2, with N the number of years (not the number of values). The results are plotted on semi-logarithmic axes with discharge on the linear axis and average recurrence interval on the log scale, as in Figure 8.6. Floods with average recurrence intervals of less than one year can then be evaluated. For example, from the partial-duration series in Figure 8.6 the flow exceeded every 4 months, on average (ARI = 0.33 yr), is about 30 m^3/s. NERC (1975) provides a more rigorous explanation of the partial-duration series.

The partial-duration flood series may be practical for studying how frequently different levels within a stream channel are inundated. Gregory and Madew (1982) recommend an evaluation of the annual exceedance series when investigating changes in flood frequency due to land-use change. Their reasoning is that the less extreme discharges may show a change in frequency due to land use, whereas the largest events may not be affected.

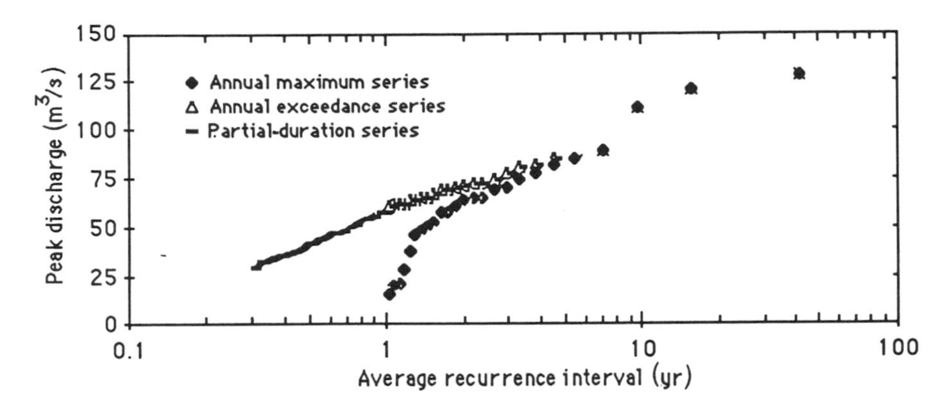

Figure 8.6. Flood data for the Acheron River at Taggerty, years 1963, 1965–88 (25 years). Shown are the annual maximum series, annual exceedance series, and partial duration series with threshold = 29 m^3/s (total of 86 values, 3.4/yr). The data are plotted on semi-logarithmic axes. The average recurrence intervals were computed from the Cunnane plotting position with α = 0.4

A frequency curve is interpreted under the assumption that it is a good representation of the population from which the observed flood data were taken. When comparing frequency curves for two different streams it should be kept in mind that the curves may differ because of actual flood-generating differences or simply due to chance, or both. Reliability can be assessed by computing confidence limits and/or standard errors of estimate. Methods will not be given here but can be found in references such as Chow (1964b), Haan (1977) and NERC (1975).

8.3 FLOW-DURATION CURVES

8.3.1 GENERAL

Flow-duration curves display the relationship between streamflow and the **percentage of time** it is exceeded. They have been used for assessing the percentage of time a streamflow will provide adequate dilution for industrial wastes or sewage, for evaluating the feasibility of hydropower stations and for investigating instream flow requirements. By evaluating these curves we can ask, for example, "what percentage of time is the daily flow on the Acheron River above 2 m^3/s (about 0.05% of the mean annual flow)?" From the daily curve in Figure 8.7 it can be seen that 90% is the correct answer. Thus, in a period of one year, about 329 days (0.9 × 365.25) will be above 2 m^3/s and 36 days below. This discharge is also called the daily Q_{90}.

As compared to a flood or low-flow frequency analysis, a flow-duration curve is derived from all the data rather than just the high or low flows. The streamflows analysed are the flows over some period of time or **duration** (e.g. one hour, one day, 10 days, 6 months, or a year). The period chosen will depend on the time interval of the available data as well as the intended use of the duration curves. If computerized methods are not available the amount of work required will also be a factor.

Daily curves will show more of the details of variation in the data than monthly or yearly ones, where extremes are smoothed out through averaging. For evaluating the flows over the length of time required for, say, the hatching of mayflies, the migration of ducks, or the germination of sedges, specific durations such as 3 days, 10 days or 2 months may be appropriate. Data can be analysed with program FLOWDUR in AQUAPAK to produce annual, monthly and daily curves as in Figure 8.7.

These data, particularly the shorter-duration data, will be serially correlated (Appendix 1) rather than being independent. Therefore a probability distribution is not fitted to the data. Rather, flow-duration curves are simply a way of representing the historical record, and provide another method for characterizing the pattern of streamflows.

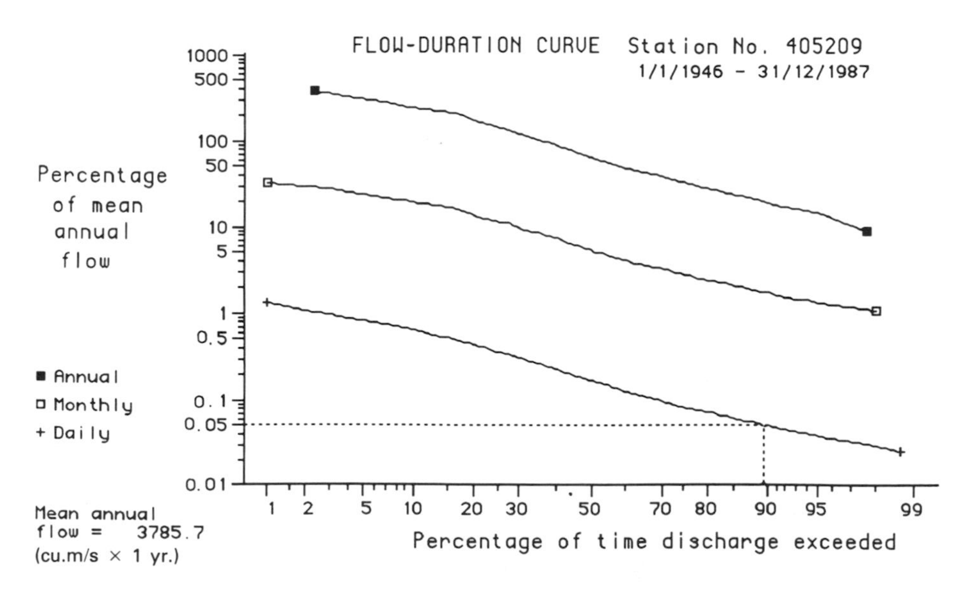

Figure 8.7. Daily, monthly and annual flow-duration curves for the Acheron River at Taggerty, based on calendar year data

8.3.2 CONSTRUCTING THE CURVES

Computationally, the technique is much like the sieving of sediment. Flows of a given duration are "sifted" into class intervals based on size. Searcy (1959) recommends 20–30 intervals. The number of items within each class interval is counted and expressed as a percentage of the total number of items "sifted" (either daily, monthly, yearly, or *n*-day flows). Beginning with the largest discharge, percentages are summed to obtain cumulative percentages of time above the lower limit of each class. As an alternative, the flows can be ranked from highest to lowest and assigned a plotting position (Table 8.2). The highest flow would have rank 1. Without a computer the latter method is more laborious.

When the cumulative frequencies have been computed the data are plotted on log-normal probability paper. This tends to linearize the relationship, which helps in defining the tail ends of the curves. For water-supply analysis the values are sometimes plotted on arithmetic paper, since the area under the curve is then equal to the total volume of water flowing by the gauging station.

It should be emphasized that the sequence of flows and the time of year in which they occur are not considered. It makes no difference whether July's data are put into the "sieves" before January's or whether the lowest flows occur in one month or are spread across all years. Because flow-duration curves are based on the whole record, the data are essentially averaged. Variability within the time period (e.g. diurnal fluctuations within a day or seasonal fluctuations within a year) will mean that actual flows within the time interval can be higher or lower. It may be desirable to supplement the "whole record" curve with duration curves for the single wettest and driest years on record to show the extremes. If the data record is of sufficient length it can also be divided into separate "samples" and flow-duration curves developed for each month or season.

In program FLOWDUR the flows are standardized as percentages of the average annual flow. This allows the comparison of curves from streams of different size. Searcy (1959) also suggests standardizing the discharges by dividing by the drainage area and expressing them in units of $m^3/s/km^2$. Class intervals are based on 1%, 2%, 5%, 10%, 20%, . . ., 80%, 90%, 95%, 98% and 99% exceedances. The number of intervals actually used in the analysis is based on the record length. For example, if a data series has less than 100 values the 1% exceedance (1/100) cannot be evaluated. In addition to a graph, FLOWDUR also generates an output file with a summary table.

8.3.3 INTERPRETATION AND INDICES

Flow-duration curves are a convenient way of portraying the flow characteristics of a stream under natural or regulated conditions. At the high flow end of the curve they can give an indication of the duration of overbank flows; at the lower end, they can be used to estimate the amount of time a particular log or rock is under water. The shapes of the curves also summarize the flow characteristics of a stream for comparison within or between catchments.

Typically, the curves flatten out as the duration increases from daily to yearly. As with a moving average (Section 4.4.2), increasing the amount of time over which the data are averaged tends to reduce the variability. The slope of the flow-duration curves reflects the catchment's response to precipitation. Stable streams such as the Acheron will tend to have monthly curves which differ little from daily ones in terms of slope. For "flashy" streams they will differ considerably.

The daily curve for flashy streams will tend to have a steep slope at the high-flow end. A flatter high-flow end is characteristic of streams with large amounts of surface storage in lakes or swamps, or where high flows have mainly resulted from snowmelt. The low-flow end of the curve is valuable for interpreting the effect of geology on low flows. If groundwater contributions are significant, the slope of the curve at the lower end tends to be flattened whereas a steep curve indicates minor baseflows (McMahon, 1976). Streams draining the same geological formations will tend to have similar low-flow duration curves (Searcy, 1959).

Flow-duration curves can provide a number of indices which characterize the stream for classification or regionalization purposes. If two stream sites are to be compared, it is important that the same period of record is used in deriving the duration curves. The median (Q_{50}) value (the flow exceeded 50% of the time) gives an "average" measure. The Q_{90} or Q_{95} values are commonly used as low-flow indices. The ratio Q_{90}/Q_{50} can be used as an index of baseflow contribution. At the high flow end of the curve a value such as the Q_{30} or Q_{10} may be of interest in indexing the length of time a floodplain is under water or as a variable which might correlate with channel dimensions. Harvey (1969), for example, suggests that the duration of floods of a given frequency generally increases in the downstream direction.

Measures of variability can be obtained from the slope of the flow-duration curves. One index, an estimate of the standard deviation of the logarithms of the streamflows ($s_{\log x}$), is given by IH (1980) as:

$$s_{\log x} = \frac{\log Q_5 - \log Q_{95}}{3.29} \tag{8.11}$$

Another measure is Lane's variability index (Lane and Lei, 1950), defined as the standard deviation of the logarithms of the Q_5, Q_{15}, Q_{25}, . . ., Q_{85} and Q_{95} values. This index is unsuitable for streams where zero flows comprise more than 5% of all the flows.

The indices apply only to the given duration, whether daily, monthly, annual, or *n*-day. For example, the Q_{90} for the annual duration curve is the flow which is lower than 90% of the annual values; the Q_{90} for the 10-day flows is the flow which is less than 90% of the 10-day flows. The standard deviation or Lane's index is also associated with a particular duration curve. Thus, the duration should be stated when citing index values.

Duration curves of sediment, turbidity, water hardness or other water-quality characteristic may also be derived if a relationship is established between streamflow and the characteristic of interest (e.g. the sediment rating curve, Section 7.5.4). A flow-duration curve must first be developed and then the discharge values are simply converted to the water quality measure by using the appropriate relationship. The amount of error depends on both the flow-duration analysis and on the correlation between discharge and the water quality measure.

In Section 8.5 methods of estimating flow-duration curves for ungauged stream sites and gauged sites with short record lengths are presented. Further information on the flow-duration curve technique is given by IH (1980) and Searcy (1959).

8.4 FLOW-SPELL ANALYSIS

Flow-spell analysis is a procedure developed by the UK Institute of Hydrology (IH, 1980) for the analysis of low-flow periods. Flow-duration curves (Section 8.3) give no information on how the low-flow days are distributed, and streams with similar flow-duration curves may be very different in the way low flows are grouped into long or short periods of time. Flow-spell analysis, in contrast, considers how long a low flow (below some threshold) has been maintained and how large a "deficit" has been built up, and thus takes into account the sequencing of flows.

The technique has been used for estimating the amount of storage needed on a catchment to maintain water supplies, and in checking the representativeness of synthetically generated streamflow time series (McMahon and Mein, 1986). In the analysis of low flows for ecological and water quality requirements it may be of more value than flow-duration curves.

A graphical description of the method is shown in Figure 8.8. Two measures are obtained from a flow-spell analysis: **spell duration** and **deficiency volume**. The spell duration is the length of time that the streamflow is continuously below a threshold and the deficiency volume is the amount of water which would be required to keep the stream at the threshold level. **Annual frequency** refers to the proportion of years in which a deficit volume or spell duration is exceeded. To analyse the spell data using this method, the *longest* spell duration or *largest* deficit volume below a given threshold is found for each year. These are the **annual spell maxima**.

In program SPELL in AQUAPAK, the threshold discharge is set at 5%,

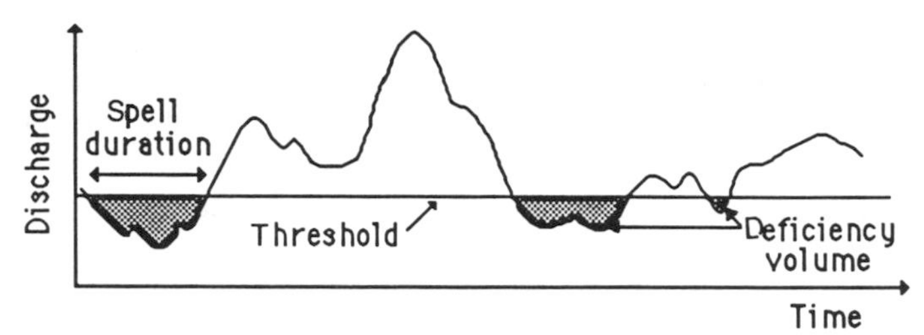

Figure 8.8. Generalized diagram of the flow-spell analysis technique. Adapted from IH (1980), by permission

10%, 20%, 40%, 60% and 80% of the mean daily flow. The results of the analysis are plotted as cumulative frequency curves. A sample plot is shown in Figure 8.9. The results are also tabulated in an output file. The spell duration is expressed in days and the deficiency volume expressed as a percentage of the mean annual flow, to allow comparisons between catchments. Mean annual flow has been expressed in the form ($m^3/s \times 1$ year) to maintain consistency with the flow duration and low-flow frequency analyses. This value can be converted to a volume in million m^3 by multiplying by 0.0864.

IH (1980) and Nathan and McMahon (1990a) recommend fitting a log-normal distribution to the spell maxima. In program SPELL, a log-normal distribution is fitted to the non-zero data with exceedance probabilities less than 80%. In the output file, the probabilities are adjusted for the reduced data set using the method described in Section 8.2.7. A sample plot is shown in Figure 8.9 for the Acheron River data.

For example, from the 20% MDF curve (20% MDF $\approx$ 2.0 m^3/s for this site) in Figure 8.9 it can be seen that a deficiency volume of 1% of the MAF (1% $\times$ 3785.7 $\times$ 0.0864 = 3.27 million m^3) is exceeded in about 12% of the years. For the spell-duration curve (also output by SPELL, but not shown) the interpretation is that a spell of a certain number of days below the threshold value is exceeded a given percentage of years.

Flow-spell annual frequency curves tend to flatten out at the higher recurrence intervals (the rarer droughts), possibly due to some limiting mechanism related to flow seasonality (IH, 1980). Regionalization procedures for improving estimates of these extremes are briefly discussed in Section 8.5.3.

8.5 EXTRAPOLATING FROM THE KNOWN TO THE UNKNOWN

8.5.1 GENERAL

In practice, gauged streamflow records will often be limited or non-existent at the site of interest, particularly on small streams. The typical problems

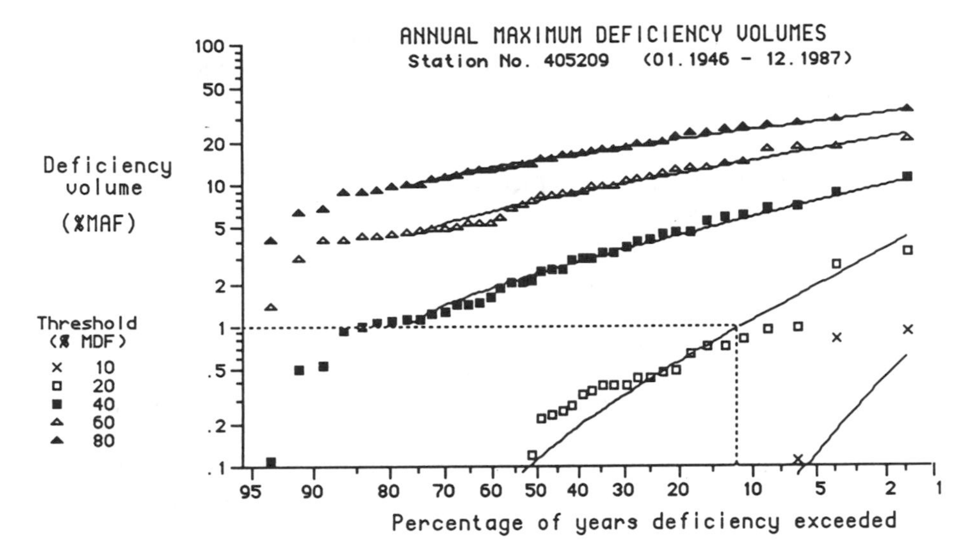

Figure 8.9. Deficiency volume analysis for the Acheron River at Taggerty, as output by program SPELL. MAF = mean annual flow (3785.7 m^3/s × 1 year for this site); MDF = mean daily flow (10.30 m^3/s). Dotted line refers to example given in text

are therefore (1) the development of streamflow estimates for sites which have no records, and (2) the extension of short-term records.

Both problems can be addressed with regression methods (Appendix 1). A regression model can be developed which relates a short-term record to the corresponding record from a long-term station to improve estimates of measures such as the mean, variance, 20-year flood, or the daily Q_{90} for the short-term station. Relationships between discharge and catchment and/or rainfall characteristics can also be developed for a region. This allows the estimation of streamflow measures for ungauged sites within the same region. To be more specific, the region must be *hydrologically homogeneous*, meaning that the catchments should be as similar as possible in terms of their water-producing characteristics (e.g. rainfall patterns, vegetal cover, land use, geology, and topography). For example, a station in the rain shadow of a mountain range should not be used for predicting streamflows on the other side (Searcy, 1959). In some techniques, large and small catchments can be grouped together but, as a rule, they behave very differently hydrologically.

Section 8.5.2 covers methods of extrapolating data from one gauged site to another and Section 8.5.3 describes regional methods. In both sections, general methods for estimating streamflow characteristics will be given first, followed by methods of deriving flood-frequency and flow-duration curves for short-term and ungauged sites. Homogeneous regions can be defined using the methods of numerical taxonomy in Section 8.6 and stream and catchment classification in Section 9.2.

8.5.2 TRANSPOSING DATA FROM GAUGING STATIONS TO LESS-ENDOWED SITES

Infilling missing data and/or extending a record

As mentioned in Section 3.2.4, simple linear regression is one method of extending or infilling missing data using records from another site. The dependent variable (Y) represents records from the short-term or intermittent site, and the independent variable (X) the long-term or "index" site. Maps showing correlation coefficients (Appendix 1) between the short-term site and surrounding stations can be developed to assist in the choice of an index station.

General guidelines for simple log-log regression (Appendix 1) should be followed in developing a relationship between the short-term and index station records. However, the relationship will not always be consistent over the entire range of data. A break in the slope may occur at a point when one stream goes dry and the other does not, or where the relative proportions of baseflow and direct runoff shift. In these instances, an "eye-fitted" curve or several line segments may be preferable to a regression line. A line of 1:1 slope is sometimes plotted with the data. For higher flows, the relationship between the streamflows at two sites tends to parallel this line (Searcy, 1960).

The resulting curve can be used to generate data for missing periods or to extend the short-term record to the length of the long-term one. A major assumption is that the relationship for the short concurrent period remains valid over the entire period of estimation. The magnitude of error can be assessed using one or more "test" sets of data which were not included in the regression analysis.

Frequency curves for floods and low flows

Flood or low-flow frequency curves can also be translated from long-record stations to produce estimates for ungauged sites and improve estimates for short-term stations. If records from several nearby stations are available, regionalization (Section 8.5.3) is a superior method. With only one other station the flood-frequency curve developed for the long-term station is transposed to the site of interest using regression methods.

To obtain information at a study site, a recording or crest gauge can be installed and operated for a few seasons or a few years to obtain a few records for correlation with the long-term station. The slope–area method can also be used to estimate the magnitudes of floods from the recent past (Section 5.8.2).

The steps for transposing frequency curves from long-term (index) stations to short-term ones are described by McMahon and Mein (1986) as follows:

(1) For both the short-term station and the index station a frequency curve (annual maximum, partial duration or low-flow frequency) is constructed for the **concurrent** period of record.
(2) From these curves, discharges are selected for several average recurrence intervals (e.g. 1.1, 1.5, 2, 5, 10, 20 and 50 years) for both stations. Discharges for the short-term station are plotted against those for the index station on log-log paper.
(3) A curve is drawn through the points, giving additional "weight" to values in the region of most interest (high, medium or low flows). The line will generally approach a 1:1 slope for the higher values.
(4) A frequency curve is then developed from the full record for the index station. Using this curve, specific discharges are again selected for several recurrence intervals and adjusted to the short-term site using the log-log relationship.
(5) A new frequency curve for the short-term site is drawn through the translated points.

To standardize the discharge values they can first be divided by catchment area. A 1:1 slope in Step 3 then indicates that both catchments are producing the same amount of runoff per unit area. Error is dependent on both the accuracy of the index frequency curve and the relationship between the sites.

The regional flood-frequency method described in Section 8.5.3 can also be abbreviated to extend or verify the flood-frequency curve at a particular station. For example, if a curve for a long-term station indicated that the 1965 maximum flood was a 50-year flood, it can also be assumed as a rough guideline that the 1965 flood at the short-term station was also 50-year. A regional flood-frequency curve (see Figure 8.11, p. 387) can also be used to improve the estimates of rarer floods at short-term stations.

Flow-duration curves

Flow-duration curves for short-term stations can be improved using records from long-term stations in the same manner as for flood-frequency curves. Again, the procedure requires a number of concurrent measurements. However, with flow-duration curves the measurements are not restricted to, for example, annual maxima or annual low flows. Spot measurements taken over a period of one or two years and a range of flows may be adequate for general estimates. Enough concurrent records should be obtained to establish a usable relationship.

The procedure is exactly the same as in the previous section except that in Step 2 the discharges are selected for a number of percentage duration values (e.g. 10, 20, 30, . . ., 80, 90). The curve drawn through the points plotted on log-log paper at Step 3 will normally be curved in the lower range due to response differences in the catchments from differences in geology and topography. For the higher range of flows the curve often tends towards a 1:1 slope as baseflow becomes insignificant. In Step 4 an improved flow-duration curve for the short-term station is obtained by selecting discharges of specific percentage duration values from the index curve and adjusting them using the relationship developed in Step 3.

To standardize the flow duration curves, the discharges can first be converted to $m^3/s/km^2$ by dividing by catchment area. This eliminates differences in mean annual flow due to catchment size. Synthetic duration curves can be approximated as a straight line on log-probability paper by developing estimates for the median flow (the Q_{50}) and the standard deviation (which defines the slope). Dingman (1978) is an additional source of information on this procedure.

8.5.3 REGIONALIZATION

The regionalization of streamflow characteristics is based on the premise that areas of similar geology, vegetation, land use and topography will respond similarly to similar weather patterns; i.e. a weather pattern which produces large floods or droughts in one catchment will likely have the same effect on nearby ones. Regional analysis methods are useful for strengthening estimates of rare events at gauged sites, both short and long term, as well as providing a means of estimating flow characteristics at ungauged sites. Additionally, they can be used to identify sites which do not "belong" in the region and to detect anomalies in gauged streamflow data.

Rough estimates of general parameters

For ungauged sites, rough estimates of the mean annual flow can be made from nearby gauged sites simply by adjusting for the difference in area, e.g.:

$$\bar{x}_1 = \bar{x}_2 \left(\frac{A_1}{A_2} \right) \tag{8.12}$$

where $\bar{x}_1$ = mean annual flow (volume units) for the ungauged site, $\bar{x}_2$ = mean annual flow (volume units) for the gauged site, and A_1 and A_2 are the areas of the ungauged and gauged catchments, respectively. More generally, this equation becomes:

$$\bar{x}_1 = \bar{x}_2 \left(\frac{A_1}{A_2}\right)^a \tag{8.13}$$

where a is generally less than 1.0. For Australian streams, McMahon (1976) gives a value of 0.65.

The coefficient of variation can be similarly estimated by adjusting for differences in the mean annual runoff:

$$C_{V_1} = C_{V_2} \left(\frac{\bar{x}_1/A_1}{\bar{x}_2/A_2}\right)^b \tag{8.14}$$

where b is generally less than 0.0. McMahon (1976) gives a value of -0.33 for Australian streams. It should be noted that in Equations 8.12 and 8.13 the ratio of the mean annual *flow* (MAF) is used (total volume over the year), whereas in Equation 8.14 mean annual *runoff* (MAR) is used (the total flow per unit area). MAF tends to increase with catchment area whereas MAR does not. Regional relationships using gauged records can be developed to obtain values of a and b.

Multiple-regression approaches

There have been a multitude of studies in which multiple regression has been applied to the prediction of discharge from catchment, channel, and rainfall characteristics. Where no flow records exist at a site, a common approach for estimating streamflow measures is to develop relationships between gauged data and measures obtained from maps or field measurements.

Some of the factors which have been considered include catchment area, channel slope, drainage density (Section 4.2), elevation, stream length, mean annual precipitation, mean summer air temperature, and percentage of the area covered by forest, swamps, lakes or permeable rock. Many of the factors are intercorrelated, and often the equation is "boiled down" to a few representative variables. Comparative studies have shown that catchment area, precipitation and geology are typically the most important characteristics.

Regression relationships can be developed between these factors and any of a number of streamflow indices derived from gauged data. These might include the mean annual flow, mean annual flood, annual coefficient of variation, flood peaks of given recurrence intervals, low-flow indices, or the Q_{50} or Q_{90} from flow-duration curves. McMahon and Mein (1986) summarize studies using regression methods in low-flow hydrology.

A regression analysis provides (1) insight into which factors have the most influence on flow attributes, and (2) an equation which allows the prediction of these flow attributes for ungauged catchments. The coefficients obtained will be specific to the hydrologically homogeneous area from which they were derived. Their reliability for predicting ungauged flow parameters can be tested by collecting a range of actual stream flow measurements at an ungauged site or by reserving the record from one or more gauging stations to act as test data sets.

For example, in a study of 81 catchments smaller than 250 km^2 in south-eastern Victoria, Australia, Gan et al. (1990b) found that a regression equation based on catchment area and mean annual rainfall explained 97% of the variance in mean annual streamflows. The equation he established was:

$$Q = 9.3 \times 10^{-6} A^{0.99} R_m{}^{1.48} \tag{8.15}$$

where Q is the mean annual streamflow (million m^3), A is the catchment area (km^2) and R_m is the mean annual rainfall (mm). Using the Marysville precipitation average from Appendix 1 (1380 mm), the equation would give a mean annual streamflow of 10.3 million m^3 for Site 3 on the Acheron River (catchment area 25.8 km^2; see Figure 4.7). This compares to a value of 41.5 million m^3 obtained using Equation 8.13 and data from the gauged site at Taggerty (mean: 327.09 million m^3, catchment area: 619 km^2) and $a = 0.65$. The lack of agreement illustrates the difficulty in estimating flows for small, ungauged streams.

Envelope curves

To examine flood-producing properties of a catchment and to estimate the maximum expected flood on ungauged streams, envelope curves are a useful tool. Within a given region, the **highest observed discharge** at all gauging stations is divided by the corresponding catchment area and then plotted against area using log-log axes. Curves which form an upper bound to the data are called "envelope curves". The graph provides a summary of the flood magnitudes experienced in a region. In Figure 8.10, points have been plotted for several stations in the vicinity of the Acheron River. As can be seen from the graph, the highest flood recorded on the Acheron River is low for its catchment size in comparison with other stations in the region. This may be due to the fact that flood-generating characteristics for the Acheron basin are different, or that none of the floods recorded during the 26 years of record have been rare events (e.g. the 100-year flood).

By labelling each point with a catchment name, trends with elevation, aspect, catchment slope and location can also be examined. Major floods recorded at longer-term stations will also reveal the effects of short record lengths. As more data are collected the envelope curves will shift upwards. Thus, the method is not satisfactory for accurate flood estimation but can be used for preliminary estimates as well as for checking whether flood estimates obtained by other methods are realistic (IEA, 1987).

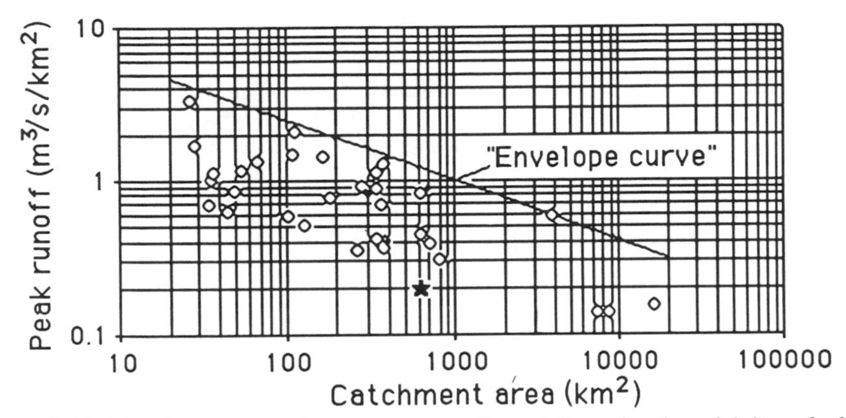

Figure 8.10. Envelope curve for stream gauging stations in the vicinity of the Acheron River. The points represent the highest instantaneous flow on record for each site. The flood for the Acheron River is shown as a star

Regional flood frequency

Combining data from a number of catchments will help to reduce the uncertainty resulting from short records. In this way, the record is effectively "lengthened" by extending it across space rather than time (Linsley et al., 1975b). Stations should be from a region which is homogeneous in terms of its flood-producing characteristics and those with very short records or regulated flows should be excluded.

Even for sites with long-term records, a regional flood-frequency analysis is recommended for improving the estimates of rare floods. For ungauged sites or stations with only a few years of record, this is essential. As mentioned previously, parameters based on higher moments such as skewness will not be good estimates of the true population parameters if they are calculated from short records. Regional skew values can be obtained by pooling information from other sites in the region. Generalized maps of skew coefficients for the LPIII distribution have also been published. Chow et al. (1988) give a map with regional skew values for the USA.

If a region is relatively homogeneous, flood-frequency curves should have approximately the same slope. Curves from several stations can thus be put on the same scale and superimposed to develop a composite regional curve. From this curve, estimates of extreme events can be made for ungauged and short-record sites. The method presented here is the Index Flood Method, given by Dalrymple (1960). Other variations on this method are given by IEA (1987) and NERC (1975).

The first step is to assemble records from a number of stations and establish a standard record length or base period. Since the common record length will usually be very short, the normal approach is to "fill in" some of the peak flows. Estimates are obtained from another longer-term station by a regression analysis of annual peak flows. A certain amount of judgement is necessary in deriving the standard base period. For example, if only one station has a record of 100 years, with all others being less than 25 years, not much is gained by using regression to put all stations on a 100-year basis. An analysis of the 100-year station alone would produce as much information.

All flood peak data are divided by an **index flood** discharge, usually the mean annual flood, to make the quantity dimensionless. The mean annual flood is commonly assumed to be the flood with an annual recurrence interval (ARI) of 2.33 years (see EVI distribution, Section 8.2.6). This can be changed to reflect the chosen probability distribution (e.g. for a normal distribution, the mean has an ARI of 2 years). Bankfull discharge (Section 7.2.5) can also be used as the index flood, which allows development of frequency curves for ungauged sites using field estimates of bankfull discharge. This is based on the assumption that bankfull flow is exceeded on different streams with about the same ARI (Leopold et al., 1964).

The values of Q/Q_{index} for each station are plotted against probability on the same graph, and a composite curve is fitted through all the points by eye, as shown in Figure 8.11. Alternatively, WMO (1983) recommends using the median value of Q/Q_{index} at several probability levels to establish the curve. For the regional curve, some departure from linearity is acceptable if it appears to improve the fit (NERC, 1975). The line should always pass through the index flood probability at a Y-value of 1.0, as shown.

The procedure is based on the assumption that all streams have the same variance in Q/Q_{index}. Floods which plots as "outliers" on individual curves will still plot as outliers on the composite curve, but regionalization will assist in obtaining better estimates of the average recurrence interval for these floods. Scatter in the data may be due to the variability of short record lengths or actual within-region differences in geology or climate. Individual curves which exhibit slopes noticeably different from the trend of the others may not belong in the region. A homogeneity test given by Dalrymple (1960) can be performed on sets of flood-frequency curves in a region to determine whether they are similar.

To apply the regional frequency curve to ungauged sites, an estimate of the index flood must first be made. If the index flood is the bankfull flow, this is done using field measurements. If it is the mean annual flood, it is made by developing a multiple-regression relationship between the mean annual flood at gauged sites and catchment and/or climatic factors. The equation is often just a simple log-log relationship between the mean annual flood and drainage area (NERC, 1975; WMO, 1983). Scatter in the plot may indicate that other factors besides area should be considered or that the stations are from a non-homogeneous region. Once an estimate of the index flood for the ungauged site is obtained, the value is multiplied by the

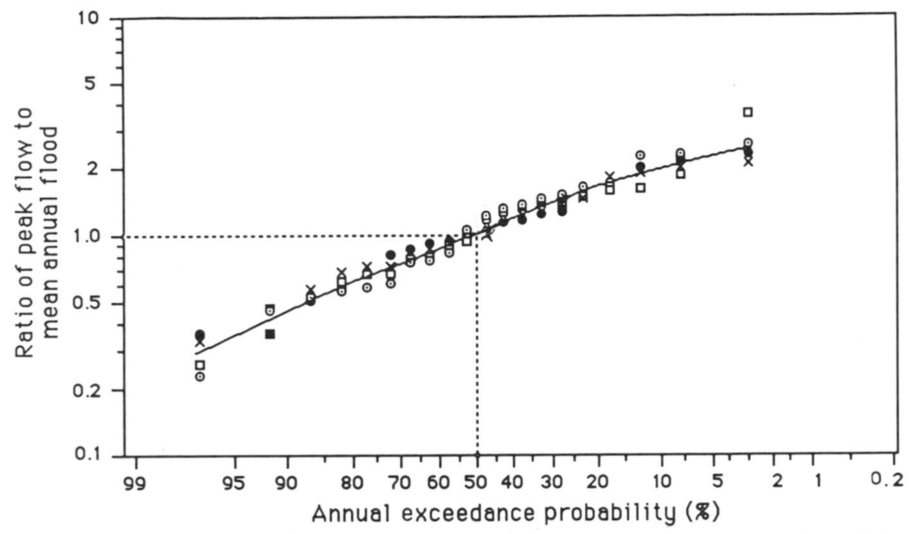

Figure 8.11. Regional flood-frequency curve derived using four stations in the vicinity of the Acheron River with similar catchment sizes. The solid dots represent the Acheron data. The fitted curve must pass through the point (index flood probability, 1.0), indicated by the dotted line. Base period is 1962–81

Q/Q_{index} ratios from the regional curve to develop a frequency curve for the ungauged site.

Regional flow-duration curves

Flow-duration curves from several stations within a region can also be put on the same scale and superimposed as for flood-frequency curves. The regional flow-duration curve allows "smoothing" of local effects such as thunderstorm-produced events which occur in one basin and not another. Again, the discharge values are divided by an "index" flow such as the median (Q_{50}) value, or by the catchment area. The individual curves are then superimposed and averaged. The regional curves will be most reliable towards the middle and less so at the extremes of low or high flow.

Depending on how the regional curve is constructed, a duration curve for an ungauged station can be developed by multiplying back by either catchment area or an estimate of the index flow. The index flow would be estimated from catchment, channel and/or climatic factors using multiple-regression methods, as for the flood-frequency curves.

Regional flow-spell curves

Nathan and McMahon (1990a) developed a method for regionalizing spell duration and deficiency volume frequency curves. Because the data are considered log-normally distributed, the frequency curves plot as straight lines on log-normal probability paper. Thus, to define a curve, only two points need to be estimated. Nathan and McMahon suggest developing regional equations using multiple-regression analysis for the prediction of 2- and 50-year events. These points are then plotted on log-normal probability paper and a straight line is drawn through them.

In their study, catchments were first divided into hydrologically homogeneous groups and regression techniques used to select and weight the most important variables. Preliminary groupings were first obtained using cluster analysis (Section 8.6.4), and Andrews curves (Section 8.6.4) were applied to fine tune the groupings. For the multiple-regression equations, they found that the most important variables were mean annual rainfall and estimated ratio of baseflow to total streamflow.

8.6 NUMERICAL TAXONOMY: MULTIVARIATE ANALYSIS TECHNIQUES

8.6.1 GENERAL

In science, classification is a fundamental principle for imparting order onto the diversity of nature: biologists classify organisms, geographers classify landscapes, and psychologists classify personalities. **Numerical taxonomy** is a mathematical procedure for classifying items based on how numerically similar they are to others within the same class and how different they are from items in other classes (e.g. bugs with six legs are insects; those with eight legs are arachnids). These procedures developed from efforts to separate categories which are *intuitively* recognized as being distinct by using more objective statistical methods. Although the statistical methods add some standardization, the final classification obtained is still largely a matter of preference.

With the increased availability of "canned" statistical software, numerical taxonomy techniques have seen wide application in biology, ecology and biogeography and in the analysis of remotely sensed spectral data. There are numerous opportunities for their use in hydrology—for example, in the classification of catchments based on their sensitivity to land modifications or of streams on the basis of their runoff characteristics or biota. However, it has only been recently that these techniques have been applied to the classification of streams and catchments. Stream and catchment classification will be covered in Section 9.2.

To develop the area of numerical classification, let us consider the classification of ice-cream flavours. Although raspberry frozen yoghurt could easily be distinguished from chocolate ice cream, finer criteria would be needed to distinguish raspberry frozen yoghurt from blackberry frozen yoghurt. Qualitative approaches might include classifying flavours according to personal taste preferences, or consulting a panel of ice-cream experts to classify and name the various flavours. To numerically classify flavours, some combination of objective measures such as ingredients or water content or melting temperature would be used to distinguish one flavour from another. Attributes can also be "weighted" according to importance; e.g. the number of walnuts may be less important to flavour than the millilitres of mint extract.

The methods which will be discussed are considered forms of **multivariate analysis**, which simply considers the interrelationships between a number of variables. Whereas multiple regression analysis results in an equation, multivariate analysis in numerical taxonomy results in a "grouping" of individual observations which have similar characteristics. Multiple regression is sometimes considered a type of multivariate analysis. Tabachnick and Fidell (1989) have written an easily followed text on multivariate analysis, from which much of the information in this section has been gleaned. Other information sources are given in Section 8.6.5.

Primary "classes" under the heading of numerical taxonomy are those of ordination and classification, which can be jointly or separately applied to data. The techniques can be based on actual measurements, presence/absence data, rankings (e.g. 1 = poor water quality, 5 = excellent), or statistical abstractions which define the similarity or dissimilarity between individuals.

8.6.2 SIMILARITY/DISSIMILARITY INDICES

Any two sets of data in continuous, discrete or attribute form can be represented by **similarity/dissimilarity indices**. A high-similarity measure (low dissimilarity) indicates "alikeness" whereas a low-similarity measure (high dissimilarity) indicates "unlikeness". For example, the amount of fruit might be used to calculate a dissimilarity measure which would indicate that tutti-frutti ice cream is much more like apricot swirl than mint chocolate chip.

One index is the r^2 value from a regression analysis (Appendix 1), where high values indicate similarity and low values indicate dissimilarity. Measures based on the geometric distance between points include (SPSS, 1986):

(1) *Euclidean distance*:

$$\text{Distance}\,(x, y) = \sqrt{[\Sigma(x_i - y_i)^2]} \qquad (8.16)$$

(2) *Squared Euclidean distance*:

$$\text{Distance}\,(x, y) = \Sigma(x_i - y_i)^2 \qquad (8.17)$$

Here, x and y might represent values of different attributes such as catchment shape and water temperature. If a graph of catchment shape versus water temperature were developed, sites which were similar would plot near each other, whereas those which were different would plot at a greater distance away. Thus, high values obtained from Equations 8.16 or 8.17 indicate dissimilarity. A pattern similarity measure was used by Haines et al. (1988) for separating river regimes into classes (Section 4.4.3), given by SPSS (1986) as:

(3) *Cosine similarity measure*:

$$\text{Similarity} = \frac{\Sigma(x_i y_i)}{\sqrt{[(\Sigma x_i^2)(\Sigma y_i^2)]}} \tag{8.18}$$

For this measure, higher values indicate similarity and lower values dissimilarity. For ecological data, Faith et al. (1987) make explicit recommendations on appropriate measures.

8.6.3 ORDINATION

Ordination is a method for investigating the structure in data. It may be based on the original data, such as ice-cream density or the presence of fine sediment in a stream, or on some index of similarity or dissimilarity. One aim of ordination is to reduce a large number of characteristics (dimensions) to a lower number of indices which still account for nearly all the variance (Pielou, 1977). The process is somewhat like the editing of technical prose, where the object is to condense the essence of a sentence into as few words as possible while still retaining all the original meaning. Redundancies in the form of intercorrelations between variables are removed.

For example, if several factors were measured in an assessment of stream physical habitat, one might ask whether they can be condensed into a small number of combined characteristics which more effectively or efficiently describe the habitat. Factors such as stream density, relief, bifurcation ratio and catchment shape might all be grouped into one “catchment morphology” index. The procedure is best used where a number of variables contribute “overlapping” information but none works well by itself. If the original variables are essentially independent, there is no need to use ordination.

Factor analysis (FA) and **principle component analysis (PCA)** are two common methods used in ordination. In both methods, variables which are correlated with each other (but not with other groups of variables) are combined. This combination of variables is called a “component” in PCA and a “factor” in FA. The factors or components are considered to reflect underlying processes which might have caused the intercorrelations. For example, stream density, relief, bifurcation ratio and catchment shape may all be related through landscape-evolution processes.

Components and factors are represented by a linear combination of variables, which forms a model much like a regression equation. The model effectively "weights" each variable according to its importance. This relationship is used to summarize patterns of intercorrelations among the variables, to define some underlying process, or to test a hypothesis about whether an underlying process exists.

For example, the hypothesis could be made that elevation, flow variability, substrate, canopy cover, water temperature and a number of other factors influence the population of mayfly larvae in streams. Ordination techniques can be used to reveal patterns and order in the mixture of measurements. These patterns might indicate that canopy cover is an unnecessary measurement, that there is a need to find an additional measure, or that substrate, water temperature, stream width and depth can be combined into a common factor which may reflect some underlying process.

The choice of whether to use PCA or FA is based on the purpose of the study. PCA is the "method of choice" for reducing a large number of variables into a smaller number of components, whereas FA is of more use in generating hypotheses about underlying processes. Cattell (1965) considers PCA to be a "closed model" in that it accounts for all the variance, whereas FA is an "open model" in which part of the variance is reserved for factors which may yet be included.

Principal component analysis (PCA)

As mentioned previously, PCA is used to develop a smaller set of components which summarize the correlations among the original variables. By collapsing several different measures onto a common axis of ordination a type of "ranking" is obtained. This ranking can then be used as an index in subsequent analyses. The groupings are also useful for exploring differences between groups; e.g. the components for trout habitat may be made up of a different set of variables than for bass habitat.

Continuing with the ice cream example, an axis representing the amount of coffee flavouring, chocolate content and weight of almonds might provide a basis for ordering (on an axis, not over a counter) the ice cream flavours chocolate chip and Swiss almond mocha. The position of a flavour on the axis indicates how "coffee-chocolatey-nuttey" it is (component 1 in Figure 8.12). Multiple axes would be needed to represent *all* flavours of ice cream; "chunkiness", "berriness", or "wateriness" being other possibilities. Each flavour would then be represented by a position (given by co-ordinates) in a "hyperspace" of multiple dimension. In Figure 8.12 two axes are shown, the second representing "wateriness".

The object of PCA is to extract the maximum amount of variance from the data set with each component. The first component extracted is a combination of variables which best separates the individual observations (i.e. "spreads them out" along an axis). It accounts for as much of the variation in the original data as possible. The second component extracts the maximum amount of remaining variance which is not correlated with the first component (orthogonal to it). Ideally, the interrelationships will be displayed using as few dimensions as will suffice (Pielou, 1977).

A problem of PCA is that components are dependent on the scale of the variable; e.g. the variable with the highest variance will dominate the first component. It is preferable to first scale variables such that all have a variance near one. Results of a PCA are normally plotted with two axes, representing pairs of components. From these plots, the distance, clustering and direction of the points relative to the axes can be examined. A large distance away from the origin along an axis indicates a close correlation with that component. Optimally, points should cluster near the end of an axis and near the origin; if clustering is not obvious, the component may not be clearly defined. If clusters do not line up on the axes, it may indicate a need to rotate the axes.

Rotation of axes is used to improve the usefulness and ease of interpretation of the solution. Varimax (orthogonal) rotation is most commonly employed, which assumes that the underlying processes influencing each component are independent. Spatially, the axes are rotated so that they more closely pass through the variable clusters, as in Figure 8.12. This allows each cluster or individual to be more easily ranked by distance along an axis.

The resulting components should also be examined in an attempt to understand the underlying, unifying principle. This is usually characterized by assigning the component a name; e.g. "richness of flavour" or "landscape form". In fact, there is no objective measure for testing how "good" the resulting PCA solution is. The final choice is up to the researcher, based

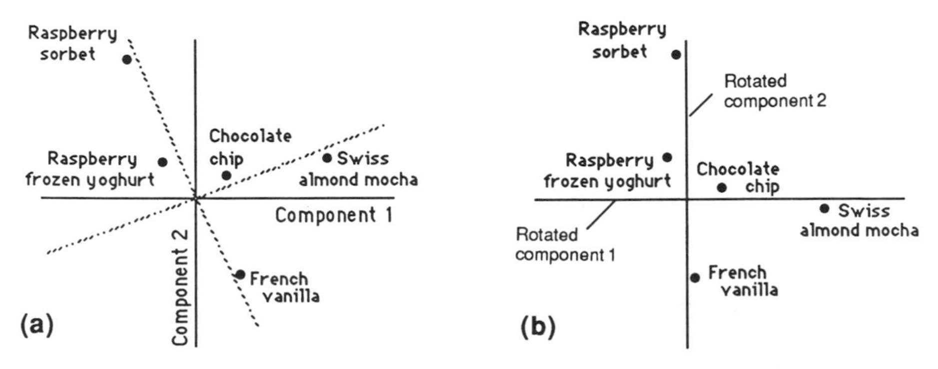

Figure 8.12. A fictitious illustration of the use of rotation in principal component analysis: (a) location of various ice cream flavours after extraction of principal components; and (b) location after rotation of axes. Component 1 might represent "coffee-chocolate-nuttiness" and component 2, "wateriness"

on its interpretability—i.e. does it "make sense"? Useful solutions ideally have a few components which are related by some common process and are highly unlike the other components.

Once obtained, principal components can be used as independent variables in a multiple-regression analysis. Because the components are uncorrelated, the regression results tend to be more stable and reliable than regressions made using the original variables (Wallis, 1965).

Factor analysis (FA)

The process of factor analysis is very similar to that of principal component analysis. However, FA differs from PCA both in theory and in which applications are most appropriate. In PCA *all* the variance in the observed variables is analysed, whereas in FA only the variance *shared* by the variables is analysed. Thus PCA is an analysis of the total variance whereas FA is one of covariance. In many statistical references the term "factor analysis" is often used for both factor and principal component analysis.

PCA is normally used for condensing the information contained in the data into indices. Factor analysis, however, might be employed to estimate the amount of variance in the data set which is accounted for by those indices; e.g. "how effective is catchment morphology at explaining the variability in endemic leech populations?" A low "score" may highlight the need to measure other variables. Thus, FA may be of use in the exploratory stages of research to identify critical variables and possibly eliminate others which do not add any more information and may be expensive to measure.

Factor analysis is a more formal statistical technique than PCA. Limitations on its use, such as the requirements of independence and normality, can be relaxed for **exploratory** uses of the technique. However, they become more critical in **confirmatory** factor analysis, where the goal is to test a hypothesis about underlying processes (Tabachnick and Fidell, 1989). Factor analysis is most appropriate for generating and testing hypotheses about underlying structures, patterns or processes, whereas PCA is more useful for summarizing the amount of information contained in the data, and is the technique more commonly used in hydrology.

An example

Principal component analysis (with Varimax rotation) was used by Hughes and James (1989) to classify streams in Victoria, Australia, into regions according to hydrological characteristics. The classification procedure produced five distinctive groups which generally corresponded with climatic regions. Sixteen variables were calculated using data from stream-gauging stations. These variables included peak and low-flow variables, mean flows, and coefficients of variation for monthly and annual streamflow data. It was found that 72.4% of the variation was accounted for by three components.

Figure 8.13 shows how the five groups "cluster" on a scatter plot in ordination space. The axes represent components 1 (monthly variability) and 2 (annual variability).

8.6.4 CLASSIFICATION

Whereas similarity measures and ordination give an indication of how closely individual data points are "related", **classification** methods group individuals (objects, traits or measures) which are "alike" into classes. "Alike" may mean close together in space or time, or in terms of having similar characteristics such as chocolate flavouring or catchment size (Sokal and Sneath, 1963). This grouping can be done by dividing the whole collection of data into increasingly finer groups, or by starting with individuals and combining and recombining them to form successively larger groups (Pielou, 1977); i.e. "splitting" versus "lumping".

Two techniques used in classification are discriminant analysis and cluster analysis. In both the aim is to predict membership in a group from a set of variables. These variables might be actual measurements, similarity/dissimilarity indices or ordination scores from a principal component analysis. The two techniques are very similar except that, in discriminant analysis, class

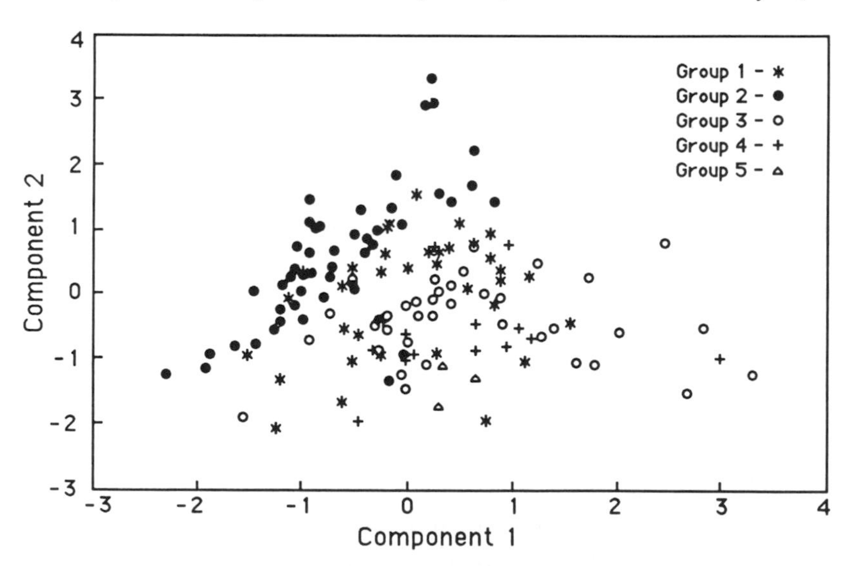

Figure 8.13. Two principal components used in a hydrological regionalization analysis on streams in Victoria, Australia. The five groups of streams shown are hydrologically distinct. Component 1 is an index of monthly variability whereas component 2 reflects annual variability. The Acheron River is in Group 4. Modified from Hughes and James (1989), by permission of *Austr. J. Mar. FW Res.*

membership is known beforehand, at least for the data which are being analysed. In cluster analysis, classes are "assembled" based on similarities among individuals.

For example, cluster analysis might be used to classify regions based on vegetation communities. In contrast, discriminant analysis would be preferred for classifying new areas as being likely to contain certain vegetation types, based on physical or meteorological or other measures. Equations would be developed from data which included observations of vegetation type; thus "class" membership of a particular area (e.g. containing spider orchids and sundews) would be known prior to analysis.

Andrews curves (Andrews, 1972) provide a third technique which can be used for classification. This method provides a means of concisely displaying the data so that classes can be grouped visually. With this method, new observations can be assigned to existing groups or identified as being unique.

Lately, another technique, multidimensional scaling, has been recommended for displaying patterns in complex ecological data. This technique will not be discussed here, and the reader is referred to Minchin (1987) for further information.

Cluster analysis

Cluster analysis (not to be confused with cluster *sampling*, introduced in Section 2.3.5) is a numerical technique for grouping the data points floating around in hyperspace into clusters of "like" members. Individual points in different groups should be dissimilar. A common technique of clustering data into similar groups is the nearest-neighbour method. Starting with single points, the groups which are closest to each other, based on a similarity/dissimilarity measure (Section 8.6.2), are combined. Each combination reduces the number of groups by one.

Cluster analysis results in a "tree diagram" or "dendrogram" (see Figure 8.15), which shows the sequence in which the groups were combined or divided. The structure of the diagram indicates the similarity between "stems". Vertically, the scale gives an indication of the amount of heterogeneity remaining or removed at each step. Different methods of analysis can create dendrograms of different structures.

Continuing the ice cream example, the top of the dendrogram might represent the class encompassing "all things that taste like ice cream" to the other extreme at the lower end where one class exists for every existing flavour. Where to "cut" the dendrogram for classification purposes is subject to one's own tastes. The final level of classification can be chosen based on (1) the number of classes to be recognized, (2) the amount of heterogeneity permitted in the final classes, or (3) the point at which reduction in heterogeneity from further subdivision becomes too small to be worth while (Pielou, 1977).

Probabilistic stopping rules can provide an objective means for deciding on the final level, but it is more often guided by practical judgement. Like PCA, interpretability is a criterion in selection of the final classification. It is best to investigate what the classes represent, and whether they match one's own perceptions and requirements. Romesburg (1984) and several of the other texts on multivariate analysis listed in Section 8.6.5 provide further information on cluster analysis.

Discriminant analysis

In discriminant analysis the objective is to predict membership in groups from a set of independent variables. For example, one might ask whether biologists, engineers, environmental scientists and geographers can be distinguished from a set of spatial perception and mathematical test scores. A significant difference between groups implies that, given certain test scores, the group in which the individual belongs can be predicted with some degree of certainty. In stream ecology an example might be the separation of "good trout habitat" from "good bass habitat" on the basis of variables such as the density of blackfly larvae, the substrate d_{85}, the mean summer water temperature, the variability of streamflow, and the percentage of streambank covered by boat docks. The classes would need to be previously established by some method such as fish sampling or angler interviews.

Discriminant analysis can also be used to assess the relative contribution of the individual variables to the prediction of group membership. For the ecological example, the main sources of discrimination between trout and bass habitat might be water temperature and substrate, with little predictability contributed by the other factors. When a classification is produced, results can be analysed to evaluate how well individual observations fit into their appropriate classes.

The combination of variables (predictors) which separate groups from each other are called **discriminant functions**. These are much like multiple-regression equations but produce a discriminant function "score". This gives an indication of the group to which an individual or stream or catchment belongs, and whether it falls solidly into one class or on the fringes of two different classes. In the previous example on using test scores, the test scores might be combined into one discriminant function score, with "low", "below average", "above average", and "high" groupings reflecting membership in various career "classes".

Several functions may be needed to separate groups; e.g. one function may separate bass habitat from trout habitat whereas another might be needed to separate brook trout habitat from brown trout habitat. The second discriminant function would separate groups on the basis of associations not used in the first function. The total number of possible "dimensions" is either one less than the number of groups or the number of predictor variables, whichever is smaller.

Analogous to multiple-regression analysis, the predictors can all be entered at once or in a stepwise fashion, with contribution to prediction of group membership evaluated as each predictor is entered. It is typical for only one or two discriminant functions to be chosen. Individual points (or group centroids) can be plotted along axes representing the discriminant function scores. A large distance between groups along an axis indicates that the function is effective in separating them.

The major questions which discriminant analysis is designed to answer are: "Can group membership be reliably predicted from the discriminant functions?" "What is the likelihood of misclassification?" "If individuals are misclassified, with what other groups are they most often confused?" As with multiple-regression analysis, the more variables used, the better the relationship; however, after a certain point, the variables are no longer adding much in the way of additional information. Tests of significance can be performed to determine how many discriminant functions are needed and the strength of relationship between class membership and the set of predictor variables. Williams (1983) discusses uses of discriminant analysis in ecology, and other references on multivariate analysis listed in Section 8.6.5 will provide further information on the technique.

Andrews curves: displaying multi-dimensional data

If only two variables are required for describing the similarity between sites, streams or catchments, then a simple two-dimensional scatter plot is sufficient for displaying groupings. Displaying these data becomes much more difficult with a larger number of variables. A graphical approach presented by Andrews (1972) provides a good method of viewing patterns of similarity or dissimilarity across multiple dimensions. A point in multi-dimensional space is represented by a curve described by the function:

$$f(t) = \frac{x_1}{\sqrt{2}} + x_2 \sin(t) + x_3 \cos(t) + x_4 \sin(2t) + x_5 \cos(2t) + \ldots \quad (8.19)$$

where x_1, x_2, . . . are the variables used to characterize a particular site. The function is plotted over the range $-\pi$ to $+\pi$ (-3.14 to $+3.14$), as in Figure 8.14. Program ANDREWS in AQUAPAK will plot Andrews curves for any number of variables.

Curves representing points which are located near one another in multi-dimensional space will look similar, whereas points which are distant will produce curves which look different. Results will depend on the order in which the variables are labelled. The first variables will be described by low-frequency components (wider "waves"). These are more readily seen than the higher-frequency components representing the latter variables. Thus, it is more useful to associate the most important variable with x_1, the second with x_2, and so on (Nathan and McMahon, 1990d). This relative importance can be determined from a stepwise multiple-regression analysis (Appendix 1). The values should also be scaled to the same order of magnitude (for example, by choosing an appropriate unit). In ANDREWS, standardization is done by subtracting the mean and dividing by the standard deviation of all observations of a given variable.

Andrews curves have great potential in stream classification (Section 9.2) since they provide a method for visual comparison of biological and/or hydrological data. Stream sites or catchments which have similar properties would produce a band of similarly shaped curves. If a curve falls outside some margin, the given site can be assigned to a different group. The curves are thus useful in evaluating the results of cluster analysis.

Importantly, group membership can also be determined for a new, unclassified site or catchment. The curve for the new site can simply be compared to bands defined for other site groupings to determine where the new site "fits". Nathan and McMahon (1990d) point out that the technique thus has an advantage over discriminant analysis for regionalization (Section 8.5.3) because it is possible to identify catchments which do not belong to *any* of the existing groups and thus would not be properly described by regional prediction equations.

Example 8.3

Three hypothetical stream sites have the following characteristics, ranked in order of importance for aquatic insect habitat. Construct and compare Andrews curves for the three sites:

Variable	Site 1	Site 2	Site 3
Catchment area (km^2)	50	65	100
Sediment d_{50} (mm)	130	100	50
C_v of annual flow (%)	28	32	52
Canopy overhang (%)	32	35	12

For Equation 8.19 catchment area would be variable x_1, sediment d_{50} would be x_2, and so on. Figure 8.14 shows the curves for the three sites. The values have been standardized using the option in ANDREWS. The curve for site 3 differs substantially from the other two curves, which are somewhat similar. Based on this interpretation, the first two sites would be grouped together and site 3 separately, based on the physical parameters used. Ideally, these would also reflect different groupings of insect species.

Case examples in classification: Example 1

Barmuta (1989) analysed patterns of benthic macroinvertebrate occurrence at one site on the Acheron River, Victoria, to test the hypothesis that distinguishable habitat "patches" would support different communities. Environmental data on velocity, substrate, temperature, DO, EC, pH and food resources as well as invertebrate populations were obtained from four riffle–pool sequences over a period of one year. Community pattern was defined by similarity of species composition among samples. Clustering techniques were first used to find groups of samples with similar faunal characteristics. Then, the discreteness of these groups was assessed using ordination. Finally, faunal patterns were related to environmental data.

The dendrogram from one cluster analysis is shown in Figure 8.15. The nine groups represent "clusters" of sites with similar faunal patterns. Groups A1–A4 generally contained fauna intolerant of fast-flowing, turbulent water, whereas groups A5–A9 had higher abundances of these organisms (e.g. hydrobiosid caddisflies, baetid mayflies). Thus the two main "stems" of the dendrogram were deemed to represent depositional (pool) and erosional (riffle) habitats.

In the ordination analysis the first axis was found to be highly positively correlated with velocity and the presence of aufwuchs ("biofilm") and negatively correlated with mean particle size, sorting coefficient and organic detritus. This reflected the difference between samples collected in pools

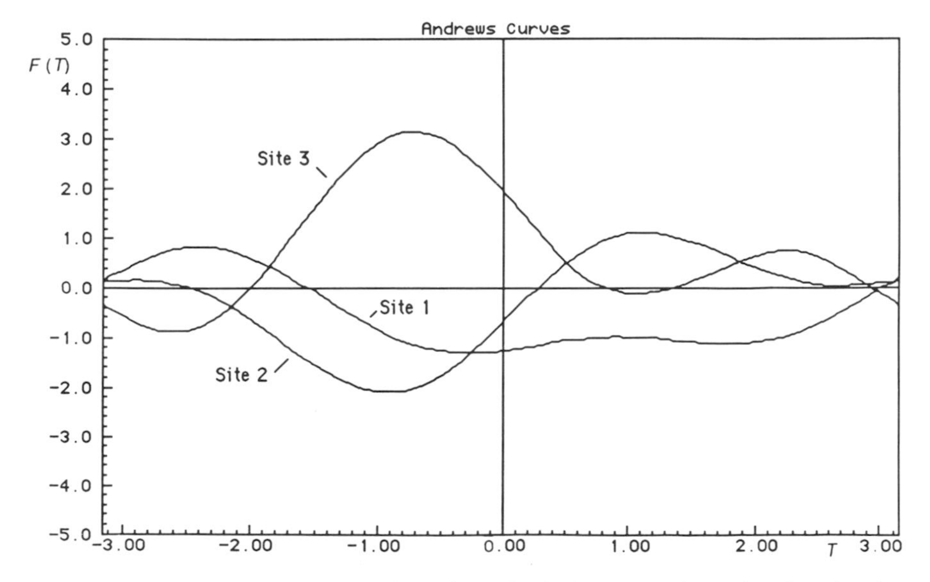

Figure 8.14. Andrews curves for three hypothetical stream sites, developed using four variables (see Example 8.3). Andrews curves are useful for displaying multi-dimensional data (for example, in the classification of streams)

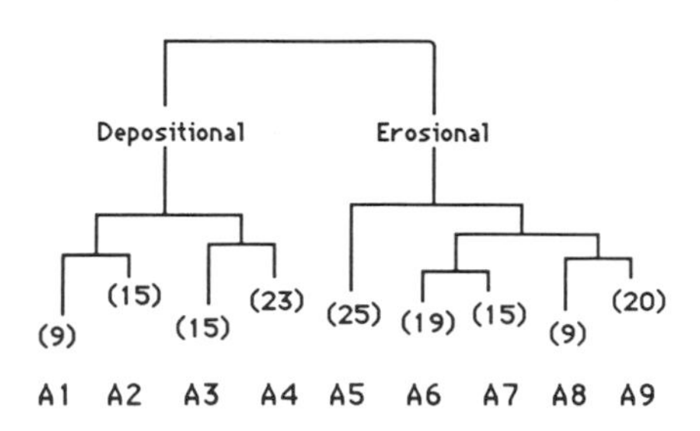

Figure 8.15. Dendrogram from a clustering procedure used by Barmuta (1989) to separate benthic macroinvertebrate samples on the basis of faunal similarity. Numbers in parentheses indicate the size of each sample group. The first "split" in the dendrogram suggested a major division into "depositional" (pool) and "erosional" (riffle) habitats. Environmental variables were insufficient for explaining the remaining divisions. Reproduced by permission of Blackwell Scientific Publications

and riffles. However, the ordination did not show clear distinctions between groups, indicating that community structure was continuous rather than composed of distinct "pool" and "riffle" unit communities. A few individual species preferred distinct habitats, particularly those whose diet was restricted to the aufwuchs on the upper surfaces of rocks, those which required fast, turbulent water for feeding or physiological requirements and conversely, those which could avoid fast-flowing water by burrowing.

The author concluded that environmental variables, at the scales measured, were insufficient to account for the patterns of species co-occurrence, and stated a need for a more precise description of near-bed flows which may be more meaningful to the species studied.

Case Example 2

In a study on clear New Zealand rivers, Jowett (1990) used both discriminant analysis and classification techniques to relate the distribution and abundance of trout to environmental factors. An experienced team of divers made observations on trout abundance by drift-diving. At each site, instream cover was evaluated and measurements were made of water depth, velocity and substrate along at least two pool–run–riffle sequences. Instream habitat and hydraulic parameters were determined from hydraulic modelling with RHYHABSIM (see Section 9.3).

Brown trout and rainbow trout populations were classified into six groups based on species, size and biomass (computed from size class and abundance). The primary division was between rainbow and brown trout, with further subdivisions based on biomass. This classification corresponded well with geographic patterns of trout distribution in New Zealand streams, with rainbow trout occurring only on the North Island.

Much like the study of Example 1, the groups were then examined to determine whether there were significant differences in hydrological, water quality, biological, instream habitat and catchment characteristics between groups. Variables were selected for discriminant analysis if the variance between groups was about twice the within-group variation. Temperature, catchment lithology, hydrological indices, instream habitat and total aquatic invertebrate biomass showed the most significant between-group differences.

A discriminant model to classify sites was developed based on eleven environmental factors. Three discriminant functions were found to be statistically significant. Variables closely related to the first two functions were minimum annual water temperature, percentage volcanic ash, ratio of mean annual low flow to median flow, square root of river gradient, percentage lake area, and physical habitat. Figure 8.16 shows the group centroids plotted using the first two functions. The resulting discriminant functions correctly predicted group membership for 48 of the 65 sites.

From the analysis it was found that lake and spring-fed rivers with stable flow regimes were likely sites for rainbow trout. Examination of misclassified sites suggested that connection of the stream to lakes might affect species distribution and abundance. Overall, the author concluded that the factors most related to the distribution of trout species were climatic whereas factors determining abundance were related primarily to instream habitat (both river morphology and streamflow magnitude and variation).

8.6.5 OTHER SOURCES OF INFORMATION ON NUMERICAL TAXONOMY

This brief introduction to numerical taxonomy is designed only to present various techniques which may be helpful in classifying streams and catchments

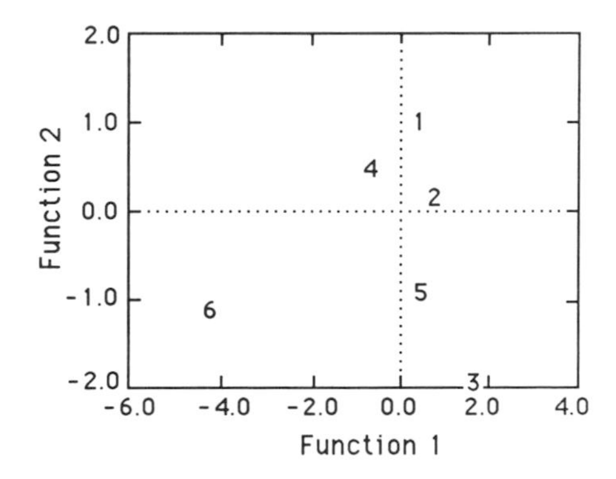

Figure 8.16. A plot of the first and second discriminant functions from a study by Jowett (1990). Riverine sites were divided into six groups based on trout species, size and abundance. The points on the plot represent group centroids. Group 6, for example, represents streams with rainbow trout. Both functions were found to be related to several environmental factors. Reproduced by permission of *NZ J. Mar. FW Res.*

and to aid in the interpretation of studies where the techniques have been used. Prior to conducting an analysis, statistical texts should be consulted for information on theoretical limitations and assumptions. A number of computer programs are available for rapid number-crunching, which lowers the "threshold" for entry into the field of numerical taxonomy. When selecting and using "canned" computer programs it is important to choose the right one, use it correctly and know how to interpret the output. Some programs provide more options and more readable documentation than others.

Pielou (1984), Sneath and Sokal (1973) and Sokal and Sneath (1963) are a few of the many references on the subject of numerical taxonomy. Especially for biologists, an article in *Scientific American* by Sokal (1966) provides interesting reading on the topic. The user's guides to commercial statistical software packages are also good sources of information on the various techniques. Other important references on multivariate analysis include Cattell (1965), Chatfield and Collins (1980), Green (1979), and Legendre and Legendre (1983).

Afifi and Clark (1984) provide a guide to multivariate analysis using computerized methods. A book on cluster analysis by Romesburg (1984) describes and compares features of commercially available packages such as SAS, BMDP and CLUSTAN. Romesburg also offers a companion statistical software package to use in conjunction with the book. Tabachnick and Fidell (1989) give a comparison of the statistical packages SPSS, BMDP, SAS and SYSTAT. They state that SYSTAT, although less comprehensive than the other packages, is available for both Macintosh and IBM-compatible microcomputers, easy to use, and ideal for fast, interactive work. Cost, reliability, and "user-friendliness" of the software, and the opinions of experienced users should serve as a guide in the selection of an appropriate package. Other references and statistical packages are given in Appendix 1.

9 "Putting It All Together": Stream Classification and Management

9.1 PUTTING THEORIES INTO PRACTICE

In previous chapters the interrelationships between ecology and hydrology have been explored at a number of scales: whole river systems, the pools, riffles and meanders of a stream reach, the flow patterns around rocks, and the lift and drag on individual organisms. Whereas scientists may study these relationships in an attempt to improve our understanding of streams and their biota, the step from research to management requires a more practical problem-solving approach; i.e. what happens if we manage a stream one way rather than another?

Streams and their floodplains have a wide range of conflicting values for recreation, wildlife habitat, drainage, water supply and residential, industrial and agricultural uses. The increasing complexity of water-resource problems and the overwhelming amount of available information has created a need for multidisciplinary teams which include both biologists and engineers. Some of the areas in which interaction is required include the classification of streams for environmental value, the development of design criteria and operating procedures for dams, the design of fishways, the simulation of field hydraulic characteristics in laboratory flumes to study flow patterns around obstacles and organisms, and the development of new biologically relevant instrumentation.

Causes of habitat loss and some solutions

In the endeavour to put freshwater sources to productive use and to tame and control floodwaters and their pathways, the consequences in terms of habitat loss have often been overlooked. It is only recently that the level of environmental awareness has reached a point where many of the modifications to streams and their catchments have been viewed unfavourably. The benefits of dam construction and channel modification in terms of water supply, hydroelectric power generation and protection of buildings and farmlands on floodplains are not to be slighted. Rather, the intent here is to describe their effects on the physical habitat and to offer suggestions for mitigating these effects.

"Disturbance" to the stream ecosystem can result from volcanic activity, wildfires, floods, and prolonged droughts as well as anthropogenic factors such as pollution or channel modification. In general, maximum biotic diversity is maintained in streams by a level of disturbance which creates environmental heterogeneity yet still allows the establishment of communities (Ward and Stanford, 1983). For example, a "patchy" substrate maintained by periodic flooding will support a more diverse biota than a silted-up streambed. Resh et al. (1988) give an extensive review of disturbance theory. In this chapter the emphasis will be on anthropogenic disturbance to the physical habitat of streams, particularly land-use changes, dam construction and channel modification.

Land-use changes can have an impact on streams by affecting runoff rates and the input rates of sediment, woody debris and chemical pollutants. A well-vegetated catchment with deep soils will absorb rainwater, releasing it slowly. If this vegetation is removed or changed, as by clearing lands for farmland, logging, or (less obviously) by grazing, changes in stream hydrographs can occur. Clearing lands for urbanization, agriculture or timber harvest is generally thought to increase peak discharges and reduce their duration.

Runoff changes due to land clearing can give rise to enlarged and/or entrenched channels. Gregory and Madew (1982), for example, suggest that urbanization can lead to increased peak flows and enlarged channels. Land-use changes can also lead to increased siltation in streams as a result of both channel bank and bed erosion and increased influxes from upland sources. Siltation can affect the survival of fish eggs and benthic invertebrates and can have an influence on aquatic and river-edge plant growth.

Changes in runoff can also lead to alterations in the stream biota which have evolved under a different flow regime. For example, a shortened spring runoff can reduce the available period for fish spawning and egg incubation (Newbury and Gaboury, 1988). Changes can also make conditions more favourable for introduced species, as Davies et al. (1988) discovered in a study on a river system in Tasmania, Australia. They found that a near-doubling in the amount of cleared land led to an increase in the mean annual flow but a decrease in the interannual variability. It was found that brown trout, *Salmo trutta L.*, had higher populations and a more stable age structure in the period following clearing. Australian streams are characterized by high variability (see Section 4.4). Since brown trout were introduced from England, the change to a less variable flow regime more like that of English streams may have had a positive effect on these fish.

Studies on the impacts of land-use change on runoff, channel morphology and stream biology are difficult since upland changes (e.g. urbanization) are often accompanied by instream changes (e.g. channelization), and because there are few "control" catchments for comparison. However, it is undeniably a good practice to leave a buffer strip of natural vegetation along streams to reduce the impact of land-use changes in the catchment (Gore and Bryant, 1988). This serves to filter out sediments from surrounding areas, slow overland flow, eliminate bank disturbance and preserve riparian vegetation.

Channel modifications have more direct impacts on streams. The impacts can occur not only in the modified reach but also in upstream and downstream sections. Channelization is typically carried out to improve drainage or flood-carrying capacity, usually leaving a smooth, trapezoidal channel with improved conveyance and more predictable hydraulic behaviour. In extreme cases the riverbed may be reduced to a concrete channel or a buried conduit.

The straightening of channels and reduction in roughness leads to greater flow velocities and thus higher erosive forces; as a result, the channel will often erode downwards or outwards. This degradation often progresses upstream as headward erosion (Section 7.2). Increases in channel erosion can create increased turbidity, and can cause sedimentation in downstream reaches where the slope decreases or where the stream enters a lake, estuary or ocean. To control erosion, regular maintenance or bank-stabilization measures must be employed which can cause frequent disturbance of stream biota (Lewis and Williams, 1984). However, stabilization works can be designed for biological habitat improvement as well as geomorphological soundness, as suggested in Section 9.4.

In terms of habitat, channelization reduces the structural diversity of streams through the elimination of meanders, smoothing of pools and riffles and irregular bank boundaries, and removal of snags and riparian vegetation. This not only reduces the total amount of stream area and shoreline length for habitation but also eliminates the natural diversity of velocity and substrate patterns. Fish no longer have backwaters, pools or low-velocity regions for refuge during high flows, and fish eggs may be swept downstream by the higher velocities (Newbury and Gaboury, 1988). Changes in hydraulic conditions selectively alter or reduce fish fauna, as increased velocities and shear stresses affect the hydrodynamics of body shape (Scarnecchia, 1988). Riffles which aerate the flow are removed, shelter in the form of undercut banks and overhanging vegetation is eliminated, and the substrate is typically more unstable, reducing benthic invertebrate production (Statzner and Higler, 1986).

The clearing of riparian vegetation during channelization can reduce food input in the form of leaf litter and affect water temperatures. The removal of snags from within the channel can release trapped sediments and change the bed topography, leading to alterations in the fish community structure

(Sullivan et al., 1987). Channelization also isolates the river from its floodplain, resulting in the loss of marshes, billabongs and their ecological diversity (Petts, 1989). The hydraulic function of the floodplain for storing, releasing and directing waters is also lost, and downstream flood peaks can be accentuated.

By employing channel designs which do not destroy the natural structural diversity or morphological processes, some of the detrimental effects of channelization can be avoided. It is important for the biologist to be involved during the design phase rather than after construction to preserve or incorporate habitat features, within engineering constraints. Practices might include minimal straightening or shaping, emulating the natural stream morphology and structure, selective removal of trees and snags, single-bank modification, and ecologically-based bank stabilization. The incorporation of environmental features into the design of flood-control channels is reviewed by Shields (1982). The use of these practices in the rehabilitation of streams is covered in Section 9.4.

Dams typically affect both the hydrology and channel morphology of the regulated stream. These impose an artificial lake environment on the stream which changes the biota from lotic to lentic and can increase water losses to evaporation and groundwater recharge.

Reservoirs are usually effective sediment traps, releasing clearer water downstream. This clearer water is considered "sediment hungry" and can cause removal of fines in the reach immediately downstream of the dam. Typically, regulation results in a reduction of peak flows which reduces the ability of the stream to carry sediment. Thus, scouring below a dam will be limited to the size range of materials which the stream can transport. These sediments may also deposit out again at some point further downstream where the sediment influxes from tributaries and upland sources cause the stream's transport capacity to be exceeded.

Flow regimes can be altered through regulation, in terms of the duration of flows of a given magnitude, the total annual discharge, flow variability, or the frequency of flood peaks. Irrigation storage may generate short-term variable flows during peak demands and constant flows otherwise, whereas hydroelectric dams can yield rapidly fluctuating flows (Walker, 1985). These changes can be quantified with statistical methods such as flow-duration curves, flow-spell analysis and flood frequency, monthly histograms, and autocorrelation methods (Chapters 4 and 8 and Appendix 1).

The altered flow regimes can influence oxygen levels, temperature, suspended solids, drift of organisms, and cycling of organic matter and other nutrients, as well as having direct impacts on biota. Sudden fluctuations in flow, for example, can wash away deposited eggs or leave fish, crustaceans and molluscs stranded out of water. Regulation can affect community composition by altering "triggers"; for example, fish activities may be synchronized to periods of low or high flow. Inundation of the floodplain may be reduced, altering the frequency with which floodplain vegetation is "watered", and reducing opportunities for fish spawning or juvenile growth in flooded backwaters (Fenner et al., 1985; Petts, 1989). Decreases in flooding of overbank areas also reduces the amount of food input to the stream (Cadwallader, 1986; Ward, 1989).

The thermal stratification which occurs in most lakes can affect the temperature of released waters, depending on the depth of the reservoir, the location of the outlet, and the time of storage. Typically, water is released from the lower depths of reservoirs. In temperate regions this regulated discharge may warm more slowly in spring and remain warmer longer into autumn, which can affect ice formation and melt (Ridley and Steel, 1975). Regulated discharges also tend to have more stable temperatures rather than going through a daily cycle of warming and cooling. The altered temperature regime can result in the elimination of many species of aquatic insects (Ward, 1984). It also affects fish species which require specific temperatures to spawn (Cadwallader, 1986). The release of colder waters from deep-release dams can therefore permit the establishment of fisheries where high temperatures had previously prevented it. However, these waters may also be anoxic as a result of decomposition processes, and may contain reduced chemical compounds which increase the total oxygen demand in the tailwaters and can be toxic to aquatic biota (Ward, 1982). Thus, large releases of these waters can lead to fish kills in downstream reaches. Aeration of the released water can be accomplished by installing air draughts in the water-release ports (Ward, 1984). In contrast, releases of warmer waters from surface layers can encourage the growth of large quantities of algae downstream (Ridley and Steel, 1975). Multiple-level outlets allow the temperature and water quality of released water to be controlled. To improve the temperature regime of released water, Ward (1984) suggests that a certain number of degree days (Section 1.3.4) should be programmed within the annual cycle of releases.

Dams act as barriers to fish migration, and there is little doubt about their effect on the production of anadromous fish such as salmon and sturgeon (Goldman and Horne, 1983). Options for providing fish passage include the addition of various "fishways". These can include fish ladders which simulate a pool-drop structure; the use of locks (either existing boat locks or locks built specifically for lifting fish); or "trap and transport" methods in which fish are trapped in a holding area at the base of a dam and taken upstream to a release site (Petts, 1989). A fish-ladder design is shown in Figure 9.1 with a vertical slot which allows fish to pass through at preferred levels and accommodates varying discharges. For this design, the slope is 1:18 (5.56%) to 1:30 (3.33%). Locks and transport methods may have an advantage for the transport of very small fish which would not be able to tolerate the higher velocities in fish ladders. A major factor in the usefulness of any of these methods is the provision of hydraulic conditions and/or fencing to guide the fish to an entryway. Other design considerations are the fish burst speed, fish endurance and number of fish which will use the structure. The design of fishways is beyond the scope of this text; however, Blake (1983), Clay (1961), Larinier (1987), and Rajaratnam and Katopodis (1984, 1988) can provide additional information. A book on designing water intake structures for fish protection has also been published by ASCE (1982).

The effect of dams on downstream channel morphology will depend on the streambed materials, the amount, size and source of suspended sediment, and the extent of alteration to the natural flow regime. Typically, regulation decreases the peak flows which affects its ability to transport sediment and causes channels to become narrower and deeper (Reiser et al., 1989). Downstream sinuosity can also be affected. Fine sediments which would have been swept away during peak flows can accumulate in the streambed gravels of regulated rivers, reducing its suitability for spawning and incubation of embryos and causing changes in the composition of benthic invertebrate communities. Williams and Wolman (1984) reviewed the results from 287 cross-sections downstream from 21 dams on alluvial rivers in the United States. On average, they found that peak discharges were reduced, although

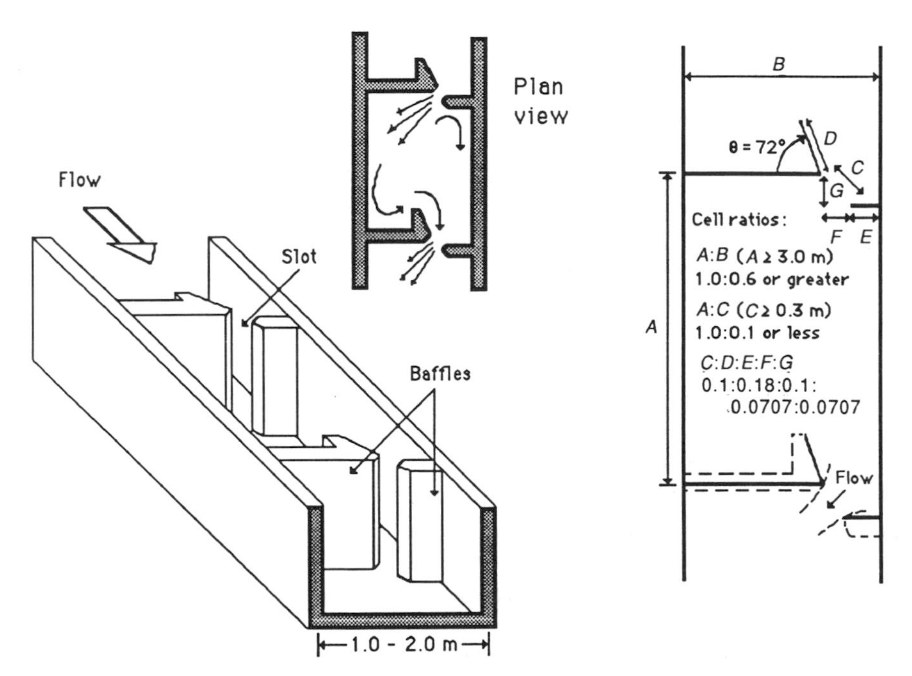

Figure 9.1. Pool and weir type fish ladder with a vertical slot. The pools provide low-velocity resting areas and the vertical slot allows fish to pass through at the preferred depth. Adapted from Mallen-Cooper (1989)

the average annual flow was increased on some rivers and decreased on others. On most of the streams the channel bed degraded in the reach immediately downstream from the dam. The effect on channel widths was variable; in some cases there was no change; in others, width increased or decreased.

If the average flow downstream of a dam is reduced it can lead to vegetation encroachment and narrowing of the channel (Rosgen et al., 1986). Species composition is also altered if non-riparian vegetation invades the shore zone (Ward, 1984). The lowered flows will also be warmer during summer months, which can cause changes in the growth of algae and aquatic plants. Reiser et al. (1989) give a graphic illustration of siltation, vegetation encroachment and channel narrowing which occurred after 85 years of water removal for transbasin diversion. Bain et al. (1988) compared fish community and habitat in two rivers: one with a natural flow regime and one which was regulated with frequent discharge fluctuations. They found that those species requiring stream margin habitat with shallow depths and slow water velocities were reduced in abundance in the regulated river.

Channel changes can often be detected several kilometres downstream from dams (Gregory and Madew, 1982). Petts (1980) suggested that channel adjustments extend downstream to a point where the reservoir's catchment becomes less than 40% of the total upstream catchment. However, the prediction of channel adjustment is difficult because of the uniqueness of each reservoir-catchment system.

"Ecohydrological" approaches, in which efforts are made to understand the biological, chemical and physical nature of streams, are needed in the analysis and solution of problems arising from land-use changes, channelization and dam construction. Karr and Dudley (1978) suggest general approaches to the improvement of biological integrity which require knowledge of:

(1) *Energy sources* (e.g. organic matter from outside the stream versus primary production within the stream);
(2) *Water quality* (including water volume and flow patterns, temperature, DO, soluble organics and inorganics, heavy metals and toxic compounds); and
(3) *Habitat structure* (e.g. substrate type, distribution of water velocity, diversity of small-scale habitats, availability of spawning, nursery and hiding places).

Some studies have been made in an attempt to relate habitat to the biota (amount, type, diversity) which can be supported by that habitat. An example is the Habitat Suitability Index (HSI) model, developed by the US Fish and Wildlife Service (Raleigh et al., 1984; Wesche et al., 1987). This involves the development of an index from 18 variables which are believed to control the carrying capacity of brown trout in streams (including water temperature, depth, cover, DO, substrate type, and baseflow as a percentage of average flow). The theory of this type of study is that if a model can be developed for predicting changes in the aquatic biota from changes in habitat then the impact of various activities (e.g. dam construction, provision of instream flows, habitat improvement) on the biota can be determined based on their potential impact on the physical and chemical character of the stream. However, biological processes such as predation and disease add another degree of complexity which make these direct predictions difficult.

Stream and catchment classification (Section 9.2) can assist in the assessment of a stream's present "state", its response to disturbance, and its ability to recover. It also enables the experience and knowledge acquired in one region to be translated to others of similar classification. For reducing disturbance due to the presence of dams, controlled releases of "instream flows" for the preservation of a more natural downstream ecosystem is an option (Section 9.3). Where stream channels have been modified, various rehabilitation measures (Section 9.4) can be applied to improve or restore stream habitats. The aim is to develop a package for the management of streams which will meet the requirements of humans and the aquatic biota—indeed, to integrate the human component into the natural ecosystem.

Sources of further information on the ecological impacts of stream modifications include Barton et al. (1972), Cheslak and Carpenter (1990), Craig and Kemper (1980), Gore and Petts (1989), Milhous (1982), Ridley and Steel (1975), Ward and Stanford (1983), and issues of the journal *Regulated Rivers*.

9.2 STREAM AND CATCHMENT CLASSIFICATION

9.2.1 GENERAL

Classification is a basic procedure for imposing order on the diversity of the real world. From dealing with the myriad of life forms which inhabit the Earth, biologists and ecologists are quite comfortable with the concept. The classification of landscapes and streams, however, is still in its infancy. Traits such as "flashy", "pristine", "regulated", "low mayfly population", and "gravel substrate" have not yet been united to form a universal classification system, although the need for one is evident.

However, before imposing a classification system on a region or a river one might want to ask whether a classification is needed. Classifications only label objects; they do not of themselves produce any information (Cowardin, 1982). Classification, then, is somewhat "in the eye of the beholder" in the sense that it is artificial. A stream will continue to be what it is, regardless of whether it is named "Cudgewa Creek" or "Bubbling Brook".

Thus, classification should have a purpose. Classification of an area for oil exploration, for example, will differ from one for earthworm potential. In the same manner, classifying a river as "wild", "scenic" or "recreational" will perhaps require different criteria than classifying it in terms of platypus habitat or archaeological interest.

One reason for classifying streams and their catchments is the identification of hydrologically homogeneous areas (Section 8.5.3). Another is the grouping of areas with similar flora or fauna so that results from a few areas can be extrapolated to all areas of the same class (e.g. as part of a stratified sampling scheme, Section 2.3.5). From a management perspective, classification is essential in designating sites for habitat preservation and/or stream rehabilitation. If regions are properly classified, it can be assumed that similar areas will respond in a like manner to similar management practices, allowing the transfer of successful practices from one area to another.

Classification is simply the grouping of objects with similar attributes. It can be "general purpose", in which classes have a large number of traits (attributes) in common, or "special purpose", in which classes share only a few attributes (Macmillan, 1987). "Desk-top" classifications of catchments from maps or satellite imagery will be sufficient for some purposes, whereas field data and close-range photography would provide the basis for more detailed classification of streams or microenvironments within a stream.

Scale plays an important role because different features are apparent when the landscape is viewed at different scales. Just as "wave-like" and "particle-like" are both useful ways of classifying the nature of light, "continuum" and "discrete segments" are both valid means of viewing a stream. For anadromous fish species the characteristics of whole river systems such as the drainage pattern classes presented in Section 4.2.4. may be important. However, for classifying the habitat of benthic invertebrates a much finer resolution at the level of microenvironments within a stream reach would be necessary. Hierarchical schemes (Section 9.2.4) which describe levels of classification nested within other levels provide an organized structure for classification across a range of scales. Again, the purpose of the classification should determine the level of detail; one should "choose a balance between being accurate but not too complicated" (Chapman et al., 1985, p. 211).

It is perhaps questionable whether classification is even appropriate for natural phenomena which vary continuously over space, such as rivers, topography, vegetation, animal communities, or soils. In community ecology, Gauch and Whittaker (1981) believe that naturally clustered data are rare and "boundaries" are thus controlled by the ecologist. Likewise, regions and rivers can be subdivided, but do the subdivisions represent distinct differences? If not, one may still wish to classify "as a convenient system of cataloguing the data" (Pielou, 1977, p. 313). When working with natural features, there will almost always be categories between categories; characteristics defying definition as either "this" or "that". Zukav (1979, p. 271) gives the following illustration:

> During the Lebanese civil war, a story goes, a visiting American was stopped by a group of masked gunmen. One wrong word could cost him his life.
> "Are you Christian or Moslem?" they asked.
> "I am a tourist!" he cried.

Proceeding with the assumption that classification provides a useful framework for understanding river processes and for identifying areas differing in terms of management needs, the actual procedure for classification is largely subject to individual requirements. Traditionally, river-classification systems have been fairly qualitative, such as Davis' (1899) famous classification of river courses into "youth", "maturity" and "old age", or the classification of "Trout", "Grayling", "Barbel" and "Bream" river zones by late-nineteenth-century German fishery biologists (Hawkes, 1975).

More recently, the power of numerical taxonomy methods (Section 8.6) provides an opportunity for developing quantitative classification systems. In developing a system suited to a particular level and purpose, physical variables such as sinuosity, catchment shape or area, stream length, mean annual flow, parameters from flow-frequency, flow-duration or flow-spell analysis, mean stream velocity, dominant particle size, water temperature or chemistry, and hydraulic variables at both macro- and micro-scales can be tested for relevance. The object is to determine which criteria are most important for distinguishing one class from another. Then traits such as "large pebble substrate", "third-order stream" and "moderate winter river regime" can be entered into a classification scheme to yield, for example, "brown trout stream".

When selecting criteria to set up classes a fundamental principle is that the causes of class differences (the "disease") are a better basis for classification than the effects ("symptoms") the differences produce (Lotspeich, 1980). In other words, brown trout streams are better classified by the factors that cause brown trout to be present rather than the fact that the trout happen to be there (which may be a function of how often the stream is stocked rather than its potential to support fish populations). These underlying factors are, however, often difficult to determine.

A plethora of ecological classification methods exist (Pielou, 1977). Aquatic ecosystems can be classified based on dominant species, indicator species, assemblages of organisms, or on the inputs and cycling of energy and nutrients. Hawkes (1975) provides a comprehensive review of classification methods used in stream ecology. She states that because of geographic differences in biota, river-classification schemes based on biota are best restricted to regions.

Wright et al. (1989), for example, describes a comprehensive nationwide research programme in the UK in which the aim was to develop a classification of polluted and unpolluted streams based on macroinvertebrate communities. A second objective was to develop methods for predicting faunal groupings based on physical and chemical variables. The method could then be used to predict which communities *should* be present at a particular site. *Absence* of a number of expected species would be reason to suspect environmental stress. Also, if changes in physical and chemical factors were anticipated (e.g. from land-use change, pollution input or river regulation), predictions could also be made of the likely impact on the macroinvertebrate fauna.

The UK programme is an example of how hydrological and ecological information can be combined in a classification scheme for environmental assessment. It illustrates the fact that physical laws apply everywhere whereas the biological communities existing within a certain range of physical conditions typically will be different from one region or continent to another. For developing a "universal" stream-classification system, then, a combination of physical variables appears to be the most promising. Biological indices also tend to be more variable and more expensive to measure. Thus, the emphasis in this section will be on classification methods based on physical rather than biological factors.

When classifying riverine systems one immediately runs into the problem of whether to classify the stream alone or as part of the catchment it drains. It has already been mentioned that when considering a stream ecosystem there is no "boundary" at the water's edge, and land and water should be considered an interacting unit. Both river- and catchment-classification methods will be described as well as hierarchical techniques in which they are combined.

9.2.2 LAND CLASSIFICATION

By using the word "land", the first step in classification has already been made, by eliminating the 75% of the Earth's surface covered by the oceans. Lands can be classified based on such factors as geology, topography, soils, climate and vegetation patterns. For example, world distribution patterns of terrestrial ecosystems are commonly classified by biogeographers into "biomes" which consider fauna and vegetation as well as climate and geology.

Maps and geographic information systems (Chapter 2) are forms of land classification. A common approach in land classification is to combine the information from several maps (digitally in GIS systems) and delineate areas sharing similar combinations of attributes. Although this "land systems mapping" approach (Gerrard, 1981) can be criticized because it implies a static relationship between features, the capabilities of geographic information systems and remote-sensing methods allow land changes to be recorded as they occur. Examples of climatic, geological and physiographical classification methods will be discussed.

Climatic classification

Since climate is considered a "master factor" in controlling watershed processes (Lotspeich, 1980), it is useful to classify an area based on its broad climate characteristics. One of the more widely used methods of climatic classification was devised by Koppen, a Russian-born German climatologist and amateur botanist (McKnight, 1990).

The Koppen system considers the association between vegetation and climate, and although classes are based on an objective, numerical appraisal of temperature and precipitation, actual class boundaries are aligned with vegetation patterns. The technique uses easily acquired statistics on mean monthly values of temperature and precipitation to separate climatic patterns into "homoclimes"—areas having broadly similar climates.

The nomenclature for labelling climate types consists of a combination of letters, each having a precise meaning; for example, Marysville (in the Acheron River basin; see Figure 4.7) is a "Cfb" climate, indicating that it has warm summers and rain all year. Definitions for each letter in the label are given by Chapman et al. (1985) as:

First letter (general categories):
A—tropical rainy climates, with no month cooler than 18°C;
C—humid warm climates, with temperatures of the coldest month less than 18°C but more than −3°C;
D—humid cool climates, with the coldest month below −3°C, and the warmest above 10°C.
E—tundra climates, with the warmest month below 10°C;
B—arid climates, where evaporation exceeds rainfall.

Second letter (refers to the season of least rainfall):
s —summer drought;
w—winter drought;
f —rain falls all year;
m—monsoonal-type rains.

The third letter "fine-tunes" the description of summer temperatures, "a" being warmer than "b", which is warmer than "c".

For "B" climates which are classified by precipitation rather than temperature, there are some exceptions to the labelling rules. The second letter can be either "S" meaning semi-arid, or "W" meaning arid, and the third letter can be "h" for hot or "k" for cold. Also, some areas may simply be labelled "H", for **H**ighland (alpine) climates. Koppen's method has been modified by geographers and climatologists to devise systems more suitable for their own purposes. Program KOPPEN in AQUAPAK will compute a Koppen classification based on the method given by Chapman et al. (1985). It requires average monthly precipitation and temperature values.

Geological classification

The geology of a region is important in riverine classification because it gives an indication of the erodibility of bedrock materials, potential for groundwater movement, surface and groundwater chemistry, landform, stream form and drainage pattern, and streambed composition. In the development of a broad classification of the ecological value of rivers and streams in Victoria, Australia, for example, Meredith et al. (1989) used rock types, grouped according to their potential erodibility, in an attempt to classify streams by substrate type. The geological data for this classification were obtained from 1:250 000 geology maps.

Substrate type in particular is frequently used by fisheries biologists and aquatic insect ecologists for classification of instream habitat. The first-order tributaries high up in a catchment will typically have a substrate derived from the geology which they flow across. However, further downstream, substrate can become quite mixed. Broad geological categories from geological maps may be of limited value for assessing substrate type because they do not normally reveal how far the materials have travelled from their parent sources. A further complication is due to the dynamic nature of streams which causes the substrate to change constantly. Thus, the age of a catchment area, its rock type, flow characteristics and fluvial processes are all considerations when attempting to predict the type of substrate material from geologic maps (S. Seymour, pers. comm., 1989). The input from an experienced geologist or geomorphologist is recommended if geology is to be included in a classification scheme as an indicator of substrate type.

Physiographical classification

"Physiography", or the description of landscape, combines both geology and climate. Since major differences in stream morphology and stream chemistry occur in different types of landscapes, physiography may be of interest in the classification of streams and river systems for biological purposes. Landscapes (large land-surface forms) and landforms (finer-scale forms, such as landslides or gravel bars) are, in general, easier to observe than ecosystem processes, and may serve as a template for designing sampling strategies for soil, vegetation and aquatic biota. Physiographical units have been used for fisheries studies and wild rivers evaluation (Macmillan, 1987). Classification by physiography is commonly used in the first steps of hierarchical systems (Section 9.2.4).

Swanson et al. (1988, p. 92) claim that understanding landscapes is crucial in understanding ecosystems at different scales in time and space since "landforms, such as floodplains and alluvial fans, and geomorphic processes, such as stream erosion and deposition, are important parts of the setting in which ecosystems develop and material and energy flows take place." They give four classes of landform effects on ecosystem processes and patterns:

Class 1: Topographic influences on ground and air temperature (e.g. the aspect of a hillslope affects solar energy exposure), moisture (e.g. orographic influences on precipitation), nutrients and other materials (such as pollutants).

Class 2: Landform effects on the flow of organisms, propagules (seeds, spores, sprouting root fragments), energy and matter (water, nutrients and organic material) through a landscape. Landscapes define gravitational gradients (e.g. for water flow) and form pathways and barriers for movement.

Class 3: Landform influences on the spatial pattern and frequency of disturbance from exterior factors such as fire, snow and wind. For example, ridges and valleys are important in controlling the spread of wildfire and the pattern of snow drifting.

Class 4: The effects of the physical dynamics of landforms (e.g. landslides, geomorphic processes) on biotic processes and features. For example, geomorphic processes affect biota in fluvial environments through their effects on the distribution of substrates, the growth of floodplains, water chemistry, and flow variability.

9.2.3 STREAM CLASSIFICATION

Numerous methods have been developed for the classification of rivers and streams based on geomorphic and hydrological characteristics such as channel shape, slope, streamflow, water quality, and substrate. For most "hydrobiological" purposes, the classification of whole river systems is not particularly useful, except perhaps in the first levels of a hierarchical scheme where one might separate, for example, "tropical", "temperate" and "arctic" river systems. A more popular approach is the use of river zones (Hawkes, 1975). Davis' method, mentioned previously, or Schumm's (1977) description of "sediment production", "transfer" and "depositional" zones (this section) would be examples of broad-scale zonation. In contrast, Vannote et al. (1980) propose that zonation is non-existent and the stream biota changes in a "continuum" along a stream in response to predictable changes in physical parameters.

Again, with a focus on physical variables, stream-classification methods can be subdivided into (1) hydrological, (2) geomorphological and sedimentological, and (3) methods based on a wide combination of physical attributes.

Hydrological classification

Streamflow is a useful measure for classification purposes because it "integrates the influences of most landscape features into a single measurable characteristic" (Likens et al., 1977, p. 27). Hughes and James (1989), for example, used streamflow indices to classify Victorian streams for "hydrobiological" purposes (see Section 8.6.3). The regime-classification method of Haines et al. (1988) mentioned in Section 4.4.3 is based on average monthly flows.

With an interest in developing objective, reproducible methods for delimiting community structure in unpolluted streams, Jones and Peters (1977) also used a flow-regime classification. They reasoned that changes in regime could be used to predict alterations in invertebrate communities. Regime was obtained by visually inspecting graphs of mean, maximum and minimum monthly flows (both mean and standard deviation) and ranking the regimes from "stable" to "spatey" (flashy).

Poff and Ward (1989) used a much more detailed analysis of streamflow patterns to group 78 streams from across the continental United States into nine stream types. They employed average daily flow values from gauging stations on streams with minimal alteration from impoundments, large diversions or channelization. Cluster analysis (Section 8.6.4) was applied to the analysis of 15 variables, including basin area, mean annual flow, annual coefficient of variation, Colwell's predictability index (Section 4.4.5), the mean number of floods per year, an index of flood predictability, an index of flood seasonality, and the average number of zero flow days per year. Highly variable and/or unpredictable flow regimes were considered to have a dominant effect on ecological patterns, whereas in more predictable flow environments the patterns were thought to be influenced more strongly by biotic interactions such as competition or predation. In most streams, both abiotic (e.g. flow regime) and biotic factors contribute to community structure. Figure 9.2 shows the conceptual stream-classification model developed, which was based on a hierarchical ranking of four components of flow regime (intermittency, flood frequency, flood predictability and overall flow predictability).

The authors give a discussion of the implications in terms of ecological patterns. For example, highly intermittent streams would tend to support fish with small body size and invertebrates with resting stages and increased dispersal capacity. In contrast, streams with high flow predictability would tend to support larger, more specialized fish and invertebrates and more long-lived species. The discussion was based on observations from a number of other studies because of the lack of data on stream organisms for the gauged streams.

Although stream discharge measures can act as surrogates for a number of related measures such as stream width, catchment area, average velocity, and catchment topography, geology and climate, they are useful only *if* discharge measurements or estimates can be obtained. "Runoff maps" do exist for various parts of the world, which can be helpful in broad-scale hydrologic classifications; however, these will most likely be inadequate at the stream reach level. With a modest amount of data collection at ungauged sites to facilitate extrapolation within a region (see Section 8.5.3) it should be possible to improve the resolution of hydrological classification schemes. However, from a practical point of view they are probably most appropriate at a regional scale.

Geomorphological/sedimentological classification

Classification systems based on stream geomorphology are popular because much, if not all, of the information can be obtained from maps and/or aerial photographs. For example, in Schumm's (1977) classification scheme (Figure 9.3) it is fairly easy to distinguish between sediment production, transfer and deposition zones from aerial photographs or casual (i.e. car-window)

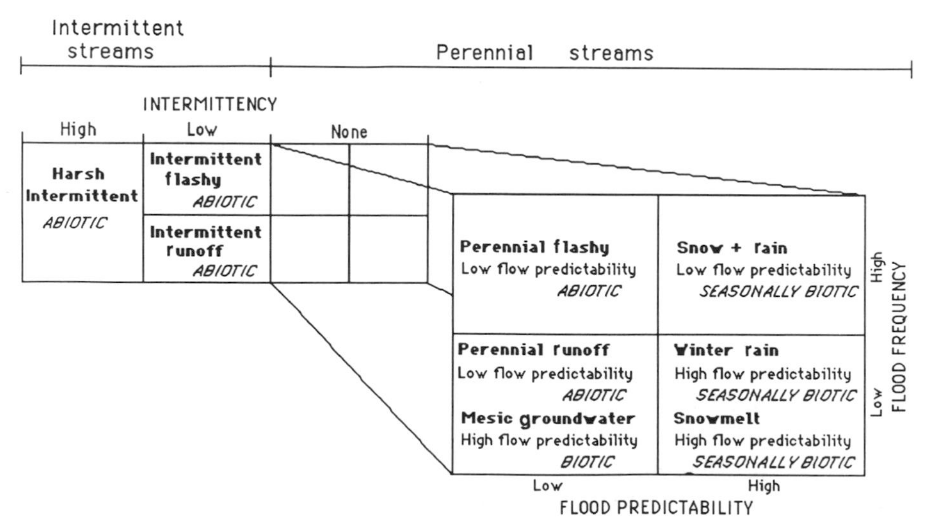

Figure 9.2. A conceptual model of stream classification based on discharge regime. The relative effects of abiotic (flow regime) and biotic (e.g. predation, competition) processes on community structure differ between the various classes, as indicated. Redrawn from Poff and Ward (1989), by permission of *Can. J. Fish. Aq. Sci.*

observations. Sediments will obviously be eroded, transported and stored within all sections of the river, and boundaries will not be clear-cut; however, one process will predominate within each zone.

Brussock et al. (1985) also present a three-zone classification method for an "ideal river" based on sedimentological settings relevant to stream biota:

Cobble and boulder-bed channel: pools form behind large boulders or large accumulations of debris; valley is generally V-shaped; streams flow across bedrock, with little floodplain development.

Gravel-bed streams: pools and riffles are more distinct and related to sinuosity. Streams flow through alluvium, with moderate to extensive floodplain development.

Sand-bed channels: channel beds are mobile at discharges less than bankfull, with the formation of ripples, dunes, etc. (see Section 7.3.4). The stream channel is easily modified by flood events and floodplains are extensive.

The three-zone classification of Davis (1891) was mentioned in Section 4.1.3 under the heading "headwater, middle-order or lowland streams?" It should be realized that his terminology should not be taken as an indication of the relative age of different parts of the landscape. For example, V-shaped valleys may actually be older geologically than floodplains formed by alluvial deposits.

Kellerhals et al. (1976) review classification schemes relevant to river engineering practices and propose a system based on the supply and variability of water and sediment, streambed materials, and the geomorphological and geological setting. They give coding sheets for summarizing descriptive field

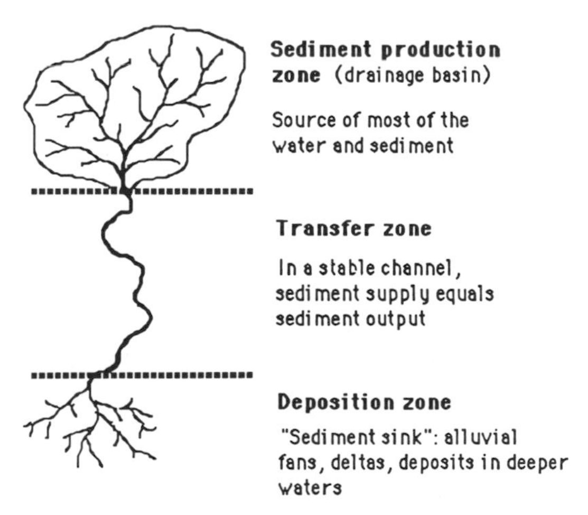

Figure 9.3. Zones of an "ideal" fluvial system. Modified from Schumm (1977), by permission of John Wiley & Sons, Inc.

data (e.g. terrain, vegetation, land use, relation of channel to valley, channel pattern, obstructions, and channel bed and bank materials). The approach is very similar to that taken in the State of the Streams survey, described in Section 4.1.4.

Rosgen (1985) developed a quantitative stream-classification system using channel morphology indices. The system, designed for application at the river reach scale, was developed from analysis of hundreds of streams from diverse hydrophysiographic regions in the United States. Identification codes are assigned to streams based on the following criteria, which can be largely determined from topographical maps and aerial photographs: channel gradient, sinuosity, width/depth ratio, dominant channel particle size, confinement of channel in valley, and landform feature/soils/stability. Stream subtype criteria which describe potential influences on channel change and thus associated changes in fisheries habitat and channel stability are also defined. The subtypes provide higher resolution which may or may not be required depending on the level of application. In use since 1978, Rosgen's classification has been utilized for a variety of purposes such as guidelines for riparian areas; estimating hydraulic roughness, shear stress and sediment supply coefficients for engineering calculations; hydraulic geometry relationships; fisheries habitat interpretation; land systems inventories; and stream-restoration guidelines. Details of the method will not be given here; however, program ROSGEN in AQUAPAK lists the stream types and subtypes.

Methods based on combinations of physical attributes

Among stream ecologists one of the more popular methods of classifying streams has been the generalized method of Illies (1961). The method, based on data from several continents and designed for worldwide use, divides streams into two zones:

Rithron—the cooler upper reaches of the stream where oxygen concentration is high, the discharge is relatively low, the current is fast and turbulent, and the bed is mostly rocks or gravel with only patches of sand and silt in pools and sheltered areas. Organisms are predominantly characteristic of colder, highly oxygenated running water, with little or no true plankton.

Potamon—the lower reaches of the stream where velocities over the riverbed are lower and the bed is mainly mud and sand. In deeper pools, dissolved oxygen and light penetration may be limited. Organisms are tolerant of warm, still water with lower oxygen levels, and there may be a rich plankton.

The point on a stream separating rithron and potamon is defined as the location where the monthly mean temperature is 20°C. For tropical rivers, Illies and Botosaneau (1963) modified the criterion to a summer maximum monthly mean of 25°C. The dividing point will lie at different altitudes depending on the general climate, with the rithron extending to lower altitudes at latitudes closer to the poles. Because a variety of factors may influence where the boundary occurs, it should not be considered precise— or fixed. Additionally, some streams will not fit into the scheme, such as those which never have a monthly mean temperature exceeding 20°C.

The general idea of Illies' classification has been found useful (Hawkes, 1975), and Hynes (1970) says that it provides a reasonable method of distinguishing between "streams" and "rivers". Hynes also points out that other researchers have further subdivided and expanded the two classes, based on "zones" which show distinct ecological structures, with high faunal diversity in the "transition" regions where the zones intersect. Lake (1982) suggests that this is perhaps only telling us that the eroding, intermediate and depositing zones of geomorphic interest (see Figure 9.3) also reflect ecological zones. Zonation is perhaps more useful for general description than for precise classification of a stream's ecology.

The work of Pennak (1971) is often cited as an example of how physical and chemical parameters are substituted for biological criteria in an attempt to develop a worldwide stream-classification method. He proposed the system for ecological use under the assumption that "widely separate streams and rivers having very similar non-biological features will usually have parallel and ecologically similar faunas" (p. 324). His system uses 13 parameters: stream width, flow, current speed, substrate, summer and winter temperatures, turbidity, total dissolved organic and inorganic matter, water hardness, dissolved oxygen, rooted aquatic plants and streamside vegetation.

Similarly, for classifying small streams in terms of their instream biota. Savage and Rabe (1979) used stream order and gradient from maps, and substrate and flow characteristics from field observations to classify stream types. They concluded that these readily identifiable physical features adequately reflected the differences in organisms present. Also, since gradient was correlated with substrate and flow characteristics, preliminary classification was possible using only map information.

Regionalization, or the classification of areas which behave in a similar fashion hydrologically, is commonly used for the purpose of extrapolating records from gauged to ungauged sites and for predicting streamflow statistics from catchment or channel characteristics (Section 8.5). It is routinely used by engineers in flood and water yield analysis. Relatively recently, numerical taxonomy techniques have been applied to regionalization. Cook et al. (1988), for example, used landscape attributes from topograpic and geological maps (relief, elevation, slope, stream counts, wetland class, vegetative cover, surface geology, etc.) to group catchments of similar streamflow in New South Wales, Australia. Nathan and McMahon (1990a) explored a number of techniques for identifying hydrologically homogeneous regions for the prediction of low-flow characteristics. Using several physical catchment variables, they applied cluster analysis, multiple regression and principle component analysis, with Andrews curves used to "fine tune" the optimum groupings (see Chapter 8), to categorize 184 catchments from south-eastern Australia into eight hydrologically homogeneous groups. The authors pointed out that regions similar in terms of hydrologic or basin characteristics may not necessarily be grouped together geographically.

9.2.4 HIERARCHICAL CLASSIFICATION

Hierarchical classification systems provide several levels of resolution at which classification can take place. They thus allow flexibility, since some applications require details which are too costly or unnecessary for others. Rosgen's (1985) classification, described in the previous section, actually has a two-tiered hierarchy. The classification of microenvironments by Davis and Barmuta (1989) can also be considered hierarchical (Section 6.4.2). The examples presented here are those which classify streams within a larger landscape, which influence water quality and quantity and energy input.

Cowardin (1982) developed a hierarchical classification scheme as part of a national inventory of wetland areas conducted by the US Fish and Wildlife Service. Wetlands are lands which are transitional between terrestrial and aquatic systems, where the water table is close to the land surface. The three-tiered structure of this classification system considered first ecological systems (e.g. marine, estuarine, riverine), second, hydrological characteristics (e.g. subtidal, perennial, intermittent), and third, vegetation type (e.g. moss-lichen wetland)—or if no vegetation was present, then substrate composition (e.g. rock bottom, reef). Additional modifying terms could be added for water regime, water chemistry and soils. Distinctions can also be made between regions even if classes are otherwise similar. Cowardin states that the correlation between soil, hydrology and vegetation is not well documented for wetlands, yet an understanding of this relationship is important for predicting the hydrologic behaviour of wetlands. It was anticipated that the system would be modified with use, and perhaps incorporated into a general site-classification system for lands.

Lotspeich and Platts (1982), drawing from the work of a number of researchers, developed a hierarchical land-classification system based on the ecosystem concept. Geology, soils and vegetation were considered landscape elements formed at different time scales. The system is based on an interpretation of how the terrain would appear in a natural state, to reduce bias from land-management practices. Categories are defined for various spatial scales from Domain (about 2000 km wide, for broad climatic separation) to Land type association (1–10 km wide, for separation of geologies, vegetation and aquatic communities). The authors considered the smaller units to be more useful for stream classification and site-level management.

Larsen et al. (1986) followed a semi-hierarchical method similar to that of Lotspeich and Platts. Classification was applied to stream habitat quality analysis, under the assumption that the physical character of the stream, and thus its biota, is controlled by the physical character of the catchment: its form, geology, soil and climate. Classification was used to delineate homogeneous "aquatic ecoregions" and to select stream sample sites for determining fish assemblages. Multivariate analysis techniques (ordination and discriminant analysis; see Section 8.6) were used to demonstrate that regional differences in fish species assemblages were related to the aquatic ecoregions. They concluded that an ecological framework could be quite useful for integrating land and water resource management.

9.2.5 SUMMARY

Classification is a subjective procedure, dependent upon its purpose and the type of data available. Clearly, there is a need for an "all-purpose" stream/catchment classification system which encompasses landscape, geology, soil, topography, temperature, climate, runoff, water quality, fauna, vegetation, aesthetics and economic and societal characteristics. A major limitation to the development of a broad classification scheme is the availability of both physical and biological data at the stream reach level, and the lack of agreement on which attributes are most important for classification. Another consideration is the dynamic nature of landscapes and streams: a frog will maintain its identity for its lifetime, whereas changes in a stream reach may cause it to slip into a different class (e.g. braided versus sinuous). In the classification of riparian and wetland areas, Gebhardt (1989, 1990) recognized the potential for change in channels, water table levels, and associated vegetation, and introduced the concept of "state" to allow process-oriented descriptions of classes.

As stream-classification methods continue to be applied and tested, progress will continue towards a universal, standardized stream-classification system. Multidisciplinary efforts are needed to determine which variables are the most critical and universal. "Lumpers" who tend to coalesce related classes and "splitters" who tend to divide them will need to reach a compromise on the actual format of the classification system. As Pielou (1977) says in reference to work in ecology, much of the literature is on testing and comparison of methods, and if the classifications are to be used, ecologists must choose one or a few methods and employ them consistently.

As in biology, classifications can be expected to change constantly within a general framework as "new species" are encountered and new observations and measurements reveal deficiencies in older systems. Hopefully, this section has illustrated some of the criteria which could serve as the basis for a joint ecological–hydrological classification system "as fluid as the medium" (Carpenter, 1928).

9.3 ENVIRONMENTAL INSTREAM FLOW REQUIREMENTS

9.3.1 GENERAL

Dams, reservoirs and diversion systems allow the modification of natural patterns of streamflow. Flood-mitigation dams, for example, reduce the peak discharges which would normally overflow the riverbanks and spill onto the floodplain. Reservoirs operated for irrigation water supply or hydropower production modify the natural flow regime through storage of water during high-runoff periods for later release when demands are highest. "Peaking power" hydroelectric production may also impose an "on–off" pattern on the natural flow regime as turbines are quickly brought on-line to supplement daily electricity requirements during peak-demand periods (Gore et al., 1989).

Concern about the impact of dam construction and flow regulation on the ecology of regulated streams has prompted efforts to quantify and preserve the flow patterns required for survival of aquatic species. Aquatic ecosystems are best preserved under natural, pristine conditions; however, modification of the natural flow regime through storage and abstraction can also be considered a requirement for survival of the modern-day human species. Conflicts over water use are common, especially where water supplies are limited as in arid areas and where demands for water supplies are increasing. Where people and nature must co-exist, compromises should be made and satisfactory balances struck between competing uses.

Instream flows are those which are retained in their natural setting, as opposed to those waters which are diverted for "offstream" users such as industry, agriculture and swimming pools. Instream waters can support economically important uses such as transportation (e.g. by barge or kayak), production of hydroelectricity, and waste disposal. Although more difficult to quantify in economic terms, instream flows are valuable for maintaining fish and wildlife habitat. Concern about these more intrinsic values has increased, as have studies for evaluating the environmental impact of water-resource projects. This concern has led to the provision of instream flows specifically for environmental purposes in a number of countries.

Instream flows provided for environmental reasons, sometimes called **environmental flows**, are designed to enhance or maintain the habitat for riparian and aquatic life. They may be provided for preserving native species of flora and fauna, maintaining aesthetic quality, maximizing the production of recreational or commercial species for harvest (e.g. geese, trout, molluscs), or protecting features of scientific or cultural interest (Kinhill, 1988).

The initial impetus for instream flow studies came from north-western North America, where salmon fisheries of significant commercial value were threatened (Nestler et al., 1989). The aim of instream flow studies is to assess the flow requirements of certain species and to develop a recommendation for flows needed to assure maintenance of the population. For indigenous species, the best model is one which mimics nature, since the biota have evolved in accordance with the historical patterns of high, low, and zero flows, and may depend on particular streamflow patterns to carry out their life cycles.

Recommendations for instream flow levels usually specify minimum flows necessary to allow the passage of fish, to provide sufficient "living space", or to ensure acceptable levels of temperature, dissolved oxygen or salinity at a particular point on a stream. The minimum flow is normally specified as an instantaneous flow rather than a daily average; i.e. the flow should never drop below the minimum at any time. Instream flow recommendations may also include artificial floods (called "flushing flows" or "channel-maintenance flows"), which are designed to remove fine materials from the streambed, scour out encroaching vegetation, or flush anoxic or highly saline waters from stratified pools. For example, Tunbridge and Glenane (1988) specified three levels of environmental flows and a flushing flow for the Gellibrand River and Estuary in Victoria, Australia:

- An *optimum environmental flow*, to allow the full production of fish, especially for recovery after a period of stress (e.g. drought, overfishing)
- A *minimum environmental flow*, which would result in little or no reduction in numbers of fish, for average rainfall years
- A *survival environmental flow*, which may cause a reduction in numbers of fish but no loss of species, for low-rainfall years, and
- A short-term *flushing flow*, to remove the salt wedge in the estuary and maintain the freshwater section of the river.

In establishing instream flow requirements the difficulty lies in deciding on how much modification of the natural flow regime is acceptable. Although there have been a number of studies on the effects of regulated flows on organisms (e.g. Ward and Stanford, 1987), a great deal of scientific uncertainty persists. One of the main difficulties in determining an instream flow requirement is this lack of quantitative data. This limitation becomes especially critical when the preservation of aquatic habitat conflicts with other uses of the water.

Managers, under pressure from the public to meet traditional as well as environmental water requirements, usually cannot wait for completion of extensive studies on species requirements in specific rivers before making decisions. Therefore, negotiation comprises much of the instream flow recommendation process. Objective, consistent methods which produce reliable estimates of habitat requirements will increase the power of a manager's arguments.

To assist in the development of instream flow recommendations, a number of numerical techniques have been developed over the past few decades. Most of the efforts have centred around the preservation of trout or salmon habitat in cold-water streams. The methods first developed were based on the judgement of biologists, but were soon followed by simple methods using some measure of the unregulated streamflow. Over time, increasingly sophisticated techniques have been derived which consider the changes in stream hydraulics at varying flow levels and the habitat requirements of species at different life stages and seasons.

Several reviews and evaluations of these techniques have been published (Kinhill, 1988; Mosley, 1983; Prewitt and Carlson, 1980; Richardson, 1986; Stalnaker and Arnette, 1976; Wesche and Rechard, 1980). The techniques fall basically into three categories: historical discharge or "rule-of-thumb", threshold, and instream habitat simulation.

9.3.2 HISTORICAL DISCHARGE OR "RULE-OF-THUMB" METHODS

These techniques utilize streamflow records only, with instream flow recommendations based on set proportions of the gauged discharge. In the simplest form, a single minimum flow value is computed. For example, the 7-day, 1-in-10-year low flow (see Section 8.2.6) is often mentioned in regard to maintenance of water quality standards (Stalnaker, 1979), but is usually considered too low for habitat maintenance (Bovee, 1982; Orth and Leonard, 1990). More sophisticated methods vary the proportion of flow retained at different times of the year. For example, to satisfy flow requirements for fish, higher proportions may be allocated during migration periods (Kinhill, 1988).

Tennant method

The Tennant method (Tennant, 1976) also referred to as the "Montana" method, is one of the most widely cited examples of this type of technique. Recommended minimum flows are based on percentages of the average annual flow, with different percentages for winter and summer months (Table 9.1). The recommended levels are based on Tennant's observations of how stream width, depth and velocity varied with discharge on 11 streams in Montana, Wyoming and Nebraska. At 10% of the average flow (the mean daily flow, averaged over all years of record), fish were crowded into the deeper pools, riffles were too shallow for larger fish to pass, and water temperature could become a limiting factor. A flow of 30% of the average flow was found to maintain satisfactory widths, depths and velocities. The choice of a maximum flow was based on the theory that prolonged large releases would result in severe bank erosion and degradation of the downstream aquatic environment. The method was designed for application to streams of all sizes, cold and warm water fish species, as well as for recreation, wildlife and other environmental resources.

One main limitation of Tennant's method is that application of the technique to other streams requires that they be morphologically similar to those for which the method was developed. The required criteria, however, are not given by Tennant, making direct transfer of the technique difficult. Field observation of the stream at the various base flow levels is recommended for verification. Also, since the method is based on the average flow it does not account for daily, seasonal or yearly flow variations. Comparison of the recommended flows with the average 10- and 30-day natural low-flow values is advisable to determine whether the flows are available naturally during low-flow periods (Wesche and Rechard, 1980). Prewitt and Carlson (1977) also recommend the examination of mean monthly flow data to check the validity of the method.

Hoppe method

Hoppe (1975) devised another rule-of-thumb method based on flow-duration curves (Section 8.3). Data from the Frying Pan River in Colorado were initially used in developing the method, which is based on trout requirements. The recommended flows are given in Table 9.2.

For this method, flow records of adequate length (20 years) are required. A daily time unit is preferable for constructing the duration curves since resolution is lost with longer time units.

The main attraction of rule-of-thumb techniques is the fact that an answer can be obtained rapidly if gauged records are available, eliminating the time and cost of field data collection. However, the techniques tend to be site- and species-specific, and for them to be applicable in other situations the relationship between habitat and discharge must be similar. As given, the methods do not require an understanding of the ecosystem; a limitation which can be significant if, for example, results from snowmelt-dominated

Table 9.1. Instream flow recommendations for fish, wildlife, recreation and related environmental resources by the Tennant method. After Tennant (1976)

Description of flows	Recommended base flow regimes October–March	April–September[a]
Flushing or maximum	200% of the average flow	
Optimum range	60–100% of the average flow	
Outstanding	40%	60%
Excellent	30%	50%
Good	20%	40%
Fair or degrading	10%	30%
Poor or minimum	10%	10%
Severe degradation	10% of average flow to zero flow	

[a] The seasons would be reversed for Southern Hemisphere streams.

Table 9.2. Instream flow recommendations based on the method of Hoppe (1975). The flushing flow is a flow maintained for 48 hours to flush fines from gravels

Description	Percentage of time flow is equalled or exceeded
Flushing flow	17
Spawning flow	40
Food production and cover flow	80

streams are applied to the flashy streams in semi-arid lands. It may be possible to develop "rule-of-thumb" recommendations for other species or streams if a relationship between habitat and discharge can be derived from the experiences of local researchers or anglers, or special field studies.

Orth and Leonard (1990) argue that simple methods requiring little or no field work are needed for basin-wide planning purposes. Using nine target fish species in four streams in Virginia, USA, they compared the results from PHABSIM (Section 9.3.4) to those from the Tennant method. It was concluded that the 10% recommendation correctly identified poor habitat conditions and the 30% recommendation corresponded with near-optimum habitat in small streams but the flow was higher than optimum for a larger stream.

9.3.3 "TRANSECT" METHODS

These methods involve the collection of field data at one or more transects (cross-sections) in a stream reach and the development of relationships between discharge and other physical variables such as wetted perimeter, water depth and velocity. Typically, data are collected at sites where the maintenance of flows is most critical. For example, fish passage may be prevented if water depth and velocity are outside certain limits. Water depth may also be important for separating islands from the mainland for waterfowl nesting (Cochnauer, 1976). Commonly, shallow riffled reaches are chosen for analysis as these areas are the first to be affected by flow alterations, and because it is reasoned that maintenance of suitable riffles will also maintain suitable pool conditions.

With some methods, the relationship between discharge and physical variables is combined with information on the physical habitat requirements of various species to develop recommendations for the optimum discharge. Examples of habitat criteria are presented in Table 9.3.

The Idaho and Washington methods will be discussed as examples of this type of technique. They can be considered precursors to the instream habitat simulation methods which follow (Section 9.3.4).

Idaho method

This method, given by Cochnauer (1976) and White (1976), was developed for large unwadable rivers, specifically the Snake River in Idaho, USA. It is based on the prediction of habitat loss at reduced discharge and the relationship of this predicted loss to the requirements of key fish species. The methodology was proposed as a starting point for developing flow recommendations rather than as a method for obtaining a rigid, absolute value.

Field data are collected only once, at the lowest practical flow, to allow evaluation of physical features. Critical areas for spawning, rearing, and passage of fish are first identified in the field. Measurements of cross-sectional profiles, water-surface elevation, velocity and substrate type are taken at between four and 100 transects in each study reach, including a transect located at the control section—the downstream section controlling the water surface profile in the reach (see Sections 5.6 and 6.6). If bridges, logs or debris dams are present, cross-sections are established 10 m above and below these controls.

A backwater curve computer program (see Section 6.6.6) is used for generating hydraulic characteristics (depth, velocity, wetted perimeter) over a range of flows. Curves are then constructed (e.g. wetted perimeter versus discharge) which are compared with known biological criteria such as those given in Table 9.3.

By comparing the simulated habitat conditions to the known requirements

Table 9.3. Depth, velocity and substrate requirements for rearing habitat of fish of southwestern Victoria. Extracted from Tunbridge (1988)

Species	Depth (m)	Velocity (m/s)	Substrate type
Blackfish	>0.20	0–0.30	All types
Brown trout (adult) (juveniles common in shallow, fast water with boulder and rubble cover)	>0.20	0–0.50	All types
Redfin and common carp (juveniles also found in faster water in gravel/rubble sites)	>1.0	0–0.20	Mud/sand
Short-finned eel	>0.20	0–0.30	All types

of three warm-water fish species (white sturgeon, smallmouth bass and channel catfish), the authors developed recommended minimum flows for passage, spawning and rearing, with allowances for later adjustments after further data analysis. Flows for passage were based on a minimum required depth. For spawning, the flow providing the maximum width for spawning (as an average across all transects) was used as a guideline for determining the minimum sustained discharge. The wetted-perimeter method (see Washington method, following) was used for determining rearing discharges.

Recommended flows were assigned by month or two-week period, with the highest flow requirement (for spawning, rearing or passing) selected for the given time period. The authors felt that the weakest point in the application of the method was the limited amount of information on the fish species.

Washington Department of Fisheries method

Collings (1972) describes this method, which was developed for salmonids. The technique uses mapping of stream reaches to determine the amount of spawning and rearing (food production) habitat at a range of discharges. To apply this technique, "usable" and "unusable" levels of depth and velocity must be known for the species of interest.

Data requirements include measurements of stream depth and velocity at a number of transects, taken at several different discharge levels. The authors recommend the selection of at least three representative study sites (rearing or spawning), with four transects at each site (10 m apart) and at least five different levels of discharge. Depths and velocities are also measured at several points between the transects to more accurately define the "iso" lines of equal depth and velocity (Figures 9.4(a) and (b)).

For each flow level, a planimetric map (Figure 9.4(c)) is drawn to display the combined distributions of depth and velocity. The total area within the study reach meeting depth and velocity criteria for spawning or rearing is measured from the maps with a planimeter and plotted against the corresponding flow level. A curve is then fitted through these points (Figure 9.4(d)). The discharge yielding the highest habitat area (the peak on the curve) is considered optimal. A discharge not less than 75% of the optimal value is taken as the recommended minimum flow.

An advantage of this method is that it yields a graph which shows the change in habitat with discharge rather than giving a single value. The effect of alternative flow regimes on habitat can thus be evaluated. In comparison with the Idaho method, this is much more labour-intensive since it requires data collection at several flow levels and hand-mapping of the results. However, the field observations at different discharges may provide invaluable information about how the stream and its inhabitants respond to changes in flow.

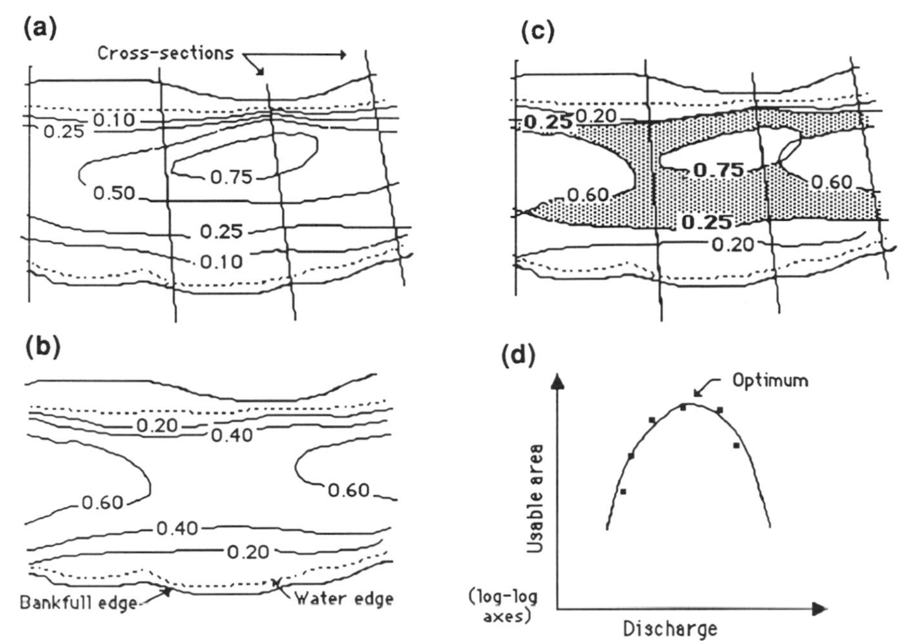

Figure 9.4. Washington method for determining preferred discharge. For this theoretical example, "preferred habitat" is: depth 0.25–0.75 m and velocity 0.20–0.60 m/s. Figures display: (a) depth contours in metres, (b) velocity contours in m/s, (c) a combination of maps (a) and (b) with shaded region showing usable area, and (d) trend-fitted curve derived from measurements (point shown) taken at several discharges. The "optimum" discharge corresponds with the greatest usable area. Adapted from Collings (1972)

Alternatively, if fish habitat criteria are not available the Washington method specifies the use of a **wetted-perimeter method**. For this, transects are located in several representative riffles, and measurements of depth and velocity are taken during at least five different flows. A plot of wetted perimeter against discharge is drawn (see Figure 9.5(b)) and the first break in slope in the curve is taken as an indication of the optimum rearing discharge. Increases in discharge above this point produce smaller changes in wetted perimeter. This technique is based on the premise that the slope breakpoint represents the quantity of water preferred by salmon.

As compared with rule-of-thumb methods, transect techniques take into consideration the habitat requirements of fish species and the availability of habitat at various discharge levels. The need for field data, however, makes the techniques more time-consuming and costly. Again, although the techniques were developed for the spawning requirements of salmon, it would be possible to modify them slightly for other species and life stages.

Kinhill (1988) argues that for almost the same amount of effort, the more sophisticated stream habitat-modelling methods (Section 9.3.4) can be

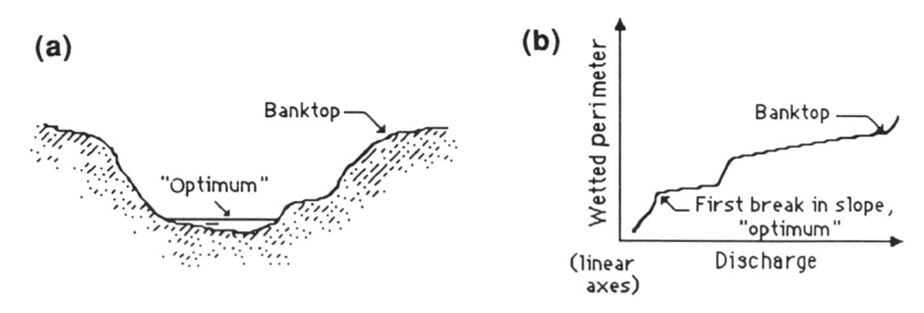

Figure 9.5. Wetted-perimeter method: (a) hypothetical channel cross-section and (b) graph of wetted perimeter versus discharge. The first breakpoint in slope is used as an index of optimum available water

applied. Transect methods are based on the optimum ranges of velocity and depth determined for a species ("usable/unusable" criteria), whereas the more sophisticated habitat-modelling methods use a continuous function to describe preferences. Transect methods may thus require less information about a species but do not allow as much flexibility in evaluating the effects of different flow levels on available habitat.

9.3.4 INSTREAM HABITAT SIMULATION METHODS

These methods consider not only how physical habitat changes with streamflow but combine this information with the habitat preferences of a given species to determine the amount of habitat available over a range of streamflows. Results are normally in the form of a curve showing the relationship between available habitat area and stream discharge. From this curve, the optimum streamflows for a number of individual species can be ascertained and the results used as a guide for recommending instream flows.

The Instream Flow Incremental Methodology (IFIM)

The IFIM is a conceptual framework for the assessment of riverine habitats. It can be thought of as a flexible set of guidelines for solving any problem involving disturbance of a riverine ecosystem. The methodology has been developed by the US Fish and Wildlife Service Cooperative Instream Flow Service Group (now the Aquatic Systems Branch of the National Ecology Research Center), and is described by Bovee (1982).

Basically, IFIM is a problem-solving tool made up of a collection of analytical procedures and computer models. It was designed as a communication link between fishery biologists, hydrologists and hydraulic engineers (Stalnaker, 1980). The aim of IFIM is normally to determine the effect of some activity such as irrigation withdrawals, dam construction or channel modification on aquatic habitat. Since each riverine system will have different sets of flora, fauna, hydrological and hydraulic characteristics, types of disturbances and regulating agencies, the IFIM allows the development of a different approach for each situation.

The IFIM is a complete thought process that begins with the structuring of a study design and the description of the present condition, and carries through to the final negotiation of a solution. "Incremental" means the slight or incremental modification of the problem or the perspective or view of the problem until a solution is found. It also refers to the ability to look at the effects of incremental changes in a variable (e.g. discharge) on available habitat. Since there will be a number of perspectives on management of water for instream uses, incrementalism is a valuable approach to problem-solving in this field. Rather than generating a single answer, the methodology produces a range of solutions which permit the evaluation of different alternatives. Two identical applications of the method can lead to different solutions simply due to different management goals.

Before an IFIM study is initiated, alternative and competing uses of the stream system should be identified. This should include consideration of which species are to be preserved and their physical habitat requirements. Is the stream to support carp or an endangered fish species? Are flushing flows needed to maintain or restore desirable streambed characteristics? What are the most critical time periods, depths, velocities and other factors for maintaining the species?

Instream flow studies should weigh the relative values of the water resource so that changes in stream habitat can be balanced with other beneficial uses. Changes in both "macrohabitat" (channel characteristics, temperature and water quality) and "microhabitat" (the distribution of hydraulic and structural features making up the living space for an organism) are considered by the IFIM (Bovee, 1982).

An application of the IFIM to a water-resource project consists of seven steps (Bovee, 1982; Nestler et al., 1989):

(1) Describe the present state of the river or system in terms of key variables (e.g. water quality, channel form or flow regime).
(2) Develop functions or mathematical expressions describing the habitat preferences of evaluation species including humans; for example, a description of how changes in the wetted perimeter affects the suitability of the reach for mayfly larvae or whitewater rafting.
(3) Develop functions or mathematical expressions integrating the macro- and microhabitat availability of the present system; for example, a function describing how wetted perimeter changes with streamflow.
(4) Incrementally change one or more variables (e.g. the flow regime or the channel dimensions) to reflect a particular management option, and

then, using relationships developed in (2) and (3), determine habitat availability under the "new" system. Options should be evaluated over a range of streamflows.

(5) Determine possible alternatives or remedial actions to correct any adverse impacts found in step 4.
(6) Repeat steps 3 and 4 to evaluate the impact of a range of management alternatives.
(7) Evaluate the alternatives in light of the various perspectives and management objectives of the organizations involved, and prepare assessments/recommendations for the project.

Step 1 involves the delineation of a study area and the collection of data. A study area may be a specific segment of a stream or a larger section of a drainage area. Reaches within the study area can be "representative" of the habitat in a particular stream and/or "critical", meaning that they either have characteristics which are in short supply within the stream or they limit or control some important species activity (such as migration or spawning). Bovee (1982, 1986), Bovee and Milhous (1978), and Milhous (1988a) give thorough instructions on site selection and the collection of both physical and biological data for use with the IFIM. A review of their work is recommended before the initiation of field studies.

Step 2 involves the selection of appropriate evaluation species and the development of species habitat criteria from the advice of experts, the literature or field studies. One or more species may be chosen based on their value as game, commercial, endangered or indicator species, or some other value. Orth (1987) recommends selection of species with narrow ranges of habitat preference since these are most sensitive to flow alterations. The time of year for which requirements are most critical should also be determined (Mosley, 1983).

Observations of the populations of a given species (at particular life stages) are made at locations having a given set of habitat characteristics (e.g. depth, velocity, cover and substrate), and are used to determine the "suitability" of a particular location for use by that species and life stage. This is expressed in the form of a habitat "preference" or "suitability" curve (Figure 9.6(b)), which provides an index of habitat suitability over a range of values of some factor such as velocity.

Curves are developed by combining population-habitat data with the availability of habitat type, and then normalizing the curve so that the resulting "suitability" index varies from 0 to 1, a value of one representing the optimal or most productive range (see Figure 9.6(b)). Ideally, criteria should be developed during different times of the year and for different life stages. Bovee (1986) presents several methods of developing habitat-suitability curves as well as procedures for evaluating their accuracy and precision. Gore and Judy (1981) and Morin et al. (1986) present preference criteria for benthic invertebrates; Bovee (1978) gives preference curves for 10 species of the family Salmonidae, and Fritschen et al. (1984) and Mosley (1983) list preference criteria for recreational uses such as wading, fishing and canoeing. The most recent version of PHABSIM (see below) will produce suitability curves directly from data files.

In step 3, the way in which microhabitat variables vary with discharge is determined. This is achieved either through field evaluation over a range of discharges or by hydraulic simulation.

Steps 4 to 6 involve iterations of the procedure to produce an array of options and their impacts on habitat availability, and step 7 is an institutional process. This could involve estimates of the impact of a specific development, a statement of optimum or minimum acceptable habitat or flow, or a recommendation of some compromise arrangement (Mosley, 1983). The final selection of instream flow levels is developed from knowledge of which levels limit the amount of usable habitat in a stream for specified species. Nestler et al. (1989) give further details on institutional analysis and methods of producing defensible results.

PHABSIM

PHABSIM (Physical HABitat SIMulation System) is a collection of computer programs which form a major component of IFIM. The assumption of PHABSIM is that aquatic species will react to changes in the hydraulic environment. Further, individual organisms will tend to select the most favourable instream conditions, but will also use less favourable ones, with preference decreasing as conditions become less favourable (Stalnaker, 1979).

The amount of physical habitat suitable for use changes with discharge, and it is this habitat–discharge relationship which is produced by PHABSIM. Programs are available in PHABSIM for modelling changes in water surface and velocity patterns with discharge (IFIM step 3, above) and for combining these relationships with habitat-suitability curves (IFIM step 2) to produce habitat–discharge relationships (IFIM step 4). The process is conceptualized in Figure 9.6.

The final curve displays the change in a composite factor, the weighted usable area (WUA), with discharge. WUA is an indicator of the net suitability of use of a given reach by a certain life stage of a certain species. At a particular stream discharge the pattern of distribution of physical habitat (depth, velocity, cover and substrate) is evaluated over the stream reach (Figure 9.6(a)). This is combined with the habitat-suitability curves (Figure 9.6(b)) to determine the WUA for that discharge. The physical habitat is redefined at each discharge and the computations repeated to obtain WUA as a function of discharge (Figure 9.6(c)).

PHABSIM, first released in 1978, has since undergone an extensive

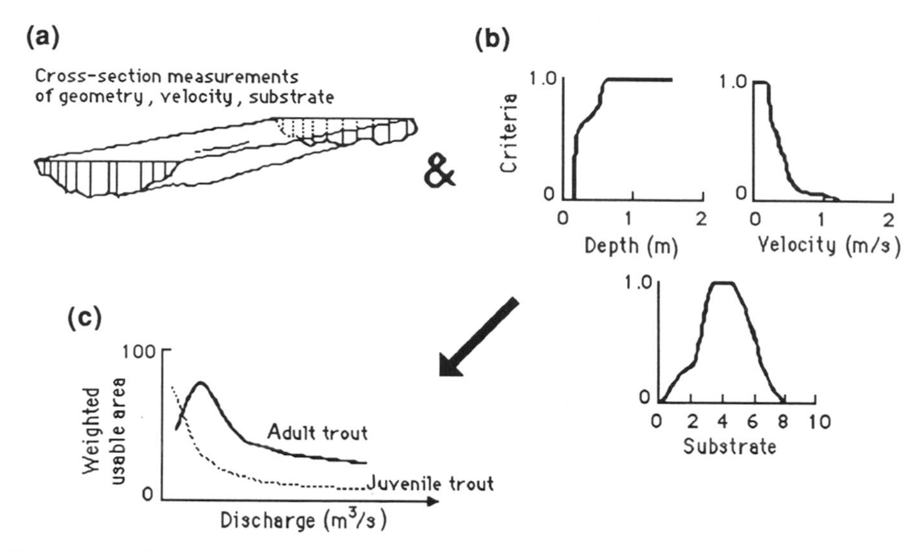

Figure 9.6. A conceptualization of the procedures in PHABSIM: a stream reach is selected and surveyed at a particular discharge (a), and the information is combined with habitat suitability curves defined for a particular species and life stage (b). The procedure is repeated for several discharges (using measured or modelled data) to obtain a curve describing the suitability of use of that reach as a function of discharge (c). Adapted from Gore and Nestler (1988), by permission of John Wiley & Sons, Ltd

evolution as the level of experience of the developers has increased and modifications have been suggested by other researchers. The most recent release of the program, PHABSIM, Version II (Milhous et al., 1989), is a collection of many separate FORTRAN programs. Gan and McMahon (1990a) extensively examined and reviewed the microcomputer version of the model, which is supplied on a set of 25 floppy diskettes (360 K each). For running PHABSIM, an AT computer with 20 K of storage memory and graphics capability is recommended. In addition, an editor with ASCII file compatibility and MS-DOS or PC-DOS, version 3.00 or later, is required (Gan and McMahon, 1990a; Milhous et al., 1989).

There are two basic components of the model: (1) hydraulic simulation and (2) habitat simulation. Simulation is carried out either by using average conditions in a stream channel or by dividing the reach into rectangular "cells" of smaller areas to more finely describe the distribution of physical habitat. Thus, WUA can be calculated either as an average value for the reach or as a summation of the individual "cell" WUAs. In the latest version of PHABSIM, options also exist for the prediction of WUV (weighted usable volume) and WUBA (weighted usable bed area—essentially, wetted perimeter). The output habitat–discharge function(s) can be subjected to further analysis using additional programs to simulate "effective habitat" and/or to generate a time series of physical or effective habitat.

Hydraulic simulation

The user must choose between several options for calculating water-surface elevations and velocities, i.e.:

(1) Stage–discharge relationship, fitted to measured data from each cross-section as a power function (see Section 5.8.1);
(2) Manning's equation (see Section 6.6.5) (in one program, Chezy's equation is an option);
(3) Standard step backwater method (see Section 6.6.6). The highly sophisticated US Army Corps of Engineers backwater curve program, HEC-2 (HEC, 1982) can also be interfaced with PHABSIM.

For most applications, calibration requires a set of water-surface elevations and velocities taken at several transects for at least one discharge. For most instream habitat work an option would be selected for calculating the water-surface elevation at a given discharge. Then the velocity distribution across a channel would be developed using Manning's equation on a cell-by-cell basis. Other options yield average stream cross-section velocities only, which may or may not be sufficiently accurate for habitat simulation. In relatively steep streams, Milhous et al. (1989) recommends a mixture of stage–discharge, Manning's equation and step-backwater calculations to determine water-surface elevations.

Habitat simulation

Habitat simulation models compute the quantity of physical habitat area in a stream reach for a given species and life stage or activity. The habitat models can be used with measured data or with data simulated by the hydraulic simulation programs. For habitat simulation, there is a choice of three different model types:

(1) Average parameter models which calculate top width, average velocity, wetted perimeter and other hydraulic parameters based on cross-section and discharge measurements. From these programs, curves can be developed to show how hydraulic habitat changes with discharge, as for the "transect methods" described in Section 9.3.3.
(2) Microhabitat simulation models which combine habitat suitability curves and hydraulic data to generate a habitat–discharge relationship.

(3) Dual flow/habitat or "effective habitat" models which consider the effects of alternate flows, or the competition between two species or life stages for space. The term "effective habitat" implies that less habitat is available for use than the computed physical habitat. For example, where two flows are of importance as in spawning flows followed by incubation flows, the program computes conditions at the two flows and two life stages in determining the amount of habitat. Alternatively, when there is competition between two species or life stages for one area, the relative preferences of one life stage/species over another can be considered. These options give the user more flexibility in describing the actual amount of habitat available for use by a species or life stage.

In support of the habitat simulation models, several programs are available for building and plotting files of habitat suitability curves. Normally, these curves are derived for depth, velocity and a channel index which considers both substrate type and cover (see Bovee, 1986).

For the simulated or measured values of depth, velocity and channel index, a preference factor ($f(v)$, $f(d)$, $f(s)$ for velocity, depth and channel index, respectively) is obtained from each of the habitat-suitability curves. These factors are weighted and the WUA for each cell is calculated by multiplying the cell area by this weighted preference factor. There are three alternative methods of weighting, described by Gan and McMahon (1990a):

(1) *Multiplication*: $f(v) \times f(d) \times f(s)$. This implies a synergistic action, whereby optimum habitat will only exist if all variables are optimum.
(2) *Geometric mean*: $(f(v) \times f(d) \times f(s))^{0.333}$. This implies a compensation effect; i.e. if two of the factors are in the optimum range, the third has little effect unless it is zero.
(3) *Minimum*: the minimum of $f(v)$, $f(d)$ or $f(s)$. This implies that the habitat is no better than its worst component.

For example, if a 5 m^2 area of streambed had these preference factors for a particular fish species: $f(v) = 0.70$, $f(d) = 0.95$, and $f(s) = 0.80$, then the weighted preference factors for the three options would be 0.53, 0.81, and 0.70, respectively, with associated WUA values of 2.7, 4.0, and 3.5 m^2. Thus, the multiplication option gives the most conservative index of suitable area, a little over half of the total area for this fish species.

In addition to velocity, depth and substrate, temperature can be considered as an additional factor if a function relating suitability of habitat and temperature is developed. The temperature factor is multiplied by the WUA for the whole reach (not each cell). This is a future option for PHABSIM.

Time-series simulation (TSLIB)

The habitat–discharge functions from PHABSIM may be combined with flow data to obtain monthly or daily habitat time series and habitat-duration curves using a set of programs, TSLIB (Time Series LIBrary), also developed by the US Fish and Wildlife Service. The procedure is shown diagrammatically in Figure 9.7. For a minicomputer the programs are transferred on 11 floppy diskettes (360 K each). These programs will:

(1) Calculate basic statistics (mean, standard deviation, log-normal statistics, lag-1 correlation coefficients) on monthly data;
(2) Generate flow-duration habitat curves for selected months; and
(3) Create monthly or annual habitat time series or effective annual habitat time series for four to seven life stages of a given species.

The habitat time series is useful for comparing pre- and post-project habitat availability. Habitat "duration curves" (see Section 8.3) can be developed for different water-allocation rules to help in selection of the "best" alternative. Periods critical to species survival during a given life stage can also be identified, as can the limiting habitat availability (i.e. physical carrying capacity) for each species and life stage. This information is particularly useful for evaluating potential changes in species composition since different species will react differently to changes in hydraulic characteristics (Milhous et al., 1989; Stalnaker, 1979).

Using average monthly flow data, Milhous (1986) found the physical habitat time series much less variable than the streamflow time series. However, at lower streamflows, a moderate reduction in flow resulted in a large reduction in physical habitat. These trends are illustrated in Figure 9.7. Milhous points out that the WUA–discharge relationship may not be the same across all seasons, and that it may not be applicable for all seasons (e.g. spawning flow requirements would not be pertinent during non-spawning seasons). Bovee (1982) gives additional information on the use of habitat time series.

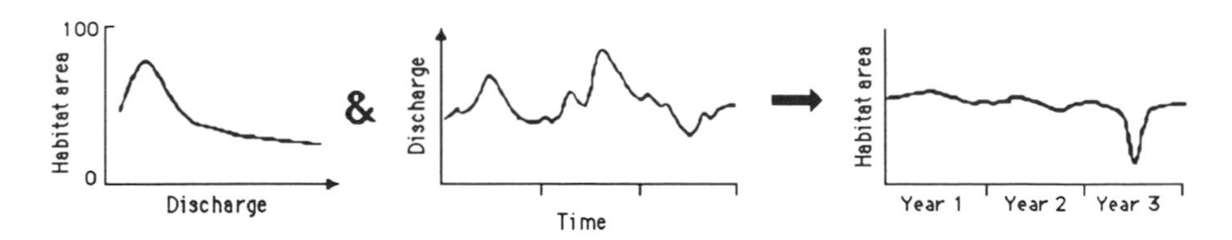

Figure 9.7. Generation of a habitat time series using PHABSIM. Adapted from Bovee (1982)

Other PHABSIM-style models

Other habitat-simulation methods have been developed, although they are basically modified versions of PHABSIM. In New Zealand, the River Hydraulic and Habitat Simulation (RHYHABSIM) model has been developed by Jowett (1989). This can be considered a highly simplified version of PHABSIM (Gan and McMahon, 1990a). In comparison to PHABSIM, RHYHABSIM is transferred as one program with test data on one 360 K floppy diskette, and will run on an IBM-compatible microcomputer (either AT or XT), with on-screen graphics.

Hydraulic simulation is carried out by first obtaining surface-water elevations at different flows using a standard step backwater curve method and then employing Manning's equation to obtain velocities within each cross-section. One set of cross-section geometry and velocity measurements are required for calibration. An option also exists for using stage–discharge relationships at individual cross-sections, which may be preferable in situations where the backwater curve method is not feasible such as during low-flow periods. Habitat simulation results in a WUA–discharge relationship as for PHABSIM, with preference factors multiplied together. RHYHABSIM has a library of habitat-suitability curves.

In Norway, Vaskinn (1985) is developing another model called RIMOS (river-modelling system). This uses some of the hydraulic simulation routines from PHABSIM; however, RIMOS does not incorporate habitat preferences. In contrast to PHABSIM and RHYHABSIM, RIMOS accepts meteorological data from which river discharges are estimated for ungauged sites. Detailed documentation on the model is available, but written in Norwegian.

Limitations of habitat-simulation models

A critical limitation on the use of habitat simulation models is the lack of well-defined habitat-suitability curves. Since these curves are essentially empirical correlations, some authors (e.g. Nestler et al., 1985) state that the curves may not be transferable from one stream to another, and indicate that site-specific curves may be preferable. However, the development of these curves is costly, with Bovee (1986) estimating a cost of US$10 000 per species, and this approach may be highly impractical for large regions. Further research is needed to develop "standard" curves for non-salmonid fish species as well as for benthic invertebrates and other stream biota and users, which consider the seasonality of habitat use and carrying capacity of the stream.

There have been a number of criticisms of both the hydraulic and habitat simulation routines of PHABSIM. For hydraulic simulation, it is important that the underlying assumptions are understood before the model is applied. Standard step backwater methods, for instance, require that the flow is steady and gradually varied. None of these criteria may be met under certain conditions, such as in steep-gradient whitewater streams, during hydropeaking operations, in flooding onto vegetated floodplains or at low flow when rocks are exposed. Osborne et al. (1988), for example, found the backwater curve model difficult to calibrate under low-flow conditions. Manning's equation also has its limitations, as discussed in Section 6.6.5. The assumption that the stage–discharge relationship can be described by a log-log function may be unrealistic for complex cross-sections. In producing a habitat time series an assumption is that the structure of the stream channel does not change under the range of flows simulated. However, channels can realistically be expected to change, both naturally and in response to flow regulation, altering the available habitat (Bleed, 1987). Bovee (1982) thus recommends that an evaluation of the channel's equilibrium conditions should be the first step of an analysis.

These limitations should be considered not only when running a hydraulic simulation model but also before collecting field data, since the individual hydraulic and habitat models have differing data requirements. However, this also allows more flexibility in applying PHABSIM since it will accept anything from a few average values to detailed microhabitat measurements.

The strongest criticism has centred on the ecological interpretation of the WUA index. Gore and Nestler (1988) review and comment on the criticisms put forward by a number of authors. The assumption is made in instream flow uses of PHABSIM that if the habitat is maintained, the fish population will be maintained. A common interpretation is that habitat (the WUA index) is a proxy for species biomass or abundance. This has led to a number of studies comparing the relationship between fish biomass and WUA, some of which discovered satisfactory relationships (Orth and Maughan, 1982; Stalnaker, 1979) and others which found that they were often uncorrelated (Irvine et al., 1987; Mathur et al., 1985; Scott and Shirvell, 1987).

Other factors such as food supply, biological interactions (e.g. competition, predation), nutrients, dissolved oxygen, presence of ice cover, temperature and flow regime (including the effect of floods) may be of greater importance than physical habitat in limiting species biomass or abundance. For example, fewer fish than expected may use the habitat during floods or if food is limited, whereas more may use a suboptimal habitat if preferred areas are in short supply, as during droughts (Mosley and Jowett, 1985). Thus, physical habitat may only be limiting during certain seasons (Orth, 1987).

Gore and Nestler (1988) found that the relationship between fish biomass and WUA was better in coldwater streams which have simpler ecosystems and more predictable hydrological regimes in comparison to more hydrologically and ecologically complex cool- or warmwater streams. Improvement in the relationship has also been observed when the channels are approximately uniform (Mosley and Jowett, 1985) and if the flows are in a steady-state condition (Gore and Nestler, 1988), thus more closely meeting the requirements for hydraulic simulation. Scott and Shirvell (1987) also observed that a better relationship was obtained when usable space was a limiting factor, such as for older age classes of fish and when the biomass was near the system's carrying capacity. If a river is carrying the maximum possible number of individuals, then reducing the amount of habitat will cause a near-immediate reduction in numbers or size of fish, whereas the same reduction at less than carrying capacity may have little effect (Tunbridge and Glennane, 1988).

It is also commonly assumed that the relationship between biomass and WUA is linear. However, the biota may show threshold responses to discharge rather than gradual, linear changes. Also, as stated by Milhous (1988), the physical habitat usually changes relatively quickly and the aquatic population responds more slowly. Therefore, a direct linear relationship between WUA and biomass would mean that the aquatic biota are able to reproduce, migrate or die at the same rate as changes in habitat—an unrealistic assumption.

Routine users of PHABSIM-type models should, therefore, realize that the method cannot be expected to predict changes in biomass or abundance on a short-term basis, or in situations where the species is limited by some factor other than physical habitat. Gore and Nestler (1988) suggest that WUA should be treated as an index of available physical habitat rather than an indicator of actual biomass or species numbers, and that this is the appropriate level of utility of PHABSIM as a management tool. Milhous (pers. comm., 1990) suggests that improvements could be made in the prediction of biomass or populations through use of the effective habitat and time-series analysis option, which allows the sequence of streamflows and the interactions between species to be addressed.

The large number of options in PHABSIM may diminish the objectivity of the results. In an evaluation of PHABSIM, Gan and McMahon (1990b) found up to an elevenfold difference in computed WUA, depending on the options selected. They concluded that this is not surprising since the options are based on different assumptions, but can lead to "results fudging". Comparisons of PHABSIM results are valid, then, only if similar options are chosen. Some US agencies have adopted a "standard" path of options for this reason (Milhous, pers. comm., 1990). RHYHABSIM, in comparison, has few options. Thus different analysts will get similar results, but the program does not provide as much flexibility.

Orth and Maughan (1982) conclude that PHABSIM is a useful tool for managing streams with altered flow regimes but is not a panacea. Biological expertise is still needed in the interpretation of results. However, the model can provide a basis upon which a biologist may apply professional judgement (Mosley and Jowett, 1985) and a methodology for comparing the relative effects of different management decisions. Milhous (1988b) stresses that the analyst(s) must first understand the stream system, including its hydraulics, hydrology and aquatic biology. He recommends for best results that both a biologist and an engineer should be involved in any application of the model.

9.3.5 ARTIFICIAL FLOODS: "FLUSHING FLOWS"

In unregulated rivers, channels are maintained by periodic high flows, often described by some dominant discharge such as "bankfull" (see Section 7.2.5). A change in the timing, frequency and magnitude of these high flows due to regulation can have an effect on channel shape and the arrangement of the streambed materials. If flows are reduced, vegetation encroachment, siltation and narrowing of the channel can occur, reducing the flood-carrying capacity of the channel as well as the amount of stream habitat. As mentioned in Section 9.3.1, an instream flow recommendation will sometimes include a provision for "flushing flows" to remove fines and restore intragravel permeability or "channel-maintenance flows" to maintain the existing channel configuration and ability to pass flood discharges. These flows are typically (1) a one-time "flush" to remove sediments (for example, those deposited during road or dam construction) or (2) a periodic maintenance flow release to prevent narrowing of the channel.

Reiser et al. (1989) give several points to be considered in the determination of flushing flow requirements: the location of sediment sources (upstream or downstream of the project), the effect of land-use changes on sediment influxes, the sensitivity of biota to sediment deposition or channel width changes, the operational requirements of the project, and the timing and hydrograph shape of natural peak flows. Historical data or data from undisturbed sites are needed to determine the unaffected state.

At present, there is a need for more quantitative information to determine if flushing flows are required and to develop recommendations for the magnitude, duration, timing and hydrograph shape of flow releases. Milhous et al. (1986) cite a need for physical process approaches. Tennant's method (Section 9.3.2), for example, was based on field observations and may only be applicable to the observed streams. As advocated by Tunbridge and Glenane (1988), experiments with flow levels may yield the best site-specific solutions.

There is also a need to take into consideration the ecological as well as morphological requirements for flooding. For example, as mentioned in Section 9.1, floods act as a "cue" to some fish species that it is time to migrate and/or spawn, and in rivers with floodplains, seasonal inundation is necessary for riparian vegetation and for juvenile fish nursery grounds. The releases of floodwaters should thus coincide with biological requirements.

Reiser et al. (1989) and Milhous (1986) both review flushing flow methods. The approaches generally fall into three categories:

(1) Hydrological—These methods use an index obtained from runoff records;

for example, the 2-day flow exceeded 17% of the time in Hoppe's method (Section 9.3.2) or the 1.5-year flood. Hydrological records (or regionalized estimates) are needed at the site to be "flushed".

(2) Morphological—Channel characteristics such as some percentage of bankfull flow (Section 7.2.5) are used to obtain a flushing flow. This discharge can be estimated at the site using the slope–area method (Section 5.8.2) or more elaborate backwater curve methods (Section 6.6.6).

(3) Sedimentological—A flushing flow is determined by using a sediment-transport equation to find the flow level needed to move a particle of a given size (Section 7.4). Knowledge of channel dimensions, slope, and substrate composition is required.

In the first two methods the selected flow level corresponds with flow levels at which satisfactory flushing takes place, as determined from observations at the actual site, in a flume study or from the literature. If operation rules permit, it may be possible to perform experiments downstream of a dam site, adjusting the flow level until flushing begins to occur. It is reasonable to assume that statistics derived from one study site can be applied to others within a hydrologically homogeneous area (Kondolf et al., 1987).

The third method is physically based, derived from sediment-entrainment theories (Section 7.4). It is normally assumed that fines will be flushed out when the threshold of motion for some percentage of the particles is reached. The Shield's equation (7.17; see Example 7.1) is most commonly used. However, as mentioned in Section 7.4, the difficulty lies in selection of a proper coefficient since the method was developed for uniform sands. The amount of shielding, packing or imbrication, or armouring must be taken into account as well as the particle size to be mobilized.

Although Hoppe's method gives a recommended duration of 48 hours the others only indicate peak-flow values. The duration of the flow and the shape of the hydrograph (Section 4.3) should be part of the design for both ecological and geomorphological reasons. For example, if the flow is dropped too quickly it can lead to sloughing of saturated streambanks (Green, 1974) or stranding of fish or other stream-margin biota. The quality of the released water should also be considered; for example, releasing large quantities of anoxic water from the lower levels of a reservoir may kill off much of the downstream aquatic life, somewhat defeating the purpose of habitat improvement. The time of release of flushing flows should preferably coincide with historical periods of high flow, since the biota will be adapted to this regime. For example, for Murray cod of the Murray River, Australia, Lloyd (1990) suggests strong flows to inundate the floodplain in late spring for at least two months, and recommends that the hydrograph should taper off slowly and be free of sudden changes to prevent stranding of floodplain fish and nesting sites.

Rosgen et al. (1986) give flushing flow recommendations developed by the US Forest Service for snowmelt-dominated, perennial, alluvial streams. Their method uses bankfull discharge as the peak flow of a synthetic "flood" hydrograph, with the rising limb approximating the natural hydrograph shape. The recession flows are dropped somewhat more quickly than natural floods under the assumption that most of the sediment transport occurs on the rising limb. The authors state that adjustments to the basic method should be made based on local conditions and professional judgement.

O'Brien (1987) gives a case study of a flushing-flow study. The purpose of this was to determine instream flow requirements for the Yampa River, Colorado, to preserve the habitat of an endangered fish species (the Colorado squawfish) which deposits its eggs in the interstitial spaces of cobble streambeds. Proposed was the concept of a minimum annual streamflow hydrograph with a peak flow period designed to preserve the existing cobble substrate and inundate the active channel to eliminate encroaching vegetation. Flume studies were used to determine the flows needed to mobilize sand trapped within a cobble bed. The author found that sand could be scoured from the bed to an approximate depth of one median cobble diameter below the cobble surface without cobble mobilization. From bedload measurements made during a high-water year, sediment load–discharge relationships (Section 7.5.4) were developed and used to calculate the "effective" discharge—the flow which transported the most sediment over a long period of time. This was calculated by multiplying the sediment transport at a given discharge with the frequency of that discharge to find the peak. The peak of the minimum streamflow hydrograph was set at this peak effective discharge, which had an average recurrence interval of about 1.5 years, and was approximately one-half of the bankfull discharge. The authors questioned whether the proposed minimum hydrograph could maintain a relatively sand-free cobble bed for long periods of time, and suggested that flows approaching natural flood peaks for a short period (e.g. 24–48 hours) might be necessary for long-term maintenance.

The best approach, then, would be to develop flushing-flow estimates using several methods, include ecological requirements for flooding such as a specified level or period of inundation, adopt a conservative figure, test it out at the site, and monitor the results to determine its effectiveness. Monitoring sediment movement can be done (1) during the flushing event with a bedload sampler or (2) after the event, using scour chains, cross-sectional surveys, bed-sediment sampling, tracer particles of various sizes, or photography (see Section 5.9). Biological monitoring should also be conducted to evaluate whether the changes in streambed composition and/or flooding of the floodplains have the desired effects on the aquatic ecosystem.

9.3.6 SUMMARY OF METHODS

All the instream flow methods will yield "answers". The choice of a method will be based on time and budget allocations and the level of competition between instream and offstream water users. With any technique, the dictum of computer modelling, "GIGO" (for Garbage In–Garbage Out) is relevant. Answers will only be as good as the quality of information examined and the skill of the user in applying the model and interpreting the results.

Proponents of the habitat-simulation methods feel these are more attuned to biological principles than the other techniques and are thus superior for evaluating biological impacts resulting from flow alterations. None of the methods can, as yet, directly address the potential changes in biomass or populations resulting from altered flow regimes. There remains a need for more research on the effects of flow regimes on biota, due to resultant changes in physical habitat as well as changes in water quality and biological interactions.

Also, as Stalnaker (1979) recommends, flow requirements should be considered dynamic. Rather than applying a set "minimum" or "optimum" flow throughout the year—every year—considerations should be given to seasonal and annual variations, with different regimes for "wet years" and "dry years". The natural regime of streams should be considered in establishing flow requirements since the resident populations have adapted to it. Especially in ephemeral streams or drought periods, plans are needed to balance short supplies between users. Periods of above-average water supply would perhaps be the most acceptable time to allocate water for artificial floods.

It should also be recognized that plants may be able to tap groundwater reserves, frogs can bury into the mud to enter a dormant phase, fish can migrate, and aquatic insects may have adult forms which are not as dependent on water for survival. Thus, an allocation of "zero flows" may be appropriate for streams which historically dry up each year.

The field of instream flow recommendation is still a dynamic, evolving area, with heated controversy and much testing and refinement of methods. IFIM can be considered the present "state-of-the-art" method for decision making. The direction of future developments appears to be towards more complex models which incorporate ecological components, with further refinement of habitat requirements. It is unlikely that a computer model will replace the need for discussion of the impacts of management options. This is particularly true in cases where value is unquantifiable, such as in the evaluation of the intrinsic value of an endangered species or the aesthetic value of a pristine brook. The best decisions on instream flow needs will be based on an environmentally sensitive evaluation of the stream system and the relative value of its uses, both economic and ecological. Mathematical methods can be important tools in this process.

Further information on the models described can be obtained from:

PHABSIM:
Aquatic Systems Branch
National Ecology Research Center
US Fish and Wildlife Service
2627 Redwing Road
Fort Collins, CO 80526-2899
USA
(this agency also offers courses in the use of PHABSIM)

RHYHABSIM:
Dr Ian G. Jowett
Freshwater Fisheries Centre
PO Box 8324
Riccarton, New Zealand

RIMOS:
Dr Kjetil A. Vaskinn
N-7034
Trondheim
Norway

9.4 STREAM REHABILITATION

9.4.1 GENERAL

To rehabilitate a stream means to return it to a healthy or improved condition. Traditionally, streams have been "improved" for flood mitigation, navigation, and channel stabilization. More recently, the emphasis has shifted towards the re-establishment of healthy aquatic habitats, which may mean improvement of the engineering works, such as the addition of fish ladders to large dams or the enhancement of habitat in channelized reaches.

Although the focus of this section is on rehabilitation, the same concepts can be applied to new river engineering works to produce a "managed river which behaves as much like a natural one as possible, which looks as aesthetically pleasing as possible and which minimizes the morphological and ecological disruption" (Brookes and Gregory, 1988, p. 158).

Since streams have been influenced by human activities for centuries, the first question is often "to what condition should they be rehabilitated?" Should they be restored to some pre-disturbance condition or should they be repaired using ecologically sound procedures to create an enhanced habitat for a different biological community (i.e. introduced species) than what was historically found in the stream? On many streams it will be difficult to know what the "normal" or "unimpaired" state is. Dahm et al. (1989) recommend that historical changes should first be examined to gain an understanding of conditions which existed before human disturbance. Lake and Marchant (1990) state a need for more ecological knowledge and for inventories to survey the extent of damage in streams. This knowledge can then serve as a guide for rehabilitation efforts.

Some forms of rehabilitation will not be feasible, and the "natural" state may not always be desirable. For example, the re-establishment of large debris dams on navigable rivers or the removal of flood-mitigation dams upstream from thriving communities would no doubt be met with strenuous resistance. There may be difficulties in rehabilitating larger streams simply because they run across political boundaries and into regions with different management schemes (Goldman and Horne, 1983).

In some situations, localized "band-aid" solutions may be adequate, such as where a streambed has been dredged for gold or a channel re-routed away from a highway. If whole-river systems are of concern, stream rehabilitation is best undertaken within a total catchment management strategy. Stream improvements should not be used as a substitute for dealing with more complex causes of stream degradation such as grazing, logging, road construction and other impacts to upland areas (Elmore and Beschta, 1987). As Platts and Rinne (1985, p. 127) state, "artificial stream improvement must not be substituted for vigorous, responsible stewardship of the surrounding watershed". The old adage, "if it ain't broke, don't fix it" also applies to stream rehabilitation. For example, streams to be maintained in pristine condition (e.g. heritage status) should not be artificially "improved" by active rehabilitation methods.

A multi-disciplinary approach is required in the rehabilitation of streams since modifications must be both biologically and hydraulically sound. Digging out a hole in the streambed may create a nice pool environment for fish refuge at low flows but it will be of little benefit if the stream's geomorphology is such that the pool will fill up with sediment during the first flood. Conversely, planting woody vegetation along streams may improve bank stability and riparian habitat, but it can also increase local flooding hazards if the channel capacity is insufficient. The hydrological character of the stream should also be taken into account as this will affect the extent, size, and longevity of rehabilitation works. Sound data on the hydraulics, hydrology, geomorphology and ecology of undisturbed stream reaches are essential for re-creating these conditions in degraded areas. Designing with nature rather than against it will improve stability and assist in the recovery of the site.

With their continual supply of water, streams have the capability to be self-cleansing. In geological time, channels will shift and settle and become recolonized by new assemblages of flora and fauna. Stream rehabilitation is a means of speeding up the natural processes and/or using artificial features temporarily to serve the function of natural ones. Ongoing maintenance should be an integral part of stream rehabilitation to assess its effectiveness, monitor vegetation growth, faunal response and bank stability, and carry out repair work after floods.

Stream rehabilitation can be either (1) passive, in which disturbance is reduced and the stream is left to heal by itself or (2) active, in which specific repair procedures are applied. An example of the first might be the protection of streambanks from grazing pressure to permit natural revegetation, and of the second, the placement of boulders or other "cover devices" in the stream. For discussion purposes, passive measures (Section 9.4.2) will include the re-establishment of native flora and fauna and the removal of non-native species. Active measures will be separated into two categories: modifications to the channel (Section 9.4.3) and structural modifications within the channel (Section 9.4.5). Artificial floods for the removal of fines and channel maintenance (Section 9.3.5) could also be considered a form of channel modification, although in this case it is actually the flow regime which is manipulated.

9.4.2 PASSIVE METHODS

A "leave it alone" approach is quite possibly the most difficult one to implement. People seem to have an innate tendency to tinker with things; to apply plasters, poultices and pills. However, there will be some cases, particularly where degradation is caused by offstream activities, where the best solution to a river management problem will be to remove the problem source and wait for the stream to rehabilitate itself. In severely degraded streams, natural morphological and biological recovery may be a very slow process, typically requiring 10–100 years (Brookes and Gregory, 1985).

"Removing a problem source" might require controlling point pollution sources by introducing wastewater treatment plants, or controlling non-point pollution sources by reducing chemical usage and erosion on range and agricultural lands. Alternatively, it could involve controlling the spread of "pest" plant or animal species. For example, in many areas, introduced game fish have displaced native species, and vigorous rehabilitation plans might include the elimination of non-native species such as carp.

In heavily grazed areas the solution may be to exclude riparian areas to domestic livestock during all or certain times of the year through management or fencing. The fencing of corridors along the stream can protect riparian plants from grazing pressure and banksides from trampling. Stock access points can be located where the animals will do the least damage to the stream and rock-lined fords used to reduce streambed disturbance. In studies on streams in the western United States, Platts and Nelson (1985) found that streams in heavily grazed areas tended to widen and become more shallow, with reductions in streamside vegetation. Their studies demonstrated that providing rest from grazing could lead to significant improvements in riparian vegetation, streambank stability and stream-channel conditions.

After the problem source is removed or controlled, natural rehabilitation can be assisted through revegetation and re-introduction of faunal species. Ideally, river management should progress to a point where a "preventive medicine" approach can be used, with problems foreseen and eliminated before they become problems.

Revegetation

Riparian vegetation provides bank stabilization, nutrient regulation, filtering of sediments, shading, nesting areas for birds, and cover for fish. Vegetation cover also has a modulating effect on stream temperatures, keeping the water cooler in summer and insulating it in winter. For example, Winegar (1977) found that heavy anchor ice appeared in a reach where brush cover was almost absent, but only thin ice formed in an adjacent ungrazed reach with good cover. Without streamside vegetation, creeks can also dry up more readily. On ten streams in Oregon good riparian management by the US Bureau of Land Management caused them to be perennial rather than intermittent (Stuebner, 1988). The improvement in vegetation made the streams act more like reservoirs, retaining the runoff and then releasing it more slowly.

As bankside protection, plants absorb the impact of ice chunks and waves. The root mats of grasses can reduce scour and the failure of undercut banks. Vegetation also reduces the velocity of water flowing through it, encouraging sediment accumulation (Lewis and Williams, 1984). Platts and Rinne (1985), for example, discuss the revegetation of point bars at meander bends (high-deposition areas) for building up banks. In contrast to inert materials (e.g. rocks and concrete), plants are self-regenerating following "structural failures". Historically, plants were routinely used for stabilizing banks and deflecting flows, with routine maintenance to monitor and modify their growth as needed. As Lewis and Williams (1984, p. 9) state, these practices are now often regarded as a "luxury", and in relying on technological advances in machinery and materials, the water industry has "abandoned its roots".

Streamside management should aim to maintain a stable, self-sustaining cover of native vegetation. As a general recommendation, indigenous species should be used wherever possible, since they are adapted to the local conditions and utilized by the local wildlife. Plants can also be selected for their value to fish, invertebrate, waterfowl or wildlife habitat, their rapid regrowth attributes, their successional standing (e.g. as a primary colonizer), their aesthetic quality, or their rareness (e.g. to re-introduce endangered species). The various "zones" of the stream (i.e. channel edges versus mid-channel; see Section 2.1.2) should also be kept in mind during the planning of revegetation work. Site preparation, plant propagation and planting procedures are best addressed on a local basis. Risser and Harris (1989) provide some valuable guidelines, and a handbook by Lewis and Williams (1984) is also highly useful. Other references are included in a book on the restoration of damaged ecosystems edited by Cairns (1988). Local government agencies may also publish revegetation guides, and extension agents can provide professional advice.

The planting of exotic species can introduce foreign organic matter into the ecosystem and change the timing and rate of processing of the material (Campbell et al., 1990). They may also outcompete and displace native riparian vegetation. Willows (*Salix* sp.) have often been used to protect banksides because they can be established easily, are quick-growing and resilient, can withstand inundation, and are dense enough to promote sediment deposition (Petersen, 1986). They can be grown from slips and cuttings; in fact, this regrowth ability may cause them to be a nuisance weed species when broken branches are washed downstream. Thus, in regions where they are not native, they should be regarded only as a temporary measure if they must be used at all.

If willows are an appropriate bank-protection measure, Strom (1962) advocates pruning them or partly cutting the stems and bending them over to keep the trees short and bushy. This improves water resistance and prevents them from becoming top-heavy. He also recommends planting them by burying cuttings horizontally in the soil to generate a mat of roots and a thicket of stems. Willow logs can be used to create "growing" groynes (see Section 9.4.3) to encourage silt deposition.

Protection and maintenance are important components of revegetation work. Seeds and/or cuttings may require watering until their roots reach the water source. Losses should be anticipated; thus allowances should be made for overplanting or replanting of the sites. The sites should be maintained to reduce weed growth, with mulching or cultivation preferred over herbicide application. Protection of the new plants with stock- and vermin-proof fencing and/or tree guards is also essential.

Consideration must be given to the hazards caused by the introduced vegetation, especially when the stream passes through private lands. The retarding effects of vegetation within a channel will slow the water and encourage silt build-up, which may be desirable from an ecological viewpoint but can lead to local flooding. If landholders living next to incised or channelized streams have become accustomed to a reduced threat from floods, they may be reluctant to allow tree planting on the streambanks for erosion control. When vegetation becomes dislodged it can lead to log jams or weed jams at bridges or other constrictions. Even in a riparian zone, the dense vegetation can pose a fire hazard during dry conditions. "Lawn mowing" levels of grazing to reduce fuel levels may be feasible in some areas during seasons when the streambanks and plants are least sensitive. It may also be a method of removing undesirable species (e.g. the use of tethered goats for blackberry control). Alternatively, mechanical cutting, thinning and removal of aquatic and bankside vegetation may be required. The re-establishment of vegetation should, therefore, be combined with good management practices both during the rehabilitation stage and afterwards.

Re-introduction of fauna

Although revegetation has received much of the emphasis in stream-rehabilitation efforts, faunal species are also part of a healthy riparian ecosystem. As with revegetation, it may be possible to speed up the natural rehabilitation process by re-introducing animals at the appropriate stage of recovery. Again, to guide rehabilitation efforts, it will be necessary to acquire knowledge of the requirements of each species and the typical population densities found in undisturbed areas.

Recolonization of invertebrates will occur rapidly if substrate and nutrient requirements are met, but for some species with limited migration capabilities, "seeding" larvae from undisturbed sites may also be feasible. Fish stocking has a long history in areas which are heavily fished or which cannot support self-sustaining populations, and an extensive literature is available on game fish species. In the re-establishment of native fish species considerations must include competition between species, predators (which may include introduced game species), and food and physical habitat requirements.

Among the higher animals, the beaver has had an active role in stream rehabilitation. Because the silt-trapping dams they construct are a natural alternative to concrete structures, beavers are being introduced into a number of Northern Hemisphere streams. The dams, which can have extreme longevity (Johnston and Naiman, 1990), help to control the downcutting of channels, bank erosion, and the movement of sediment downstream. They also create pools which are important for fish habitat.

Dahm et al. (1989) document the degradation of a stream in New Mexico, USA, following more than a century of heavy grazing, logging and beaver trapping. Expansive meadows, most likely formed and maintained by beaver activity, were greatly reduced in extent, and the vegetation adjacent to the stream became dominated by xeric (dry-adapted) species as the water table dropped. Since acquisition of the land by the US Forest Service, stream reaches have been rehabilitated through riparian plant re-introduction, bank stabilization and the addition of large woody debris. Beavers naturally recolonized an upper tributary, stabilizing the upper catchment with a network of dams and ponds. On another tributary, beavers were introduced as part of a stream ecology study.

Planning ahead: "preventive" measures

In the planning phase of water-resource projects it is a routine procedure to evaluate several alternatives in terms of cost, effectiveness and environmental impact. The substitution of passive methods for active ones may reduce the need for later rehabilitation works—much like a "preventive medicine" approach. As an example, many of the water-supply catchments near Melbourne, Australia, are closed to all uses (i.e. logging, recreation and subdivision). This has maintained a high water quality and has reduced the need for expensive purification.

If feasible, alternative solutions can replace the need for engineering works such as flood-mitigation dams and channelized streams. With proper planning, lands in the floodplain can be set aside for restricted activities (e.g. parks, golf courses, car parks), where adjustments of the channel and overbank flows do not pose serious problems.

In streams where channel modifications are required for flood mitigation, measures can be taken to reduce the impact on the aquatic environment. Brookes (1989), Brookes and Gregory (1988) and Shields (1982) discuss environmental features for flood channels. These include reducing the amount of channel straightening or shaping, selectively leaving mature trees and other bank-stabilizing vegetation, and minimizing the removal of large rocks, snags and aquatic plants from the stream. Retaining the existing features is preferable to "fixing" the stream after disturbance.

For many streams, passive methods will be the only possible solution simply for economic reasons. Without help, however, some degraded streams may take centuries to recover. The rehabilitation process can often be sped up through active intervention. Active restoration methods may be more popular (and thus more likely to be funded) simply because the effects are more immediate and more visible—they give the impression of "doing something".

9.4.3 CHANNEL MODIFICATIONS

In comparison to passive methods, active methods for rehabilitating streams are more immediate but will usually be much more costly. The purpose of channel modification and instream structures (Section 9.4.4) is to restore diversity of physical habitat to streams which have been modified or degraded. Meanders, pools and riffles, islands, billabongs and side channels contribute to biological richness by providing richness of channel habitat: turbulent and still water, shade and sun, sand and mud, eroding cliffs and point bars (Lewis and Williams, 1984). As channelization often reduces the total channel length and irregularity of the stream margins, the restoration of a more natural meandering, rough-boundary form will also provide more surface area and total amount of habitat.

The principles discussed might be applied to the rehabilitation of streams which have been straightened, desnagged and otherwise channelized, to those where the natural channel structure has been totally destroyed by mining or dredging, or in the design of a new channel when a stream must be rerouted to avoid highways or strip mines. A possible application of a number of these techniques is illustrated in Figure 9.8.

As mentioned in the previous section, proper planning during the design of engineering works on streams can minimize the need for later rehabilitation work and reduce the amount of channelization work required. Permanent structures are installed with the intent of locking the stream channel into a fixed condition, neglecting the fact that alluvial streams naturally adjust to changes in flow and sediment load (Elmore and Beschta, 1987). As stated by Lewis and Williams (1984, p. 6), "much expense and environmental degradation is involved in forcing a river to flow where it is put".

Stabilized, smooth channels with trapezoidal sides and uniform slopes act much more predictably than natural ones. It is thus much easier to describe the behaviour of water flows with simple backwater curves and Manning's equation in channelized reaches. In natural streams, there is much more variability, and simplifying assumptions must be made in order to apply

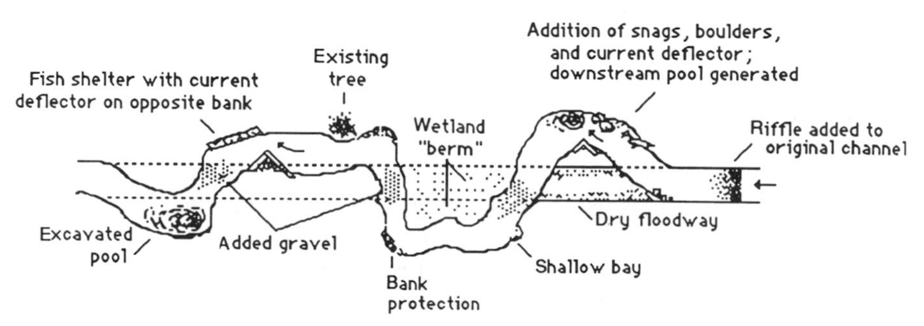

Figure 9.8. A theoretical example of how a straightened, channelized reach (dotted line) might be rehabilitated using techniques described in this chapter

techniques which were developed for uniform conditions. This is where the experience and "gut understanding" of river managers is especially valuable. There is great opportunity for advances to be made in the modelling of water behaviour in natural streams if we can move away from thinking of them as "canals".

It is important to have an understanding of the hydraulic, sediment transport and morphological characteristics of the stream. Knowledge pertaining to the stability of a stream reach can be crucial to the success or failure of rehabilitation works. Is the stream aggrading or degrading? What is the natural pool–riffle ratio or frequency of debris dams? Is this a braided or meandering stream? Predevelopment surveys and aerial photographs of the stream will be of the most help in designing rehabilitation works. If these are not available then characteristics of unaltered streams or reaches with similar runoff rates can be used.

The uniqueness of each situation will require individual solutions. No one design will work universally. Because of this, an experimental attitude should be taken in approaching each site. Processes can be investigated on a small scale with flume studies or field tests. Follow-up studies to evaluate the biological and hydraulic effectiveness of the works is needed so that bad concepts can be discarded and effective solutions promoted. Successful demonstration sites can help to "sell" these ideas to both private landholders and public agency decision makers.

The designs given for both channel modification and instream structures are primarily for use on low- and middle-order streams. On larger streams they become major engineering works. In all cases, interaction between biologists and engineers during the design, construction, and monitoring phases can lead to improved designs which are effective from both standpoints. For example, channels designed for flood conveyance can be modified to include the "biological component" by allowing for the resistance effects of vegetation. Techniques in this section include the establishment of natural channel dimensions, bank-stabilization measures, the re-establishment of natural pool–riffle sequences and/or meander patterns, and the reconstruction of floodways.

Channel dimensions

If a streambed has become widened or entrenched it will change the frequency of overbank flows, modify water table levels, influence water temperatures, and affect riparian vegetation growth. One goal of stream rehabilitation may be to re-establish a more natural stream geometry.

Determining the "best" channel dimensions can be guided by the natural characteristics of streams in hydrologically similar settings, such as the width: depth ratio. For example, on the Mink River, an alluvial stream in Manitoba, Canada, Newbury and Gaboury (1988) suggest that although narrow, deep channels provide the best hydraulic radius for conducting flood flows, natural cross-sections tended to be wider and shallower with a general width-to-depth ratio of about 15:1.

Measurements should be taken in both pools and riffles. If there are no suitable, unaltered streams for comparison, hydraulic geometry relationships (Section 7.2.6) obtained from upstream, unaltered tributaries can be used to estimate the "unaffected" downstream dimensions (Newbury, 1989). For example, a reach of the Steavenson River, a tributary of the Acheron River (see Figure 4.7), was diverted into a straightened channel with a width of approximately 3.5 m. The catchment area for this site is about 140 km^2, and from the hydraulic geometry relationships (Figure 7.9) it can be seen that the width should be closer to 10 m.

Under natural conditions, shorelines are seldom smooth. The natural variability in width, depth, and bed topography should therefore be worked into the design. Small-scale diversity should also be a design consideration, with efforts made to keep channel edges "rough". Lewis and Williams (1984), for example, discuss the addition of shallow bays to the outside of meanders (see Figure 9.8). These create pond-like conditions which support a rich population of algae, plankton, water beetles and pond skaters, and may protect fish fry from predation by larger fish. Shallow bays can also function as stock watering points.

Reinstating meanders

Meandering channel forms are described in Section 7.3.1. In comparison to a straight channel, the presence of meanders reduces the slope of the stream and thus its velocity and sediment-transport capacity. The reinstatement of meanders can accompany stream-rehabilitation works to increase both the quality and quantity of habitat. The slope and bed materials of the rehabilitated channel should be compatible with those of upstream and downstream reaches. For example, if roughness is too low or the channel too straight, water will exit the reach with excess energy which can cause scour downstream and headcutting upstream (see discussion of Equation 7.6).

General relationships between meander parameters and other hydrological or geomorphic measures have been developed (Hasfurther, 1985; Morisawa, 1985). In re-establishing meanders a general rule of thumb is to simply set them five to seven stream-widths apart, one-half of the meander wavelength (Leopold et al., 1964; see Table 7.1). It should be realized, however, that this rule may not be applicable to all streams. It is preferable to develop relationships for a particular region by studying the meander patterns on aerial photographs of undisturbed sections.

It should also be realized that it is normal for meander spacings to have a large amount of variability. There are few streams which look like perfect sine waves; instead, waves are superimposed on waves at different scales, with short, tight meanders and irregular shorelines superimposed on a larger snake-like form. This variability should be worked into a design by taking into account the location of trees, boulders and variations in soil or substrate.

Hasfurther (1985) summarizes techniques for using meander parameters in channel design. He gives four methods of meander design:

(1) The carbon-copy technique: The meander is reconstructed exactly to its pre-disturbance form. This assumes other factors affecting stream patterns (e.g. discharge, bed materials) remain the same. A variation on this technique is to "carbon-copy" the pattern of a similar reach in an undisturbed section.
(2) Empirical relationships: Meander parameters are developed for specific regions and/or stream sites within a region and applied to the affected site.
(3) "Natural" approach: The stream is allowed to seek its own path. The disadvantage of this method is that it may take a long time for the stream to reach a stable form, and high erosion rates and sediment movement may occur.
(4) A "systems" approach: This is the approach advocated by Hasfurther, and includes an analysis of undisturbed meanders, an evaluation of the geomorphology of the disturbed area and consideration of the interaction between the stream and the surrounding areas.

Equation 7.6 should be used as a guideline for modifying the design if streambed conditions in the new channel (e.g. sediment size) are different from those in the reaches used for guidance on meander design. Some self-adjustment of the channel after construction should be expected; however bank stabilization should be considered in areas where excessive adjustment of the channel is undesirable.

In river engineering work for improvement of flood conveyance, meanders are often eliminated. Lewis and Williams (1984) suggest an alternative method of cutting a bypass channel across the meander neck and leaving the meanders as refuge areas for wildlife. The original meandering channel continues to carry some portion of the flow, with the bypass channel conveying the peak discharges (Figure 9.9). This improves flood conveyance and minimizes erosion within the meander. The bypass channel can be designed to carry a small proportion of the flow year-round or to remain dry until floods occur. Dry bypass channels may require maintenance by grazing or mowing to keep plant cover under control, unless the resistance effects of vegetation are included in the design. Engineering input is needed in selecting the stage at which water enters a dry bypass channel, as poor design may lead to scour or silt deposition at the entryway. Bank protection at the entryway and exit may thus be required. As shown in Figure 9.9(b), weirs or headgates can also be built for a more precise control of flow into the bypass or into the meander.

Horner and Welch (1982) give a case study from the Pilchuck River in Washington, USA, in which a segment of the stream channel was reconstructed to go around a new section of highway. The original bends in the river reach were replaced with an S-shaped meander which retained the original stream length. The bases of the banks were stabilized with rock, and grasses and trees were planted along the banks and tops of the slopes. Substrate material of a variety of sizes was placed on the bed of the new channel to create habitat diversity, and large rocks were used for directing the streamflow to scour out pools.

The new channel was left dry for about a year before the streamflow was gradually diverted into it during the following summer. There was a relatively rapid recovery of the bed topography and substrate to that of the original

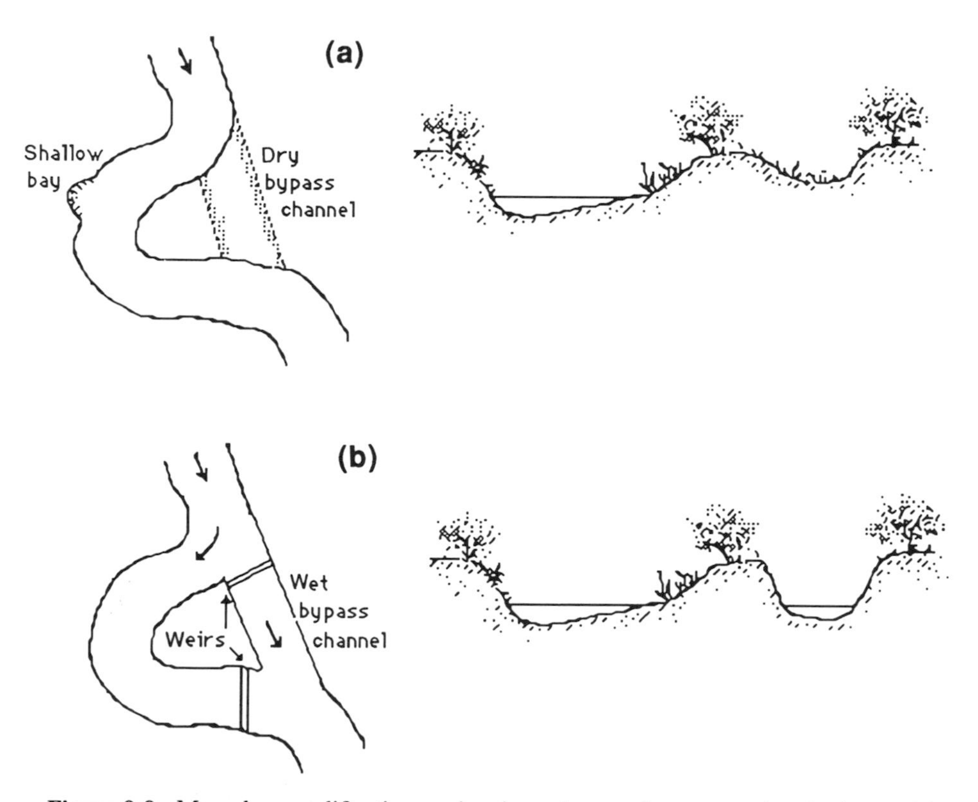

Figure 9.9. Meander modifications, showing plan and cross-sectional views: (a) shallow bay and dry bypass channel and (b) wet bypass channel with optional weirs for control of flows. Adapted from Lewis and Williams (1984), by permission of the Royal Society for the Protection of Birds (RSPB)

channel, accompanied by a rapid development of extensive and diverse macroinvertebrate and fish populations. Monitoring of fish and invertebrates was conducted over a 5-year period, and showed no deterioration in size, diversity or quantity in the reconstructed channel.

On private lands, especially where the stream divides two properties, changing the stream course may meet with resistance because it can mean the loss of land to one landholder and a gain to the other. Thus, the reinstatement of meanders may require negotiation and sufficient proof of potential benefits.

Re-establishing pool–riffle morphology

A meander design would be incomplete without considering the "vertical meanders"—the pool–riffle sequences. Section 7.3.2 describes these features and their value to the instream biota. Energy can be dissipated in the turbulent flow between and over the larger rocks of riffles. Thus, pool–riffle structures can be added to straightened reaches to dissipate the energy which would have been lost over a longer meandering reach.

As with meander re-establishment, historical or regional patterns or rules of thumb (e.g. five to seven channel widths) can guide the placement of riffles. Leopold et al. (1964) cite a study where a stream was dredged by a dragline, but the operator was instructed to leave piles of gravel at about five to seven stream widths apart. These "riffles" smoothed out over a few flood seasons and then remained stable.

A line through the crests of the riffles should follow the general slope of the channel. Since these will act as "control sections" (see Sections 5.6 and 6.6) their effect on flood flows should be considered when deciding on an appropriate height. The slope of the downstream end of the riffle, the "tailrace", should be similar to that in natural channels, and allow fish passage. For example, on the third-, fourth- and fifth-order streams surveyed in the Acheron River basin the average riffle slope was 2.3%, or about 1:43.

Pools and riffles can be established by (1) excavation, (2) placement of gravel, cobbles and/or boulders to form riffles and an upstream pool, (3) building low weirs or debris dams, or (4) strategically placing instream structures so the stream will "do the work" to scour out pools and create gravel bars. Again, knowledge of morphologic processes is necessary to prevent excavated pools from filling up on the next flood event or riffles from becoming buried in mud. Following natural pool–riffle patterns in the design, with allowances for local variation such as the presence of large boulders or snags and massive roots of bankside trees, can help to ensure stability and longevity of the rehabilitated form.

The first three alternatives will produce immediate effects and may thus be looked upon more favourably by landholders and funding agencies than the fourth. With the first method it may be possible to use material excavated from a pool to build a riffle if the particle sizes are not too small for riffle-living invertebrates; otherwise, it might be employed to create levees or earthen current deflectors (Section 9.4.4). Allowing the stream to do its own excavation work, however, has its advantages in cost as well as in the distribution of substrate materials. The stream will be more selective about what it transports than a scoop shovel.

Lewis and Williams (1984) give an example of reinstatement works performed on a channelized stream in Wales which had previously been a salmonid spawning and rearing area. Spawning riffles were created by removing the fine gravel substrate to a depth of 0.3 m and replacing it with a suitable rubble mixture. Stabilizing boulders were placed at the lower end of the riffle to prevent downstream movement of the gravel. The riffles were combined with groynes, instream placement of scattered rocks, revegetation and bank stabilization.

The use of natural channel characteristics in stream-rehabilitation works is described by Newbury and Gaboury (1988). They give a case study from the Mink River in Manitoba, Canada, in which pool–riffle structures were used to improve stability and fish habitat in a channelized stream. The stream had been channelized and straightened from a tightly meandering form to increase the channel flood capacity. The channelized section was narrower and deeper than the natural channel, and continued to degrade over a period of 30 years, causing the banks to slump and steepen.

The degraded reaches were selected for rehabilitation in 1985–6. A series of riffles was constructed in each reach using cobbles and boulders collected from surrounding fields (Figure 9.10). The downstream tailrace was designed to dissipate energy in the reach; its low slope was consistent with natural riffles which did not obstruct upstream fish passage. A "V"-shape crest concentrated low flows in the centre of the channel.

Riffles were spaced approximately six times the predevelopment bankfull width, creating pools which overlapped part of the next upstream riffle. After a bankfull flood, none of the riffles were displaced. A scour hole formed in the pool downstream of each riffle. In comparison, bank erosion and slumping were severe in the unrehabilitated sections. Walleye, a game fish species, were observed using the eddies created both upstream and downstream of the reconstructed riffles.

Pool–riffle structures can also be used to dissipate energy and control headward erosion as a "soft engineering" solution, with environmental benefits. For example, on the Bunyip River in Southern Victoria, Australia, channelization resulted in headward erosion which contributed excessive sediment to Western Port Bay and threatened an upstream water supply pipeline. An approach involving the use of five rock drop structures and floodplain remoulding was chosen over other options, including a single large concrete drop, because it caused minimal adverse impact, provided

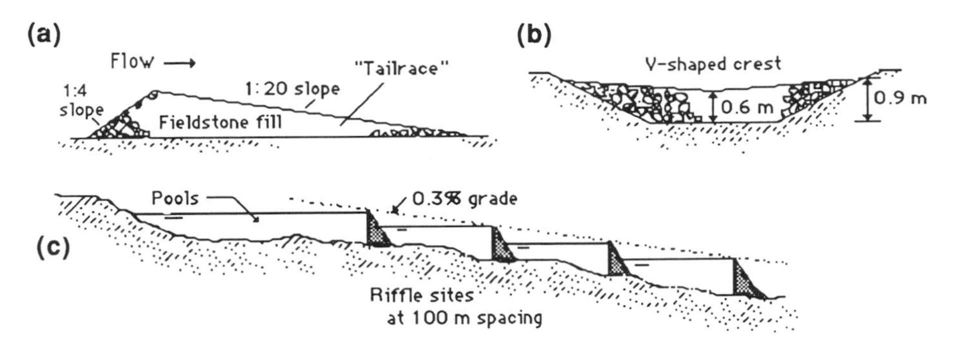

Figure 9.10. Reconstructed riffle design used by Newbury and Gaboury (1988) on the Mink River in Manitoba, Canada: (a) cross-sectional view through riffle, (b) front view, and (c) riffle placement, showing pools formed behind each riffle. Modified from Newbury and Gaboury (1988), by permission of *Can Water Resour. J.*

fish passage, and resulted in cost savings. Angular granite rocks of varying sizes up to 1.5 m were used to create the structures, which each had a 1:3 (33%) upstream slope and a 1:12 (8.3%) downstream slope. Together, the five structures of varying heights took out 8 m of head over about 700 m of stream length (S. Seymour, pers. comm., 1990).

Pool–cascade structures (artificial weirs)

The pool–cascade forms of high mountain streams can also serve as a template for rehabilitation designs. In these streams a "plunge pool" absorbs the energy of the cascading water, also oxygenating it through turbulent action. Artificial weirs can replace natural boulder structures in rehabilitation works. The stable surface can form a site for colonization by mosses, lichens and encrusting algae, which in turn supply food or habitat for organisms higher in the food chain.

Weirs are probably best used in environments where debris dams or large piles of boulders would have occurred naturally; i.e. small, steep-gradient headwater streams. Design of the weir should be consistent with the stream's biology and hydrology. For example, it should not form a barrier to migrating fish species during critical times of the year. One option is to design the weir such that it is underwater during high flow to permit upstream migration, with a notch to concentrate low flows over a "spillway" into a downstream pool which can provide refuge from summer droughts. Weirs may be made of:

(1) Stone walls: Natural stones with crevices and rough surfaces for invertebrate shelter are preferable to smooth bricks, and are aesthetically

more pleasing. The stones can simply be stacked and held in place by stakes of metal or wood, or mortared together for a more permanent structure.

(2) Gabions: Gabions are wire mesh baskets filled with cobbles and gravel from the streambed. They are relatively low-cost and do not require heavy equipment. Their best application may be in lowland gravel-bed streams where larger rocks are uncommon. Their initial artificial appearance is improved when they become silted and overgrown. However, they can become aesthetically unattractive and a hazard when the wire mesh rusts and begins to break up.

(3) Logs: As shown in Figure 9.11, different configurations can be used depending on the log size, stream size and desired weir height. More elaborate structures can be fashioned log-cabin style to create boxes or "cribs" which are then filled with stones (Strom, 1962). Logs will preferably be obtained from local downed trees, although planks and other commercial timbers such as railroad sleepers ("ties") can be used. Weirs can also be created using some of the "wicker" structures of woven branches as described in the section on bank stabilization.

A review of a number of successes and failures of these structures is given by Wesche (1985). He recommends that a good location for weir placement is in a straight, narrow reach at the lower end of a sharp break in gradient. Both ends of the dam should be anchored into stable banks and the base sunk into the streambed. Weirs may require bank-stabilization works to prevent erosion around the plunge pool and end-cutting around the weir. The addition of rocks below the overfall can help to dissipate some of the erosive energy. In larger streams, weir design should be done with care to avoid creating "roller" currents downstream which are hazardous to swimmers and boaters who can be pulled under by the strong currents.

Wesche (1985) recommends that the low-flow spillway should be located

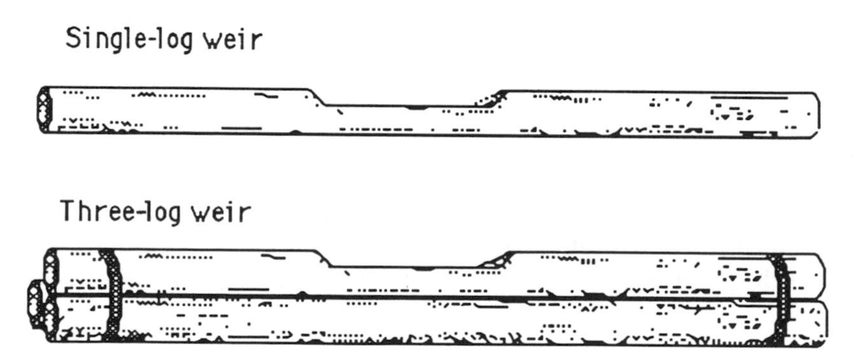

Figure 9.11. Log weir designs. Redrawn from Wesche (1985), by permission of Butterworth–Heinemann

near the thalweg line of the channel to preserve the natural flow path. Substrate materials can also be packed against the upstream face of the structure to create an upward tapering "ramp" (as in a beaver dam or riffle) to help to stabilize the weir and to direct flows smoothly over the top. Deposition of sediment upstream of the structure in subsequent years will further stabilize the weir. The gradual loss of the upstream pool should be considered part of the natural evolution of the streambed.

Floodbanks, floodways and floodplains

Sandbags filled and placed in front of buildings, the levees alongside irrigation channels, and the dikes in Holland are all methods of preventing water from spreading into areas where it is not wanted. Levees or floodbanks are probably one of the oldest methods of stopping rivers from flooding adjacent lands. For protection against a flood of a given magnitude, the height of the floodbanks will depend on the distance they are set away from the stream, since this defines the amount of area available for floodwaters: high banks are needed close to the stream; lower banks further away. Levees have disadvantages because they disrupt the natural function of the floodplain in directing waters into the main channel, and in large floods when the levees are overtopped, they can actually prolong the flooding duration. If they are built close to the channel they can also result in the loss of wetland areas of ecological value (Ward, 1989).

From an ecological standpoint it is preferable to set levees further apart to preserve more of the floodplain habitat and to provide a buffer strip to reduce chemical and sediment input from lands which are farmed, inhabited, mined or logged. On the areas between the levees, land uses should be adapted to the flood-prone conditions, perhaps by being expendable or intermittent-use (e.g. cycle paths or seasonal grazing or agriculture).

Traditionally, materials for the construction of levees have been dredged from the stream itself, sometimes as part of a design to create a deeper, straighter flood channel. The use of smaller levees further from the stream would reduce construction costs since less material is required and would reduce impacts to the stream by discouraging the use of stream materials.

Multi-stage channels (Figure 9.12) are a common engineering practice. Usually, the "normal" lower flows are contained within a relatively narrow channel, with the higher flows carried by the wider, leveed floodplain. Concrete-lined examples of this design are common in metropolitan areas. In a more natural environment the equivalent practice is to leave the stream undisturbed and cut "berms" adjacent to the stream to increase the flood capacity of the immediate floodplain. Land use, geology and amount of flood mitigation required will determine the height, width and location of the berm(s). Lewis and Williams (1984) recommend cutting berms on one side only (sides may be alternated) to reduce disturbance of bank vegetation.

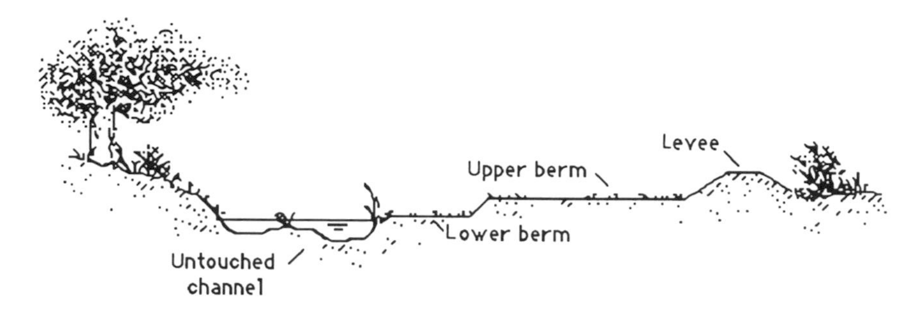

Figure 9.12. Multi-stage channel design. Berms have been excavated and the material used to form a levee. The lower berm would be inundated several times each year, whereas the upper one might only be inundated once every 3–5 years. Modified from Lewis and Williams (1984), by permission of the RSPB

With this method not only is most of the original channel habitat preserved but the low-lying berm may function as a wetland or a damp meadow habitat. Vegetation on the berms may require periodic pruning unless the resistance effect of plants has been included in the design.

For streams which have already been converted to a trapezoidal floodway Bovee (1982) offers some habitat improvements which can be made without affecting the flood-carrying capacity. Even in these constricted channels, meandering and reworking of the bed materials will still occur, creating microhabitat patterns. Figure 9.13 illustrates the addition of cutout areas, boulders and vegetation to provide sites of lower velocity during flood flows. Bovee states that it is important to leave some unaltered trapezoidal sections to act as hydraulic controls.

Bank stabilization

Bank slope and stability have an influence on channel form, vegetation growth and habitat for bank-living species such as otters, platypus and kingfishers. Bank stabilization is a common engineering practice, the rock rip-rap on the face of dams or the sides of channels being an example. The purpose is to reduce erosion which can lead to the failure of a dam, the

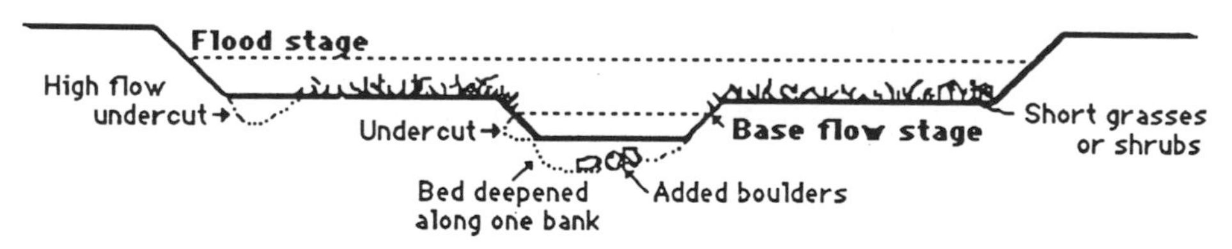

Figure 9.13. Modifications of flood channels for habitat improvement. Adapted from Bovee (1982)

slumping of banks, the gradual migration of a channel or the sudden "breakaway" or "avulsion" of a river into a different course. Bank revegetation (Section 9.4.2) is one stabilization measure; however, other protective works may be required until the vegetation becomes well established. Fencing areas to exclude stock is also an effective practice in bank rehabilitation.

The methods included in this section are those which predominantly use natural materials which are more environmentally compatible and aesthetic than nylon, steel, used tyres, concrete slabs and plastic. These applications are sometimes termed "soft engineering" or "biotechnical engineering" (Lewis and Williams, 1984). Their construction will normally be less costly; however, they are more labour-intensive and may be shorter-lived than more conventional engineering works. They also require more monitoring and maintenance. Anchoring of the works is important because if they drift away they can create flooding or erosion hazards elsewhere or endanger bridges. The benefit, perhaps unmeasurable in economic terms, comes in the improvement of bankside habitat and water quality.

An important element in the design of stabilization works for habitat rehabilitation is non-uniformity. This preserves the natural variability, providing different surfaces and textures for colonization by a diverse array of organisms. Bank protection should ideally use local materials for the same reason, and to reduce costs. Flexible structures (e.g. plants) are preferable in some cases since they will absorb the erosive energy rather than reflect it to some other location; although in other cases this will be the intention (see current deflectors, below).

Bank battering or shaping to some standard slope (e.g. 1 in 2) is often a part of the channelization process. In severely entrenched streams this technique is performed to smooth down canyon-like banksides, improving access to the river for stock and anglers and facilitating revegetation efforts. Material from the upper part of the bank is piled at the foot of the cliff to form the slope. Tamping or battering the soil can help to prevent later slumping or erosion of saturated, loose soils. If lower soil layers are of low fertility it is advisable to set the topsoil aside until the slope is formed, then spread it across the surface. This is primarily an engineering practice, and expertise is required to determine the proper bank angle and degree of compaction.

Protection of the re-formed bank with mulch or cut brush weighted down with rocks will help to prevent rainfall-induced erosion until the vegetation takes hold. The toe of the slope may also require protection with rocks or logs to prevent scour by the stream. Again, diversity is important in restoring "naturalness". Creative "sculpting" of the channel in harmony with geomorphic processes might leave undercut banks and some cliff faces, preserve existing trees or shrubs, and even include a few flat ledges on the slopes for angler and wildlife access.

Bank-protection works are applied where bank erosion is likely to be highest, as on the outside of meander bends, on the bank across from a current deflector, or on newly constructed bank or levee slopes. The selective use of bank protection only where it is most needed can provide stability at a lower cost as well as maintaining habitat diversity. Some of the discussed bank protection designs are shown in Figure 9.14. The longevity of the works will depend on what kind of materials are used; for example, Australian red gum timbers can last centuries. Texts on river engineering such as Petersen (1986) can be consulted for additional information. Engineering assistance should be obtained in the design and construction of larger works.

Stone "rip-rap" or "beaching"

Lining the banks with stone is feasible if a local source of rock is available and equipment is available for transporting and depositing it. The pits in the stones and the nooks and crannies between them increase the amount of area for colonization by aquatic insects. Plants can take root between the rocks and rock ledges can be used as resting and roosting sites by waterfowl. Stonework is a relatively permanent option if the stones used are sufficiently large that they do not wash away during high flows. More angular stones will interlock and are thus more stable. The methods of Section 7.4 can be used as a guideline for determining the appropriate rock size. Placing a layer of fine gravel or stone under the larger rip-rap material will help to prevent erosion of finer bank materials from beneath the rocks. To avoid undermining at the toe of the rip-rap, Strom (1962) describes the use of stone "aprons" which extend well out into the riverbed. An extra pile of rocks can be added to the edge of the apron which fall in and continue protecting the bed if it degrades (Figure 9.14(a)).

Dikes, groynes or revetments

These structures extend out from a streambank into the flow (Figure 9.14(b)). They are used to reduce the speed or change the direction of the stream's current and thus reduce the erosive force on the banks. There are many designs, from impenetrable earth and rock dikes or "cribs" of logs and stones, to timber planks or logs bolted to posts (see "Weirs", above). With proper design silt will be deposited in the quieter water between the groynes. Vegetation can be planted in these silt beds which will eventually form floodplain benches. Normally, a series of groynes are used along a bank. As a rule of thumb, a groyne will protect three to five times its own length of bank (Strom, 1962).

Vertical timber posts

Wooden posts or piles driven vertically into the streambed can be used to protect the lower part of a bank, for example where a vertical cliff on the outside of a meander bend is to be preserved (Figure 9.14(c)). Some bird species use these vertical cliffs for nesting. Lewis and Williams (1984) recommend that at least half of the post should be driven into the ground against the base of the cliff and local materials banked against the area behind the posts to prevent washout.

Woody debris

In areas where tree pruning, snag removal and brush cutting must be carried out these materials can be "recycled" to the stream in the form of bank protection. Debris can also be used to slow the water, trap silt and sediment and permit the growth of river-edge plants. Cut logs and branches can be secured against a bank for protection or for filling scour holes. Large trees should be tethered with the trunk end upstream. Stakes driven into the streambed or heavy-gauge wire or cables looped around the debris and attached to firmly embedded stakes on the banktops may be necessary to hold the materials in place until anchored by siltation and vegetation growth.

"Wicker" spiling and hurdles

Smaller materials can be "woven" together for bank protection. Spiling (Figure 9.14(d)) can be used to protect the base of steep banks and to create dikes on smaller streams. Stakes are driven into the ground and branches woven between them. Alternatively, posts can be staggered in two rows and brush piled between them. The distance between the posts should be consistent with the length and thickness of the "weaving" materials. As with timber posts (above), the area behind the spiling should be backfilled. If appropriate for the region the use of willow stakes can create a semi-permanent, living form of bank protection.

Hurdles are woven of smaller branches and pegged in place against banks of milder slopes. These function as a "mulch" through which plants grow, and degrade at about the time that the plants become established. Hazel, which will not grow from cuttings, is the usual construction material in the UK (Lewis and Williams, 1984). Hurdles are only suitable for small streams or low-turbulence reaches.

9.4.4 INSTREAM STRUCTURES

Like channel modifications, instream structures are intended to create physical diversity in the stream. Considerations include biological requirements and

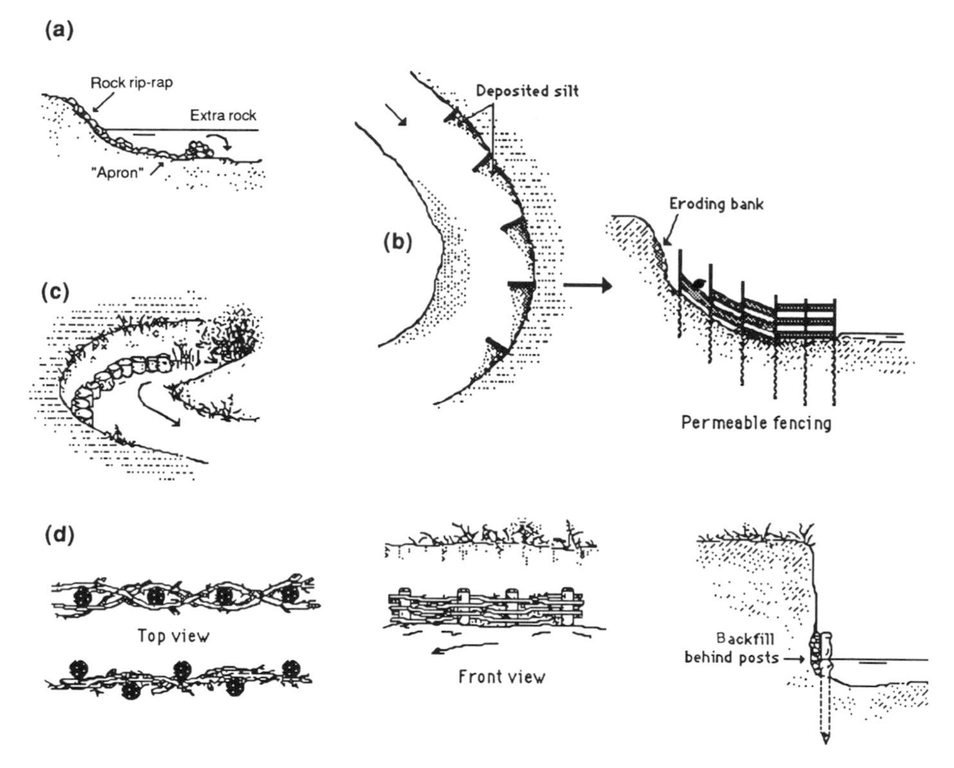

Figure 9.14. Bank-protection measures: (a) stone "rip-rap", (b) dikes or groynes, (c) vertical timber posts, and (d) wicker spiling. Adapted from Lewis and Williams (1984), by permission of the RSPB, and Strom (1962)

the geomorphologic and hydrological character of the stream. For example, it would be relatively useless to install fish shelters on flashy streams where they might either be swept away by floods or left sheltering a dry streambed.

Current deflectors

The purpose of these structures is to alter the direction of flow; for example, to divert water away from erodible banks (see Dikes, above), or to direct the current towards key locations such as fish shelters. Flow can also be directed to put the stream's energy to work in widening meanders or in creating, maintaining or enhancing the cirulation in pools. Increased turbulence leads to the creation of a scour pool along the edge of the deflector—either upstream or downstream, depending on its orientation. Deflectors may be used to increase velocity and/or restrict the channel width for low flows by using them in pairs. The structure itself adds to the bank length, increasing the amount of margin for biological activity, and the pools created provide shelter for fish during both high and low flows.

Placement and alignment of the structures should be compatible with the channel form and composition. In braided streams with large gravel bedloads, for example, the technique would be useless. The five to seven channel-width "rule of thumb" can be used in preliminary designs, although, again, local terrain irregularities and good judgement should guide the final placement. The observation of Lisle (1986) (see Section 7.3.3) that relatively permanent pools and bars were correlated with the presence of both upstream and downstream "natural deflectors" warrants closer inspection. Similar studies in a particular region would perhaps assist the placement of deflection structures on streams requiring rehabilitation. These should include observations of sand/silt deposits and pool locations near snags, boulders, and large bankside trees.

Current deflectors must be designed and installed carefully to prevent erosion and removal by floods. Erosion of the bank opposite a deflector may occur, necessitating bank protection. Height, distance into the flow, shape, and angle are all important design considerations. Bovee (1982) mentions that the US Federal Highway Administration recommends that deflectors should extend no more than halfway across a channel and be no taller than 0.5 m. Wesche (1985) recommends a height of no more than 0.15–0.3 m above the low-flow elevation.

Deflectors can be composed of natural materials (rocks, tree trunks, logs) or concrete or wire-mesh gabions. They are relatively inexpensive and easy to construct, and can be modified to suit the individual stream site. The use of a triangular form (Figure 9.15) rather than a narrow peninsula can reduce erosion of the adjacent bank during high flows. The configuration of point bars might serve as a good template for design. Vegetation of the top surface and addition of fish shelters at the downstream edge can further improve fish habitat (Wesche, 1985). The base of the current deflector should be buried low enough to reduce undermining and the ends well secured into the bank.

Wesche (1985) gives details on the construction of deflectors for fish-habitat enhancement and a number of examples where deflectors were applied both successfully and unsuccessfully. An increase in production of benthic invertebrates and fish was reported in several studies. Failures were mostly due to flood damage and/or improper design.

Boulder placement

The placement and arrangement of boulders in streams (Figure 9.16) can create diverse flow conditions, shelter for fish, amphibians, invertebrates and mammals, and "patchiness" of streambed substrates. In smaller disturbed

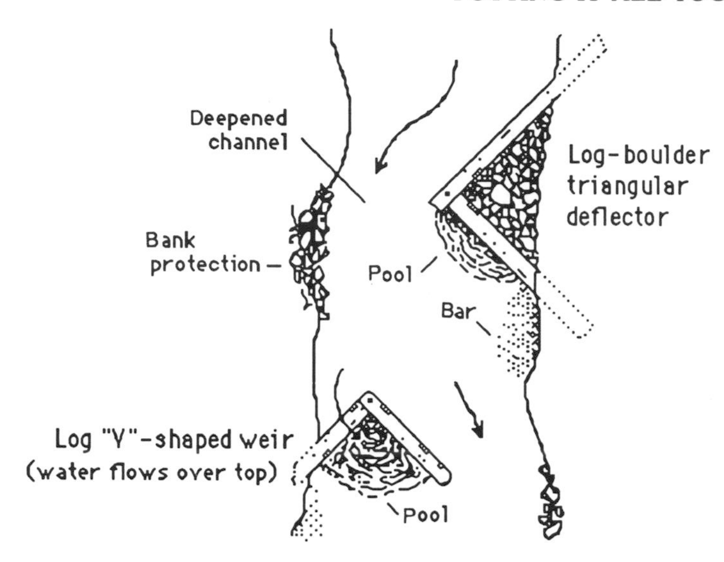

Figure 9.15. Current deflector designs. Adapted from Koehn (1987) and Wesche (1985), by permission of Butterworth–Heinemann

stream sites such as power-line crossings or disused road crossings, rocks can be added or rearranged to improve fish migration through the reach at low flows. The streambed topography is altered not only by the objects but also by scour which can occur around the sides or under these boulders and deposition downstream. As obstructions, boulders cause energy to be dissipated through turbulence, and can be used to slow the total velocity through a reach (e.g. in a fishway) (Newbury, 1989).

Boulders can be arranged to create specific local flow variations preferred by individual organisms (Nowell and Jumars, 1984). With some "playing in the water", the wake region behind one rock can be connected with that of the next rock downstream, to provide a continuous region of low velocity (Newbury, 1989) as shown in Figure 9.16.

In larger streams, boulders should be carefully placed, preferably based on the combined judgement of a biologist and a geomorphologist or engineer. Another consideration is whether the stream is to be used for boating (Wesche, 1985). If the placements are to be relatively permanent an estimate should be made of the size of rock which will move at high flows (see Section 7.4). Embedding the boulders in the streambed will reduce their tendency to roll away. If the object is to re-create a natural boulder pattern, natural arrangements should first be surveyed by observation or by formal methods (Section 5.2) to determine the mean and variability of boulder spacings. A design should be worked out in advance, especially where larger boulders will need to be moved with heavy equipment.

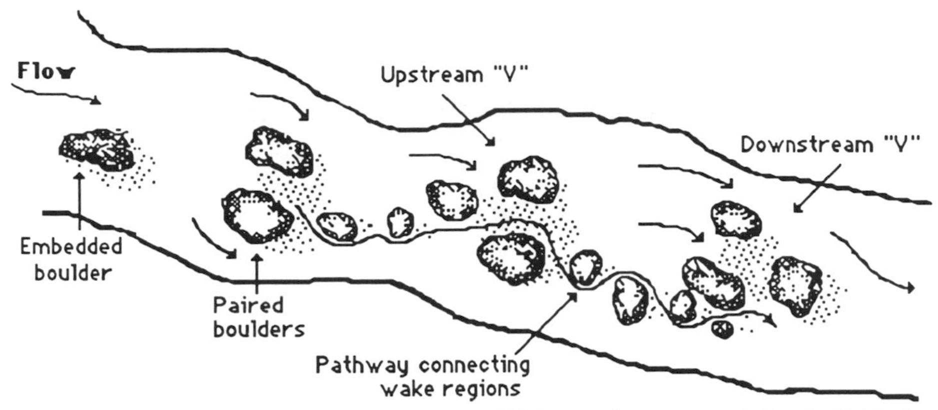

Figure 9.16. Boulder placement patterns. Wake regions created by individual boulders can be connected to provide a continuous path of low velocity between boulder clusters. Adapted from Wesche (1985), by permission of Butterworth–Heinemann

If boulders are not readily available it may be possible to "mould" them out of a mixture of concrete and local gravels or cobbles. The wet mixture can be poured into burlap (hessian) bags, shaped and left to dry. Colonization of the surface by algae should soon eliminate the artificial look. Submerged log sections are another alternative.

Fish shelters

Fish shelters (Figure 9.17) are substitutes for natural overhanging banks. These provide shade, shelter from swift flows and protection from predators. Lewis and Williams (1984) state that these shelters can be built above or below water level or made to float on the water surface. Covers of brush, either tied to the bank or woven like the hurdles mentioned in the previous section, are a simple form. A platform of split logs or planks built out from the bank and secured to the streambed by posts or reinforcing rods ("rebar") is another design. Bovee (1982) recommends that pilings (support columns) should be about 2 m long for low-gradient streams. Concrete footings can be added to the pilings for more permanence. Earth, rock and sod can be added to the top for additional weight, to improve aesthetics and to provide opportunities for overhanging vegetation to become established.

Bovee (1982) states that fish shelters should be located in the erosional parts of the channel such as the outsides of meander bends, since they may be lost to siltation in depositional areas. However, a close look at small undamaged streams will reveal that undercuts can occur almost anywhere along the length of a stream reach.

Fish shelters should be installed when the flow is low, both to facilitate

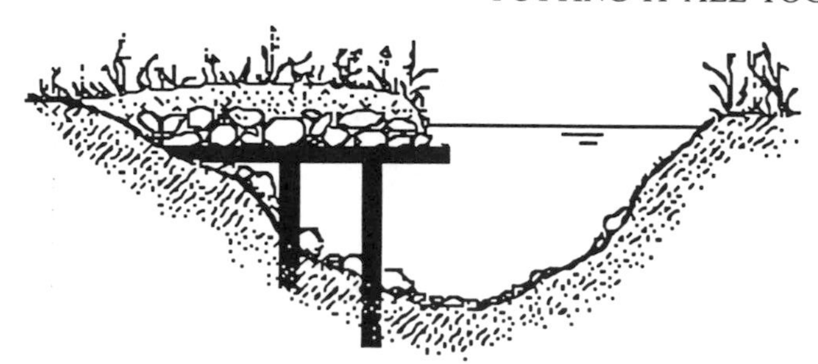

Figure 9.17. Fish shelter design. Redrawn from Bovee (1982), after White and Brynildson (1967)

construction and to ensure that the sheltered region contains water. Boulders can be placed under the overhang to enhance cover and protect the bank against erosion (Wesche, 1985). Fish shelters can also be attached to other structures such as current deflectors or weirs. Alternatively, current deflectors can be installed on the opposite bank to improve circulation of water under the shelter and lessen sediment deposition.

Although a fish shelter will form an obstruction at high flow, the narrow width should not increase flood hazards significantly. The main concern is their tendency to float or wash away during high flows. Because of this susceptibility they should perhaps be designed to break up during floods so that they do not create problems downstream. As Bovee (1982) mentions, the failed structure itself may not form a large part of the debris load moving down a river, but if blockage problems occur, the structure can receive a disproportionate part of the blame.

In a study on the Ovens River in Victoria, Australia, Koehn (1987) reported a ninefold increase in the number of native blackfish (*Gadopsis bispinosis*) after the addition of artificial habitat. A low (1 m height) V-shaped weir composed of six telegraph poles was installed near the edge of the stream (see lower part of Figure 9.15). A 24 m downstream reach was also "seeded" with large rocks (0.35–1.1 m in diameter). The previously uniform water flows were broken up into regions of both higher and lower velocity, and the log weir produced a pool 1.4 m deep within the "V". The weir was found to be quite sturdy and self-cleaning. It survived several high water flows and did not lead to an increase in channel width. However, fine sediment accumulated around the large rocks, increasing the streambed elevation and reducing the amount of fish cover as compared with the initial condition.

Stream rehabilitation is still a relatively new field in which there is room for more experimentation and refinement of methods. Documentation is needed on both successes and failures. As stressed throughout this section, each stream site should be considered unique. The geomorphology, fluctuations in water level, native plants and animals, bank soil types and channel bed topography must all be taken into account. The continuing interaction of biologists, geomorphologists and engineers is needed to develop guidelines for stream-rehabilitation works which are sufficiently versatile to accommodate a wide range of situations and flexible enough to incorporate improvements suggested by research and experimentation.

Appendix 1 Basic Statistics

A1.1 SUMMARY STATISTICS

Summary statistics describe properties of sample data. For this summary, $x_1, x_2, x_3, \ldots, x_n$ represent sample data values, where n values have been selected from a population of all possible values of x. The symbols Σ and Π indicate summation and multiplication, respectively, of values from $i = 1$ to n. Many of the statistics given are calculated by program STATS in AQUAPAK.

Measures of central tendency

Arithmetic mean or average, $\bar{x}$

$$\bar{x} = \frac{\Sigma x_i}{n} = \frac{x_1 + x_2 + \ldots + x_n}{n} \tag{A1.1}$$

Geometric mean, $\bar{x}_g$

$$\bar{x}_g = (\Pi x_i)^{1/n} = \sqrt[n]{(x_1 * x_2 * \ldots * x_n)} \tag{A1.2}$$

For calculating $\bar{x}_g$, Armour et al. (1983) recommend taking the mean of the logarithms of the sample values, then taking the antilog of the mean:

$$\bar{x}_g = \text{antilog}\left(\frac{\Sigma(\log x_i)}{n}\right) \tag{A1.3}$$

Median

If a data set is ranked in order of magnitude, the **median** is the middle value; e.g. for the values 1, 3, 4, 6, 8, the median is 4.

Mode

The **mode** is the sample value which occurs most often in the data set; e.g. for the values 1, 1, 2, 2, 3, 3, 3, 4, 5, 5, the mode is 3. If a sample has two modes, it is **bimodal**; if more, it is **multimodal**.

Measures of spread

Range

$$\text{Range} = (x_{max} - x_{min}) \tag{A1.4}$$

where x_{max} is the highest sample value and x_{min} the lowest.

Root mean square (RMS)

$$\text{RMS} = \sqrt{\left[\frac{\Sigma(x_i - \bar{x})^2}{n}\right]} \tag{A1.5}$$

The RMS is used as a measure of turbulence intensity with x = velocity (Section 6.4.1).

Variance, s^2

$$s^2 = \frac{\Sigma(x_i - \bar{x})^2}{n - 1} \quad \text{or} \quad \frac{\Sigma(x_i^2) - \left[\frac{(\Sigma x_i)^2}{n}\right]}{n - 1} \tag{A1.6}$$

The second version is preferred for computational purposes.

Standard deviation, s

The standard deviation, s, is the positive square root of the variance.

Coefficient of variation, C_v

The coefficient of variation is a dimensionless measure of variability commonly used in hydrology:

$$C_v = s/\bar{x} \tag{A1.7}$$

Index of variability, I_v

$$I_v = \sqrt{\left[\frac{\Sigma(\log x_i - \overline{\log x})^2}{n - 1}\right]} \tag{A1.8}$$

where $\overline{\log x}$ is the mean of the logarithms of the x values. I_v is often used to characterize the year-to-year variability of peak flood flows. Baker (1977) also refers to this as the "flash flood index". Streams with high values of I_v are more likely to have flash flood behaviour. Those with a high variability of peak flows may also have lower species diversity and abundance than streams with only moderately variable peak regimes (Ward and Stanford, 1983).

Measures of shape

Skewness (C_s)

$$C_s = \frac{n\Sigma(x_i - \bar{x})^3}{(n-1)(n-2)(s^3)} \tag{A1.9}$$

$$\text{or} \quad \frac{n(\Sigma x_i^3 - 3\bar{x}\Sigma x_i^2 + 2n\bar{x}^3)}{(n-1)(n-2)(s^3)}$$

This measure describes the lack of symmetry in a set of data. A positive value of C_s means the data are skewed to the right,  and a negative one that the data are skewed to the left,  . Streamflow data typically show a positive skewness, with the skewness decreasing as the time interval increases from daily to annual.

Kurtosis (C_k)

$$C_k = \frac{n^2\Sigma(x_i - \bar{x})^4}{(n-1)(n-2)(n-3)(s^4)} \tag{A1.10}$$

Kurtosis describes the "peakedness" or "flatness" of the data distribution. For a normal distribution (Section A1.2), the kurtosis is 3.

A1.2 DESCRIBING DISTRIBUTIONS OF DATA

Measurement variables can be either **continuous** or **discrete**. Data which can take on any value or any value within a certain range (e.g. 0 to 100) are considered continuous. Examples of continuous variables are channel width and depth, stream temperature, streamflow, and substrate particle diameters. Discrete data, in constrast, have a limited number of possible values. Examples are counts of rocks, plants or fish, or quality ratings (e.g. 5 = excellent, 1 = poor). Continuous data are presented in discrete form due to measurement limitations (e.g. measured to the nearest 1 mm).

If a large amount of data is collected (either continuous or discrete), the values can be summarized by a **frequency distribution**. To accomplish this, the samples are ranked from smallest to largest and then grouped into classes. **Frequency** refers to the number in each class. Frequency data can be plotted as a **histogram** (Figure A1.1(a)), where the height of each column represents the frequency and the class intervals are given on the x-axis.

If the frequencies are divided by a factor such that the area under the curve equals one, then the values become **relative frequencies** (Figure A1.1(b)). These can also be expressed as **percentages** if multiplied by 100. By progressively summing up these values, a **cumulative frequency distribution** can be developed, as shown in Figure A1.1(c). The sediment data of Figure 5.32 are presented in this form.

When the number of values increases and the size of the class interval decreases, the profile of the frequency distribution becomes smoother. At the limit, where the number of values approaches infinity and the class interval approaches zero, the resulting curve becomes a **probability distribution** or **probability density function (pdf)**, as shown in Figure A1.1(d). The function describing the curve is denoted $f(x)$, or "function of x". Probability density functions for standard theoretical distributions (e.g. normal, log-normal, etc.) are usually defined by mathematical equations. In a pdf, the function $f(x)$ must be ≥ 0 for all x, and the area under the graph of $f(x)$ is equal to 1. For the probability distribution of Figure A1.1(d) the probability that x takes on a value between a and b (written "$P(a < x < b)$") is equal to the shaded area under the curve. A probability of zero means "impossible" whereas a probability of 1 means "certain". The probability that x is *exactly* a or *exactly* b is zero thus, we usually use the symbols $>$ and $<$ rather than $\leq$ and $\geq$.

Another, more useful form of the probability distribution is the **cumulative distribution function (cdf)**. The probabilities are cumulated, and cumulative values are plotted against their associated x values, as shown in Figures A1.1(e) and (f). The probability becomes the **probability of non-exceedance**, $F(x)$, or the **probability of exceedance**, $F'(x)$, depending on the direction in which the values are cumulated and the intended use of the curves. $F'(x)$ is sometimes called the **inverse cumulative distribution function (icdf)**. Thus, in Figure A1.1(e) the probability that $x < 1800$ mm ("$P(x < 1800)$") is about 0.95, whereas in Figure A1.1(f) the probability that $x \geqslant 1800$ mm ("$P(x \geqslant 1800)$") is about 0.05. As is apparent, $F(x) = 1 - F'(x)$.

Cumulative distributions are commonly plotted with the x and y axes reversed, using a probability scale based on a theoretical probability distribution (e.g. normal). Data which are described by this distribution will then plot as a straight line. Commercial probability papers are available for a number of distributions. The large sediment data of Figure 5.33, the flood-frequency data of Figure 8.4 and the flow-duration data of Figure 8.7 are plotted in the cumulative form on log-normal probability axes. It should be noted that Figures 5.33 and 8.7 are cumulative frequency plots, whereas in

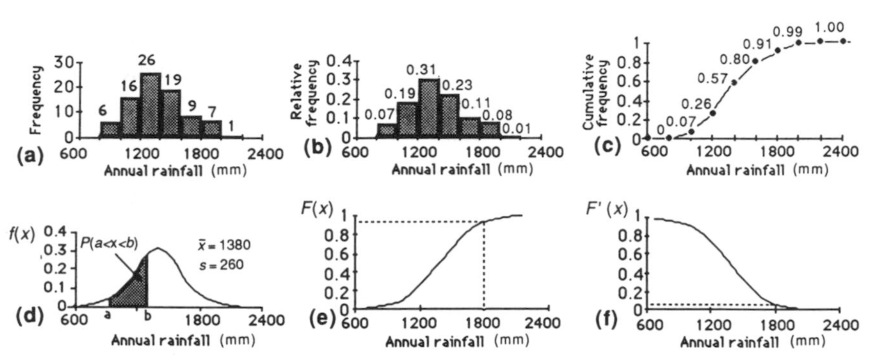

Figure A1.1. Marysville, Victoria, annual rainfall, presented as (a) frequency histogram, (b) relative frequency histogram, (c) cumulative frequency distribution, (d) fitted probability distribution (normal), (e) cumulative distribution function, and (f) inverse cumulative distribution function. $f(x)$ = probability, $F(x)$ = probability of non-exceedance, and $F'(x)$ = probability of exceedance

Figure 8.4 a theoretical probability distribution has been fitted to the data. The first summarizes the data, the second allows inferences to be made about the underlying population.

A basic assumption in statistics is that sample data can be described by probability distributions. For example, in a flood-frequency analysis, a distribution is fitted to the data to improve estimates of, say, 100-year floods. If the fitted distribution does not truly represent the population of all possible sample values, then results can be misleading.

Distributions used for discrete data include the binomial, negative binomial, and Poisson distributions. These will not be discussed here, but are described in many basic statistics texts. The **normal probability distribution** ("bell-curve") is one of the most widely used models for both discrete and continuous variables. Other distributions are described in Section 8.2.6. The normal distribution is defined by its parameters μ (mean) and σ^2 (variance). In this distribution, 68.3% of the values lie within $\pm 1\sigma$, 95.4% within $\pm 2\sigma$, and 99.7% within $\pm 3\sigma$. Normal probability tables give values of the standard normal distribution, for which $\mu = 0$ and $\sigma^2 = 1$. To use the table, sample values are standardized with the transformation:

$$Z = \frac{x - \bar{x}}{s} \tag{A1.11}$$

where Z is the standardized variate. Thus, if the data of Figure A1.1 are assumed to be represented by a normal distribution, then we would expect the annual precipitation to be between 860 mm and 1900 mm about 95% of the time (solving for x with $\bar{x} = 1380$, $s = 260$ and $Z = \pm 2$).

The normal distribution is particularly important because a number of statistical methods require that the data are normally distributed, e.g. t-tests, analysis of variance, and correlation analysis. Sometimes **transformations** can be applied to the data (e.g. log, square root, etc.) to satisfy this requirement. Even when variables are not normally distributed, sums and averages of the variables may be. According to the **central limit theorem**, if samples are selected randomly from a given distribution then the distribution of the sample means tends towards the normal distribution. The more values included in the sample means, the better the approximation. Thus, many variables have distributions which can be fitted closely with a normal distribution, such as measurement errors, heights, weights and other physical characteristics. For rainfall and streamflow data, for example, the distributions tend towards normality as the time base is increased; e.g. average monthly rainfall and yearly runoff may be approximately normally distributed. As a rule of thumb, if $n > 30$, the central limit theorem can be applied (conveniently similar to the number of values in monthly totals).

A1.3 JUST A MOMENT

Moments provide a way of quantifying location, spread and shape of probability distributions. "Moment" is a term from the bending and twisting branch of engineering mechanics, defined as distance times force. For example, if a bag of 500 "Newton's apples" of 1 N each were placed on the end of a 3 m diving board, the applied moment would be 1500 N·m. A related concept, torque, explains why less force is needed to move a locknut if a longer-handled wrench is used.

In statistics, infinitely small strips called "elemental areas" and distances along the x-axis replace the physical measures of force and distance (see Figure A1.2). For continuous variables, moments are computed from calculus formulae. "Distances" are measured in relation to either the origin ($x = 0$) or the mean ($x = \mu$). The two types of moments are called "moments about the origin" and "moments about the mean" ("central moments"), respectively.

The mean, μ, is the same as the first moment *about the origin*. If the x-axis were "balanced" on a wedge as shown in Figure A1.2 the downward "forces" on either side would be in balance and the point of balance would be the mean. The first moment *about the mean* is therefore zero.

The second moment is calculated in the same manner except that the distances are *squared*. For the third moment, the distances are cubed, and so on. To generalize the situation, the rth moment about the mean (m_r) for a continuous distribution can be described by

$$m_r = \int_{-\infty}^{\infty} (x - \mu)^r f(x) \, dx \qquad \text{(A1.12)}$$

Here, x is the distance along the x-axis from the origin, μ is the population

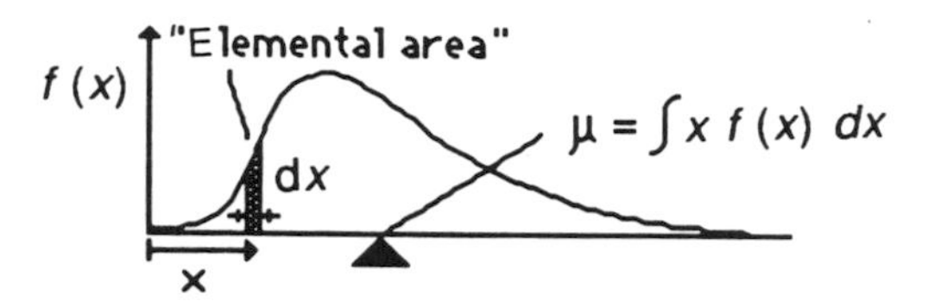

Figure A1.2. Illustration of the first moment about the origin, the mean (μ), for a continuous probability distribution. One half of the area under the curve, $f(x)$, lies to either side of the mean. The sum of all "elemental areas" is equal to one

mean, and $f(x)$ is the probability density function. To compute moments about the origin (m'_r), the term μ is deleted from Equation A1.12.

For sample data, Equation A1.12 becomes:

$$M_r = \frac{\Sigma(x_i - \bar{x})^r}{n} \tag{A1.13}$$

where $\bar{x}$ is now the sample mean. To calculate the rth sample moment about the origin (M'_r), $\bar{x}$ is deleted from Equation A1.13.

Probability distributions can be fitted to data by approximating the moments of the distribution by sample moments. This is called the **method of moments**, and is mentioned in Section 8.2.5. Sample moments are related to the sample statistics, mean, variance, skewness, and kurtosis, as follows:

$$\bar{x} = M'_1 \text{ (the first moment about the origin)} \tag{A1.14}$$

$$s^2 = \frac{n}{n-1}(M_2) \tag{A1.15}$$

$$C_s = \frac{n^2 M_3}{(n-1)(n-2)s^3} \tag{A1.16}$$

$$C_k = \frac{n^3 M_4}{(n-1)(n-2)(n-3)s^4} \tag{A1.17}$$

A1.4 CORRELATION AND REGRESSION: HOW CLOSELY IS THIS RELATED TO THAT?

A1.4.1 GENERAL

Correlation and regression are well-worn and well-proven methods used in hydrology and ecology, and a multitude of examples can be found in the literature. For example, regression relationships can be developed to extend the record at one gauging station from data obtained at another, to predict streamflow from catchment characteristics and precipitation or to determine which physical habitat factors are most closely related to the abundance of organisms.

The objective may be (1) to simply determine whether certain variables are related or (2) to develop a relationship so that one variable can be predicted from one or more other variables. The first process is one of correlation and the second, regression. The two have important distinctions although the methods of analysis are much the same. Haan (1977) and Holder (1985) are useful references on the use of these statistical methods in hydrology.

Correlation

This is a means of defining the strength of the relationship between two variables, e.g. substrate size and benthic populations of blackfly larvae. A sample **correlation coefficient** is a quantitative measure of this "strength", based on sample data. For this example the sample data might be the substrate d_{50} and numbers of larvae per square metre. A close association between two variables does not imply cause-and-effect or a common influencing factor. The fact that substrate size and numbers of blackfly larvae are correlated does not mean that changing the number of blackfly larvae on the streambed will affect the substrate size or vice versa.

The sample correlation coefficient (r) for n pairs of (x, y) data is calculated as follows:

$$r = \frac{n\Sigma x_i y_i - (\Sigma x_i)(\Sigma y_i)}{\sqrt{[n\Sigma x_i^2 - (\Sigma x_i)^2]}\sqrt{[n\Sigma y_i^2 - (\Sigma y_i)^2]}} \tag{A1.18}$$

The value of r lies between -1 and $+1$, inclusive. As a rule of thumb, the correlation is weak if $0 \le |r| \le 0.5$, strong if $0.8 \le |r| \le 1$ and moderate otherwise (Devore, 1982).

To illustrate how r is interpreted, let the graphs of Figure A1.3 represent paired measurements of frequency of frog croaking (y-axis) and air temperature (x-axis). A value of r near $+1$ means that frogs croak more frequently at higher temperatures and less often at low ones. Conversely, a value of r near -1 means that frogs croak more often as the temperature goes down. An r value of zero would imply that the correlation between the two variables is non-existent or not linear. Thus, r is a measure of the degree of linearity of the relationship. Clearly, in the fourth graph, there is a relationship (a reasonable one for this example) even though $r \approx 0$.

A high correlation between two variables when in fact none exists is called **spurious correlation**. For example, if the data forms clusters of two or more groups such as in the last graph of Figure A1.3, a large correlation may result solely due to the heterogeneity of the data. Another potential source of spurious correlation is the use of ratios with a common element, e.g. the correlation of X_1/X_2 with X_1, or X_1/X_3 with X_2/X_3. Plots of these variables may show a well-defined relationship with a narrow scatter of points due only to the presence of the common element. This type of

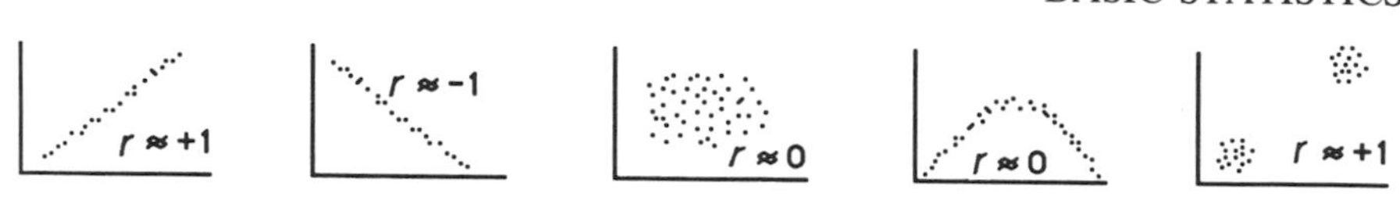

Figure A1.3. Interpretation of the correlation coefficient, r

spurious correlation is quite common, and often goes undetected. Examples in this text include the C_v versus MAR relationship (Figure 4.19(b)), where mean annual flow is a common factor, and the sediment rating curve (Figure 7.22), where discharge is a common factor. As Benson (1965) points out, the relationship is not invalid as long as interpretations are made in terms of the ratios and not the original factors. For example, the sediment-rating curve should be used only to estimate sediment discharge, not sediment concentration. Thus, care should be exercised in interpreting the results of scatter plots or regression relationships when common elements are involved.

If sample values for the two variables are assumed to be selected from a bivariate ("three-dimensional") normal distribution, and free of measurement error, then one can perform tests of hypotheses about whether the population correlation coefficient is significantly different from zero, and construct confidence limits. Neglecting these requirements can bias results. Correlation coefficients can have very large confidence limits for sample sizes of 30 or less; thus for small samples, it would be splitting hairs to say that r values of 0.701 and 0.702 have a meaningful difference. For the same reason, correlation coefficients should only be reported to two significant figures.

Regression

In correlation, the "end product" is the correlation coefficient. In **regression**, the "end product" is a **regression equation** which describes one variable, the dependent (Y) variable, as a function of one or more independent (X) variables. A coefficient is also obtained, which in this case describes how well the equation fits the data. The term "regression" was first used by Francis Galton to explain the phenomenon that men's heights "regressed" back towards the average. From a relationship developed between father's and son's heights, he found that if a father was taller than average, his son would also be expected to be taller than average, but not by as much. Similarly, the son of a shorter-than-average father would be expected to be shorter than average, but taller than his father (Devore, 1982).

The deviations of observed values (y_i) from values predicted using the regression equation ($\hat{y}_i$) are the error terms or **residuals**, $\epsilon_i = y_i - \hat{y}_i$. If the regression equation is to be accompanied by confidence intervals or tests of significance, the required assumptions about the residuals are that they:

- Are random and independent

- Are normally distributed
- Have constant variance ("homoscedastic")

Plots of the residual values provide a means of visually assessing whether the assumptions are met, and how well the equation fits the data. The recommended procedure is to develop a **scatter plot** of ϵ_i versus y_i. Plotting the predicted values, $\hat{y}_i$, against y_i is also advisable. Statistical computer packages normally have an option for plotting residuals, and users' manuals will provide information on interpreting the patterns. Various statistical tests of hypothesis are also available for testing the validity of the assumptions. Details will not be given here, but can be found in statistical texts (see Section A1.5).

Regression techniques can be used despite violations of these assumptions, but the resulting equation should be applied only over the range of data selected. If the assumptions are not met, the correlation coefficient will change over the range of data, making it unsuitable as a measure of adequacy of the regression.

When nothing is known about the relationship between variables, a "first-try" approach is to simply assume that the relationship is linear. A linear relationship works well for some variables such as the relationship between annual rainfall and annual runoff. Data are often **transformed** to linearize the relationship or to achieve constant variance or normality of the residuals. Transforming the data by taking logarithms is by far the most common approach in natural sciences. Examples are the stream-order relationships in Figure 4.12, the hydraulic geometry relationships in Figure 7.9 and the relationship of fish height to length (Armour et al., 1983). Relationships based on transformed data are considered **non-linear**.

It should be noted that the correlation coefficient from an analysis of transformed data applies *only* to the transformed data. Therefore, it is not "legal" to compare correlation coefficients from regressions on original, log-transformed and otherwise-transformed data to determine which has the "best fit". This should, instead, be judged visually or with the assistance of various tests to ascertain whether the transformation does in fact improve the linearity of the model and/or the residual patterns.

Regression and correlation are powerful methods for examining relationships between variables. As with any procedure, it is important to understand the assumptions and limitations. A general caution about the use of regression methods is well summed up in a phrase by Haan (1977, p. 218):

> The ready availability of digital computers and library regression programs has led many to collect data with little thought, throw it into the computer, and hope for a model. This temptation must be avoided.

A1.4.2 SIMPLE LINEAR REGRESSION AND LEAST SQUARES FITTING

In simple linear regression the fitted line is described by an equation of the form:

$$Y = aX + b \tag{A1.19}$$

or

$$Y = \beta_0 + \beta_1 X + \epsilon \tag{A1.20}$$

depending on whether one's preferences are algebraic or statistical. The second form will be used in this discussion. In Equation A1.20, Y is the dependent variable (the one which is to be predicted) and X is the independent variable (the "predictor"). The coefficients β_0 and β_1 are called **regression coefficients**. In this model, β_1 represents the slope of the line and β_0 the point where the line crosses the Y-axis (the "Y-intercept"). A scatter plot of the observed data (y_i versus x_i) is always advisable as a first step to investigate whether Y increases or decreases with X, whether the relationship is strong, moderate or non-existent, its linearity, the presence of "outliers" (points which considerably depart from the general trend), and the general range of X and Y (Haan, 1977).

When fitting a line through the data the objective is to make the deviation between the observed points and the fitted line small. This is usually accomplished by using the principle of **least squares**. A least squares fit of the equation to the data minimizes the sum of the squares of the deviations between the observed and predicted values of Y. This deviation is the "error" or "residual" term, and is illustrated for one point in Figure A1.4.

By using calculus, equations are derived for β_0 and β_1 such that the sum of the squares of the errors becomes a minimum. These equations are:

$$\beta_1 = \frac{n\Sigma x_i y_i - (\Sigma x_i)(\Sigma y_i)}{n\Sigma x_i^2 - (\Sigma x_i)^2} \tag{A1.21}$$

and

$$\beta_0 = \bar{y} - \beta_1 \bar{x} \tag{A1.22}$$

where x_i and y_i are now actual observations, n is the number of data pairs, and $\bar{x}$ and $\bar{y}$ are mean values. From Equation A1.22 it can be seen that this solution must pass through the point ($\bar{x}$, $\bar{y}$).

In Figure A1.4, 40 years of annual runoff data from the Acheron River at Taggerty (the dependent variable) have been plotted against the annual precipitation data at Marysville (the independent variable). A regression line fitted by least squares has been drawn through the points. The resulting equation allows runoff to be predicted if rainfall is known.

The total sum of the squares of the Y values (SST), Σy_i^2, can be partitioned into three components:

Sum of squares due to the regression (SSR), $\Sigma(\hat{y}_i - \bar{y})^2$;
Sum of squares due to the mean (SSM), $(\Sigma y_i)^2/n = n\bar{y}^2$;
Sum of squares due to the error or residual terms (SSE), $\Sigma(y_i - \hat{y}_i)^2$.

Thus, SST = SSM + SSR + SSE. A derivation can be found in Haan (1977).

The SSE is a measure of the variability of the y values about the fitted line. It is what is minimized in the least squares procedure. The ratio of the sum of squares due to the regression (SSR) to the total sum of squares adjusted for the mean (SST − SSM) is commonly termed the **coefficient of determination** or r-squared (r^2) value:

$$r^2 = \frac{\text{SSR}}{\text{SST}-\text{SSM}} = \frac{\Sigma(\hat{y}_i - \bar{y})^2}{\Sigma(y_i - \bar{y})^2} \tag{A1.23}$$

The coefficient of determination is interpreted as the proportion of variation in Y explained by the regression equation. If thc model were a "perfect fit" to the data, the r^2 value would be 1, whereas if none of the variation in the Y values were explained by the regression equation, the r^2 value would be 0. Thus, the closer the value of r^2 is to 1, the better the fit. The square root of r^2, r, is a type of correlation coefficient which describes the correlation between the values of y_i and $\hat{y}_i$.

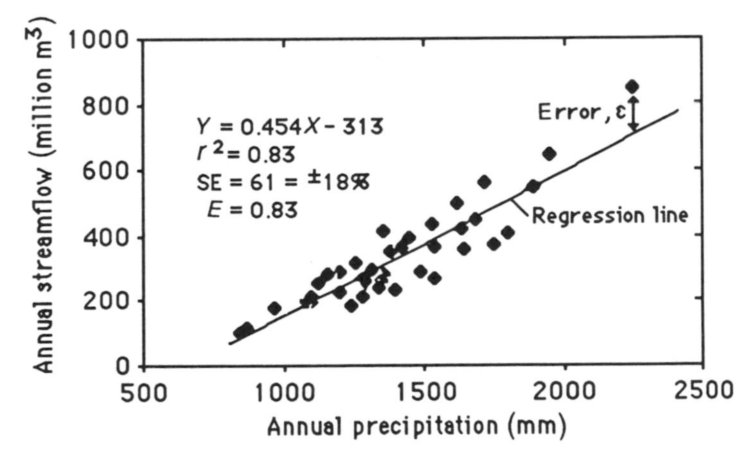

Figure A1.4. Simple linear regression analysis of annual streamflow data from the Acheron River at Taggerty versus annual precipitation data recorded at Marysville (see map of Figure 4.7). Shown are the scatter plot of the actual data, the fitted regression line, regression equation, and values of r^2, SE (standard error) and E (coefficient of efficiency)

The **standard error** or **standard error of estimate** is preferred over r^2 as a measure of reliability of the regression equation. It is equal to the standard deviation of the residuals about the regression line. The spread is assumed to be normally distributed and of constant variance over the range of data. An unbiased estimate for this standard error (SE) term is (Haan, 1977):

$$\mathrm{SE} = \sqrt{\frac{\Sigma(y_i - \hat{y}_i)^2}{(n-2)}} \tag{A1.24}$$

The $(n - 2)$ term indicates that two degrees of freedom are "lost" by estimating β_0 and β_1 (Holder, 1985). If the standard error is squared, the term SE^2 is often called the **mean squared error (MSE)**. If the regression equation explains a large part of the variation in Y, then SE will be much less than the standard deviation of the y_i values. Because SE has the same units as Y, it is sometimes expressed as a percentage of $\bar{y}$, as in Figure A1.4. When data are log-transformed, SE can be converted back to the original scale to give a more useful description of the spread about the regression line. For log-log or semi-log regressions (with y the logarithmic variable), these percentages are calculated as:

$$\mathrm{SE\ (lower)} = \left(\frac{1}{\mathrm{antilog}(\mathrm{SE}')} - 1\right) \times 100 \tag{A1.25}$$

and

$$\mathrm{SE\ (upper)} = (\mathrm{antilog}(\mathrm{SE}') - 1) \times 100 \tag{A1.26}$$

where SE' is the standard error calculated on the log-transformed scale (this is the value normally given on a computer output). For example, if a log-log (or semi-log) regression (base-10) gives $\mathrm{SE}' = 0.135$, then $\mathrm{antilog}(\mathrm{SE}') = 1.36$, giving $\mathrm{SE} = -27\%, +36\%$. Both SE and r^2 are usually listed with the regression equation when reporting the results of a regression analysis.

Another useful measure is the **coefficient of efficiency**, E. The coefficient of efficiency describes the degree to which observed and predicted values agree (Aitken, 1973). The term is computed from the equation:

$$E = \frac{\Sigma(y_i - \bar{y})^2 - \Sigma(y_i - \hat{y}_i)^2}{\Sigma(y_i - \bar{y})^2} \tag{A1.27}$$

The value of the coefficient of efficiency is always less than or equal to 1. In program REGRESS in AQUAPAK the values used in Equation A1.27 are on the original scale rather than the transformed one. This makes E a better measure for computing the fit of various models than r^2.

In obtaining a least squares estimate, no assumptions about the distribution of error terms are required. They enter the picture, however, when confidence intervals and tests of hypothesis are of interest. Haan (1977) and Holder (1985) and most statistics texts offer procedures for calculating confidence intervals about the regression line and for testing the significance of the parameters β_0 and β_1. Procedures are also available for testing whether two regression lines are significantly different. For example, in the double-mass curve analysis (Section 3.2.3) a test can be performed to determine whether an observed change in slope is significant. Confidence bands expand at both the high and low end of the range of data. Also, the relation between X and Y (or their transformed values) may be linear only over the range studied. For these reasons, extrapolation of a regression equation beyond the range of X is discouraged.

It is also worth noting that the regression line for Y versus X is not the same as that for X versus Y. For example, in Figure A1.4 the inverse relationship could *not* be found by solving the regression equation for X. A new relationship must be developed by regressing precipitation versus runoff (which, in this example, makes little physical sense). The r^2 value will be identical but the equation will be different. However, the equation will still pass through the point defined by the means of the two variables.

As Holder (1985) points out, the least squares method is an impartial, objective method of fitting a line through the data. For some purposes, a critical assessment by eye which gives more weight to some values (e.g. measurements which were known to be of high quality) and less to others may yield a line which is more useful than a least squares fit. Therefore, the intended use of the equation and the consequences of poor predictions should always be considered when using regression methods.

A1.4.3 MULTIPLE REGRESSION

The "simple" in simple linear regression simply means that only one independent variable, X, is involved. Often, a phenomenon is "dependent" on several factors. For example, the number of a fish in a stream reach may depend on the velocity of the stream, area of bank overhang, substrate type, amount of aquatic vegetation, populations of aquatic insects and other edible organisms, amount of shade, temperature, flow duration, and other factors which may or may not be measurable. A regression model for predicting the number or biomass of fish might include some or all of these influencing factors.

The general form of the multiple regression model is:

$$Y = \beta_0 + \beta_1 X_1 + \beta_2 X_2 + \ldots + \beta_p X_p + \epsilon \qquad \text{(A1.28)}$$

where Y is the dependent variable, X_1 to X_p are independent variables, β_0 to β_p are parameters of the model and ϵ is the error term, as described for Equation A1.20. Data may consist of actual measurements (e.g. stream width)—or "dummy" variables can be included to represent presence/absence criteria or a qualitative "1 to 10" rating, as described by Holder (1985).

Equation A1.20 describes a straight line. In multiple regression, the equation describes a "hyperplane" of p dimensions, which becomes more difficult to visualize or plot when $p > 2$. Andrews (1972) presents a method of displaying the observations of a number of variables on a single graph. This allows visual comparison of the sets of observations and provides a concise summary of the data. Andrews curves are described in Section 8.6.4. Two-dimensional scatter plots can still be developed for evaluating residual patterns.

To obtain the parameters β_0 to β_p a number of observations are made of all the variables. This number must be at least as great as the number of parameters, $p + 1$, and preferably three to four times as large or more (Haan, 1977). For example, if one were trying to predict the density of Chironomid larvae in stream sediments from (1) the average size of the sediment grains, (2) percentage organic matter, (3) stream velocity and (4) depth, then at least five concurrent observations of each variable (and preferably 20 or more) would be needed to solve for the parameters β_0 to β_4.

This can be done using a least squares procedure which, in this case, minimizes the deviation of points from the "hyperplane". The procedure requires the manipulation of matrices containing the observations, and is best carried out with the assistance of commercial statistical software packages. A method of hand calculation is given by Riggs (1968). A large number of significant figures are carried throughout the computations to prevent large roundoff errors. However, in reporting results, the number of significant figures should be reduced to three or four.

The **multiple coefficient of determination** (r^2) obtained from a multiple regression analysis is the proportion of the total sum of squares (corrected for the mean) explained by the regression equation. This r^2 value also has a range of 0 to 1, as does its positive square root, the **multiple correlation coefficient** (r). The coefficient of efficiency can be calculated with Equation A1.27 and interpreted as for simple linear regression.

An unbiased estimate of the variance of the error terms is normally given in the computer output as the "expected mean square". The positive square root of this value is the **standard error** (SE) of the multiple regression equation.

The r^2 value will never decrease as more values are added; thus a high r^2 may be obtained simply by using a large number of independent variables of dubious value. In many studies, these variables are often intercorrelated. If two people always give the same advice then nothing is gained by listening to both of them. By the same token, there is no need to include two intercorrelated variables in an equation when one will do. The final choice of variables may be based on which are the easiest and least expensive to measure.

The r^2 value is not necessarily the best indicator of how well the regression equation fits the data. It is preferable to observe the effect of additional variables on SE, since the equation with the smallest SE will have the narrowest confidence intervals (Haan, 1977). If the addition of variables causes SE to increase, this should be an "alert" signal that these variables are not adding any more information to the equation and can be left out. The values of SE and r^2 can be plotted—graphically or mentally—and the number of variables selected by noting where the graph reaches a "point of diminishing returns", as shown in Figure A1.5.

One of the first steps in developing an equation is to look over the **correlation matrix**, a table giving the correlation coefficients (r values) between each variable and every other variable. A correlation matrix is normally available as an option in multiple regression programs. Variables should be selected which are the most correlated with the dependent variable and the least correlated with each other. The equation can then be modified by adding or subtracting variables by trial and error, assisted by judgement about which variables make the most practical sense.

A formal, systematic procedure is provided by "all possible regressions" and "stepwise regression" methods, which will be discussed. Statistical texts will yield additional approaches. With any approach, the results should always be tempered with knowledge about the phenomena of interest.

In **all-possible regressions**, regression equations are developed with every possible combination of the independent variables. If all regression equations are of the form of Equation A1.28 then 2^p regressions would need to be developed. This may be a simple matter if, for example, only three independent variables are considered, requiring only eight equations. A final equation would be selected based on the r^2, SE and relevance of the

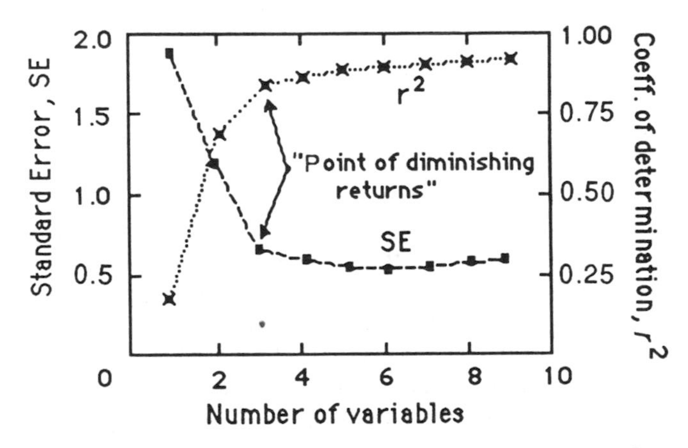

Figure A1.5. Illustration of the effect of additional variables on r^2 and SE in a multiple-regression equation

independent variables. As p increases, however, the number of possibilities increases exponentially; e.g. for $p = 10$, 1024 regressions would be needed. Thus, this method is not practical for a large number of variables.

Stepwise regression is one of the more common approaches. The regression equation is "built", one step—one variable—at a time. The variable added at each step is the one which explains the largest amount of the remaining unexplained variation (Haan, 1977). Thus, at the first step the variable added is the one with the highest correlation with the dependent variable. After each step, all variables which have been included are re-examined to see if they are still significant. Non-significant variables can be discarded. The procedure is repeated until all variables have been "tested" for possible contribution to the equation. The final equation should be composed of a "best" set of significant variables and the variables left out should be insignificant. To make this an objective procedure, the level of significance is set before the regression procedure is carried out. Statistical references should be consulted for methods of testing significance.

After an equation is selected, a scatter plot of the observed and predicted values should be examined to check for numerical errors, outliers, violations of assumptions about the error terms, and the appropriateness of the model. These points should ideally plot evenly about a 1:1 line. Residuals can also be plotted against Y or any of the X's or against time to examine trends.

In some cases, a transformation is needed to linearize a model so that a least squares approach can be taken. Scatter plots of the residuals against individual X variables can identify which variables need transformation. In some cases it may be possible to improve the model by including cross-product terms or ratios (e.g. X_1X_2 or X_1/X_2). Care should be taken to avoid introducing spurious correlation (Section A1.4.1) (for example, by including both the ratio X_1/X_2 and X_1 or X_2 in the relationship). Assumptions about the distribution of errors are made relative to the transformed scale; i.e. the errors in the transformed domain should be of constant variance, random, and normally distributed.

The resulting equation is empirical, i.e. not necessarily based on physical reality. In most cases the units on either side of the equation are different. A little "applied common sense" (Holder, 1985) is valuable when investigating whether the equation correctly mimics the real situation. Regression equations will also change slightly as more data are added. For example, snowmelt forecast equations are often "updated" each year after both winter snowpack and spring runoff measurements have been obtained.

In deriving a multiple regression equation, simplicity is the goal. The final equation should contain just a few meaningful variables which explain most of the variability in the data. They should all contribute to the regression unless there is some other logical or intuitive reason for retaining them. It is also helpful to evaluate the usefulness of the final equation with a "test set" of observed values which were not used in its development.

A1.4.4 AUTOCORRELATION AND INTERCORRELATION

Autocorrelation

A look at daily streamflow data plots (e.g. Figure 3.1) will reveal that runoff does not occur in a totally random fashion. High flows tend to follow other high flows and low flows also tend to group together. For stable streams, this will be more true than for flashy ones. Seasonal precipitation patterns, the carryover effects of soil moisture, and storage of water in reservoirs or in the channel, can all create this **persistence** in the record. Persistence actually has the effect of increasing the variability of the mean over the period of record, meaning that more years of data are required to accurately estimate the mean than if the data were truly random.

Autocorrelation or **serial correlation** refers to the correlation of data series with the same data series "shifted" by some time interval. Correlations are computed for a given **lag time**, k, to obtain an **autocorrelation coefficient**, $r(k)$. For example, the correlation between today's streamflow and yesterday's (or between this month's and last month's) would be a lag-1 autocorrelation coefficient, $r(1)$.

Figure A1.6(a) diagrammatically illustrates a lag-1 autocorrelation, with the arrows indicating the data "pairs" used in the correlation analysis. The number of pairs of observations decreases as k increases, as can be seen by comparing Figures A1.6(a) and A1.6(b). Autocorrelations can also be computed in a "circular" manner, where the shifted data are wrapped around to start over at the beginning, as in Figure A1.6(c). The correlation coefficient should only be computed for k much less than the number of items in the data series, n. The upper limit of k is usually taken as $n/4$. For a series of length 100, this would mean that only lag-1 through lag-25 autocorrelations would be performed.

The values of $r(k)$ can range from -1 (high values correlate with low values) to $+1$ (high values correlate with high values and low with low). The value of $r(0)$ will always be $+1$; i.e. an observation is perfectly correlated with itself. A value of $r(k)$ significantly different from zero means that part of the information contained in each observation is already known through its correlation with a preceding observation (Haan, 1977).

Lag-1 autocorrelation coefficients are usually of the most interest in hydrology and are considered a measure of persistence. For daily streamflows, a high positive value of $r(1)$ implies that streamflows are similar to those from the previous day. For annual streamflows, $r(1)$ will not be significantly different than zero for most streams. For monthly flows, however, it will nearly always be positive and significant (McMahon and Mein, 1986). Large, stable streams will have a high correlation between consecutive days and thus a high daily $r(1)$ as compared to flashy streams. Jones and Peters (1977), for example, used the average of the 1-, 2-, 3-, 4- and 5-day lag autocorrelation coefficients as a measure of the instability of the flow regime

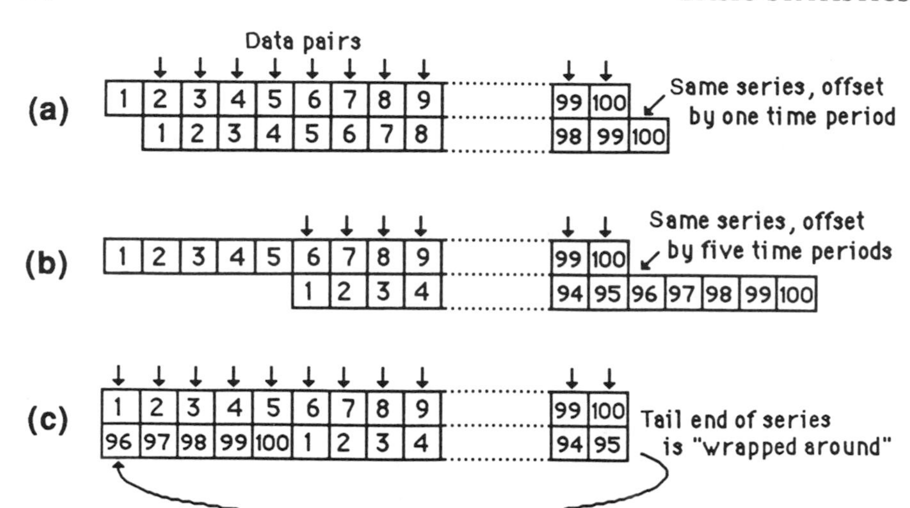

Figure A1.6. Diagrammatic illustration of (a) lag 1, (b) lag 5, and (c) "circular" lag 5 autocorrelations, with $n = 100$

in stream classification. Negative lag-1 coefficients are unusual and difficult to explain in hydrology, implying that a large flow is followed by a small one and vice versa.

The lag-k autocorrelation coefficient can be calculated from the equation (Wallis and O'Connell, 1972):

$$r(k) = \frac{\sum_{i=1}^{n-k} (x_i - \bar{x})(x_{i+k} - \bar{x})}{\sum_{i=1}^{n} (x_i - \bar{x})^2} \tag{A1.29}$$

Program AUTOCORR in AQUAPAK will compute $r(k)$ for a given lag, k, by either the "offset" or "circular" method (Figure A1.6).

If the autocorrelation coefficient is calculated for a range of k values, a graph can be constructed by plotting $r(k)$ against k. This graph is called a **correlogram**. Examples are shown in Figure A1.7. A data series which is entirely random will yield a graph similar to Figure A1.7(a), with the autocorrelation decreasing quickly from 1.0 to near-zero. Figure A1.7(b) represents a data series with a cyclical or periodic component. Figure A1.7(c) shows the correlogram for daily streamflow data from the Acheron River, $k = 0$ to $k = 730$ (2 years).

Assumptions required in computing a serial correlation coefficient are that observations are equally spaced in time and that the statistical properties do not change with time. Correlation methods are also based on the

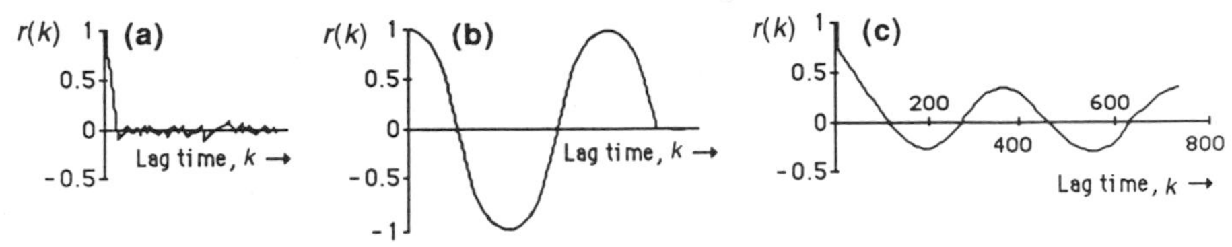

Figure A1.7. Correlograms showing the relationship between the lag-k autocorrelation coefficient, $r(k)$, and the lag period, k, for (a) a random data series, (b) a cyclical data series, and (c) daily streamflow data from the Acheron River at Taggerty (lag-0 to lag-730)

assumption that the data are independent and composed of random observations (Haan, 1977). This is paradoxical since, by definition, correlated data are not random. Therefore, statistical tests are used to evaluate whether the deviation of the coefficient from zero is too large to have occurred by chance. Raudkivi (1979) gives a simple approximation for the confidence limits on the autocorrelation coefficient as $\pm 2/\sqrt{n}$ (at the 0.05 significance level). If the computed value of $r(k)$ lies outside these limits, it is considered significantly different than zero, indicating persistence in the data. A more precise equation can be found in statistical texts.

Intercorrelation

Intercorrelation or **cross-correlation** was mentioned as a consideration in reducing the number of significant variables in a multiple regression equation. The implication is that if variables are intercorrelated, they do not add as much information as they would if they were totally independent. Thus, a few carefully selected independent variables provide as much information as several intercorrelated ones.

For the same reason, a few independent measurements of rainfall or streamflow or stream length from different sites will provide as much information about the "true" mean and variance as several intercorrelated measurements. Raudkivi (1979) gives the following equation for computing a regional mean intercorrelation coefficient ($\bar{r}$):

$$\bar{r} = \frac{2}{n(n-1)} \sum_{j=1}^{n-1} \sum_{i=j+1}^{n} r_{ij} \tag{A1.30}$$

where r_{ij} represents the correlation coefficient calculated between stations i and j and n is the total number of stations. Alexander (1954) suggests that the number of independent stations (n') required to give the same variance as some number n of intercorrelated stations is given by:

$$n' = \frac{n}{[1 + \bar{r}(n-1)]} \tag{A1.31}$$

where n is the number of stations having a mean intercorrelation coefficient of $\bar{r}$ and n' the number of equivalent independent stations which would provide as much information about the variance. It can be shown that as n becomes large, n' approaches $1/\bar{r}$. As an example, if 100 stations have an average correlation of 0.25, n' is 3.9. Thus, only about four independent stations would be needed to provide as much information as the original 100. A logical conclusion is that only a few independent stations should be installed rather than many correlated ones. However, "overlap" of stations is often desirable for other reasons such as the estimation of missing data.

Thus, data from the same station may be serially correlated and those between stations intercorrelated. Intercorrelation actually has the effect of *improving* the amount of information about the mean. For example, having many rainfall gauges in an area which report similar values strengthens the evidence that mean values are correct. Serial correlation, in comparison, tends to *decrease* the amount of information about the mean (Haan, 1977). For example, if streamflows occurred randomly about a mean value the mean daily flow for that year could probably be estimated after only a few months, but because of persistence in the data due to seasonality, we must wait until the end of the year.

A1.5 FURTHER INFORMATION ON STATISTICS

Statistical texts which can be consulted for more information include Devore (1982), Dunteman (1984), Elliot (1977), Green (1979), Haan (1977), Holder (1985), Sokal and Rohlf (1969), Steel and Torrie (1960), Tabachnick and Fidell (1989), Wardlaw (1985), Yevjevich (1972), and Zar (1974).

Commercial computer packages for statistical analysis include BMDP (Dixon, 1984), MINITAB (Ryan et al., 1985), SAS (SAS, 1987), SPSS (SPSS, 1986), and SYSTAT (Wilkinson, 1986). BMDP, MINITAB, SPSS, SAS, and SYSTAT are also available for both mainframe and IBM-compatible microcomputers. In addition, SYSTAT is also available for Macintosh microcomputers. Woodward et al. (1988) provides a guide to different statistical packages for the microcomputer. A text written by Kvanli (1988) describes the use of SPSS, SAS and MINITAB, and Afifi and Clark (1984) provide a reference on multivariate analysis using the packages BMDP, SAS, and SPSS-X. Green (1979) also lists several statistical packages used by environmental biologists, recommending SAS.

Appendix 2 About AQUAPAK

AQUAPAK is a package of stand-alone computer programs, written in FORTRAN, which supplement this text. A file editor of some type (e.g. word processor) is also needed for editing and/or creating data files. To run a program on the Macintosh, double "click" on its icon; on IBMs and compatibles, type the name of the program. All the programs are interactive to some degree (i.e. require answers typed in from the keyboard). For the Macintosh, the f77.rl file (run-time library) must be placed in the system folder. This file is the intellectual property of ABSOFT Corporation (Copyright © Absoft Corp 1988, Michigan, USA), and is distributed under licence.

A listing of the programs, their function, and the form of input data, is given overleaf in Table A2.1 (see also Section 3.2.2). A help program, 0HELP, is also included with the package; in addition, some programs offer "help" options which provide additional information.

Copies of this software can be obtained from the Centre for Environmental Applied Hydrology, Department of Civil and Agricultural Engineering, University of Melbourne, Parkville, Victoria 3052, Australia.

Suggestions for improvements to the software are welcome, and should be directed to the authors at the address given above.

Table A2.1. Listing of programs in supplementary software package, AQUAPAK

Program		Form of input data[a]		
Name	Description	From terminal	Std. data file (D=daily, M=monthly, Y=yr)	Free Format (≥ 1 column)
0HELP	Information on programs			
ANDREWS	Andrews curves			√
AUTOCORR	Lag-k autocorrelation		D,M, or Y	√
BACKWATR	Water surface profiles	√		
COLWELL	Colwell's indices		M	
CONVERT	Converts units (e.g. cfs to m^3/s)		D,M	
DMASS	Double-mass curve		M,Y	
FLOODFRQ	Flood-frequency analysis			√
FLOWDUR	Flow-duration curves		D,M	
FROUDE	Froude numbers	√		
HISTO	Histograms; river regime		M,Y	
INPUT	Data input into files (D,M,Y or 1 col.)	√		
KOPPEN	Koppen classification	√		
MANNING	Manning's equation	√		
PLOTFLOW	Plots streamflow time series		D	
QCALC	Discharge from velocity–area	√		
REGRESS	Simple regression	√	D,M, or Y	
REYNOLDS	Reynolds numbers	√		
ROSGEN	Rosgen's stream classification			
SPELL	Flow-spell analysis		D	
SPHERE	Particle sphericity	√		
STATS	Basic statistics	√	D,M, or Y	√
SUMUP	Produces monthly/yearly files		D,M	
TRANSFRM	Transformations (e.g. $\log_{10}$, $\log_2$)		D,M, or Y	
WEIBULL	Low-flow frequency		D	
XSECT	Cross-section data analysis	√		√

[a] Standard data files must be in a specified format, as described by the 0HELP program. Free-format files have one or more columns of data, and are created with a file editor. The first two lines of either standard or free-format files are reserved for information (e.g. station name, catchment area, dates). The following is a listing of the sample data files included; ZANDREWS.DAT and ZFLOOD.DAT are free-format files; ZXSECT.DAT was created with program XSECT.

ZACHERON.DA	Std daily data file
ZACHERON.MN	Std monthly data file
ZACHERON.YR	Std yearly data file
ZANDREWS.DAT	Sample file for ANDREWS
ZFLOOD.DAT	Sample file for FLOODFRQ
ZMARYPPT.YR	Yearly precipitation data
ZTEST.MN	Sample data file to use with ZACHERON.MN in DMASS and REGRESS
ZXSECT.DAT	Sample data file for XSECT

References

Ackers, P., White, W.R., Perkins, J.A. and Harrison, A.J.M. (1978). *Weirs and Flumes for Flow Measurement*, John Wiley, Chichester (as cited by Shaw, 1988).

Afifi, A.A. and Clark, V. (1984). *Computer-aided Multivariate Analysis*, Lifetime Learning Publications, Belmont, California.

Aitken, A.P. (1973). 'Assessing systematic errors in rainfall-runoff models', *J. of Hydrol.*, **20**, 121–36.

Alexander, G.N. (1954). 'Some aspects of time series in hydrology', *J. Inst. of Engineers* (Australia), September, 196 (as cited by Haan, 1977).

Allan, J.D. (1984). 'Hypothesis testing in ecological studies of aquatic insects', in *The Ecology of Aquatic Insects* (Eds V.H. Resh and D.M. Rosenberg), pp. 484–507, Praeger, New York.

Allen, T. (1981). *Particle Size Measurement*, 3rd edn, Chapman and Hall, New York (as cited by Klute, 1986).

Amoros, C., Roux, A.L., Reygrobellet, J.L., Bravard, J.P. and Pautou, G. (1987). 'A method for applied ecological studies of fluvial hydrosystems', *Regulated Rivers,* **1**, 17–36.

Andrews, D.F. (1972). 'Plots of high dimensional data', *Biometrics*, **28**, 125–36.

Andrews, E.D. (1983). 'Entrainment of gravel from naturally sorted riverbed material', *Geol. Soc. of Amer. Bull.*, **94**, 1225–31.

APHA (1985). *Standard Methods for the Examination of Water and Wastewater*, 16th edn, American Public Health Association, American Water Works Association, and Water Pollution Control Federation, Washington, DC.

Armour, C.L., Burnham, K.P. and Platts, W.S. (1983). 'Field methods and statistical analyses for monitoring small salmonid streams', *FWS/OBS-83/33*, US Fish and Wildlife Service, Washington, DC.

ASCE (1982). *Design of Water Intake Structures for Fish Protection*, American Society of Civil Engineers, New York.

AWRC (1984). 'Workshop on surface water resources data', *AWRC Conference Series no. 10*, Canberra, 23–25 November 1983, Australian Water Resources Council, Canberra.

Bagnold, R.A. (1966). 'An approach to the sediment transport problem from general physics', US Geol. Survey Prof. Paper 442–I, USGS, Washington, DC.

Bagnold, R.A. (1980). 'An empirical correlation of bedload transport rates in flumes and natural rivers', *Proc. Roy. Soc. London,* **A372**, 453–73.

Bain, M.B., Finn, J.T. and Booke, H.E. (1988). 'Streamflow regulation and fish community structure', *Ecology,* **69**, 382–92.

Baker, M.B. Jr (1984). 'Changes in streamflow in an herbicide-treated pinyon-juniper watershed in Arizona', *Water Resour. Res.*, **20**, 1639–42.

Baker, V.R. (1977). 'Stream-channel response to floods, with examples from central Texas', *Geol. Soc. of Amer. Bull,* **88**, 1057–71.

Barmuta, L.A. (1989). 'Habitat patchiness and macrobenthic community structure in an upland stream in temperate Victoria, Australia', *Freshwater Biol.,* **21**, 223–36.

Barnes, H.H., Jr (1967). 'Roughness characteristics of natural channels', US Geol. Survey Water-Supply Paper 1849, USGS, Washington, DC.

Barnes, H.H., Jr and Davidian, J. (1978). 'Indirect methods', in *Hydrometry: Principles and Practice* (Ed. R.W. Herschy), pp. 149–204, John Wiley, New York.

Barnes, R.K. and Mann, K.H. (1980). *Fundamentals of Aquatic Ecosystems*, Blackwell Scientific, Oxford.

Barnett, V. (1982). *Elements of Sampling Theory*, Hodder and Stoughton, London.

Barrett, E.C. and Curtis, L.F. (1982). *Introduction to Environmental Remote Sensing*, 2nd edn, Chapman and Hall, London.

Barton, J.R., Peters, E.J., White, D.A. and Winger, P.V. (1972). *Bibliography on the Physical Alteration of the Aquatic Habitat (Channelization) and Stream Improvement*, Brigham Young University Publications, Provo, Utah.

Bayly, I.A.E. and Williams, W.D. (1973). *Inland Waters and their Ecology*, Longman Australia, Victoria.

Beaumont, P. (1975). 'Hydrology', in *River Ecology* (Ed. B.A. Whitton), pp. 1–38, Blackwell Scientific, Oxford.

Bell, F.C., and Vorst, P.C. (1981). 'Geomorphic parameters of representative basins and their hydrologic significance', Australian Water Resources Council Technical Paper No. 58, Australian Government Publishing Service, Canberra.

Bencala, K.E. (1984). 'Interactions of solutes and streambed sediment. 2. A dynamic analysis of coupled hydrologic and chemical processes that determine solute transport.' *Water Resour. Res.,* **20**, 1804–14.

Bencala, K.E., Kennedy, V.C., Zellweger, G.W., Jackman, A.P. and Avanzino, R.J. (1984). 'Interactions of solutes and streambed sediment. An experimental analysis of cation and anion transport in a mountain stream', *Water Resour. Res.,* **20**, 1797–1803.

Benson, M.A. (1965). 'Spurious correlation in hydraulics and hydrology', *J. Hyd. Div., Proc. ASCE,* **HY4**, 35–42.

Benson, M.A. (1968). 'Uniform flood-frequency estimating methods for federal agencies', *Water Resour. Res.,* **4**, 891–908.

Beschta, R.L. (1983). 'Channel changes following storm-induced hillslope erosion in the Upper Kowai Basin', Torlesse Range, New Zealand, *J. of Hydrol. (NZ),* **22**, 93–111.

Beschta, R.L. and Platts, W.S. (1986). 'Morphological features of small streams: significance and function', *Water Resour. Bull.,* **22**, 369–79.

Best, G.A., and Ross, S.L. (1977). *River Pollution Studies*, Liverpool University Press, Liverpool.

Beven, K.J., and Callen, J.L. (1979). 'HYDRODAT—a system of FORTRAN computer programs for the preparation and analysis of hydrological data from charts', *Technical Bulletin No. 23*, British Geomorphological Research Group, Geo Abstracts Ltd, Norwich.

Bhowmik, N.G. and Adams, J.R. (1986). 'The hydrologic environment of Pool 19 of the Mississippi River', *Hydrobiologia,* **136**, 21–30.

Binder, R.C. (1958). *Advanced Fluid Mechanics*, Prentice-Hall, Englewood Cliffs, New Jersey.

Binns, N.A. and Eiserman, F.M. (1979). 'Quantification of fluvial trout habitat in Wyoming'; *Trans. Amer. Fish. Soc.,* **108**, 215–28 (as cited by Wesche and Reschard, 1980).

Birnbaum, Z.W. (1952). 'Numerical tabulation of the distribution of Kolmogorov's

statistic of finite sample size', *J. of the American Statistical Assn,* **47**, 425–41.
Blackman, D.R. (1969). *SI Units in Engineering*, Macmillan, Melbourne.
Blake, R.W. (1983). *Fish Locomotion*, Cambridge University Press, London.
Bleed, A.S. (1987). 'Limitations of concepts used to determine instream flow requirements for habitat maintenance', *Water Resour. Bull.,* **23**, 1173–8.
Blyth, K. and Rodda, J.C. (1973). 'A stream length study', *Water Resour. Res.,* **9**, 1454–61.
Bohn, C. (1986). 'Biological importance of streambank stability', *Rangelands,* **8**, 55–6.
Bolemon, J. (1989). *Physics: an Introduction,* 2nd edn, Prentice-Hall, Englewood Cliffs, New Jersey.
Bone, Q. and Marshall, N.B. (1982). *Biology of Fishes*, Blackie, Glasgow.
Bovee, K.D. (1974). 'The determination, assessment, and design of "instream value" studies for the Northern Great Plains region'; *No. Great Plains Res. Prog.* (as cited by Stalnaker and Arnette, 1976).
Bovee, K.D. (1978). 'Probability-of-use criteria for the Family Salmonidae', Instream Flow Information Paper No. 4, *FWS/OBS-78/07*, US Fish and Wildlife Service.
Bovee, K.D. (1982). 'A guide to stream habitat analysis using the Instream Flow Incremental Methodology', Instream Flow Information Paper 12, *FWS/OBS-82/26*, Co-operative Instream Flow Group, US Fish and Wildlife Service, Office of Biological Services.
Bovee, K.D. (1986). 'Development and evaluation of habitat suitability criteria for use in the Instream Flow Incremental Methodology', Instream Flow Information Paper 21, *US Fish Wildl. Serv. Biol. Rep. 86*, US Fish and Wildlife Service.
Bovee, K.D. and Milhous, R. (1978). 'Hydraulic simulation in instream flow studies: theory and techniques', Instream Flow Information Paper 5, *FWS/OBS-78/33*, US Fish and Wildlife Service, Office of Biological Services.
Boyer, M.C. (1964). 'Streamflow measurement', in *Handbook of Applied Hydrology* (Ed. V.T. Chow), Section 15, McGraw-Hill, New York.
Brakensiek, D.L., Osborn, H.B. and Rawls, W.J. (co-ordinators) (1979). *Field Manual for Research in Agricultural Hydrology*, Agriculture Handbook 224, USDA.
Branson, F.A., Gifford, G.F., Renard, K.G. and Hadley, F.F. (1981). *Rangeland Hydrology*, 2nd edn, Kendall/Hunt, Iowa.
Brater, E.F. and King, H.W. (1976). *Handbook of Hydraulics*, McGraw-Hill, New York.
Bren, L.J., O'Neill, I. and Gibbs, N.L. (1988). 'Use of map analysis to elucidate flooding in an Australian riparian river red gum forest', *Water Resour. Res.,* **24**, 1152–62.
Brewer, R. (1964). *Fabric and Mineral Analysis of Soils*, John Wiley, New York.
Briggs, D. (1977a). *Sources and Methods in Geography: Sediments*, Butterworths, London.
Briggs, D. (1977b). *Sources and Methods in Geography: Soils*, Butterworths, London.
Brinker, R.C. and Wolf, P.R. (1977). *Elementary Surveying*, 6th edn, Harper and Row, New York.
Brizga, S.O. and Finlayson, B.L. (1990). 'Channel avulsion and river metamorphosis: the case of the Thompson River, Victoria, Australia', *Earth Surf. Proc. and Landforms,* **15**, 391–404.
Brookes, A. (1989). 'Alterative channelization procedures'; in *Alternatives in Regulated River Management* (Eds J.A. Gore and G.E. Petts), pp. 139–62, CRC Press, Boca Raton, Florida.
Brookes, A. and Gregory, K. (1988). 'Channelization, river engineering and geomorphology', in *Geomorphology in Environmental Planning* (Ed. J.M. Hooke), pp. 145–67, John Wiley, Chichester.

Brown, A.L. (1971). *Ecology of Fresh Water*, Heinemann Educational, London.
Brussock, P.P., Brown, A.V. and Dixon, J.C. (1985). 'Channel form and stream ecosystem models', *Water Resour. Bull.*, **21**, 859–66.
Brusven, M.A. (1977). 'Effects of sediments on insects', in *Transport of granitic sediments in streams and its effects on insects and fish, Wildlife and Range Expt. Sta. Bull. 17D* (Ed. L. Kibbee), p. 43, USDA Forest Service, Univ. Idaho, Moscow, Idaho (as cited by Bovee, 1982).
BS (1973). Methods of measurement of liquid flow in open channels, BS 3680, British Standards Institution, London (publication dates vary as updates are common).
BS (1975). Methods of testing soils for civil engineering purposes, BS 1377, British Standards Institution, London.
Budyko, M.I. (1980). *Global Ecology*, Progress Publishers, Moscow.
Bunn, S.C. and Boughton, W. (1990). 'Ecological significance of streamflow patterns in some Australian rivers', *Verh. Int. Verein. Limnol.*, **24**, Proc. of SIL Congress, Munich, August 1989, (in press).
Cadwallader, P.L. (1986). 'Flow regulation in the Murray River System and its effect on the native fish fauna', in *Stream Protection: The Management of Rivers for Instream Uses* (Ed. I.C. Campbell), pp. 115–34, Water Studies Centre, Chisholm Institute of Technology, East Caulfield, Australia.
Cairns, J. (Ed.) (1988). *Rehabilitating Damaged Ecosystems*, CRC Press, Boca Raton, Florida.
Campbell, I.C., James, K.R. and Edwards, R.T. (1990). 'Farming and streams—impact, research, and management', in *Proc. of the Conference, The State of Our Rivers*, Aust. Nat. Univ., September 1989.
Carpenter, K.E. (1928). *Life in Inland Waters*, Sidgwich and Jackson, London (as cited by Hawkes, 1975).
Carson, M.A. and Griffiths, G.A. (1987). 'Bedload transport in gravel channels', *J. of Hydrol. (NZ)*, **26**, 1–151.
Carter, D.J. (1986). *The Remote Sensing Sourcebook: A guide to remote sensing products, services, facilities, publications and other materials*, McCarta Ltd, London.
Cattell, R.B. (1965). 'Factor analysis: an introduction to essentials, 1. The purpose and underlying models', *Biometrics*, **21**, 190–215.
Chapman, D., Codrington, S., Blong, R., Dragovich, D., Smith, T.L., Linacre, E., Riley, S., Short, A., Spriggs, J. and Watson, I. (1985). *Understanding Our Earth*, Pitman Publishing, Victoria.
Chatfield, C. and Collins, A.J. (1980). *Introduction to Multivariate Analysis*, Chapman and Hall, London.
Cheslak, E., and Carpenter, J. (1990). 'Compilation report on the effects of reservoir releases on downstream ecosystems', *Rept. No. REC-ERC-90-1*, US Bureau of Reclamation, Denver, Colorado.
Ching, F.S. (1967). *Basic Sampling Methods*, University of Singapore.
Chow, V.T. (1959). *Open-Channel Hydraulics*, McGraw-Hill, New York.
Chow, V.T. (Ed.) (1964a). *Handbook of Applied Hydrology*, McGraw-Hill, New York.
Chow, V.T. (1964b). 'Statistical and probability analysis of hydrologic data. Part I. Frequency analysis', in *Handbook of Applied Hydrology* (Ed. V.T. Chow), pp. 8-1 to 8-42, McGraw-Hill, New York.
Chow, V.T., Maidment, D.R. and Mays, L.W. (1988). *Applied Hydrology*, McGraw-Hill, New York.
Church, M.A. (1974). *Electrochemical and Fluorometric Tracer Techniques for Streamflow Measurements*, Geo. Abstracts Ltd, for the British Geomorphic Research Group.

Church, M.A. (1978). 'Palaeohydrological reconstructions from a Holocene valley fill', in *Fluvial Sedimentology* (Ed. A.D. Miall), pp. 743–72, Can. Soc. Petroleum Geologists Memoir 5, Calgary, Alberta.

Church, M.A., McLean, D.G. and Wolcott, J.F. (1987). 'River bed gravels: sampling and analysis', in *Sediment Transport in Gravel-bed Rivers* (Eds C.R. Thorne, J.C. Bathurst and R.D. Hey), pp. 43–88, John Wiley, Chichester.

Clark, J.W., Viessman, W. Jr and Hammer, M.J. (1977). *Water Supply and Pollution Control*, IEP, New York.

Clay, C.H. (1961). *Design of Fishways and other Fish Facilities*, Dept of Fisheries of Canada, Queen's Printer, Ottawa, Canada.

Cochnauer, T. (1976). 'Instream flow techniques for large rivers', in *Instream Flow Needs* (Eds J.F. Orsborn and C. Allman), pp. 387–99, Vol. II, Amer. Fish. Soc.

Cochran, W.G. (1977). *Sampling Techniques*, 3rd edn, John Wiley, New York.

Cochran, W.G. and Cox, G.M. (1957). *Experimental Designs*, 2nd edn, John Wiley, New York.

Coehn, J. (1988). *Statistical Power Analysis for the Behavioral Sciences*, 2nd edn, Lawrence Erlbaum, Hillsdale, New Jersey.

Coleman, N.L. and Alonso, C.V. (1983). 'Two-dimensional channel flows over rough surfaces', *J. Hyd. Eng.,* **109**, 175–88.

Collings, M. (1972). 'A methodology for determining instream flow requirements for fish', in *Proc. of Instream Flow Methodology Workshop*, pp. 72–86, Washington Dept of Ecology, Olympia (as cited by Wesche and Rechard, 1980).

Colwell, R.K. (1974). 'Predictability, constancy, and contingency of periodic phenomena', *Ecology,* **55**, 1148–53.

Colwell, R.N. (Ed.) (1983). *Manual on Remote Sensing*, 2nd edn, American Society of Photogrammetry, Falls Church, Virginia.

Cook, B.G., Laut, P., Austin, M.P., Body, D.N., Faith, D.P., Goodspeed, M.J. and Srikanthan, R. (1988). 'Landscape and rainfall indices for prediction of streamflow similarities in the Hunter Valley, Australia', *Water Resour. Res.,* **24**, 1283–98.

Corbett, D.M. (1962). 'Stream-gaging procedure', US Geol. Survey Water-Supply Paper 888, USGS, Washington, DC.

Costa, J.E. (1987). 'Interpretation of the largest rainfall-runoff floods measured by indirect methods on small drainage basins in the conterminous United States', *Proc. of the US–China Bilateral Symp. on the Analysis of Extraordinary Flood Events*, October 1985, Nanjing, China (as cited by Jarrett and Malde, 1987).

Cowan, W.L. (1956). 'Estimating hydraulic roughness coefficients, *Agricultural Engineering,* **37**, 473–5.

Cowardin, L.M. (1982). 'Wetlands and deepwater habitats: a new classification', *J. Soil and Water Conserv.,* **37**, 83–5.

Cracknell, A. and Hayes, L. (Eds). (1988). *Remote Sensing Yearbook 1988/89*, Taylor and Francis, London.

Craig, R.F. (1983). *Soil Mechanics*, Van Nostrand Reinhold, Wokingham.

Craig, J.F. and J.B. Kemper (Eds) (1980). *Regulated Rivers: Advances in Ecology*, Plenum Press, New York.

Csermak, B. and Rakoczi, L. (1987). 'Erosion and sedimentation', in *Applied Surface Hydrology* (Ed. O. Starosolszky), pp. 760–807, Water Resour. Publ., PO Box 2841, Littleton, Colorado.

CSU (1977). *Lecture Notes on River Mechanics*, Colorado State University, Fort Collins, Colorado.

Cummins, K.W. (1986). 'Riparian influence on stream ecosystems', in *Stream Protection: The Management of Rivers for Instream Uses* (Ed. I.C. Campbell), pp. 45–55, Water Studies Centre, Chisholm Institute of Technology, East Caulfield, Australia.

Cunnane, C. (1978). 'Unbiased plotting positions—a review', *J. of Hydrol.*, **37**, 205–22.

Curran, P.J. (1985). *Principles of Remote Sensing*, Longman Scientific and Technical, Harlow.

Dackombe, R. and Gardiner, V. (1983). *Geomorphological Field Manual*, George Allen & Unwin, London.

Dahm, C.N., Sedell, J.R. and Triska, F.J. (1989). 'A historical look at streams and rivers in North America', Technical Information Workshop, Stream Rehabilitation and Restoration, 37th Annual No. Am. Benth. Soc. Meeting, April 1989.

Daily, J.W. and Harleman, D.R.F. (1966). *Fluid Dynamics*, Addison-Wesley, Reading, Massachusetts.

Dalrymple, T. (1960). 'Flood-frequency analyses', *Manual of Hydrology:* Part 3. *Flood-Flow Techniques*, US Geol. Survey Water-Supply Paper 1543-A, USGS, Washington, DC.

Dalrymple, T. and Benson, M.A. (1967). 'Measurement of peak discharge by the slope–area method', *US Geol. Survey Techniques of Water-Resources Investigations*, Book 3, Chap. A2, USGS, Washington, DC.

Davidian, J. (1984). 'Computation of water-surface profiles in open channels', *US Geol. Survey Techniques of Water-Resources Investigations*, Book 3, Chapter A15, USGS, Washington, DC.

Davies, P.E., Sloane, R.D. and Andrew, J. (1988). 'Effects of hydrological change and the cessation of stocking on a stream population of *Salmo trutta L.*', *Aust. J. Mar. Freshwater Res.*, **39**, 337–54.

Davis, J.A. (1986). 'Boundary layers, flow microenvironments and stream benthos', in *Limnology in Australia* (Eds P. De Deckker and W.D. Williams), pp. 293–312, CSIRO Australia, Melbourne, Victoria.

Davis, J.A. and Barmuta, L.A. (1989). 'An ecologically useful classification of mean and near-bed flows in streams and rivers', *Freshwater Biol.*, **21**, 271–82.

Davis, W.M. (1899). 'The geographical cycle', *Geographical J.*, **14**, 481–504.

Dawdy, D.A. and Vanoni, V.A. (1986). 'Modeling alluvial channels', *Water Resour. Res.*, **22**, 71S–81S.

Dawson, F.H. and Charlton, F.G. (1988). 'Bibliography on the hydraulic resistance or roughness of vegetated watercourses', *Occ. Publ. No. 25*, Freshwater Biol. Assoc., Ambleside, Cumbria.

Day, P.R. (1965). 'Particle fractionation and particle-size analysis', in *Method of Soil Analysis*, Part I (Eds C.A.B. et al.), pp. 545–67, Agronomy 9, Amer. Soc. of Agronomy, Soil Sci. Soc. of America, Madison, Wisconsin.

DeDeckker, P. and Williams, W.D. (1986). *Limnology in Australia*, CSIRO Australia, Melbourne, Victoria.

Denny, M.W. (1988). *Biology and the Mechanics of the Wave-Swept Environment*, Princeton University Press, Princeton, New Jersey.

Denny, M.W., Daniel, T.L. and Koehl, M.A.R. (1985). 'Mechanical limits to size in wave-swept organisms', *Ecological Monographs,* **55**, 69–102.

Devore, J.L. (1982). *Probability and Statistics for Engineering and the Sciences*, Brooks/Cole, Monterey, California.

De Wiest, R.J.M. (1965). *Geohydrology*, New York (as cited by Gregory and Walling, 1973).

Dingman, S.L. (1978). 'Synthesis of flow duration curves for unregulated streams in New Hampshire', *Water Resour. Bull.*, **14**, 1481–1502.

Dingman, S.L. (1984). *Fluvial Hydrology*, W.H. Freeman, New York.

Dixon, W.J. (Ed.) (1985). *BMDP Statistical Software: 1985 Printing*, University of California Press, Berkeley.

Douglas, J.F., Gasiorek, J.M. and Swaffield, J.A. (1983). *Fluid Mechanics*, Pitman, London.

Drummond, R.R. (1974). 'When is a stream a stream?' *Prof. Geographer,* **26**, 34–7.

Dunne, T. and Leopold, L.B. (1978). *Water in Environmental Planning*, W.H. Freeman, San Francisco.

Dunteman, G.H. (1984). *Introduction to Multivariate Analysis*, Sage Publications, Beverly Hills, California.

Elliot, J.M. (1977). *Some Methods for the Statistical Analysis of Samples of Benthic Invertebrates*, Sci. Publ. No. 25, Freshwater Biological Assn, Ferry House, UK.

Elmore, W. and Beschta, R.L. (1987). 'Riparian areas: perceptions in management', *Rangelands,* **9**, 260–65.

Emmett, W.W. (1975). 'The channels and waters of the Upper Salmon River area, Idaho', US Geol. Survey Prof. Paper 870-A, USGS, Washington, DC (as cited by Dunne and Leopold, 1978).

Emmett, W.W. (1977). 'Measurement of bed load in rivers', IAHS Publ. No. 133, *Erosion and Sediment Transport Measurement*, Symposium, Florence (as cited by Csermak and Rakoczi, 1987).

Emmett, W.W., Burrows, R.L. and Chacho, E.F. Jr (1989). 'Gravel transport in a gravel-bed river, Alaska', paper presented at spring meeting of Am. Geophys. Union, May 1989, Baltimore, Maryland.

Everest, F.H., McLemone, C.E. and Ward, J.V. (1980). 'An improved tri-tube cryogenic gravel sampler', *Res. Note PNW-350*, USDA Forest Service, Pac. NW Forest and Range Expt. Sta., Portland, Oregon (as cited by Hamilton and Bergersen, 1984).

Everitt, B.L. (1968). 'Use of cottonwood in the investigation of the recent history of a flood plain', *Am. J. Sci.,* **206**, 417–39 (as cited by Schumm, 1977).

Faith, D.P., Minchin, P.R. and Belbin, L. (1987). 'Compositional dissimilarity as a robust measure of ecological distance', *Vegetation,* **69**, 57–68.

Fenner, P., Ward, W.B. and Patton, D.R. (1985). 'Effects of regulated water flows on regeneration of Fremont cottonwood', *J. of Range Management,* **38**, 135–8.

Finlayson, B. (1979). 'Electrical conductivity: a useful technique in teaching geomorphology', *Journal of Geography in Higher Education,* **3**, 68–87.

Finlayson, B. (1981). 'The analysis of stream suspended loads as a geomorphological teaching exercise', *Journal of Geography in Higher Education,* **5**, 23–35.

Finlayson, B. and Statham, I. (1980). *Hillslope Analysis*, Butterworths, London.

Folk, R.L. (1980). *Petrology of Sedimentary Rocks*, Hemphill Publishing, Austin, Texas.

Fontaine, T.D. III and Bartell, S.M. (Eds) (1983). *Dynamics of Lotic Ecosystems*, Ann Arbor Science, Michigan.

Fortner, S.L. and White, D.S. (1988). 'Interstitial water patterns: a factor influencing the distribution of some lotic aquatic vascular macrophytes', *Aquatic Botany,* **31**, 1–12.

Foth, H.D. (1978). *Fundamentals of Soil Science*, John Wiley, New York.

Fresenius, W., Quentin, K.E. and Schneider, W. (Eds) (1987). *Water Analysis*, Springer-Verlag, Berlin.

Fritschen, J.A., Milhous, R.T. and Nestler, J. (1984). 'Measuring resource potential for river recreation', in *Proceedings of the National River Recreation Symposium* (Eds J.S. Popadic, D.I. Butterfield, D.H. Anderson, and M.R. Popadic), pp. 484–94, Baton Rouge, Louisiana.

Gan, K.C. and McMahon, T.A. (1990a). *Comparison of Two Computer Models for Assessing Environmental Flow Requirements*, Report produced for the Department of Water Resources, Victoria, Centre for Environmental Applied Hydrology, University of Melbourne, Australia.

Gan, K.C. and McMahon, T.A. (1990b). 'Variability of results from the use of PHABSIM in estimating habitat area', *Regulated Rivers,* **15** (in press).

Gan, K.C., McMahon, T.A. and Finlayson, B.L. (1989). 'Fractal dimensions and lengths of rivers in Southeast Australia', Unpublished report, Centre for Environmental Applied Hydrology, University of Melbourne, Australia.

Gan, K.C., McMahon, T.A. and Finlayson, B.L. (1990a). 'Analysis of periodicity in streamflow and rainfall data by Colwell's indices', *J. of Hydrol.* (in press).

Gan, K.C., McMahon, T.A. and O'Neill, I.C. (1990b). 'Errors in estimated streamflow parameters and storages for ungauged catchments', *Water Resour. Bull.,* **26**, 443–50.

Gan, K.C., McMahon, T.A. and O'Neill, I.C. (1991). 'Transposition of monthly streamflow data to ungauged catchments', *J. of Hydrol.* (in press).

Gauch, H.G. Jr and Whittaker, R.H. (1981). 'Hierarchical classification of community data', *J. of Ecology,* **69**, 537–57.

Gebhardt, K.A., Bohn, C., Jensen, S. and Platts, W.S. (1989). 'Use of hydrology in riparian classification', Riparian Resource Management Workshop, 8–11 May 1989, pp. 53–9, Billings, Montana.

Gebhardt, K.A., Leonard, S., Staidl, G. and Prichard, D. (1990). 'Riparian and wetland classification review', *Tech. Ref. 1737-5*, US Dept of the Interior, Bur. of Land Mgmt, Denver, Colorado.

Gee, G.W. and Bauder, J.W. (1986). 'Particle-size analysis', in *Methods of Soil Analysis, Part I. Physical and Mineralogical Methods—Agronomy Monograph no. 9* (2nd edn). (Ed. A. Klute), pp. 383–411, Amer. Soc. of Agronomy, Soil Sci. Soc. of America, Madison, Wisconsin.

Gerhart, P.M. and Gross, R.J. (1985). *Fundamentals of Fluid Mechanics*, Addison-Wesley, Reading, Massachusetts.

Gerrard, A.J. (1981). *Soils and landforms: an integration of geomorphology and pedology*, George Allen & Unwin, London.

Gippel, C.J. (1989). *The use of turbidity instruments to measure stream water suspended sediment concentration*, Monograph Series No. 4, Dept of Geography and Oceanography, University College, Australian Defence Force Academy, Canberra.

Gleick, J. (1988). *Chaos: Making a New Science*, Viking, New York.

Goldman, C.R., and Horne, A.J. (1983). *Limnology*, McGraw-Hill, New York.

Goldstein, R. (1989). 'Power and sample size via MS/PC-DOS computers'; *Am. Stat.,* **43**, 253–60.

Golterman, H.L., Clymo, R.S. and Ohnstad, M.A.M. (1978). *Methods for Physical and Chemical Analysis of Fresh Waters*, Blackwell Scientific, Oxford.

Gordon, A.D. (1981). *Classification*, Chapman and Hall, London.

Gore, J.A. (1978). 'A technique for predicting in-stream flow requirements of benthic macroinvertebrates', *Freshwater Biol.,* **8**, 141–51.

Gore, J.A. (1985). 'Mechanisms of colonization and habitat enhancement for benthic macroinvertebrates in restored river channels', in *The Restoration of Rivers and Streams* (Ed. J.A. Gore), pp. 81–101, Butterworths, Boston, Massachusetts.

Gore, J.A. (Ed.) (1985). *The Restoration of Rivers and Streams*, Butterworths, Boston, Massachusetts.

Gore, J.A. and Bryant, F.L. (1988). 'River and stream restoration', in *Rehabilitating Damaged Ecosystems*, Vol. I (Ed. J. Cairns), pp. 23–38, CRC Press, Boca Raton, Florida.

Gore, J.A. and Judy, R.D. Jr (1981). 'Predictive models of benthic macroinvertebrate density for use in instream flow studies and regulated flow management', *Can. J. Fish. Aquat. Sci.,* **38**, 1363–70.

Gore, J.A. and Nestler, J.M. (1988). 'Instream flow studies in perspective', *Regulated Rivers,* **2**, 93–101.

Gore, J.A., Nestler, J.M. and Layzer, J.B. (1989). 'Instream flow predictions and management options for biota affected by peaking-power hydroelectric operations',

Regulated Rivers, **3**, 35–48.

Gore, J.A., and Petts, G.E. (Eds) (1989). *Alternatives in Regulated River Management*, CRC Press, Boca Raton, Florida.

Goudie, A. (1981). *Geomorphological Techniques*, George Allen & Unwin, London, edited for the British Geomorphological Research Group.

Graf, W.H. (1971). *Hydraulics of Sediment Transport*, McGraw-Hill, New York.

Gray, L.J. (1981). 'Species composition and life histories of aquatic insects in a lowland sonoran desert stream', *The American Midland Naturalist,* **106**, 229–42.

Grayson, R.B., Barling, R.D. and Ogleby, C.L. (1988). 'A comparison between photogrammetry and a surface profile meter for the determination of surface topography for micro-erosion measurement', Conf. on Ag. Engineering, Hawkesbury, 25–30 September 1988, *I.E. Aust. Natl Conf. Publication No. 88/12*, pp. 264–8.

Green, D.C. (1974). *The stability of the banks of the Mitta Mitta River during fluctuating discharges due to operation of Dartmouth power station*, Report No. H.I.1., State Electricity Comm. of Victoria.

Green, R.H. (1979). *Sampling Design and Statistical Methods for Environmental Biologists*, Wiley-Interscience, New York.

Gregory, K.J. (1976). 'The determination of river channel capacity', *New England Research Series in Applied Geography*, No. 42, University of New England, Armidale, New South Wales, Australia.

Gregory, K.J. (Ed.) (1983). *Background to Palaeohydrology: A Perspective*, John Wiley, Chichester.

Gregory, K.J. and Madew, J.R. (1982). 'Land use change, flood frequency and channel adjustments', in *Gravel-Bed Rivers* (Eds R.D. Hey, J.C. Bathurst and C.R. Thorne), pp. 757–81, John Wiley, Chichester.

Gregory, K.J. and Walling, D.E. (1973). *Drainage Basin Form and Process*, Edward Arnold, London.

Griffiths, J.C. (1967). *Scientific Method in Analysis of Sediments*, McGraw-Hill, New York (as cited by Goudie, 1981).

Gumbel, E.J. (1963). 'Statistical forecast of droughts', *Bull. IASH,* **1**, 5–23 (as cited by Shaw, 1988).

Guy, H.P. (1969). 'Laboratory theory and methods for sediment analysis', *Techniques of Water-Resources Investigations of the USGS*, Book 5, Chapter C1, USGS, Washington, DC.

Haan, C.T. (1977). *Statistical Methods in Hydrology*, The Iowa State University Press, Ames, Iowa.

Haan, C.T. and Read, H.R. (1970). 'Prediction of monthly, seasonal and annual runoff volumes for small agricultural watersheds in Kentucky', *Bull. 711*, Kentucky Ag. Expt. Sta., Univ. of Kentucky, Lexington, Kentucky (as cited by Haan, 1977).

Hadley, R.F. and Schumm, S.A. (1961). 'Sediment sources and drainage basin characteristics in upper Cheyenne River basin', US Geol. Surv. Water-Supply Paper 1531-B, USGS, Washington, DC.

Hagen, V.K. (1989). 'Flood flow frequency, PC application of Bulletin 17B', paper presented at IAHS Conf., 10–19 May 1989, Baltimore, Maryland.

Haines, A.T., Finlayson, B.L. and McMahon, T.A. (1988). 'A global classification of river regimes', *Applied Geography,* **8**, 255–72.

Hamilton, K. and Bergersen, E.P. (1984). *Methods to Estimate Aquatic Habitat Variables*, prepared by Colorado Cooperative Fishery Research Unit, Colorado State University, for the Bureau of Reclamation, Denver, Colorado.

Harrison, S.S. and Reid, J.R. (1967). 'A flood-frequency graph based on tree-scar data', *No. Dakota Academy of Science Annual Proc.,* 23–33.

Hart, B.T., Bailey, P., Edwards, R., Hortle, K., James, K., McMahon, A., Meredith, C. and Swadling, K. (1990a). 'A review of the salt sensitivity of the Australian freshwater biota', *Hydrobiologia* (in press).
Hart, B.T., Bailey, P., Edwards, R., Hortle, K., James, K., McMahon, A., Meredith, C. and Swadling, K. (1990b). 'Effects of salinity on river, stream and wetland ecosystems in Victoria, Australia', *Wat. Res.* (in press).
Harvey, A.M. (1969). 'Channel capacity and the adjustment of streams to hydrologic regime', *J. of Hydrol.,* **8**, 82–98.
Hasfurther, V.R. (1985). 'The use of meander parameters in restoring hydrologic balance to reclaimed stream beds', in *The Restoration of Rivers and Streams* (Ed. J.A. Gore), pp. 21–40, Butterworths, Boston, Massachusetts.
Hawkes, H.A. (1975). 'River zonation and classification', in *River Ecology* (Ed. B.A. Whitton), pp. 312–74, Blackwell Scientific, Oxford.
Hawkins, R.H. (1975). 'Acoustical energy output from mountain stream channels', *J. Hyd. Div., Proc. ASCE,* **101**, 571–5.
HEC (1982). *HEC-2 Water Surface Profiles Users Manual*, Hydrologic Engineering Center, US Army Corps of Engineers.
Helley, E.J. and Smith, W. (1971). 'Development and calibration of a pressure-difference bedload sampler', *US Geol. Survey Open-File Report*, USGS, Washington, DC.
Henderson, F.M. (1966). *Open Channel Flow*, Macmillan, New York.
Herschy, R.W. (Ed.) (1978). *Hydrometry, Principles and Practices*, John Wiley, Chichester.
Herschy, R.W. (1985). *Streamflow Measurement*, Elsevier Applied Science, London.
Higgins, A.L. (1965). *Elementary Surveying*, Longmans, Green and Co., London.
Hino, M., Nadaoka, K., Hironaga, K. and Muramoto, T. (1986a). 'Development of beam-scan type laser-doppler anemometry for measurement of nearly-instantaneous and continuous velocity profiles', in *Flow and Waves, No. 1, Report of the Hydraulic Engineering Laboratory*, pp. 3–6, Tokyo Institute of Technology.
Hino, M., Nadaoka, K., Kobayashi, T., Hironaga, K. and Muramoto, T. (1986b). 'Flow structure measurement by the beam scan type LDV. Fluid Dynamics Research 1, in *Flow and Waves, No. 1, Report of the Hydraulic Engineering Laboratory*, pp. 177–90, Tokyo Institute of Technology.
Hjulstrom, F. (1939). 'Transportation of detritus by moving water', in *Recent Marine Sediments, a Symposium* (Ed. P.D. Trask), pp. 5–31, American Assn of Petroleum Geologists, Tulsa, Oklahoma.
Hogan, D.L. and Church, M. (1989). 'Hydraulic geometry in small, coastal streams: progress toward quantification of salmonid habitat', *Can. J. Fish. Aquat. Sci.,* **46**, 844–52.
Holder, R.L. (1985). *Multiple Regression in Hydrology*, Institute of Hydrology, Wallingford.
Holeman, J.N. (1968). 'The sediment yield of major rivers of the world', *Water Resour. Res.,* **4**, 737–47.
Hooke, J.M. (1986). 'The significance of mid-channel bars in an active meandering river', *Sedimentology,* **33**, 839–50.
Hooke, J.M. and Kain, R.J.P. (1982). *Historical Change in the Physical Environment: a guide to sources and techniques*, Butterworth Scientific, Guildford.
Hoppe, R.A. (1975). 'Minimum streamflows for fish', Paper distributed at Soils Hydrology Workshop, USFS, Montana State University, 26–30 January 1976. Bozeman, Montana (as cited by Wesche and Reschard, 1980).
Horner, R.R. and Welch, E.B. (1982). *Impacts of Channel Reconstruction in the Pilchuck River*, Washington State Dept of Transportation, Olympia, Washington.
Horowitz, A.J., Rinella, F.A., Lamothe, P., Miller, T.L., Edwards, T.K., Roche, R.L. and Rickert, D.A. (1989). 'Cross-sectional variability in suspended sediment

and associated trace element concentrations in selected rivers in the US', in *Sediment and the Environment* (Eds R.F. Hadley, and E.D. Ongley), pp. 57–66, Proc. of Baltimore Symposium, May 1989, IAHS Publ. no. 184.

Horton, R.E. (1932). 'Drainage basin characteristics', *Trans. Am. Geophys. Union,* **13**, 350–61.

Horton, R.E. (1945). 'Erosional development of streams and their drainage basins: hydrophysical approach to quantitative morphology', *Bull. Geol. Soc. Am.,* **56**, 275–370.

Horwitz, R.J. (1978). 'Temporal variability patterns and the distributional patterns of stream fishes', *Ecological Monographs,* **48**, 307–21 (as cited by Resh *et al.*, 1988).

Hosking, J.R.M. (1989). 'The theory of probability weighted moments', *Res. Rep. RC 12210 (#54860)*, IBM Research Division, T.J. Watson Research Center, Yorktown Heights, NY 10598.

Hosking, J.R.M., Wallis, J.R. and Wood, E.F. (1985). 'An appraisal of the regional flood frequency procedure in the UK Flood Studies Report', *Hydro. Sci. J.,* **30**, 85–109.

Hubbell, D.W. (1964). 'Apparatus and techniques for measuring bedload', US Geol. Survey Water Supply Paper 1748, USGS, Washington, DC.

Hughes, J.M. and James, B. (1989). 'A hydrological regionalization of streams in Victoria, Australia, with implications for stream ecology', *Aust. J. of Mar. and Freshwater Res.,* **40**, 303–26.

Hughes, R.M. and Omernik, J.M. (1983). 'An alternative for characterizing stream size', in *Dynamics of Lotic Ecosystems* (Eds T. D. Fontaine III, and S. M. Bartell), pp. 87–101, Ann Arbor Science, Michigan.

Hupp, C.R. and Bryan, B.A. (1989). 'A dendrogeomorphic approach to flash-flood frequency estimation', paper presented at spring meeting of Am. Geophys. Union, May 1989, Baltimore, Maryland.

Hurlbert, S.J. (1984). 'Pseudoreplication and the design of ecological field experiments', *Ecological Monographs,* **54**, 187–211.

Hynes, H.B.N. (1970). *The Ecology of Running Waters*, Liverpool University Press, Liverpool.

IAC (1982). 'Guidelines for determining flood flow frequency', *Bulletin 17B*, Interagency Advisory Committee on Water Data, Hydrology Sub-committee, Office of Water Data Coordination, USGS, Washington, DC (available from Office of Water Data Coordination, US Geological Survey, Reston, VA 22092, USA).

ICA (1984). *Basic Cartography for Students and Technicians*, Vol. 1, International Cartographic Association, UK.

IEA (1987). *Australian rainfall and runoff, a guide to flood estimation,* **1**, The Institution of Engineers, Australia, Barton, ACT.

IH (1980). *Low Flow Studies: (1) Research Report; (2) Flow Duration Curves Estimation; (3) Flow Frequency Curves Estimation; (4) Catchment Characteristics Estimation Manual*, Institute of Hydrology, Wallingford.

Illies, J. (1961). 'Versuch einer allgemein biozönotischen Gliederung der Fliessgewässer', *Int. Revue ges. Hydrobiol.,* **46**, 205–13 (as cited by Hawkes, 1975).

Illies, J. and Botosaneanu, L. (1963). 'Problémes et méthodes de la classification et de la zonation écologique des eaux corantes, considerées surtout du point de vue faunistique', Mitt. int. Verein. theor. andew., *Limnol.*, **12**, 1–57 (as cited by Hawkes, 1975).

Irvine, J.R., Jowett, I.G. and Scott, D. (1987). 'A test of the instream flow incremental methodology for underyearling rainbow trout, *Salmo gairdnerii*, in experimental New Zealand streams', *NZ J. of Mar. and Freshwater Res.,* **21**, 35–40.

Jarrett, R.D. (1984). 'Hydraulics of high-gradient streams', *J. Hyd. Eng.*, **110**, 1519–39.

Jarrett, R.D. (1985). 'Determination of roughness coefficients for streams in Colorado', *US Geol. Survey Water-Resources Investigations Report 85-4004*, USGS, Lakewood, Colorado.

Jarrett, R.D. (1987). 'Errors in slope–area computations of peak discharges in mountain streams', *J. of Hydrol.*, **96**, 53–67.

Jarrett, R.D. (1988). *Hydrologic and hydraulic research in mountain rivers*, prepared for International Workshop on Hydrology of Mountainous Areas, Strbske Pleso, Czechoslovakia, 6–11 June 1988.

Jarrett, R.D. and Malde, H.E. (1987). 'Paleodischarge of the late Pleistocene Bonneville flood, Snake River, Idaho, computed from new evidence', *Geol. Soc. of Am. Bull.*, **99**, 127–34.

Jarrett, R.D. and Petsch, H.E. Jr (1985). 'Computer program NCALC user's manual—verification of Manning's roughness coefficient in channels', *US Geol. Survey Water-Resources Investigations Report 85-4317*, USGS, Lakewood, Colorado.

Jarvis, R.S. and Woldenberg, M.J. (Eds) (1984). *River Networks*, Hutchinson Ross, Pennsylvania.

Jenkinson, A.F. (1955). 'The frequency distribution of the annual maximum (or minimum) values of meteorological elements', *Quart. Jour. Roy. Met. Soc.*, **81**, 158–71.

Jennings, J.N. (1967). 'Topographical maps and the geomorphologist', *Cartography*, **6**, 73–80.

Johnson, C.W., Engelman, R.L., Smith, J.P. and Hanson, C.L. (1977). 'Helley–Smith bedload samplers', *Proc. ASCE Hyd. Div.*, **103**, 1217–21.

Johnson, C.W., Gordon, N.D. and Hanson, C.L. (1985). 'Northwest sediment yield analysis by the MUSLE', *Trans. ASAE*, **28**, 1885–95.

Johnston, C.A. and Naiman, R.J. (1990). 'Aquatic patch creation in relation to beaver population trends', *Ecology*, **71**, 1617–21.

Jones, H.R. and Peters, J.C. (1977). *Physical and biological typing of unpolluted rivers*, Technical Report TR 41, Medmenham Laboratory, UK.

Jones, P. (1983). *Hydrology*, Basil Blackwell, Oxford.

Jowett, I.G. (1989). *RHYHABSIM Computer manual*, Freshwater Fisheries Centre, Riccarton, New Zealand.

Jowett, I.G. (1990). 'Factors related to the distribution and abundance of brown and rainbow trout in New Zealand clear water rivers', *NZ J. of Mar. and Freshwater Res.*, **24**, 429–40.

Jowett, I.G. and Duncan, M.J. (1990). 'Flow variability in New Zealand rivers and its relationship to in-stream habitat and biota', *NZ J. of Mar. and Freshwater Res.*, **24**, 305–17.

Jowett, I.G. and Richardson, J. (1989). 'Effects of a severe flood on instream habitat and trout populations in seven New Zealand rivers', *NZ J. of Mar. and Freshwater Res.*, **23**, 11–17.

Karr, J.R. and Dudley, D.R. (1978). 'Biological integrity of a headwater stream: evidence of degradation, prospects for recovery', in *Environmental Impact of Land Use on Water Quality, Final Report on the Black Creek Project* (Eds J. Lake and J. Morrison), US Envt. Prot. Agency, Chicago, Illinois.

Kellerhals, R. and Church, M. (1989). 'The morphology of large rivers: characterization and management', in *Proc. of the International Large River Symp.* (Ed. D. P. Didge), pp. 31–48. Can. Spec. Publ. Fish. Aquat. Sci. 106.

Kellerhals, R., Church, M. and Bray, D.I. (1976). 'Classification and analysis of river processes', *J. Hyd. Div., ASCE*, **102**, 813–29.

Kinhill (1988). *Techniques for determining environmental water requirements—a*

review, Technical Report Series, Report No. 40, Kinhill Engineers Pty Ltd, a report to the Department of Water Resources, Victoria.

Klemes, V. (1986). 'Dilettantism in hydrology: transition or destiny?', *Water Resour. Res.*,**22**, 177S–188S.

Klute, A. (Ed.) (1986). *Methods of Soil Analysis, Part 1: Physical and Mineralogical Methods. Agronomy Monograph no. 9*, 2nd edn, Amer. Soc. of Agronomy, Soil Sci. Soc. of America, Madison, Wisconsin.

Knapp, B. (1979). *Elements of Geographical Hydrology*, George Allen & Unwin, London.

Knighton, D. (1984). *Fluvial Forms and Processes*, Edward Arnold, London.

Koehn, J. (1987). 'Artificial habitat increases abundance of two-spined blackfish (*Gadopsis bispinosis*) in Ovens River, Victoria', *Tech. Report Series No. 56*, Arthur Rylah Inst. for Envt Res., Dept. of Cons., For. and Lands, Victoria, Australia.

Kolata, G. (1985). 'Prestidigitator of digits', *Science 85,* **6**, 66–72.

Kondolf, G.M., Cata, G.F. and Sale, M.J. (1987). 'Assessing flushing-flow requirements for brown trout spawning gravels in steep streams', *Water Resour. Bull.,* **23**, 927–35.

Konijn, H.S. (1973). *Statistical Theory of Sample Survey Design and Analysis*, North-Holland, Amsterdam.

Koppen, W. (1936). 'Das geographische System der Klimate, Vol. 1, part C', in *Handbuch der Klimatologie* (Eds W. Koppen and R. Geiger), Borntraeger, Berlin.

Korte, V.L. and Blinn, D.W. (1983). 'Diatom colonization on artificial substrata in pool and riffle zones studied by light and scanning electron microscopy', *J. Phycol.,* **19**, 332–41.

Kotwicki, V. (1986). *Floods of Lake Eyre*, Engineering and Water Supply Dept, Adelaide, South Australia.

Krumbein, W.C. (1941). 'Measurement and geological significance of shape and roundness of sedimentary particles', *J. of Sedimentary Petrology,* **11**, 64–72.

Kvanli, A.H. (1988). *Statistics. A Computer Integrated Approach*, West Publishing, St Paul, Minnesota.

Lake, P.S. (1982). 'The 1981 Jolly Award address, Ecology of the macroinvertebrates of Australian upland streams—a review of current knowledge', *Bull of the Australian Society for Limnology,* **8**, 1–15.

Lake, P.S., Barmuta, L.A., Boulton, A.J., Campbell, I.C. and St Clair, R.M. (1985). 'Australian streams and Northern Hemisphere stream ecology: comparisons and problems', *Proc. Ecol. Soc. Aust.,* **14**, 61–82.

Lake, P.S. and Marchant, R. (1990). 'Australian upland streams: ecological degradation and possible restoration', *Proc. Ecol. Soc. Aust.,* **16**, 79–91.

Lane, E.W. (1955). 'The importance of fluvial morphology in hydraulic engineering', *ASCE Proc.,* **81**, No. 745.

Lane, E.W. and Borland, W.M. (1951). 'Estimating bedload', *Amer. Geophys. Union Trans.,* **32**, 121–23 (as cited by Morisawa, 1985).

Lane, E.W. and Lei, K. (1950). 'Stream flow variability', *Trans. ASCE,* **115**, 1084–1134.

Langbein, W.B. and Iseri, K.T. (1960). 'General introduction and hydrologic definitions', *Manual of Hydrology:* Part 1. *General Surface-Water Techniques*, US Geol. Survey Water-Supply Paper 1541-A, USGS, Washington, DC.

Langbein, W.B. and Leopold, L.B. (1964). 'Quasi-equilibrium states in channel morphology', *Amer. Jour. Sci.* **262**, 782–94.

Langbein, W.B. and Schumm, S.A. (1958). 'Yield of sediment in relation to mean annual precipitation', *Am. Geophys. Union Trans.,* **39**, 1076–84.

LaPerriere, J.D. and Martin, D.C. (1986). 'Simplified method of measuring stream slope', *Cold Regions Hydrology Symp.*, pp. 143–5, American Water Resources Assn, July 1986.

Larinier, M. (1987). 'Fishways: principles and design criteria', *La Houille Blanche,* **1/2**, 51–7.

Larsen, D.P., Omernik, J.M., Hughes, R.M., Rohm, C.M., Whittier, T.R., Kinney, A.J., Gallant, A.L. and Dudley, D.R. (1986). 'Correspondence between spatial patterns in fish assemblages in Ohio streams and aquatic ecoregions', *Environmental Mgmt,* **10**, 815–28.

Legendre, L., and Legendre, P. (1983). *Numerical Ecology*, Elsevier, Amsterdam.

Leopold, L.B. (1959). 'Probability analysis applied to a water-supply problem', *US Geol. Survey Circular 410*, USGS, Washington, DC.

Leopold, L.B. (1960). 'Ecological systems and the water resource', *Part D, US Geological Survey Circular 414, Conservation and Water Management*, US Geol. Survey, Washington, DC.

Leopold, L.B. and Emmett, W.W. (1976). 'Bedload measurements, East Fork River, WY', *Nat. Acad. of Sci. Proc.,* **73**, 1000–4.

Leopold, L.B. and Langbein, W.B. (1960). *A Primer on Water*, USGS, Washington, DC.

Leopold, L.B. and Maddock, T., Jr (1953). 'The hydraulic geometry of stream channels and some physiographic implications', Geol. Survey Prof. Paper 252, USGS, Washington, DC.

Leopold, L.B. and O'Brien Marchand, M. (1968). 'On the quantitative inventory of the riverscape', *Water Resour. Res.,* **4**, 709–17.

Leopold, L.B., Wolman, M.G. and Miller, J.P. (1964). *Fluvial Processes in Geomorphology*, W.H. Freeman, San Francisco.

Lewis, D.W. (1984). *Practical Sedimentology*, Hutchinson Ross, USA.

Lewis G. and Williams, G. (1984). *Rivers and Wildlife Handbook—A Guide to Practices Which Further the Conservation of Wildlife on Rivers*, Royal Soc. for the Prot. of Birds and Royal Soc. for Nature Conservation, UK.

Likens, G.E., Bormann, F., Pierce, R.S., Eaton, J.S. and Johnson, N.M. (1977). *Biogeochemistry of a Forested Ecosystem*, Springer-Verlag, New York.

Lillesand, T.M. and Kiefer, R.W. (1987). *Remote Sensing and Image Interpretation*, 2nd edn, John Wiley, New York.

Limerinos, J.T. (1970). 'Determination of the Manning coefficient from measured bed roughness in natural channels', US. Geol. Survey Water-Supply Paper 1898-B, USGS, Washington, DC (as cited by Jarrett, 1985).

Linsley, R.K., Kohler, M.A. and Paulhus, J.L.H. (1975a). *Applied Hydrology*, McGraw-Hill, New Delhi.

Linsley, R.K., Kohler, M.A. and Paulhus, J.L.H. (1975b). *Hydrology for Engineers*, 2nd edn, McGraw-Hill, New York.

Lisle, T.E. (1986). 'Stabilization of a gravel channel by large streamside obstructions and bedrock bends, Jacoby Creek, northwestern California', *Geol. Soc. of Am. Bull.*, **97**, 999–1011.

Lisle, T.E. (1987). 'Using "residual depths" to monitor pool depths independently of discharge', *Res. Note PSW-394*, Pac. SW For. and Range Expt. Sta., US Forest Service, Berkeley, California.

Lloyd, L. (1990). 'Conservation significance of wetlands in the Murray–Darling system: the assignment of conservation values', presented at the Fenner Conference, September 1989, Canberra, Australia (in press).

Lo, C.P. (1986). *Applied Remote Sensing*, Longman Scientific and Technical, Harlow.

Lotspeich, F.B. (1980). 'Watersheds as the basic ecosystem: this conceptual framework provides a basis for a natural classification system', *Water Resour. Bull.,* **16**, 581–6.

Lotspeich, F.B. and Reid, B.H. (1980). 'Tri-tube freeze core procedure for sampling stream gravels', *Progr. Fish Cult.*, **42**, 96–9 (as cited by Hamilton and Bergersen, 1984).

Lotspeich, F.B. and Platts, W.S. (1982). 'An integrated land–aquatic classification system', *Nth. Am. J. Fisheries Mgmt*, **2**, 138–49.

Low, J.W. (1952). *Plane Table Mapping*, Harper and Brothers, New York.

Macmillan, L.A. (1986). 'Criteria for evaluating streams for protection', in *Stream Protection: The Management of Rivers for Instream Uses* (Ed. I.C. Campbell), pp. 199–233, Water Studies Centre, Chisholm Institute of Technology, East Caulfield, Australia.

Macmillan, L.A. (1987). *Assessing the nature conservation value of rivers and streams with particular reference to the rivers of East Gippsland, Victoria*, Master of Appl. Science thesis, Dept of Chemistry and Biology, Chisholm Institute of Technology, Melbourne, Victoria.

Maitland, P.S. (1978). *Biology of Fresh Waters*, Blackie, Glasgow.

Mallen-Cooper, M.G. (1989). 'Fish passage in the Murray–Darling Basin', *Proc. of the Workshop on Native Fish Management*, pp. 123–36, Murray–Darling Basin Comm., Canberra, 16–17 June 1988.

Manly, B.F.J. (1986). *Multivariate Statistical Methods: A Primer*, Chapman and Hall, London and New York.

Mark, D.M. (1983). 'Relations between field-surveyed channel networks and map-based geomorphometric measures, Inez, Kentucky', *Ann. of the Assn. of Am. Geog.*, **73**, 358–72.

Mark, H.B., Jr and Mattson, J.S. (1981). *Water Quality Measurement*, Marcel Dekker, New York.

Mathur, D., Bason, W.H., Purdy, E.J. Jr and Silver, C.A. (1985). 'A critique of the Instream Flow Incremental Methodology', *Can. J. Fish. Aquat. Sci.*, **42**, 825–31.

Maude, S.H. and Williams, D.D. (1983). 'Behavior of crayfish in water currents: hydrodynamics of eight species with reference to their distribution patterns in southern Ontario', *Can. J. Fish. Aquat. Sci.*, **40**, 68–77.

McKnight, T.L. (1990). *Physical Geography*, 3rd edn, Prentice-Hall, Englewood Cliffs, New Jersey.

McMahon, T.A. (1976). 'Preliminary estimation of reservoir storage for Australian streams', *Civil Engineering Trans.*, Institution of Engineers, Australia, **CE18**, 55–9.

McMahon, T.A. (1982). *Hydrological Characteristics of Selected Rivers of the World*, UNESCO, Paris.

McMahon, T.A. (1986). 'Hydrology and management of Australian streams', in *Stream Protection: The Management of Rivers for Instream Uses* (Ed. I.C. Campbell), pp. 23–44, Water Studies Centre, Chisholm Institute of Technology, East Caulfield, Australia.

McMahon, T.A. and Diaz Arenas, A. (1982). 'Methods of computation of low streamflow', *Studies and Reports in Hydrology*, **36**, UNESCO, Paris.

McMahon, T.A., Finlayson, B.L., Haines, A. and Srikanthan, R. (1987). 'Runoff variability: a global perspective', in *The Influence of Climate Change and Climatic Variability on the Hydrologic Regime and Water Resources*, pp. 3–11, Proc. of the Vancouver Symposium, August 1987, IAHS Publ. No. 168.

McMahon, T.A. and Mein, R.G. (1986). *River and Reservoir Yield*, Water Resources Publications, PO Box 2841, Littleton, Colorado.

Megahan, W.F. (1982). 'Channel sediment storage behind obstructions in forested drainage basins draining the granitic bedrock of the Idaho Batholith', in *Sediment Budgets and Routing in Forested Draining Basins* (Eds F. J. Swanson, R.J. Janda, T. Dunne, D.N. Swanston), pp. 114–21, Gen. Tech. Rep. PNW-141, US Dept of Agriculture, Forest Service, Portland, Oregon.

Meredith, C., Goss, H. and Seymour, S. (1989). 'Nature conservation values of the rivers and catchments of Gippsland', *Report No. 44, Water Resource Management Report Series*, Department of Water Resources, Victoria, Australia.

Milhous, R.T. (1982). 'Effect of sediment transport and flow regulation on the ecology of gravel-bed rivers', in *Gravel-bed Rivers* (Eds R.D. Hey, J.C. Bathurst, C.R. Thorne), John Wiley, Chichester.

Milhous, R.T. (1986). 'Development of a Habitat Time Series', *Journal of Water Resources Planning and Management,* **112**, 145–8.

Milhous, R.T. (1988a). 'Hydraulics for physical habitat studies of streams', *US Fish Wild. Serv. Biol. Rep. 88*, US Fish and Wildlife Service, Washington, DC.

Milhous, R.T. (1988b). 'The physical habitat–aquatic population relationship', presented at the 1988 meeting of the Colorado–Wyoming Section of the American Fisheries Society (unpublished).

Milhous, R.T. and Bradley, J.B. (1986). 'Physical habitat simulation and the moveable bed', in *World Water Issues in Evolution* (Eds M. Karamoz, G.R. Baumli and W. J. Brick), pp. 1976–83, Water Forum '86, ASCE, New York.

Milhous, R.T., Updike, M.A. and Schneider, D.M. (1989). 'Physical Habitat Simulation System reference manual—Version II', Instream Flow Info. Paper No. 26, Biol. Rpt 89(16), National Ecology Research Center, US Fish and Wildlife Service, Fort Collins, Colorado.

Milhous, R.T., Wegner, D.L. and Waddle, T.J. (1981). *User's guide to the physical habitat simulation system, FWS/OBS-81/43*, US Fish and Wildlife Service, Office of Biological Services.

Minchin, P.R. (1987). 'An evaluation of the relative robustness of techniques for ecological ordination', *Vegetation,* **69**, 89–108.

Minshall, G.W. (1984). 'Aquatic insect–substratum relationships', in *Ecology of Aquatic Insects* (Eds V.M. Resh and D.M. Rosenberg), pp. 358–400, Praeger, New York.

Mitchell, P. (1990). *The Environmental Condition of Victorian Streams*, Dept of Water Resources, Victoria, Australia.

Molloy, D.P. and Struble, R.H. (1988). 'A simple and inexpensive method for determining stream discharge from a streambank', *J. of Freshwater Ecology,* **4**, 477–81.

Morin, A., Harper, P.P. and Peters, R.H. (1986). 'Microhabitat-preference curves of blackfly larvae (Diptera: Simuliidae): a comparison of three estimation methods', *Can. J. Fish. Aquat. Sci.,* **43**, 1235–41.

Morisawa, M.E. (1958). 'Measurement of drainage basin outline form', *Jour. Geol.,* **66**, 587–91.

Morisawa, M.E. (1968). *Streams, their Dynamics and Morphology*, McGraw-Hill, New York.

Morisawa, M.E. (1985). *Rivers, Form and Process, Geomorphology Texts 7*, Longman, London.

Morris, H.M. (1955). 'Flow in rough conduits', *Trans. ASCE,* **120**, 373–98.

Morris, H.M. (1961). 'Design methods for flow in rough conduits', *Trans. ASCE,* **126**, 454–90.

Mosley, M.P. (1981). 'Semi-determinate hydraulic geometry of river channels, South Island, New Zealand', *Earth Surf. Proc. and Landforms,* **6**, 127–37.

Mosley, M.P. (1983). 'Flow requirements for recreation and wildlife in New Zealand rivers—A review', *J. of Hydrol. (NZ),* **22**, 152–74.

Mosley, M.P. and Jowett, I.G. (1985). 'Fish habitat analysis using river flow simulation' *NZ J. of Mar. and Freshwater Res.,* **19**, 293–309.

Moss, B. (1988). *Ecology of Fresh Waters, Man and Medium*, 2nd edn, Blackwell Scientific, Oxford.

Muschenheim, D.K., Grant, J. and Mills, E.L. (1986). 'Flumes for benthic ecologists: theory, construction and practice', *Mar. Ecol. Prog. Ser.,* **28**, 185–96.

Nathan, R.J. and McMahon, T.A. (1990a). 'The estimation of low flow characteristics and yield from small ungauged rural catchments', *AWRAC Research Project 85/105*, Dept of Civil and Agricultural Engineering, University of Melbourne, Victoria.

Nathan, R.J. and McMahon, T.A. (1990b). 'Evaluation of automated techniques for baseflow and recession analysis', *Water Resour. Res.* (in press).

Nathan, R.J. and McMahon, T.A. (1990c). 'Practical aspects of low flow frequency analysis', *Water Resour. Res.* (in press).

Nathan, R.J. and McMahon, T.A. (1990d). 'Identification of homogeneous regions for the purposes of regionalisation', *J. Hydrol.* (in press).

NERC (1975). *Flood Studies Report*, 5 volumes, Natural Environment Research Council, London.

Nestler, J.M., Milhous, R.T. and Layzer, J.B. (1989). 'Instream habitat modeling techniques', in *Alternatives in Regulated River Management* (Eds J. A. Gore and G. E. Petts), pp. 295–315, CRC Press, Boca Raton, Florida.

Newbury, R.W. (1984). 'Hydrologic determinants of aquatic insect habitats', in *Ecology of Aquatic Insects* (Eds V. M. Resh and D. M. Rosenberg), pp. 323–57, Praeger, New York.

Newbury, R.W. (1989). 'Habitats and Hydrology of Streams', workshop conducted by the Centre for Environmental Applied Hydrology, Buxton, Victoria, 15–17 November 1989.

Newbury, R. W. and Gaboury, M. (1988). 'The use of natural stream characteristics for stream rehabilitation works below the Manitoba escarpment', *Canadian Water Resour. J.,* **13**, 35–51.

Nezu, I. and Rodi, W. (1986). 'Open-channel flow measurements with a laser-doppler anemometer', *J. Hyd. Eng.,* **112**, 335–55.

Nordin, C.F., Jr (1985). 'The sediment loads of rivers', in *Facets of Hydrology*, Vol. II (Ed. J. C. Rodda), Chapter 7, John Wiley, Chichester.

Nordin, C.F. Jr and Richardson, E.V. (1971). 'Instrumentation and measuring techniques', in *River Mechanics* (Ed. H. W. Shen), Chapter 14, Fort Collins, Colorado.

Northcote, K.H. (1979). *A Factual Key for the Recognition of Australian Soils*, Rellim Technical Publications, Adelaide, South Australia.

Nowell, A.R.M. and Jumars, P.A. (1984). 'Flow environments of aquatic benthos', *Ann. Rev. Ecol. Syst.,* **15**, 303–28.

O'Brien, J.S. (1987). 'A case study of minimum streamflow for fishery habitat in the Yampa River', in *Sediment Transport in Gravel-bed Rivers* (Eds C.R. Thorne, J.C. Bathurst, and R.D. Hey), pp. 921–46, John Wiley, Chichester.

Orth, D.J. (1987). 'Ecological considerations in the development and application of instream flow-habitat models', *Regulated Rivers*, **1**, 171–81.

Orth, D.J. and Leonard, P.M. (1990). 'Comparison of discharge methods and habitat optimization for recommending instream flows to protect fish habitat', *Regulated Rivers,* **5**, 129–138.

Orth, D.J. and Maughan, O.E. (1982). 'Evaluation of the Incremental Methodology for recommending instream flows for fishes', *Trans. Am. Fish. Soc.,* **111**, 413–45.

Osborne, L.S., Wiley, M.J. and Larimore, R.W. (1988). 'Assessment of the Water Surface Profile model: accuracy of predicted instream habitat conditions in low-gradient warmwater streams', *Regulated Rivers,* **2**, 619–31.

Overmire, T.G. (1986). *The World of Biology*, John Wiley, New York.

Park, C.C. (1977). 'World-wide variations in hydraulic geometry exponents of stream channels: an analysis and some observations', *J. of Hydrol.,* **33**, 133–46.

Parker, G., Klingeman, P.C. and McLean, D.G. (1982). 'Bedload and size distribution in paved gravel-bed streams', *J. Hyd. Div., Proc. ASCE,* **108(HY4)**, 544–71.

Pennak, R.W. (1971). 'Toward a classification of lotic habitats', *Hydrobiologia,* **38**, 321–334.

Pepi, D. (1985). *Thoreau's Method: a Handbook for Nature Study*, Prentice-Hall, Englewood Cliffs, New Jersey.

Petersen, M.S. (1986). *River Engineering*, Prentice-Hall, Englewood Cliffs, New Jersey.

Petts, G.E. (1980). 'Long-term consequences of upstream impoundment', *Environ. Conserv.,* **7**, 325–32 (as cited by Walker, 1985).

Petts, G.E. (1989). 'Perspectives for ecological management of regulated rivers', in *Alternatives in Regulated River Management* (Eds J.A. Gore and G.E. Petts), pp. 3–24, CRC Press, Boca Raton, Florida.

Petts, G.E. and Foster, I. (1985). *Rivers and Landscape*, Edward Arnold, London.

Pielou, E.C. (1977). *Mathematical Ecology*, John Wiley, New York.

Pielou, E.C. (1984). *The Interpretation of Ecological Data*, John Wiley, New York.

Pizzuto, J.E. (1986). 'Flow variability and the bankfull depth of sand-bed streams of the American Midwest', *Earth Surf. Proc. and Landforms,* **11**, 441–50.

Platts, W.S. (1979). 'Relationships among stream order, fish populations, and aquatic geomorphology in an Idaho river drainage', *Fisheries,* **4**, 5–9.

Platts, W.S., Armour, C., Booth, G.D., Bryant, M., Bufford, J.L., Cuplin, P., Jensen, S., Lienkaemper, G.W., Minshall, G.W., Monsen, S.B., Nelson, R.L., Sedell, J.R. and Tuhy, J.S. (1987). *Methods for evaluating riparian habitats with applications to management*, Gen. Tech. Rep. INT-221, USDA Forest Service, Intermountain Research Station, Ogden, Utah.

Platts, W.S., Megahan, W.F. and Minshall, G.W. (1983). *Methods for evaluating stream, riparian, and biotic conditions*, Gen. Tech. Rep. INT-138, USDA Forest Service, Intermountain Forest and Range Expt Station, Ogden, Utah.

Platts, W.S. and Nelson, R.L. (1985). 'Stream habitat and fisheries response to livestock grazing and instream improvement structures, Big Creek, Utah', *J. of Soil and Water Cons.,* **40**, 374–9.

Platts, W.S. and Penton, V.E. (1980). 'A new freezing technique for sampling salmonid redds', Res. Pap. INF-248, USDA Forest Service, Intermountain Forest and Range Expt Station, Ogden, Utah.

Platts, W.S. and Rinne, J.N. (1985). 'Riparian and stream enhancement management and research in the Rocky Mountains', *No. Am. J. of Fisheries Mgmt.,* **5**, 115–25.

Poff, N.L. and Ward, J.V. (1989). 'Implications of streamflow variability and predictability for lotic community structure: a regional analysis of streamflow patterns', *Can. J. Fish. Aquat. Sci.,* **46**, 1805–18.

Porterfield, G. (1972). 'Computation of fluvial-sediment discharge', *Tech. of Water-Resources Investigations of the USGS*, Chapter C3, USGS, Washington, DC.

Powell, F.C. (1982). *Statistical Tables for the Social, Biological and Physical Sciences*, Cambridge University Press, Cambridge.

Prestegaard, K.L. (1989). 'Selective unravelling and the downstream fining of streambed material', paper presented at spring meeting of Am. Geophys. Union, May 1989, Baltimore, Maryland.

Prewitt, C.G. and Carlson, C.A. (1980). 'An evaluation of four instream flow methodologies', *Biological Sciences Series*, Number 2, US Bureau of Land Management, Denver, Colorado.

Purcell, E.M. (1977). 'Life at low Reynolds number', *Am. J. of Physics,* **45**, No. 1, 3–11.

Pusey, B.J. and Arthington, A.H. (1990). 'Limitations to the valid application of the instream flow incremental methodology (IFIM) for determining instream flow

requirements in highly variable Australian lotic environments', *Verh. Int. Verein. Limnol.*, **24**, Proc. of SIL Congress, August 1989, Munich (in press).

Rajaratnam, N. and Katopodis, C. (1984). 'Hydraulics of Denil fishways', *J. Hyd. Div., ASCE,* **110**, 1219–33.

Rajaratnam, N. and Katopodis, C. (1988). 'Plunging and streaming flows in pool and weir fishways'; *J. Hydr. Eng., ASCE,* **114**, 939–44.

Raudkivi, A.J. (1967). *Loose Boundary Hydraulics*, Pergamon Press, Oxford.

Raudkivi, A.J. (1979). *Hydrology: An Advanced Introduction to Hydrological Processes and Modelling*, Pergamon Press, Oxford.

Reeve, R.C. (1986). 'Water potential: piezometry', in *Methods of Soil Analysis*, Part 1: *Physical and Mineralogical Methods, Agronomy Monograph no. 9,* 2nd edn (Ed. A. Klute), pp. 545–61, Amer. Soc. of Agronomy, Soil Sci. Soc. of America, Madison, Wisconsin.

Reiser, D.W., Ramey, M.P. and Wesche, T.A. (1989). 'Flushing flows', in *Alternatives in Regulated River Management* (Eds J. A. Gore and G. E. Petts), pp. 91–135, CRC Press, Boca Raton, Florida.

Resh, V.H., Brown, A.V., Covich, A.P., Gurtz, M.E., Li, H.W., Minshall, G.W., Reice, S.R., Sheldon, A.L., Wallace, J.B. and Wissmar, R.C. (1988). 'The role of disturbance in stream ecology', *J. N. Am. Benthol. Soc.,* **7**, 433–55.

Resh, V.M., and Rosenberg, D.M. (Eds) (1984). *Ecology of Aquatic Insects*, Praeger, New York.

Reynolds, O. (1883). 'An experimental investigation of the circumstances which determine whether the motion of water shall be direct or sinuous, and the law of resistance in parallel channels', *Trans. Roy. Soc. Lond.*, 935–82.

Richards, K.S. (1976). 'The morphology of riffle–pool sequences', *Earth Surf. Proc.,* **1**, 71–88.

Richards, K.S. (1982). *Rivers, Form and Process in Alluvial Channels*, Methuen, London.

Richards, K.S. (Ed.) (1987). *River Channels: Environment and Process*, Basil Blackwell, Oxford.

Richardson, B.A. (1986). 'Evaluation of instream flow methodologies for freshwater fish in New South Wales', in *Stream Protection: The Management of Rivers for Instream Uses* (Ed. I. C. Campbell), pp. 143–67, Chisholm Institute of Technology, Victoria, Australia.

Ridley, J.E. and Steel, J.A. (1975). 'Ecological aspects of river impoundments', in *River Ecology* (Ed. B.A. Whitton), pp. 565–87, Blackwell Scientific, Oxford.

Ridley, S.J. (1972). 'A comparison of morphometric measures of bankfull', *J. Hydrol.* **17**, 23–31.

Riggs, H.C. (1976). 'A simplified slope area method for estimating flood discharges in natural channels', *Jour. Research US Geol. Survey,* **4**, 285–91.

Risser, R.J. and Harris, R.R. (1989). 'Mitigation for impacts to riparian vegetation on western montaine streams', in *Alternatives in Regulated River Management* (Eds J.A. Gore and G.E. Petts), pp. 235–50, CRC Press, Boca Raton, Florida.

Roberson, J.A. and Crowe, C.T. (1990). *Engineering Fluid Mechanics*, 2nd edn, Houghton Mifflin, Boston, Massachusetts.

Romesburg, H.C. (1984). *Cluster Analysis for Researchers*, Lifetime Learning Publications, Belmont, California.

Rosgen, D.L. (1985). 'A stream classification system', in *Riparian Ecosystems and their Management, Interagency North American Riparian Conference, Gen Tech. Rept. ROM-120*, pp. 91–5, Rocky Mt. For. and Range Expt. Sta., USDA Forest Service, Fort Collins, Colorado.

Rosgen, D.L., Silvey, H. L. and Potyondy, J.P. (1986). 'The use of channel maintenance flow concepts in the Forest Service', *Hydrological Science and Technology: Short papers,* **2**, 19–26, American Institute of Hydrology.

Ryan, B.F., Joiner, B.L. and Ryan, T.A. Jr. (1985). *Minitab Handbook*, 2nd edn, PWS Publishers, Boston, Massachusetts.

SAS (1987). *SAS User's Guide: Statistics, Version 6 Edition*, SAS Institute, Cary, North Carolina.

Savage, N.L. and Rabe, F.W. (1979). 'Stream types in Idaho: an approach to classification of streams in natural areas', *Biol. Conserv.*, **15**, 301–15.

Scarnecchia, D.L. (1988). 'The importance of streamlining in influencing fish community structure in channelized and unchannelized reaches of a prairie stream', *Regulated Rivers,* **2**, 155–66.

Scheaffer, R.L., Mendenhall, W. and Ott, L. (1979). *Elementary Survey Sampling*, 2nd, edn, PWS Publishers, Massachusetts.

Scheidegger, A.E. (1965). 'The algebra of stream-order numbers', US Geol. Survey Prof. Paper 525B, pp. 187–9, USGS, Washington, DC.

Schlichting, H. (1961). 'Boundary layer theory', in *Handbook of Fluid Dynamics* (Ed. V. L. Streeter), Chapter 9, McGraw-Hill, New York.

Schumm, S.A. (1954). 'The relation of drainage basin relief to sediment loss', *Internat. Assoc. Sci. Hyd. Pub.*, **36**, 216–19 (as cited by Gregory and Walling, 1973).

Schumm, S.A. (1956). 'Evolution of drainage systems and slopes in badlands at Perth Amboy, New Jersey', *Geol. Soc. Amer. Bull.*, **67**, 597–646.

Schumm, S.A. (1977). *The Fluvial System*, John Wiley, New York.

Scott, D. and Shirvell, C.S. (1987). 'A critique of the Instream Flow Incremental Methodology and observations on flow determination in New Zealand', in *Regulated Streams—Advances in Ecology* (Eds J.B. Kemper and J. Craig), pp. 27–44, Plenum Press, New York.

Searcy, J.K. (1959). 'Flow-duration curves', *Manual of Hydrology: Part 2. Low-Flow Techniques*, US Geol. Survey Water-Supply Paper 1542-A, USGS, Washington, DC.

Searcy, J.K. (1960). 'Graphical correlation of gaging-station records', Geol. Survey Water-Supply Paper 1541-C, USGS, Washington, DC.

Searcy, J.K. and Hardison, C.H. (1960). 'Double-mass curves', Geological Survey Water-Supply Paper 1541-B, USGS, Washington DC.

Selby, M.J. (1985). *Earth's Changing Surface, An Introduction to Geomorphology*, Oxford University Press, Oxford.

Shaw, E.M. (1988). *Hydrology in Practice*, 2nd edn, Van Nostrand Reinhold International, Wokingham.

Shields, F.D. Jr (1982). 'Environmental features for flood control channels', *Water Resour. Bull.*, **18**, 779–84.

Shields, N.D. (1936). 'Anwendung der ahnlickeit Mechanik und der Turbulenzforschung auf die Geschiebelerwegung', *Mitt. Preoss Versuchanstalt fur Wasserbau und Schiffbau,* **26**.

Shreve, R.L. (1967). 'Infinite topologically random channel networks', *J. of Geology,* **75**, 178–86.

Silvester, N.R. and Sleigh, M.A. (1985). 'The forces on microorganisms at surfaces in flowing water', *Freshwater Biol.*, **15**, 433–48.

Simons, D.B. and Richardson, E.V. (1961). 'Forms of bed roughness in alluvial channels', *J. Hyd. Div., ASCE,* **87**, 87–105.

Singhall, H.S.S. *et al.* (1977). 'Sediment sampling in rivers and canals', *Erosion and Sediment Transport Measurement, Symposium, IAHS Publ. No. 133*, Florence (as cited by Starosolszky, 1987).

Smart, P.L. (1984). 'A review of the toxicity of twelve fluorescent dyes used for water tracing', *NSS Bulletin,* **46**, 21–33.

Smith, D.G. (1976). 'Effect of vegetation on lateral migration of anastomosed channels of a glacier meltwater river', *Bull. of the Geol. Soc. of Am.*, **87**, 857–6

(as cited by Knighton, 1984).
Smith, D.I. and Stopp, P. (1978). *The River Basin*, Cambridge University Press, Cambridge.
Smith, I. R. (1975). *Turbulence in Lakes and Rivers*, Scientific Publication No. 29, Freshwater Biological Association, UK.
Sneath, P.H.A. and Sokal, R.R. (1973). *Numerical Taxonony*, W.H. Freeman, San Francisco.
Sokal, R.R. (1966). 'Numerical taxonomy', *Scientific American,* **165**, 106–16.
Sokal, R.R. and Rohlf, F.J. (1969). *Biometry*, W.H. Freeman, San Francisco.
Sokal, R.R. and Sneath, P.H.A. (1963). *Principles of Numerical Taxonomy*, W.H. Freeman, San Francisco.
SPSS (1986). *SPSS^x User's Guide, Edition 2*, SPSS Inc., McGraw-Hill, New York.
Srikanthan, R. and McMahon, T.A. (1981). 'Log Pearson III distribution—an empirically-derived plotting position', *J. of Hydrol.,* **52**, 161–3.
Stalnaker, C.B. (1979). 'The use of habitat structure preferenda for establishing flow regimes necessary for maintenance of fish habitat', in *The Ecology of Regulated Streams* (Eds J.V. Ward and J.A. Stanford), pp. 321–37, Plenum Press, New York.
Stalnaker, C.B. and Arnette, J.L. (1976). 'Methodologies for the determination of stream resource flow requirements: an assessment', prepared for US Fish & Wildlife Service by Utah State University, Logan, Utah.
Starosolsky, O. (Ed.) (1987). *Applied Surface Hydrology*, Water Resources Publications, Littleton, Colorado.
Statzner, B. (1988). 'Growth and Reynolds number of lotic macroinvertebrates: a problem for adaptation of shape to drag', *Oikos,* **51**, 84–7.
Statzner, B., Gore, J.A. and Resh, V.H. (1988). 'Hydraulic stream ecology: observed patterns and potential applications', *J. N. Am. Benthol. Soc.,* **7**, 307–60.
Statzner, B. and Higler, B. (1986). 'Stream hydraulics as a major determinant of benthic invertebrate zonation patterns', *Freshwater Biol.,* **16**, 127–39.
Statzner, B. and Holm, T.F. (1982). 'Morphological adaptations of benthic invertebrates to stream flow—An old question studied by means of a new technique (Laser doppler anemometry)', *Oecologia,* **53**, 290–2.
Statzner, B. and Holm, T.F. (1989). 'Morphological adaptation of shape to flow: microcurrents around lotic macroinvertebrates with known Reynolds numbers at quasi-natural flow conditions', *Oecologia,* **78**, 145–57.
Statzner, B. and Müller, R. (1989). 'Standard hemispheres as indicators of flow characteristics in lotic benthos research', *Freshwater Biol.,* **21**, 445–60.
Steel, R.G.D. and Torrie, J.H. (1960). *Principles and Procedures of Statistics with Special Reference to the Biological Sciences*, McGraw-Hill, New York.
Stelczer, K. (1987). 'Physical, chemical and biological properties of water', in *Applied Surface Hydrology* (Ed. O. Starosolszky), pp. 150–74, Water Resources Publications, Littleton, Colorado.
Stephens, P.S. (1974). *Patterns in Nature*, Little Brown, Boston.
Strahler, A.N. (1952). 'Hypsometric (area–altitude) analysis of erosional topography', *Bull. Geol. Soc. Am.,* **63**, 1117–42.
Strahler, A.N. (1964). 'Geology. Part II. Quantitative geomorphology of drainage basins and channel networks', in *Handbook of Applied Hydrology* (Ed. V.T. Chow), pp. 4–39 to 4–76, McGraw-Hill, New York.
Streeter, V.L. (Ed.) (1961). *Handbook of Fluid Dynamics*, McGraw-Hill, New York.
Streeter, V.L., and Wylie, E.B. (1979). *Fluid Mechanics,* 7th edn, McGraw-Hill International, Sydney.
Strickler, A. (1923). 'Beitrage zur Frage der Geschwindigheits-formel und der Rauhigkeiszahlen für Strome, Kanale und Geschlossene Leitungen', *Mitteilungen*

des Eidgenössischer Amtes fur Wasserwirtschaft, no. 16, Bern (as cited by Dackombe and Gardiner, 1983).

Strom, H.G. (1962). *River Improvement and Drainage in Australia and New Zealand*, State Rivers and Water Supply Comm., Victoria, Australia.

Stuebner, S. (1988). 'Riparian renewal on range', *Idaho Statesman*, 8 May.

Sullivan, K., Lisle, T.E., Dolloff, C.A., Grant, G.E. and Reid, L.M. (1987). 'Stream channels: the link between forests and fishes', in *Streamside Management: Forestry and Fishery Interactions* (Eds E.O. Salo and T.W. Cundy), pp. 39–97, Institute of Forest Resources, University of Washington, Seattle.

Swanson, F.J., Kratz, T.K., Caine, N. and Woodmansee, R.G. (1988). 'Landform effects on ecosystem patterns and processes', *Bioscience,* **38**, 92–8.

Tabachnick, B.G. and Fidell, L.S. (1989). *Using Multivariate Statistics*, 2nd edn, Harper and Row, New York.

Tennant, D.L. (1976). 'Instream flow regimens for fish, wildlife, recreation and related environmental resources', *Fisheries,* **1**, 6–10.

Thomas, I.L., Benning, V.M. and Ching, N.P. (1987). *Classification of Remotely Sensed Images*, Adam Hilger, Bristol.

Thorne, C.R., Bathurst, J.C. and Hey, R.D. (Eds) (1987). *Sediment Transport in Gravel-Bed Rivers*, John Wiley, Chichester.

Todd, D.K. (1959). *Ground Water Hydrology*, John Wiley, New York.

Townsend, C.R. (1980). *The Ecology of Streams and Rivers*, Institute of Biology, Studies in Biology 122, London.

Trieste, D.J. and Jarrett, R.D. (1987). 'Roughness coefficients of large floods', in *Irrigation and Drainage Division Specialty Conference Proceedings* (Eds L.G. James and M.J. English), pp. 32–40, ASCE, New York.

Triska, F.J., Kennedy, V.C., Avanzino, R.J., Zellweger, G.W. and Bencala, K.E. (1989a). 'Retention and transport of nutrients in a third-order stream: channel processes', *Ecology,* **70**, 1877–92.

Triska, F.J., Kennedy, V.C., Avanzino, R.J., Zellweger, G.W. and Bencala, K.E. (1989b). 'Retention and transport of nutrients in a third-order stream in northwestern California: hyporheic processes', *Ecology,* **70**, 1893–1905.

Tunbridge, B.R. (1988). 'Environmental flows and fish populations of waters in the South-Western Region of Victoria', *Tech. Rep. Series No. 65*, Arthur Rylah Inst. for Envt Res., Dept Cons., Forests and Lands, Victoria, Australia.

Tunbridge, B.R. and Glenane, T.J. (1988). 'A study of environmental flows necessary to maintain fish populations in the Gellibrand River and estuary', *Tech. Rep. Series No. 25*, Arthur Rylah Inst. for Envt Res., Dept of Cons., Forests and Lands, Victoria, Australia.

Uhlmann, D. (1979). *Hydrobiology*, John Wiley, Chichester.

UNESCO (1970). *International Legend for Hydrogeological Maps*, Int. Assoc. of Sci. Hydrology and Int. Assoc. of Hydrogeologists, Cook, Hammond and Kell Ltd, England.

Uren, J. and Price, W.F. (1984). *Calculations for Engineering Surveys*, Van Nostrand Reinhold, Wokingham.

USBR (1977). *Design of Small Dams*, US Bureau of Reclamation, Denver, Colorado.

USEPA (1987). *A Guide to the Sampling and Analysis of Water and Wastewater*, US Environmental Protection Agency.

Van Haveren, B.P. (1986). *Water Resource Measurements: A Handbook for Hydrologists and Engineers*, American Water Works Assn, Denver, Colorado.

Vannote, R.L., Minshall, G.W., Cummins, K.W., Sedell, J.R. and Cushing, C.E. (1980). 'The river continuum concept', *Can. J. Fish. Aquat. Sci.,* **37**, 130–37.

Vaskinn, K.A. (1985). 'Fysisk beskrivende vassdragsmodell', *Report A1*, Norsk Hydroteknisk Laboratorium, Trondheim, Norway.

Vennard, J.K. (1961). 'One-dimensional flow', in *Handbook of Fluid Dynamics* (Ed.

V.L. Streeter), Section 3, McGraw-Hill, New York.
Vennard, J.K. and Street, R.L. (1982). *Elementary Fluid Mechanics*, 6th edn, John Wiley, New York.
Vogel, S. (1981). *Life in Moving Fluids, The Physical Biology of Flow*, Princeton University Press, New Jersey.
Vogel, S. (1988). *Life's Devices: The Physical World of Animals and Plants*, Princeton University Press, New Jersey.
Wadell, H. (1932). 'Volume, shape, and roundness of rock particles', *Jour. Geol.*, **40**, 443–51 (as cited by Krumbein, 1941).
Wadell, H. (1933). 'Sphericity and roundness of rock particles', *Jour. Geol.*, **41**, 310–31 (as cited by Krumbein, 1941).
Walker, K. F. (1985). 'A review of the ecological effects of river regulation in Australia', *Hydrobiologia*, **125**, 111–29.
Walling, D.E. and Kleo, A.H.A. (1979). 'Sediment yields of rivers in areas of low precipitation: a global view', in *The Hydrology of Areas of Low Precipitation*, Proc. of IAHS Symposium, Canberra, December 1979, IAHS-AISH Publ. 128, 479–93 (as cited by Knighton, 1984).
Wallis, J.R. (1965). 'Multivariate statistical methods in hydrology—a comparison using data of known functional relationship', *Water Resour. Res.*, **1**, 447–61.
Wallis, J.R. and O'Connell, E. (1972). 'Small sample estimates of ρ1', *Water Resour. Res.*, **8**, 707–12.
Ward, J.V. (1982). 'Ecological aspects of stream regulation: responses in downstream lotic reaches', *Water Pollution and Management Reviews (New Delhi)*, **2**, 1–26.
Ward, J.V. (1984). 'Ecological perspectives in the management of aquatic insect habitat', in *The Ecology of Aquatic Insects* (Eds V.H. Resh and D.M. Rosenberg), pp. 558–77, Praeger, New York.
Ward, J.V. (1989). 'Riverine–wetland interactions', in *Freshwater Wetlands and Wildlife, 1989 Conf. 8603101, DOE Symposium Series No. 61* (Eds R.R. Sharitz and J.W. Gibbons), pp. 385–400, US DOE Office of Scientific and Technical Information, Oak Ridge, Tennessee.
Ward, J.V. and Stanford, J.A. (Eds) (1979). *The Ecology of Regulated Streams*, Plenum Press, New York.
Ward, J.V. and Stanford, J.A. (1983). 'The intermediate-disturbance hypothesis: an explanation for biotic diversity patterns in lotic ecosystems', in *Dynamics of Lotic Ecosystems* (Eds T.D. Fontaine and S.M. Bartell), pp. 347–56, Ann Arbor Science, Michigan.
Ward, J.V. and Stanford, J.A. (1987). 'The ecology of regulated streams: past accomplishments and directions for future research', in *Regulated Streams—Advances in Ecology* (Eds J.F. Craig and J.B. Kemper), pp. 391–409, Plenum Press, New York.
Wardlaw, A.C. (1985). *Practical Statistics for Experimental Biologists*, John Wiley, Chichester.
Webb, P.W. (1984). 'Form and function in fish swimming', *Scientific American*, **251**, 58–68.
Welch, P.S. (1935). *Limnology*, McGraw-Hill, New York.
Wesche, T.A. (1985). 'Stream channel modifications and reclamation structures to enhance fish habitat', in *The Restoration of Rivers and Streams* (Ed. J.A. Gore), pp. 103–59, Butterworths, Boston, Massachusetts.
Wesche, T.A., Goertler, C.M. and Hubert, W.A. (1987). 'Modified habitat suitability index model for brown trout in southeastern Wyoming', *N. Am. J. of Fisheries Mgmt*, **7**, 232–7.
Wesche, T.A. and Rechard, P.A. (1980). 'A summary of instream flow methods for fisheries and related research needs', *Eisenhower Consortium Bulletin 9*, US Govt Printing Office, Washington, DC.

Wetmore, S.H., Mackay, R.J. and Newbury, R.W. (1990). 'Characterization of the hydraulic habitat of *Brachycentrus occidentalis*, a filter feeding caddisfly', *J. No. Am. Benth. Soc.*, **9**, 157–69.

White, D.S., Elzinga, C.H. and Hendricks, S.P. (1987). 'Temperature patterns within the hyporheic zones of a northern Michigan river', *J. No. Am. Benthol. Soc.*, **6**, 85–91.

White, F.M. (1986). *Fluid Mechanics*, 2nd edn, McGraw-Hill, New York.

White, R.G. (1976). 'A methodology for recommending stream resource maintenance flows for large rivers', in *Proc. of the Symp. and Spec. Conf. on Instream Flow Needs* (Eds J.F. Orsborn and C.H. Allman), Vol. II, pp. 367–86, Amer. Fish. Soc., Bethesda, Maryland.

White, R.G. and Cochnauer, T. (1975). *Stream Resource Maintenance Flow Studies*, Dept of Fish and Game and Idaho Coop., Fishery Research Unit, Idaho.

White, R.J. and Brynildson, O.M. (1967). 'Guidelines for management of trout stream habitat in Wisconsin', *Wisconsin Dept Nat. Res. Tech. Bull. 39* (as cited by Bovee, 1982).

Whitton, B.A. (Ed.) (1975). *Studies in Ecology, Vol. 2: River Ecology*, Blackwell Scientific, Oxford.

Wiesner, C.J. (1970). *Climate, Irrigation and Agriculture*, Angus & Robertson, London (as cited by Shaw, 1988).

Wilkinson, L. (1986). *SYSTAT: The System for Statistics*, SYSTAT, Inc., Evanston, Illinois.

Williams, B.K. (1983). 'Some observations on the use of discriminant analysis in ecology', *Ecology*, **64**, 1283–91.

Williams, G.P. and Wolman, M.G. (1984). 'Downstream effects of dams on alluvial rivers', US Geol. Survey Prof. Paper 1286, USGS, Washington, DC.

Wilson, E.M. (1969). *Engineering Hydrology*, Macmillan, London.

Wilson, J.F. Jr, Cobb, E.D. and Kilpatrick, F.A. (1984). 'Flurometric procedures for dye tracing', *Open File Report 84–234*, USGS, Washington, DC.

Winegar, H.H. (1977). 'Camp Creek channel fencing plant, wildlife, soil and water response', *Rangeman's Journal*, **4**, 10–12 (as cited by Platts and Rinne, 1985).

WMO (1980). *Manual on Stream Gauging*, Volume 1: *Fieldwork*, and Volume II: *Computation of Discharge*, Operational Hydrology Report No. 13, WMO No. 519, Secretariat of the World Meteorological Organization, Geneva, Switzerland.

WMO (1981). *Guide to Hydrological Practices*, Volume I: *Data Acquisition and Processing*, WMO No. 168, Secretariat of the World Meteorological Organization, Geneva, Switzerland.

WMO (1983). *Guide to Hydrological Practices*, Volume II: *Analysis, Forecasting and Other Applications*, WMO No. 168, Secretariat of the World Meteorological Organization, Geneva, Switzerland.

Wolman, M.G. (1955). 'The natural channel of Brandywine Creek, Pennsylvania', US Geol. Survey Prof. Paper 271, USGS, Washington, DC (as cited by Harvey, 1969).

Woodward, W.A., Elliot, A.C., Gray, H.L. and Matlock, D.C. (1988). *Directory of Statistical Microcomputer Software*, Marcel Dekker, New York.

Woodyer, K.D. (1968). 'Bankfull frequency in rivers', *J. of Hydrol.*, **6**, 114–42.

Woolhiser, D.A. and Saxton, K.E. (1965). 'Computer program for the reduction and preliminary analyses of runoff data', *ARS 41-109*, USDA Agricultural Research Service.

Wright, J.F., Armitage, P.D., Furse, M.T. and Moss, D. (1989). 'Prediction of invertebrate communities using stream measurements', *Regulated Rivers*, **4**, 147–55.

Yalin, M.S. (1972). *Mechanics of Sediment Transport*, Pergamon Press, Oxford.

Yalin, M.S. and Karahan, E. (1979). 'Inception of sediment transport', *J. Hyd. Div., ASCE,* **105**, 1433–43.
Yang, C.T. (1973). 'Incipient motion and sediment transport'. *J. Hyd. Div., Proc. ASCE,* **99(HY10)**, 1679–1704.
Yevjevich, V.M. (1972). *Probability and Statistics in Hydrology*, Water Res. Publ., Fort Collins, Colorado.
Zar, J.H. (1974). *Biostatistical Analysis*, Prentice-Hall, Englewood Cliffs, New Jersey.
Zellweger, G.W., Avanzino, R.J. and Bencala, K.E. (1989). 'Comparison of tracer-dilution and current-meter discharge measurements in a small gravel-bed stream, Little Lost Man Creek, California', *Water-Res. Invest. Rep. 89-4150*, US Geol. Survey, Menlo Park, California.
Zingg, T. (1935). 'Beitrag zur Schotteranalyse', *Schweiz. Min. u. Pet. Mitt.*, **15**, 39–140 (as cited by Krumbein, 1941).
Zukav, G. (1979). *The Dancing Wu Li Masters*, Bantam Books, Toronto.

Index